SOIL CONSERVATION SERVICE CURVE NUMBER (SCS-CN) METHODOLOGY

Water Science and Technology Library

VOLUME 42

The titles published in this series are listed at the end of this volume.

SOIL CONSERVATION SERVICE CURVE NUMBER (SCS-CN) METHODOLOGY

by

SURENDRA KUMAR MISHRA
Hydrologic Design Division,
National Institute of Hydrology,
Roorkee, Uttaranchal, India

and

VIJAY P. SINGH
Department of Civil and Environmental Engineering,
Louisiana State University, Baton Rouge, U.S.A.

KLUWER ACADEMIC PUBLISHERS
DORDRECHT / BOSTON / LONDON

A C.I.P. Catalogue record for this book is available from the Library of Congress.

ISBN 1-4020-1132-6

Published by Kluwer Academic Publishers,
P.O. Box 17, 3300 AA Dordrecht, The Netherlands.

Sold and distributed in North, Central and South America
by Kluwer Academic Publishers,
101 Philip Drive, Norwell, MA 02061, U.S.A.

In all other countries, sold and distributed
by Kluwer Academic Publishers,
P.O. Box 322, 3300 AH Dordrecht, The Netherlands.

Printed on acid-free paper

Printed in the Netherlands.

Dedicated

to

our families:

SKM: Rekha, Shivangi and Surabhi

VPS: Anita, Vinay and Arti

CONTENTS

PREFACE

The Soil Conservation Service (SCS) curve number (CN) method is one of the most popular methods for computing the runoff volume from a rainstorm. It is popular because it is simple, easy to understand and apply, and stable, and accounts for most of the runoff producing watershed characteristics, such as soil type, land use, hydrologic condition, and antecedent moisture condition. The SCS-CN method was originally developed for its use on small agricultural watersheds and has since been extended and applied to rural, forest and urban watersheds. Since the inception of the method, it has been applied to a wide range of environments. In recent years, the method has received much attention in the hydrologic literature.

The SCS-CN method was first published in 1956 in Section-4 of the National Engineering Handbook of Soil Conservation Service (now called the Natural Resources Conservation Service), U. S. Department of Agriculture. The publication has since been revised several times. However, the contents of the methodology have been nonetheless more or less the same. Being an agency methodology, the method has not passed through the process of a peer review and is, in general, accepted in the form it exists. Despite several limitations of the method and even questionable credibility at times, it has been in continuous use for the simple reason that it works fairly well at the field level.

Recent contributions have significantly enhanced the understanding of the SCS-CN method and consequently its application potential. In the simplest form, the fundamental proportionality concept of the method relates the two orthogonal hydrological processes of surface water and ground water and the other hypothesis relates to the atmospheric process. Qualitatively, the method broadly integrates all the three major processes of the hydrologic cycle; and therefore it can form one of the fundamental concepts of hydrology. Thus, there is a need to have another look at the SCS-CN method and highlight its potential for applications to perform hydrological tasks other than those originally intended. This text book is aimed at presenting an up-to-date account of the SCS-CN method and clarify its potential for practical applications.

The subject matter of the book is divided into nine chapters. Chapter 1 presents a brief introduction of rainfall-runoff modeling and presents elements of catchment, precipitation, interception, surface detention and depression storage, evaporation, infiltration, runoff, and the runoff hydrograph. The chapter is concluded with a discussion of the scope of the SCS-CN concept in hydrology. Providing a historical background of the SCS-CN method, Chapter 2 discusses the factors affecting the curve number (CN), the determination of CN, the use of NEH-4 tables, sensitivity analysis, advantages and limitations of the SCS-CN method, and application to distributed watershed modeling.

An analytical derivation of the SCS-CN method is presented in Chapter 3. It focuses on an analytical derivation of the Mockus and other methods, generalization of the Mockus method, implication of the generalized Mockus method, characteristics of the SCS-CN and Mockus methods, functional behavior of the existing and modified SCS-CN methods, significance of the proportional equality, antecedent moisture conditions, and the SCS-CN concept as an alternative to power law. Chapter 4 discusses a determination of 'S' using the volumetric concept encompassing an analytical derivation, verification of the existing AMC criteria, determination of S, use of NEH-4 tables and advantages and limitations of the modified model.

Chapter 5 deals with determination of 'S' using physical principles, involving Fokker-Planck equation of infiltration, description of S, S/P relations for the modified model and determination of D_s from universal soil loss equation. Simulation of infiltration and runoff hydrographs is dealt with in Chapter 6, with particular emphasis on SCS-CN-based infiltration and runoff models and application of infiltration and runoff models. Chapter 7 presents long-term hydrologic simulation. It presents hydrologic models of Williams and LaSeur, Hawkins, Pandit and Gopalkrishnan, and Mishra and others. Then, it discusses rainfall-excess computation, soil moisture budgeting, catchment routing, and baseflow computation. The chapter is concluded with a discussion of the application of the modified SCS-CN method and variations thereof.

Chapter 8 dwells upon transport of pollutants in urban watersheds. It discusses heavy metals, metal partitioning, metal transport, SCS-CN analogy for metal partitioning, application of wave analogy, application of SCS-CN analogy to metal partitioning in rainfall-runoff environment, snowmelt environment and riverflow environment, PCN-based characterization of media, and determination of annual pollutant loads. The last Chapter 9 deals with sediment yield. It presents an analytical derivation of an SCS-CN based sediment yield model and discusses its coupling with the Universal Soil Loss Equation. The chapter is concluded with a discussion of its application to several watersheds.

It is hoped that this book will be useful to agricultural scientists, agricultural and civil engineers, environmental engineers, forest and range scientists, as well as watershed managers. It will also be useful to college students and faculty members engaged in environment and water related studies.

This book would not have been possible were it not for the efforts and contributions of the USDA-SCS scientists and engineers who conceived the basic ideas in the first place and wrote the NEH-Section 4. The authors are therefore deeply grateful to the USDA-SCS. The authors also had many fruitful and enlightening discussions with their students, colleagues and friends working in the area of agricultural hydrology and watershed management and are thankful to them for these discussions. Finally, the authors would like to express their gratitude to their families for their support and understanding throughout this book project. Without their support the book would not come to fruition and it is therefore dedicated to them.

S. K. Mishra
National Institute of Hydrology
Roorkee, Uttaranchal, India

V. P. Singh
Louisiana State University
Baton Rouge, Louisiana, U. S. A.

List of Symbols

a	coefficient of K-θ, regression equations, and Singh and Yu model;
a_1	coefficient of ψ-θ relation;
A	catchment or plot area,
	a parameter of the Green-Ampt model,
	a hydrologic soil group;
A′	albedo;
A″	a constant of the Philip equation;
ABFI	antecedent baseflow index;
AMC	antecedent moisture condition;
AMC I	AMC corresponding to dry condition;
AMC II	AMC corresponding to normal condition;
AMC III	AMC corresponding to wet condition;
API	antecedent precipitation index;
A_T	total catchment area;
A_w	total wetted area of the catchment;
b	parameter of the Mockus (1949) method,
	a parameter rainfall depth-duration and regressions equations,
	an exponent of the modified Kostiakov model;
B	a parameter of the Green-Ampt model,
	top width of the channel,
	a hydrologic soil group;
B′	Bowen ratio;
BD	bulk density;
c	wave celerity,
	a coefficient of equation (1.72);
$\hat{c}$	dimensionless wave celerity;
C	runoff factor,
	a coefficient of the Meyer equation,
	a parameter of the Green-Ampt model,
	an empirical parameter,
	Chezy coefficient,
	Courant number,
	weir coefficient,
	a hydrologic soil group;
CEC	ratio of cation-exchange capacity of clay to the % clay;
C_H	a nondimensional quantity in Jensen-Haise method;
c_k	kinematic wave celerity;
CN	curve number;
$\overline{CN}$	an average value of CN;
CN_o	curve number at time t = 0;
CN_t	curve number at time t;
C_t	a coefficient related to catchment storage in equation (1.77);
C_T	temperature coefficient;
d	drainage density,

	Brooks-Corey pore-size distribution index;
D	moisture diffusivity,
	duration of unit hydrograph,
	cell Reynold's number,
	a hydrologic soil group;
DCIA	directly connected impervious area;
DRH	direct runoff hydrograph;
D_s	soil depth above impeding layer;
e	void ratio,
	an exponential;
E	potential evaporation rate or volume,
	error;
e_a	vapour pressure of the overlying air;
E_a	mass transfer evaporation;
E_n	net evaporation rate;
e_o	saturation vapour pressure of the air at the mean monthly air temperature;
E_p	pan evaporation;
ER	effective rainfall;
ERH	effective rainfall hyetograph;
e_s	saturation vapour pressure;
e_2 and e_1	saturation vapour pressure at the mean maximum and minimum temperatures, respectively;
f	infiltration rate;
f_c	minimum infiltration rate;
f_o	initial infiltration rate;
f(u)	a function of horizontal wind speed;
F	cumulative infiltration;
F_d	cumulative dynamic portion of total infiltration;
F_c	cumulative static portion of total infiltration;
g	gravitational acceleration;
g()	some function;
G	specific gravity of the medium;
h	gauge height or flow depth;
H	total potential head;
h_b	Brooks-Corey bubling pressure head;
H_c	capillary potential at the wetting front;
h_e	Green-Ampt effective wetting front suction;
h_m	matric suction;
h_o	reference gauge height;
h(t)	instantaneous unit hydrograph;
i	rainfall intensity,
	an imaginary unit;
I	inflow,
	net inflow of water infiltrated into the ground,
	heat index in Tornthwaite formula,
	percent imperviousness;
I_a	initial abstraction;

i_e effective rainfall intensity;

I_j and I_{j-1} indices for jth and (j-1)th days, respectively, in equation (1.82);

i_o uniform rainfall intensity;

IUH instantaneous unit hydrograph;

j an integer;

J annual temperature efficiency index;

k hydraulic conductivity,
coefficient of rainfall depth-duration-frequency relation;

K ratio of evaporating foliage surface to its horizontal projection,
storage coefficient,
a recession factor in equation (1.82);

K_c compactness ratio;

K_f form ratio;

K_h hydraulic conductivity;

$\overline{K}_h$ average hydraulic conductivity;

K_{ho} hydraulic conductivity at θ_o;

$K_{h(sat)}$ saturated hydraulic conductivity;

K_p pan coefficient;

L total water loss,
length of the spillway crest;

$\subseteq$ heat of vaporization;

L_c length from outlet to the centroid of the watershed;

L_i interception loss;

M exponent of K-θ and rainfall intensity-duration relations,

m_1 exponent of ψ-θ power relation,

M antecedent moisture;
a nondimensional number;

n soil porosity,
Manning's roughness,
number of reservoirs in Nash model;

N an integer,
an empirical parameter;

NEH-4 National Engineering Handbook, Section 4;

O outflow;

O_c controlled reservoir outflow;

O_s uncontrolled reservoir outflow;

p hydraulic pressure,
atmospheric pressure,
ratio of mean daytime hours for a given month to the total daytime hours;

P cumulative rainfall depth;

$\overline{P}$ an average value of cumulative rainfall depth;

P_a annual rainfall;

pC percent clay;

PD particle density;

PD_d number of people per hectare;

P_e cumulative effective rainfall depth;

PET	potential evapotranspiration;
pOM	percent organic matter;
pS	percent sand;
P_5	5-d antecedent rainfall;
q	discharge;
Q	cumulative direct surface runoff;
$\overline{Q}$	an average value of the cumulative direct surface runoff;
Q_a	annual runoff, net energy advected into the water body;
Q_b	long wave radiation loss by the water body;
Q_e	energy expended in the evaporation process, equilibrium flow rate;
Q_{ex}	extra-terrestrial radiation;
Q_h	sensible heat transferred from water body to the atmosphere by convection and conduction;
Q_i	incoming energy;
q_o	reference discharge per unit width;
Q_o	energy expenditure;
Q_p	peak flow;
Q_s	global radiation;
Q_t	increase in energy stored in the water body;
q	incremental direct runoff;
r	radius of the capillary tube;
R	hydraulic radius of the channel cross-section;
r_a	aerodynamic resistance to vapor and heat transfer;
RH	relative humidity;
r_s	bulk surface resistance, a function of leaf area index;
S	potential maximum retention, storage space available in the soil column for water retention;
S_{abs}	absolute potential maximum retention;
S^2_{CN}	variance of CN;
SCS	Soil Conservation Service;
S_d	depression storage;
SH	S-hydrograph;
S_i	interception storage depth;
S^2_λ	variance of λ;
SMI	soil moisture index;
S_o	potential storage space available in the soil column for water retention;
s_o	channel bed slope;
S^2_Q	variance of Q;
S_r	degree of saturation;
$S_{r(atm)}$	atmospheric degree of saturation;
S_s	soil sorptivity;
S_T	total storage space available for water loss;
S_v	storage capacity of vegetation for the projected area of the canopy;

SVL	soil-vegetation-land use complex;
S_x^2	variance of x;
t	time coordinate, storm duration;
$\hat{t}$	non-dimensional time coordinate;
t_c	time of concentration, a time parameter;
t_e	effective storm duration;
t_p	time to ponding;
t_r	time of rise;
T	time period, mean daily temperature;
T_a	overlying air temperature;
T_b	time-base of an infiltration decay curve;
T_L and t_l time lag;	
T_{max} and T_{min}	maximum and minimum temperatures, respectively;
T_N	temperature of the Nth month;
T_s	storm duration, water surface temperature;
T_x	intercept of temperature axis;
T_1, $T_{2...}$ T_7	time lags;
U	daily consumptive use factor;
UH	unit hydrograph;
v	Darcy velocity, uniform flow velocity;
v_o	mean flow velocity;
V_a	volume of air;
$V_{a\,(atm)}$	volume of atmospheric air;
V	total volume of soil column, reservoir storage;
V_v	volume of voids;
V_s	volume of solids, equivalent depth of depression storage;
V_w	volume of water;
$V_{w\,(evap)}$	volume of evaporated water;
W	wind speed, an index of rainfall abstractions;
W_1	wind speed at 1 m height;
x	independent variable of regression equation, space coordinate; x_i and x_j ith and jth variables of x;
$\overline{x}$	an average value of variable x;
y	dependent variable of regression equation;
Y	slope of the watershed;
z, and $\hat{z}$	space coordinates, ^ stands for non-dimensional;
α	angle of contact of the meniscus, infiltration decay factor

	a coefficient of the modified Kostiakov model;
β	a coefficient of equation (1.69), decay constant of depression storage;
$\hat{\beta}$	dimensionless frequency factor;
δ	logarithmic decrement;
∂	a partial differential operator;
Δ	slope of the curve of saturated vapour pressure against temperature, change in the variable;
ε	a measure of accuracy, weighting factor in Muskingum equation;
Φ	total potential head;
Φ_1	hydrostatic pressure immediately below the interface;
Φ_2	hydrostatic pressure at free water surface;
φ	an index of rainfall abstractions;
γ	psychrometric constant;
Γ	Gamma function;
ℓ	heat of vaporization, hydraulic length of the watershed;
ℓ_o	mean overland flow length;
λ	initial abstraction coefficient;
$\overline{\lambda}$	an average value of the initial abstraction coefficient;
μ	dynamic viscosity;
θ	moisture content;
θ'	perturbation of θ about θ_o;
θ_l	θ at any time level;
θ_o	initial moisture content;
θ_{sat}	θ at saturation;
θ_T	moisture content at time, $t = T$
ς	intrinsic sorptivity;
ρ	water or fluid density;
ρ_a	air density;
ψ	negative pressure;
ψ_e	Green-Ampt effective wetting front suction;
ψ_o	negative pressure at θ_o;
$\overline{\psi}$	average negative pressure;
ψ_l	h at any time level;
σ	surface tension;
$\hat{\sigma}$	dimensionless wave number;
Ω	non-dimensional wave amlitude;
χ	characteristic depth.

Subscripts I, II, III correspond to AMC I, AMC II, and AMC III, respectively;
Subscripts A, B, C, D correspond to hydrologic soil group A, B, C, D, respectively;
Superscript * represents normalised values of the variable.

CHAPTER 1

INTRODUCTION

Hydrology is the science of water that deals with the space-time-frequency characteristics of the quantity and quality of the waters of the earth with respect to their occurrence, distribution, movement, storage, and development. Hydrology plays a fundamental role in addressing a range of issues related to environmental and ecological management and societal development. Central to addressing these issues is rainfall-runoff modeling which, in particular, is used in water resources assessment, flood and drought mitigation, and water resources planning and management. In general, rainfall-runoff modeling is basic to design of a wide variety of hydraulic structures, environmental impact assessment, evaluation of the impact of climate change, irrigation scheduling, flood forecasting, planning of tactical military operations, augmentation of runoff records, pollution abatement, watershed management, and so on.

1.1 Rainfall-Runoff Modeling

The history of rainfall-runoff is believed to have begun about 325 years ago with the report on the quantitative measurements in hydrology published by P. Perreault in 1674 (Linsley, 1982). By comparing the measured annual rainfall (P_a) and the estimated annual streamflow (Q_a) of the Seine River near Paris, Perreault described a functional relationship as $Q_a = P_a/6$. In the context of modern hydrology, the development of this P_a-Q_a relation is very primitive, but it was a major finding of the time. Those days, it was commonly believed that rainfall was not sufficient to supply flow in streams. It is worth noting that the concept of computing runoff as a percentage of rainfall is still in use after 325 years.

The importance of rainfall-runoff modeling has long been recognized. A comprehensive literature search will undoubtedly disclose the existence of several hundred (perhaps thousand) "models" and the number of unpublished models probably exceeds the published ones (Linsley, 1982). Most of the historical developmental works in the field of rainfall-runoff modeling took place in the first half of the 20th century. The decade of the 1930's experienced an outburst on all fronts of hydrology. Tools developed in the nineteen thirties and forties to solve practical problems laid the foundation for present developments. The Second World War during 1940-1945 brought a temporary setback to hydrological advances. The subsequent period from 1945 to 1950 was a period of re-learning and catching up. As such, the period of the 1940's was a period of pause and consolidation. Another significant fall out of the war was that it created a tool that caused a quantum jump in mathematical analyses in hydrology. As a result of the advances of the 1930's and 1940's, hydrology reached a stage where the ability to mathematically state the problem exceeded the ability to solve it. Digital revolution that began in the 1960's and the ensuing advancements in computing facilities provided a range of modeling tools.

The basic form of the available rainfall-runoff models can, in general, be expressed in the form of a water budget for a given time period as:

$$Q = P - L \tag{1.1}$$

where Q is the amount of runoff, P is the amount of rainfall, L is the amount of hydrologic abstractions, such as interception, depression storage, surface detention, infiltration, evaporation, and evapotranspiration. The effect of infiltration and antecedent moisture on a rainfall-runoff event is widely recognized. A number of methods or models have been developed to simulate the rainfall-generated runoff. One of the widely used methods is the Soil Conservation Service Curve Number (SCS-CN) (SCS, 1956) method which constitutes the theme of this book. A complete discussion of the SCS-CN method, its rationale, physical significance of its parameters, its extension for short-term and long-term hydrologic simulation, and its application to environmental problems is presented.

The major elements of the rainfall runoff process, as embodied in the SCS-CN method, are (a) catchment characteristics; (b) precipitation; (c) hydrologic abstractions, including interception, surface depression, infiltration, evaporation, evapotranspiration; and (d) runoff. A brief description of these elements is in order.

1.2 Catchment Characteristics

All rainfall-runoff processes take place within a catchment and it is therefore taken as a spatial unit for all hydrologic analyses. The terms catchment, watershed, basin, and drainage area are used interchangeably in hydrology and will so be used in this book. Some important catchment characteristics include length, width, area, shape, slope, relief, linear measures, and drainage patterns. There are many other characteristics, as for example, soil type, vegetative cover, land use, and hydrologic condition, which significantly affect the rainfall-runoff process. These are described in Chapter 2.

1.2.1 CATCHMENT LENGTH, WIDTH, AND SLOPE

The rainfall-runoff process occurs on the surface of the watershed, which is represented by plane in modeling. The geometry of a plane is characterized by the topography of the watershed, such as length and width. It can be horizontal or inclined, convex or concave, and its length may not necessarily be greater than the width. Five geometric elements can be used to represent a watershed: plane, triangular section, converging section, diverging section, and channel. These elements may be arranged to provide an almost perfect representation of a watershed. The length of flow is taken to represent the length of the plane along the direction of flow. The land area, length of flow, and slope are sufficient to define a plane (Singh, 1996) and, in turn, the catchment.

1.2.2 CATCHMENT AREA

The area of a catchment or drainage area is perhaps the most important property of the catchment. The potential runoff volume is determined by the size of the catchment, provided the storm covers the whole area. The watershed divide does not strictly coincide with the topographic divide due to the effect of groundwater or subsurface flow. Because of the former being less tractable, the latter is preferred in hydrologic analyses.

1.2.3 CATCHMENT SHAPE

The horizontal projection of a catchment outlines its shape. A quantitative measure of the shape is provided by the form ratio defined as

$$K_f = \frac{A}{\ell^2} \tag{1.2}$$

where K_f is the form ratio and ℓ is the hydraulic length of the catchment, measured along the longest watercourse. It is noted that the area and length are given in consistent units, such as square kilometers and kilometers, respectively. The shape is alternatively measured by the compactness ratio expressed as

$$K_c = \frac{0.282 \times \text{Perimeter}}{A^{1/2}} \tag{1.3}$$

where K_c is the compactness ratio and perimeter is the catchment perimeter. Again, the perimeter and area are given in consistent units, for example, in kilometer and square kilometer, respectively.

1.2.4 CATCHMENT RELIEF

Relief is the elevation difference between two reference points. The maximum catchment relief is the elevation difference between the highest point on the catchment divide and the catchment outlet. The relief ratio is the ratio of the maximum catchment relief to the catchment's longest horizontal straight distance measured in the direction parallel to the main watercourse. The relief ratio is a measure of the catchment slope and, therefore, affects soil erosion.

1.2.5 LINEAR MEASURES

Linear measures are used to describe the one-dimensional features of a catchment. The catchment length or hydraulic length is the length measured along the main watercourse which is the central and largest watercourse of the catchment and the one conveying runoff to the outlet. The length to the catchment centroid is the length from the outlet to the catchment centroid measured along the main watercourse. The total stream length is the sum of the lengths of all streams. The drainage density of a catchment is defined as the ratio of the total stream length to the catchment area.

The mean overland flow length is approximately equal to the half of the mean distance between stream channels. Therefore, it can be approximated as one-half of the reciprocal of the drainage density. Expressed mathematically,

$$\ell_o = \frac{1}{2d} \tag{1.4}$$

where ℓ_o is the mean overland flow length and d is the drainage density.

1.2.6 DRAINAGE PATTERNS

Drainage patterns in catchments vary widely. An intricate pattern of a catchment is an indicator of high drainage density. The drainage patterns, which are recognizable on aerial photographs, reflect geologic, soil, and vegetative effects and are often related to hydrologic properties, such as runoff response, annual water yield, or sediment yield.

1.3 Precipitation

Precipitation is one of the primary components of the hydrologic cycle. In general, it represents the amount of moisture that reaches the earth from atmosphere and occurs on the land surface in the common form of rain, snow, drizzle, glaze, sleet, and hail. Rain is the major form of precipitation in countries, such as India, and this book primarily deals with the rainfall-generated runoff. The term rainfall is characterized by water droplets of size 0.5-6 mm. The drops larger than 6 mm tend to break into drops of smaller sizes during their fall. The rate at which rain occurs is denoted by rainfall intensity which is classified as light if it is less than or equal to 2.5 mm/hr, moderate if it is between 2.5 and 7.5 mm/hr, and heavy if it is greater than 7.5 mm/hr.

1.3.1 QUANTITATIVE DESCRIPTION OF RAINFALL

A rainfall event or storm describes a period of time having measurable and significant rainfall that is preceded and followed by a period of no measurable rainfall. The time elapsed from the start to the end of the rainfall event is known as the rainfall duration. The rainfall duration is generally measured in hours. However, it may be of the order of several minutes for small catchments and it may be of the order of days for large catchments.

The rainfall depth and duration vary widely depending on the geographic location, climate, and the time of the year. The average rainfall intensity is described as the ratio of rainfall depth to rainfall duration. The rainfall intensity varies widely in time and space and its local or instantaneous values may be very different from the spatial and temporal averages. Typically, rainfall intensities range from 0.1 to 30 mm/hr, but can be as large as 150 to 350 mm/hr in extreme cases.

Rainfall frequency refers to the average time elapsed between occurrences of two rainfall events of the same depth and duration. The actual elapsed time, however, varies widely and is, therefore, interpreted statistically. For example, if at a certain location, a 100-mm rainfall event lasting for 6 hours occurs on the average once in 50 years, the 100-mm, 6-hour rainfall frequency for this location will be 1 in 50 years, 1/50, or 0.2. The inverse of the rainfall frequency is known as the return period or the recurrence interval that, for example, is equal to 50 years for the frequency of 1 in 50 years.

Generally, the larger rainfall depths are associated with longer return periods and vice versa. For a long return period, a long historical record of rainfall is needed to ascertain the statistical properties of the distribution of the annual maximum rainfall. Due to the lack of long rainfall records, extrapolations are usually necessary to estimate rainfall depths associated with long return periods. Such extrapolations involve a certain amount of risk. When the risk involves human life, the concepts of rainfall frequency and return period are no longer considered adequate for design purposes. Instead, a reasonable maximization of the meteorological factors associated with extreme rainfall is used. It leads to the concept of probable maximum precipitation (PMP). For a given geographic location, catchment area, event duration, and time of the year, PMP is theoretically the greatest depth of precipitation. In flood hydrology, PMP is used as a basis for computation of the probable maximum flood (PMF). However, for certain projects, a precipitation depth less than PMP may be warranted for economic reasons. It leads to the concept of standard project storm (SPS), which is taken as a certain percentage of the applicable PMP and is used to compute the corresponding standard project flood (SPF).

1.3.2 TEMPORAL AND SPATIAL VARIATION OF RAINFALL

The rainfall intensities for events of shorter duration (1 hour or less) can usually be expressed in terms of the average values. However, for longer events, instantaneous values of the rainfall intensity are likely to become more important, especially for determination of the flood peak. The temporal distribution depicts the variation of rainfall depth (or intensity) within the storm duration. It can be expressed in either discrete or continuous form. The discrete form is referred to as a hyetograph that represents a histogram of rainfall depth (or rainfall intensity) as ordinates. The continuous form is the temporal rainfall distribution, a function describing the rate of rainfall accumulation with time.

Rainfall varies not only temporally but also spatially, implying that the same amount of rain does not fall uniformly over the entire catchment. Isohyets are used to depict the spatial variation of rainfall. An isohyet is a contour line showing the locus of equal rainfall depth. An individual storm may have a spatial distribution or pattern in the form of concentric isohyets of approximately elliptic shape, leading to the formation of a storm eye that depicts the center of the storm. In general, storm patterns are not static, rather move gradually in the direction of prevailing winds. For regional rainfall mapping, isohyets are commonly referred to as isopluvials. These maps show contours of equal rainfall depth, applicable for a range of durations, frequencies, and catchment sizes. For large catchments, intense storms (or thunderstorms) may cover only a fraction of the whole area, yet they may lead to severe flooding in certain localized areas. Thunderstorms are more prevalent in arid and semi-arid regions and cause flash flooding. The role of thunderstorms in determining the flood potential of large basins is usually assessed on an individual basis.

1.3.3 AVERAGE RAINFALL OVER AN AREA

Rainfall is measured using raingauges. During a given storm, it is likely that the depth measured by two or more raingauges will not be the same. Therefore, it is often necessary to determine the spatial average of the rainfall depth over the catchment. A number of methods are available to compute the average rainfall. Three of the most popular methods are (1) arithmetic average, (2) Theissen polygon, and (3) isohyetal. In the arithmetic average method, rainfall depths during a storm measured by the raingauges located within the catchment are averaged to obtain the average rainfall amount. The Theissen polygon method computes the areally weighted rainfall. The polygons drawn for each raingauge describe the area of influence for the raingauge. In the isohyetal method, the locations of the raingauges are plotted on a scale map of both the catchment and the surrounding area. Each station's rainfall is used to draw isohyets in a manner similar to that topographic maps are prepared. The isohyetal method is regarded as more accurate than other methods, especially in the regions where orographic effects play a significant role in influencing the rainfall variability. The Theissen polygon method is more accurate than the arithmetic average method, particularly where rainfall significantly varies over the watersehd.

Example 1.1: Using the arithmetic average, Theissen polygon, and isohyetal methods, determine the average rainfall for a catchment (area = 28.6 sq. km) as shown in Figure 1.1 having 7 raingauge stations, A to G, with rainfall values as:

Station→	A	B	C	D	E	F	G
Rainfall (cm) →	3.2	3.0	2.6	2.4	2.5	2.0	1.8

Solution:

(a) Arithmetic average method

The sum of the measured values of rainfall at seven stations (A to G) is computed as 17.5 cm, which when divided by the number of raingauge stations (= 7) yields the arithmetic average rainfall value. Thus,

Arithmetic average rainfall = 17.5/7 = 2.50 cm.

(b) Theissen polygon method

In this method, the area of influence under each raingauge station is determined by drawing perpendicular bisectors to the sides of triangles formed by joining the stations with straight lines, as shown in Figure 1.1a. The measured areas of each raingauge station are shown in Table 1.1. The product of these areas with their corresponding rainfall values is then determined as 71.28 cm-km^2, which when divided by the total area (= 28.60 sq. km) yields the Theissen weighted rainfall as 71.28/28.60 = 2.49 cm.

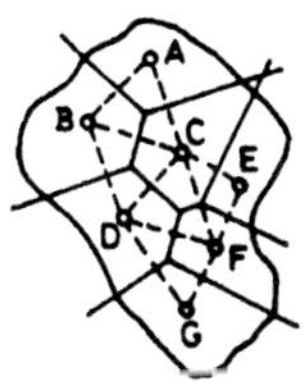

(a) Theissen polygons

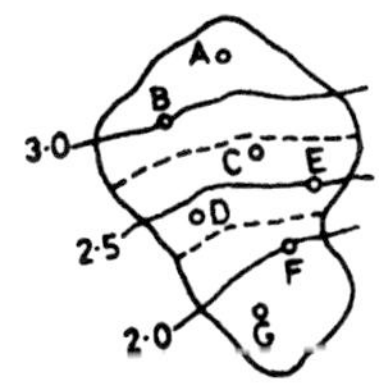

(b) Isohyetal method

Figure 1.1. Methods for computing the average rainfall over an area. A, B, C... are raingauge stations.

Table 1.1 Computation of the average rainfall using the Theissen polygon method

Station	Rainfall (cm)	Area (sq. km)	Rainfall x Area (sq. km)
A	3.2	3.28	10.50
B	3.0	5.26	15.78
C	2.6	4.01	10.43
D	2.4	4.54	10.90
E	2.5	3.16	7.90
F	2.0	3.71	7.42
G	1.8	4.64	8.35
Sum		28.60	71.28

Table 1.2 Computation of average rainfall using the isohyetal method

Rainfall (cm)	Area (sq. km)	Rainfall x Area (sq. km)
2.0	11.32	22.64
2.5	8.31	20.78
3.0	8.97	26.91
Sum	28.60	70.33

(c) Isohyetal method

Using the given rainfall data of 7 raingauge stations, isohyets (contours of equal rainfall depth) of 2.0, 2.5, and 3.0 cm of rainfall are drawn as shown in Figure 1.1b. Then, the area of influence for each isohyet is determined using a planimeter. The areas corresponding to 2.0, 2.5, and 3.0 cm isohyets are 11.32, 8.31, and 8.97 sq. km, respectively. Further procedure for weighting the rainfall is the same as for the Theissen polygon method, as shown by the computations in Table 1.2. Accordingly,

Average rainfall = 70.33/28.60 = 2.46 cm.

1.3.4 RAINFALL STORM ANALYSIS

Rainfall storms are generally analyzed for the rainfall depth and duration, intensity and duration, intensity-duration-frequency, depth and catchment area, and depth-area-duration.

(i) *Rainfall Depth and Duration*

The rainfall depth increases with storm duration and is described mathematically as

$$P = at^n \tag{1.5}$$

where P is the rainfall depth, t is the storm duration, a is a coefficient, and n is an exponent, and a positive number less than 1. Typically, n varies from 0.2 to 0.5, indicating that the storm depth increases at a lesser rate than the storm duration. By analyzing storm data on a regional or local basis, equation (1.5) can be used to predict the storm depth as a function of storm duration. The applicability of such an equation is, however, limited to the regional or local conditions for which it was derived. For extreme rainfall events (world's greatest observed rainfall events) given in Table 1.3, the resulting enveloping curve can be described as (Figure 1.2):

$$P = 4.8685\, t^{0.4888} \tag{1.6}$$

with the coefficient of determination equal to 0.996. In equation (1.6), P is in centimeters and t is in minutes.

A less accurate, but adequate, predictive relation can be given as:

$$P = \frac{a\,t}{b+t} \tag{1.7}$$

where $a = 4.54 \times 10^3$ cm and $b = 2.39 \times 10^5$ minutes. Equation (1.7) fits the data with a coefficient of determination (r^2) of 0.96. Figure 1.2 compares the predicted rainfall values from both equations (1.6) and (1.7). It is seen that low rainfall depths are less accurately predicted by equation (1.7) than by equation (1.6), which, however, over-predicts high rainfall depths. Equation (1.6) remains unchanged if the highest two rainfall points are ignored whereas the constants of equation (1.7) change and the coefficient of determination improves. The former is somewhat insensitive to very high rainfall values whereas the latter is sensitive. It is because the trend of equation (1.6) is governed by low rainfall values, in contrast with equation (1.7). Thus, the extrapolation of equation (1.6) may result in over-prediction of high values and the trend of equation (1.7) passing through the middle of the highest two rainfall values exhibits a mean trend. The advantage of equation (1.7) is that it is dimensionally homogeneous. The significance of such a relation and its parameters with regard to the SCS-CN method are further discussed in Chapter 3.

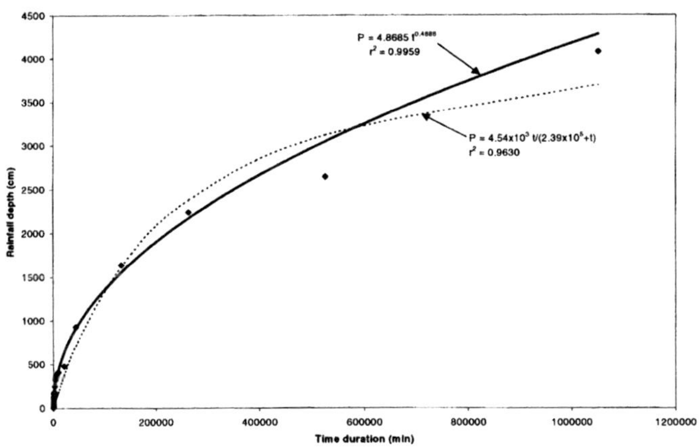

Figure 1.2. Rainfall depth-duration relation for the world's greatest observed rainfall events.

Table 1.3 Observed extreme rainfall events of the world (Data source: Ponce (1989))

Location	Date	Rainfall depth (cm)	Duration
Barot, Guadeloupe	Nov. 26, 1970	3.8	1 min
Fussen, Bavaria	May 25, 1920	12.6	8 min
Plumb Point, Jamaica	May 12, 1916	19.8	15 min
Holt, MO	June 22, 1947	30.5	42 min
Rockport, WV	July 18, 1889	48.3	2 hrs 10 min
D'Hanis, TX (17 mi NNW)	May 31, 1935	55.9	2 hrs 45 min
Smethport, PA	July 18, 1942	78.2	4 hrs 30 min
Belouve, Reunion	Feb. 28, 1964	108.7	9 hrs
Belouve, Reunion	Feb. 28-29, 1964	134.0	12 hrs
Belouve, Reunion	Feb. 28-29, 1964	168.9	18 hrs 30 min
Cilaos, Reunion	Mar. 15-16, 1952	187.0	24 hrs
Cilaos, Reunion	Mar. 15-17, 1952	250.0	2 days
Cilaos, Reunion	Mar. 15-18, 1952	324.0	3 days
Cherrapunji, India	Sept. 12-15, 1974	372.1	4 days
Cilaos, Reunion	Mar. 13-18, 1952	385.4	5 days
Cilaos, Reunion	Mar. 13-19, 1952	405.5	6 days
Cilaos, Reunion	Mar. 12-19, 1952	411.0	7 days
Cherrapunji, India	June 24-30, 1931	479.8	15 days
Cherrapunji, India	July 1861	930.0	31 days
Cherrapunji, India	May-July 1861	1637.0	3 months
Cherrapunji, India	Apr.-Sept. 1861	2245.0	6 months
Cherrapunji, India	Aug. 1860-July 1861	2646.0	1 year
Cherrapunji, India	1860-1861	4077.0	2 years

(ii) *Storm Intensity and Duration*

Storm intensity and duration are inversely related. Division of equation (1.5) by t yields the average (uniform) rainfall intensity, i_o, as follows:

$$i_o = at^{n-1} \tag{1.8}$$

or

$$i_o = \frac{a}{t^m} \tag{1.9}$$

where m = 1 − n. Since n is less than 1, m is also less than 1. Similarly, i_o can be derived from equation (1.7) as:

$$i_o = \frac{a}{b+t} \tag{1.10}$$

where a and b are constants which can be determined by regression analysis. A general form of the intensity-duration relation combining the features of equations (1.5) and (1.9) is frequently given as

$$i_o = \frac{a}{(b+t)^m} \tag{1.11}$$

For b = 0, equation (1.11) reduces to equation (1.5) and for m = 1, the former reduces to equation (1.10).

(iii) *Depth-Duration-Frequency*

It is often necessary for midsize catchments to determine several depth-duration curves, each for a different frequency or return period. Depth-duration-frequency (DDF) curves are plotted with duration on the abscissa, depth on the ordinate, and frequency (or alternatively, return period) as a third variable. The return period is attached with the constant 'a' in equation (1.5) as

$$a = kT^n \tag{1.12}$$

where k is the coefficient, n is the exponent, and T is the return period. Coupling of equation (1.12) with equation (1.5) leads to a relation for the depth-duration-frequency as:

$$P = k(Tt)^n \tag{1.13}$$

Similarly, a relation of the form of equation (1.7) can be obtained as:

$$P = \frac{kT^n t}{b + t} \tag{1.14}$$

A general form of equation (1.14), similar to equation (1.11), can be given as:

$$P = \frac{kT^n t}{(b + t)^m} \tag{1.15}$$

The DDF curves can be constructed using frequency analysis of the recorded rainfall data. The steps are as follows:

(i) Select the duration of rainfall, such as 5, 10, 20, 30, 60, or 100 min.

(ii) For the duration selected, compute the annual maximum rainfall depth for each year of record. If there are 30 years of data, then there will be 30 annual maximum depths associated with the selected duration.

(iii) Fit an appropriate frequency curve to these values. The methods of frequency analysis are described elsewhere (Singh, 1992).

(iv) Obtain from the fitted frequency curve the values of rainfall depth for the selected periods such as 5, 10, 20, 30, 50, 80, and 100 years. Tabulate these values.

(v) Repeat steps (ii) and (iii) for different durations.

(vi) Rearrange the data as rainfall depth against duration for various return periods or frequencies.

(vii) For a selected frequency, plot on log-log graph paper rainfall depth values on the ordinate versus duration on the abscissa, and fit a smooth curve or connect the points linearly. Do the same for other frequencies. These plots will be a family of DDF curves.

Example 1.2: The maximum annual 1-, 2-, 6-, 12-, and 24-hr rainfall depths recorded during 1961-1999 at the Baton Rouge airport are given in Table 1.4. Using these data, construct rainfall depth-duration-frequency curves for 2-, 5-, 25-, 50-, and 100-yr return periods.

Solution: For frequency analysis, Weibull's plotting position formula [= m/(N+1), where m is the rank and N is the total number of years (= 39 in this example)] (see Chapter 3) was used to estimate the probability of exceedance for the maximum rainfall depth of each year and each duration. The inverse of the probability of exceedance yields the return period in years. The variation of the maximum rainfall depth with the return period is shown in Figure 1.3 for 1, 2, 6, 12, and 24 hours durations. Then, lognormal distributions are fitted to these curves:

$$y = a\,\mathrm{Ln}(x) + b$$

where y is the maximum rainfall depth (inch), x is the return period (year), and a and b are the distribution parameters, shown in Table 1.5 for all durations. The lognormal

Table 1.4 Maximum rainfall depths for selected durations (Example 1.2)

Year	Maximum rainfall depth (inch)				
	1-hr	2-hr	6-hr	12-hr	24-hr
1961	1.75	2.65	4.00	4.10	4.51
1962	2.25	3.35	5.45	6.30	6.30
1963	1.45	1.98	2.06	2.06	2.06
1964	1.58	2.20	3.60	4.18	4.84
1965	1.63	2.40	2.48	3.49	4.15
1966	1.37	2.71	2.72	3.17	3.65
1967	2.42	3.35	6.97	11.31	11.99
1968	2.35	2.36	2.37	2.78	2.78
1969	2.52	3.33	4.52	4.53	4.53
1970	0.97	1.18	1.88	2.00	2.00
1971	1.63	2.21	2.45	2.45	3.05
1972	2.41	2.84	3.68	4.63	4.85
1973	1.91	3.09	4.46	4.91	6.31
1974	1.78	2.11	2.31	2.40	2.53
1975	1.87	2.33	3.35	3.45	4.26
1976	1.90	2.74	2.79	2.99	2.99
1977	2.05	2.34	3.12	4.24	5.51
1978	2.08	2.19	3.20	3.58	3.58
1979	1.22	1.94	2.56	4.16	5.24
1980	1.26	1.80	2.57	3.84	4.69
1981	0.70	0.97	2.05	2.05	2.47
1982	2.10	3.80	5.81	7.05	7.22
1983	1.08	1.29	3.62	4.54	5.11
1984	2.05	3.35	4.24	4.52	4.70
1985	1.81	2.30	3.84	4.20	4.32
1986	2.20	3.28	3.53	3.53	3.53
1987	3.52	4.68	5.05	7.88	7.94
1988	1.90	2.20	3.24	4.23	4.25
1989	1.37	2.41	5.09	6.16	6.16
1990	2.13	3.26	3.28	3.85	4.98
1991	3.13	3.80	4.35	4.38	4.38
1992	1.43	1.93	2.71	4.71	5.44
1993	1.36	1.83	3.31	6.32	9.01
1994	1.27	2.29	2.63	2.63	4.24
1995	1.13	1.68	2.90	6.27	6.40
1996	1.56	2.55	3.80	4.28	4.41
1997	2.62	3.24	3.84	4.01	4.01
1998	1.57	2.39	4.27	4.40	4.40
1999	1.65	1.87	2.21	2.44	3.23

distribution appears to fit the plots of Figure 1.3 well because the values of the coefficient of determination r^2 vary from 0.91 to 0.98. These r^2-values are quite high to indicate satisfactory fits. Here, it is appropriate to mention that any other distribution could also be attempted.

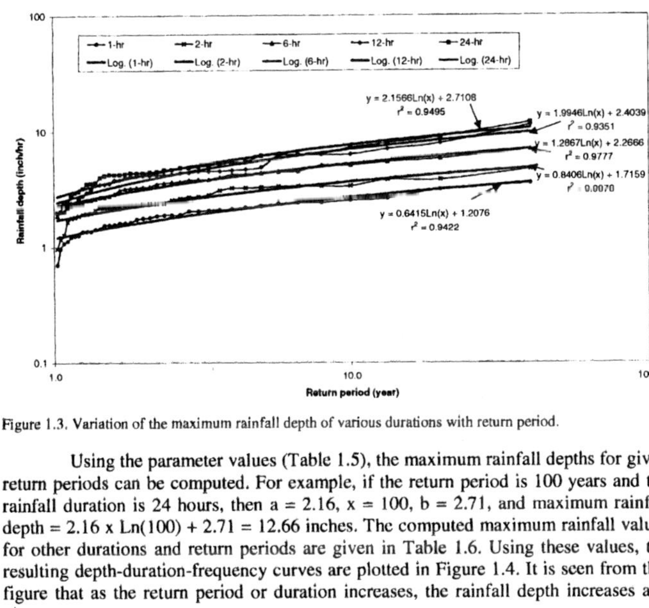

Figure 1.3. Variation of the maximum rainfall depth of various durations with return period.

Using the parameter values (Table 1.5), the maximum rainfall depths for given return periods can be computed. For example, if the return period is 100 years and the rainfall duration is 24 hours, then a = 2.16, x = 100, b = 2.71, and maximum rainfall depth = 2.16 x Ln(100) + 2.71 = 12.66 inches. The computed maximum rainfall values for other durations and return periods are given in Table 1.6. Using these values, the resulting depth-duration-frequency curves are plotted in Figure 1.4. It is seen from this figure that as the return period or duration increases, the rainfall depth increases and vice versa.

(iv) *Intensity-Duration-Frequency*

Similar to the depth-duration-frequency curves, the intensity-duration-frequency (IDF) curves are plotted with duration on the abscissa, intensity on the ordinate, and frequency (or alternatively, return period) as a third variable. Coupling of equation (1.12) with equation (1.11) leads to an intensity-duration-frequency relation as:

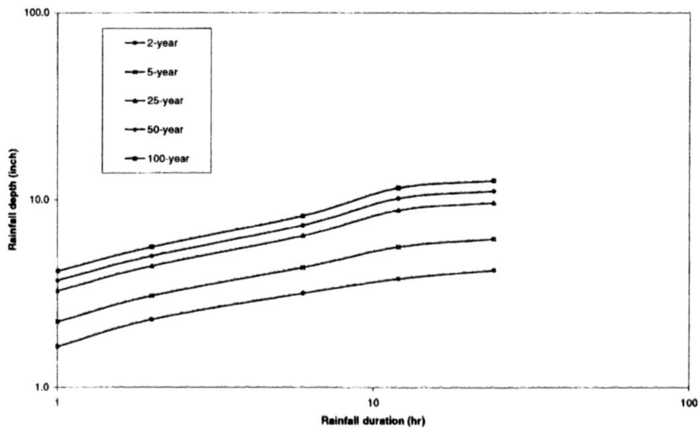

Figure 1.4. Rainfall depth-duration-frequency plots for the Baton Rouge Airport.

Table 1.5 Parameters of the fitted lognormal distribution (Example 1.2)

Parameters/coefficient of determination	Values for various durations				
	1-hr	2-hr	6-hr	12-hr	24-hr
'a'	0.64	0.84	1.29	1.99	2.16
'b'	1.21	1.72	2.27	2.40	2.71
r^2	0.94	0.91	0.98	0.94	0.95

Table 1.6 Computation of the maximum rainfall depths (inches) of given duration for various return periods (Example 1.2)

Duration (hr)	Return period				
	2-year	5-year	25-year	50-year	100-year
1	1.65	2.24	3.27	3.71	4.16
2	2.30	3.07	4.42	5.01	5.59
6	3.16	4.35	6.42	7.32	8.21
12	3.78	5.60	8.81	10.18	11.56
24	4.21	6.19	9.66	11.16	12.66

$$i_o = \frac{kT^n}{(b+t)^m} \tag{1.16}$$

For a given catchment, the values of constants k, b, m, and n are evaluated from measured data or local experience. Typical values of these constants for a few places in India are given in Table 1.7. These values are based on the reported studies of the Central Soil and Water Conservation Research and Training Institute, Dehradun, India. The values of Table 1.7 correspond to rainfall intensity, i_o, measured in cm/h; duration, t, in hr; and return period, T, in years.

Extreme point rainfall values of different durations and for different return periods have been evaluated by the India Meteorological Department and isopluvial maps covering the entire country have been prepared. These are available for rainfall durations of 15 min, 30 min, 45 min, 1 h, 3 h, 6 h, 9 h, 12 h, 15 h and for return periods of 2, 5, 10, 25, 50, and 100 years. Isopluvial maps showing storm depths (or intensity) for a range of durations, frequencies, and catchment areas are available for the entire United States (NWS, 1973, 1977; USDC, 1955, 1960, 1962).

Table 1.7 Typical values of constants in equation (1.16)

Place	'k'	'n'	'b'	'm'
Bhopal	6.93	0.189	0.50	0.878
Nagpur	11.45	0.156	1.25	1.032
Chandigarh	5.82	0.160	0.40	0.750
Bellary	6.16	0.694	0.50	0.972
Raipur	4.68	0.139	0.15	0.928

Example 1.3: Determine 1-hr and 6-hr maximum rainfall depths and corresponding intensities having return periods of 2, 10, 20, 50, and 100 years that are likely to occur in Chandigarh (India).

Solution: Here, t = 1 hr and 6 hrs. For Chandigarh, the values of the constants of equation (1.12) are: k = 5.82, n = 0.160, b = 0.40, and m = 0.75. As an example, the rainfall intensity for t = 1 hr and T = 2 years is computed from equation (1.14) as:

$$i_o = \frac{kT^n}{(b+t)^m} = \frac{5.82 x 2^{0.160}}{(0.40+1)^{0.75}} = 5.05 \text{ cm/hr}$$

The corresponding maximum rainfall depth is: 5.05 x 1 = 5.05 cm. Other computations are given in Table 1.8.

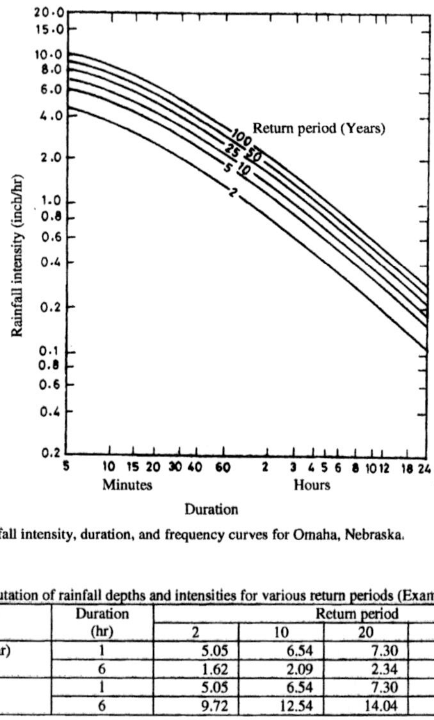

Figure 1.5. Rainfall intensity, duration, and frequency curves for Omaha, Nebraska.

Table 1.8 Computation of rainfall depths and intensities for various return periods (Example 1.3)

Rainfall	Duration (hr)	Return period				
		2	10	20	50	100
Intensity (cm/hr)	1	5.05	6.54	7.30	8.46	9.45
	6	1.62	2.09	2.34	2.70	3.02
Depth (cm)	1	5.05	6.54	7.30	8.46	9.45
	6	9.72	12.54	14.04	16.20	18.12

The construction of IDF curves is exactly similar to that of DDF curves, except for that the former utilizes the maximum rainfall intensity, and latter the maximum rainfall depth. From a given maximum rainfall depth and its duration, the maximum rainfall intensity can be computed by dividing the maximum rainfall depth by the associated duration. The resulting plots will be a family of IDF curves, as shown in Figure 1.5 for Nebraska.

Example 1.4: Construct rainfall intensity-duration-frequency curves for 2-, 5-, 25-, 50-, and 100-yr return periods using the rainfall data of Example 1.2.

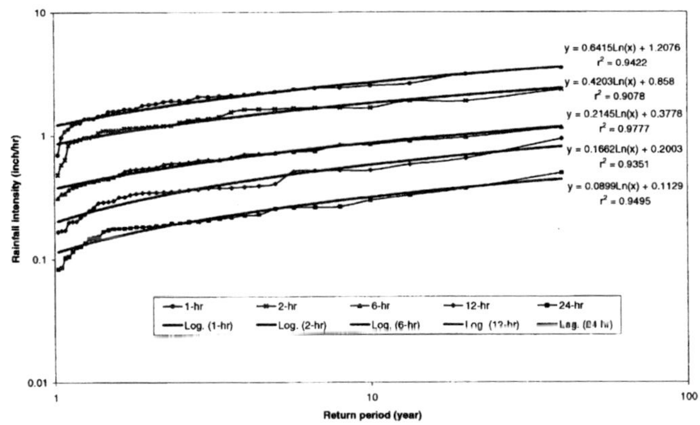

Figure 1.6. Variation of maximum, various durations rainfall intensity with return period.

Solution: For construction of IDF curves, the maximum rainfall intensities of various durations can be computed from the maximum rainfall depth and duration. Similar to Example 1.2, frequency analysis was made using the Weibull's plotting position formula. Variation of the maximum rainfall intensity with the return period is depicted in Figure 1.6 for 1, 2, 6, 12, and 24 hours durations. Again, the lognormal distribution is fitted to these curves: y = a Ln(x) + b , where y is the maximum rainfall intensity (inch/hr), x is the return period (year), and a and b are the distribution parameters, given in Table 1.9 for all durations. Following the method of Example 1.2, the lognormal distribution fits well as reflected by the high r^2–values.

Similar to the DDF analysis in Example 1.2, the maximum rainfall intensities for given return periods were computed using the parameter values shown in Table 1.9. The resulting intensity-duration-frequency curves are plotted in Figure 1.7. It is seen from this figure that as the return period or duration increases, the rainfall intensity decreases and vice versa.

(v) *Rainfall Depth-Area-Duration*

Generally, the greater the catchment area, the smaller is the spatially averaged storm depth. This variation of storm depth with catchment area has led to the concept of the point depth, defined as the storm depth associated with a given point area. A point area is the smallest area below which the variation of storm depth with catchment area is assumed to be insignificant. It is taken of the order of 25 sq. km (Ponce, 1989). The point depth applies for all areas less than the point area. For larger areas, a reduction in the point depth is necessary to account for the decrease of the storm depth with catchment area. This depth reduction is accomplished with a depth-area reduction chart, a function relating the catchment area (abscissa) to the point depth percentage (ordinate). The storm duration is a curve parameter in a depth-area reduction chart.

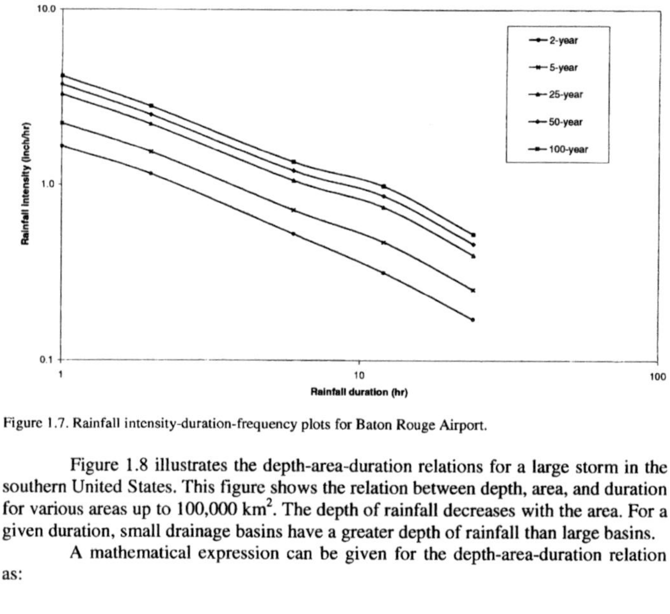

Figure 1.7. Rainfall intensity-duration-frequency plots for Baton Rouge Airport.

Figure 1.8 illustrates the depth-area-duration relations for a large storm in the southern United States. This figure shows the relation between depth, area, and duration for various areas up to 100,000 km^2. The depth of rainfall decreases with the area. For a given duration, small drainage basins have a greater depth of rainfall than large basins.

A mathematical expression can be given for the depth-area-duration relation as:

$$P = \frac{a\,t}{(A + b\,t)^m} \tag{1.17}$$

where P is the rainfall depth (cm), t is the time duration (hr), A is the catchment area (sq. km), and a, b, and m are the parameters determined from regression analysis. The fitting of data given in Table 1.10 yields a = 4.31x10^3, b = 2.77x10^2, and m = 0.803 with the coefficient of determination (r^2) equal to 0.917. The computed and observed values are compared in Figure 1.9. In this figure, the line of perfect fit shows a 1:1 relation between the observed and computed values. Apparently, most of the data points hover around the line of perfect fit, showing a satisfactory fit.

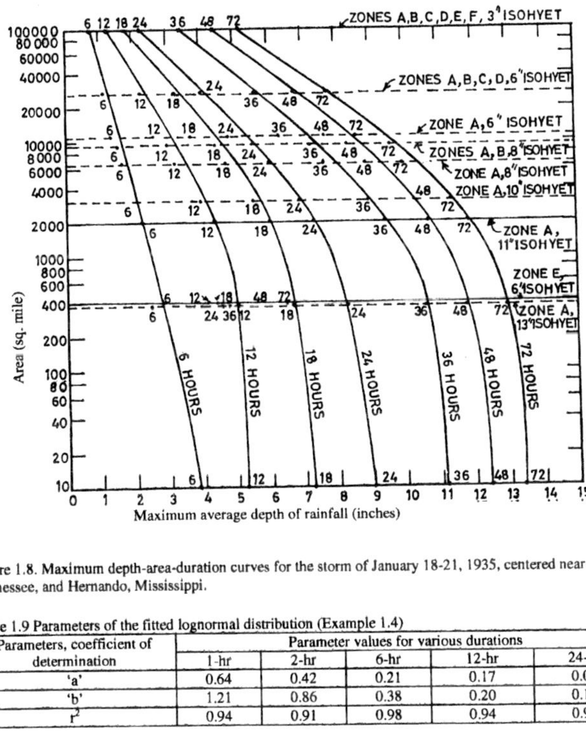

Figure 1.8. Maximum depth-area-duration curves for the storm of January 18-21, 1935, centered near Bolivar, Tennessee, and Hernando, Mississippi.

Table 1.9 Parameters of the fitted lognormal distribution (Example 1.4)

Parameters, coefficient of determination	Parameter values for various durations				
	1-hr	2-hr	6-hr	12-hr	24-hr
'a'	0.64	0.42	0.21	0.17	0.09
'b'	1.21	0.86	0.38	0.20	0.11
r^2	0.94	0.91	0.98	0.94	0.95

Example 1.5: Determine the 1-d maximum rainfall depth for catchment areas of 250 and 5000 sq. km located in the United States and compare with the observed rainfall values given in Table 1.10.

Solution: Given: t = 1 day = 24 hrs and A = 250 and 5000 sq. km. For the United States, constants of equation (1.17) are: $a = 4.31 \times 10^3$, $b = 2.77 \times 10^2$, and m = 0.803. Therefore, for A = 250 sq. km, the maximum rainfall depth is computed as:

$$P = \frac{at}{(A + bt)^m} = \frac{4.31 \times 10^3 \times 24}{(250 + 277 \times 24)^{0.803}} = 85.54\,\text{cm}$$

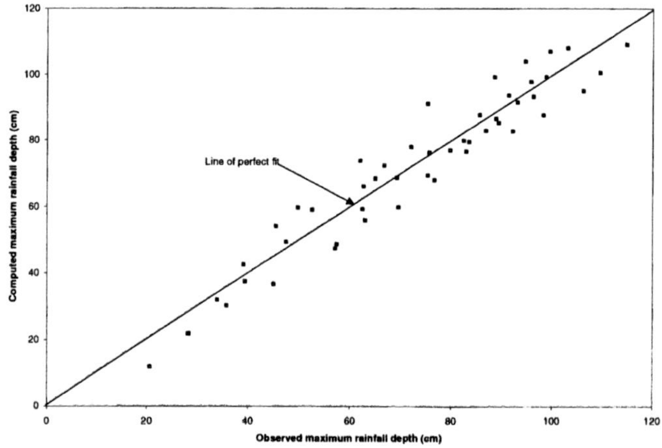

Figure 1.9. Comparison between observed and computed values of maximum rainfall depths.

Table 1.10 Maximum rainfall depth-area-duration data for the united states (source: ponce, 1989) (Average depth is given in centimeters and storm is indicated by superscripted letter)

Area (sq. km)	Duration (hr) 6	12	18	24	36	48	72
25	62.7^{a}	75.7^{b}	92.2^{c}	98.3^{c}	106.2^{c}	109.5^{c}	114.8^{c}
250	49.8^{b}	66.8^{c}	82.5^{c}	89.4^{c}	96.3^{c}	98.8^{c}	103.1^{c}
500	45.5^{b}	65.0^{c}	79.8^{c}	86.9^{c}	93.2^{c}	95.8^{c}	99.6^{c}
1250	39.1^{b}	62.5^{c}	75.4^{c}	83.0^{c}	88.9^{c}	91.4^{c}	94.7^{c}
2500	34.0^{b}	57.4^{c}	69.6^{c}	76.7^{c}	83.6^{c}	85.6^{c}	88.6^{c}
5000	28.4^{b}	45.0^{c}	57.1^{c}	63.0^{c}	69.3^{c}	72.1^{c}	75.4^{c}
12500	20.6^{b}	28.2^{b}	35.8^{b}	39.4^{c}	47.5^{d}	52.6^{d}	62.0^{d}

Storm	Date	Storm center
'a'	July 17-18, 1942	Smethport, PA
'b'	September 8-10, 1921	Thrall, TX
'c'	September 3-7, 1950	Yankeetown, FL
'd'	June 27-July 1, 1899	Hearne, TX

As seen from Table 1.10, the corresponding observed maximum rainfall depth is 89.4 cm. Thus, equation (1.17) underpredicts by 4.32%.

Similarly, for A = 5000 sq. km, P can be computed as 56.17 cm. The corresponding maximum rainfall depth (Table 1.10) is 63.0 cm. Thus, equation (1.17) underpredicts the depth by 10.84%. Thus, the error in prediction increases as catchment area increases.

The depth-area-duration (DAD) analysis is another way of portraying the reduction of storm depth with area, with duration as a third variable. To construct a DAD chart, two methods are used: (1) the mass-curve method and (2) the incremental isohyetal method. Each method portrays the time division of rainfall differently. The mass-curve method involves construction of mass curves for individual stations, an average mass curve for the entire area, and the construction of one isohyetal map for the total storm rainfall. When rainfall measurements are available at different times of the day, as is the case in the United States, construction of mass curves is fairly easy.

The incremental isohyetal method involves construction of a number of isohyetal maps. If the isohyetal pattern is simple and rainfall readings are available at a uniform time each day, then this method is relatively simple. In many cases, especially in developing countries, rainfall is recorded and reported once a day, and the standard durations are, therefore, integral multiples of a day such as 2 days, 3 days, and so on; and hence, the incremental isoyetal method may be preferable.

The following steps are involved in a DAD analysis:

(i) The storm to be analyzed is assigned a definite beginning and a definite ending to ensure the same time interval for all raingauges.

(ii) For each time period during the total storm duration, rainfall amounts are computed for each raingauge and tabulated.

(iii) Separate isohyetal maps are prepared for the area for each time period, for example, 6 hours, 12 hours, 24 hours, etc., using the rainfall amounts obtained in step (ii).

(iv) The isohyetal maps are analyzed next. The isohyetal map of the time period (say, 6 hours) leading to the maximum values at the storm centers and over substantial areas around these centers is selected. Since the maximum rainfalls of the time period may not occur simultaneously over all areas, it is advisable to investigate other isohyetal maps of the time period and then establish the maximum rainfall of the selected time period for areas of all sizes.

(v) Steps (ii) to (iv) are repeated for another time period, say, 12 hours. The maximum rainfall for this period may or may not include the maximum rainfall of the previous smaller time period. Similar calculations are done for other durations.

(vi) The depth-area calculations are made next. The isohyetal map of the lowest time interval is analyzed first. This map is divided into zones to represent principal rainfall centers. Beginning with the central isohyet (higher) value in each zone, the area encompassed by each isohyet is planimetered and then the net area between each pair of isohyets is determined.

(vii) In the storm-centered DAD analysis, the last enclosing isohyet is taken as boundary, whereas the watershed boundary is used in watershed-centered analysis.

(viii) The area enclosed by two consecutive pairs of isohyets is multiplied by the average value of these isohyets for determining the total volume. The volumes are then divided by the corresponding catchment areas to obtain the maximum rainfall depths for each area of different size.

(ix) The maximum average depths corresponding to various area sizes are plotted on a semi-log paper using the log scale for area, and a smooth curve is drawn enveloping all the points on the graph. This is the depth-area curve of the maximum rainfall for the selected time period.

Similarly, depth-area curves are prepared for other time periods. Taken together, the depth-area curves of various durations are the depth-area-duration curves, and the whole procedure to construct the DAD curves is known as the DAD analysis.

Example 1.6: Hourly rainfall for a storm in July 1964 at several stations in Western Ontario, Canada, are given in Table 1.11. Determine the maximum depth-area curve for 6-, 12-, and 24-h duration for the portion of the storm covering the South Branch of the Thames River watershed shown in Figure 1.10.

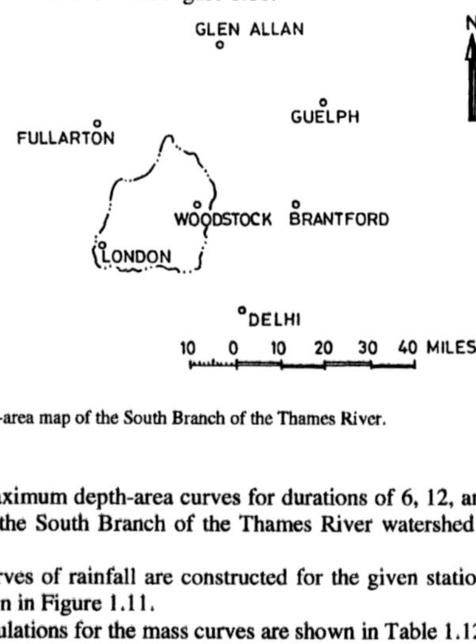

Figure 1.10. Drainage-area map of the South Branch of the Thames River.

Solution: The maximum depth-area curves for durations of 6, 12, and 24 hours for the portion covering the South Branch of the Thames River watershed are determined as follows:

(i) Mass curves of rainfall are constructed for the given stations for July 12 and 13, shown in Figure 1.11.

(ii) The calculations for the mass curves are shown in Table 1.12.

(iii) Determine those 6-h, 12-h, and 24-h periods from the given data for which the rainfall will be maximum over the watershed. This can be determined in two ways.

(a) Plot the average mass curve for the seven stations and then pick up the 6-, 12-, and 24-h periods of the maximum rainfall. This average mass curve (of course, unweighted) is shown in Figure 1.12. Computations are shown in Table 1.12 for this unweighted average mass curve. The 6-, 12-, and 24-h periods of the maximum rainfall are shown in the figure. These are as follows:

6-h period of maximum rainfall = after 2100, July 12, to 0300, July 13

12-h period of maximum rainfall = after 1600, July 12, to 0400, July 13

24-h period of maximum rainfall = 0400, July 12, to 0400, July 13

(b) Determine the 6-, 12-, and 24-h periods of the maximum rainfall as follows. Hourly rainfall values are summed up for the seven station hourly totals. These totals are summed up and cumulated for every 6-, 12-, and 24-h period throughout the period of rainfall, as shown in Table 1.13. Then mark the periods of maximum rainfall. The value of the maximum rainfall for 6 h = 6.29 inches; the corresponding period of 6 hours follows from 2100, July 12, to 0300, July 13. The maximum rainfall for 12 h = 9.13 inches; the corresponding period follows from 1600, July 12, to 0400, July 13, 1964. The maximum rainfall for 24 hours = 11.38 inches; the corresponding period is from 0400, July 12, to 0400, July 13, 1964.

(iv) For these periods, the maximum rainfall is found for each station. The calculated values are shown in Table 1.14 and the maximum average rainfall values for these are shown in Figure 1.13.

(v) The maximum rainfall values for 6, 12, and 24 hours, given in Table 1.14, are then used for drawing isohyets.

(vi) For isohyetal maps corresponding to these periods for the watershed area, a topographic map is used. The relevant portion of the complete map is traced on a separate sample paper. Figure 1.10 shows the drainage area for the South Branch of the Thames River watershed.

(vii) Isohyets are drawn at the interval of 0.1 inch on the drainage-area map for 6-, 12-, and 24-h periods separately, as shown in Figures 1.14 to 1.16.

(viii) The drainage area of the South Branch of the Thames River watershed is planimetered and is found to be 520 sq. miles.

(ix) The average rainfall over the watershed for these periods is determined by the isohyetal method. Relevant calculations are shown in Tables 1.15 to 1.17.

(x) The maximum average rainfall over the watershed for the given period is thus determined.

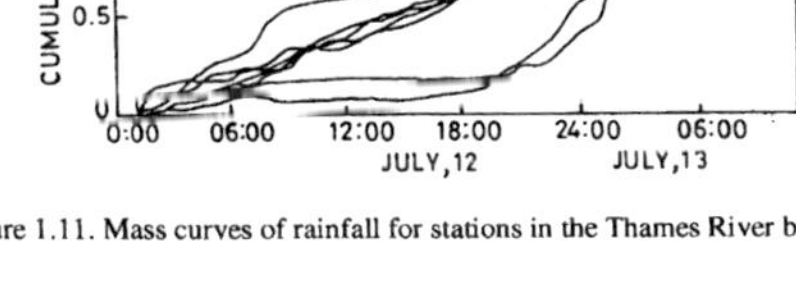

Figure 1.11. Mass curves of rainfall for stations in the Thames River basin.

(xi) The maximum depths of rainfall are determined over areas of various sizes. These values are tabulated in Tables 1.18 to 1.20 respectively for durations of 6, 12, and 24 hours. The maximum depth-area-duration curves are shown in Figure 1.17.

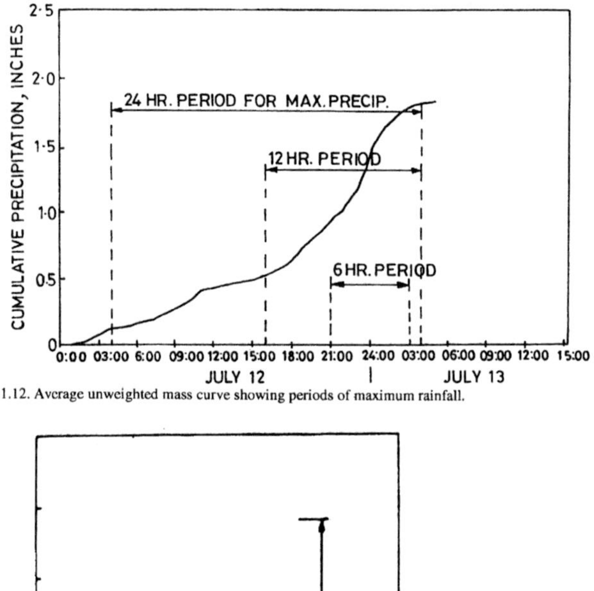

Figure 1.12. Average unweighted mass curve showing periods of maximum rainfall.

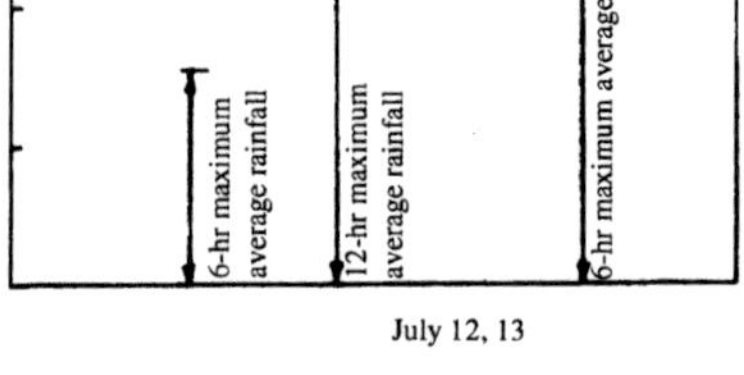

Figure 1.13. Maximum average rainfall for 6-hr, 12-hr, and 24-hr periods.

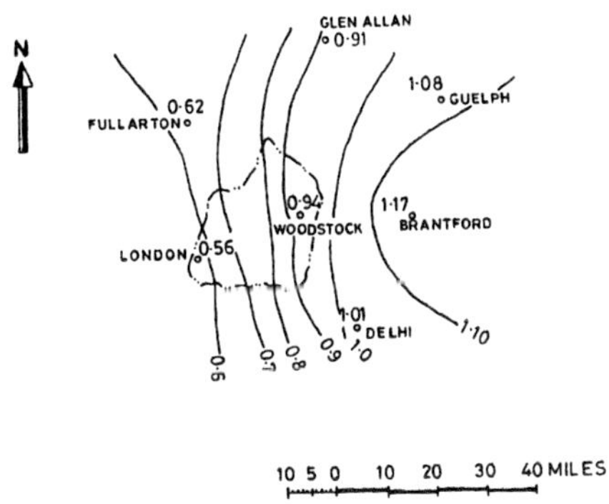

Figure 1.14. Isohyetal map for 6-hr maximum rainfall.

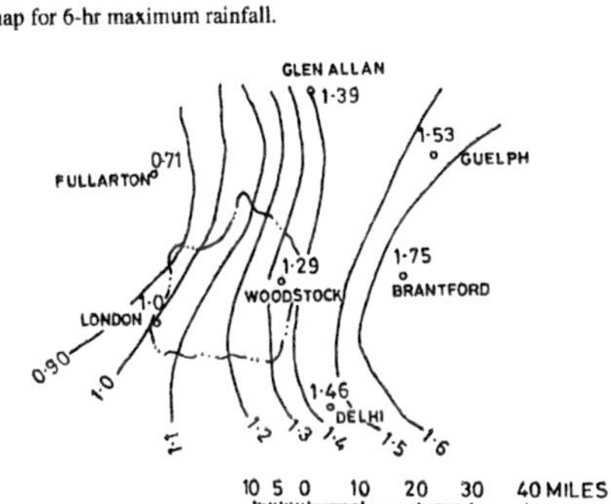

Figure 1.15. Isohyetal map for 12-hr maximum rainfall.

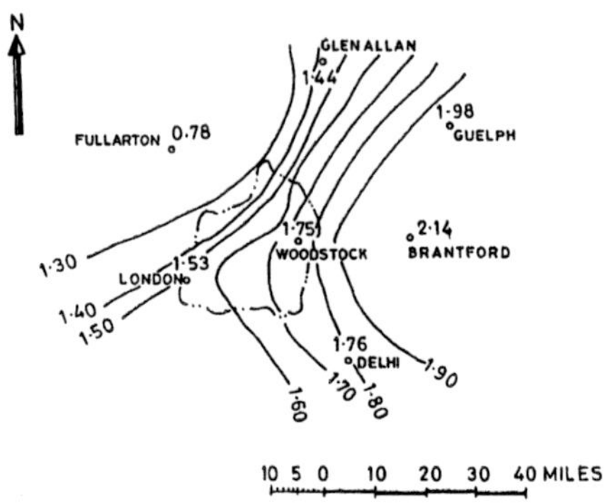

Figure 1.16. Isohyetal map for 24-hr maximum rainfall.

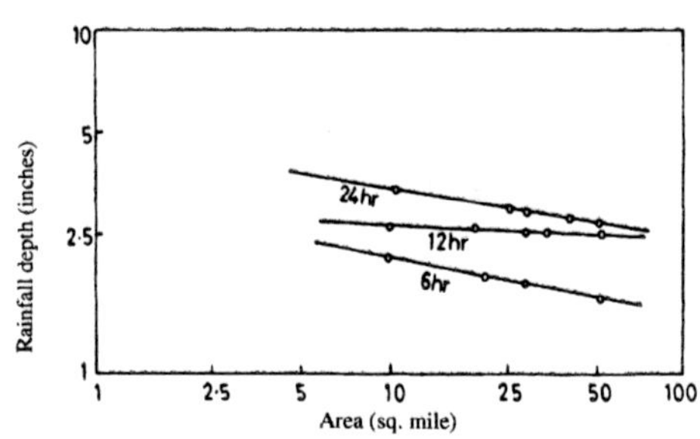

Figure 1.17. Maximum rainfall depth-area curves for 6, 12, and, 24 hours.

Table 1.11 Hourly rainfall in inches for Western Ontario for the rain storm of July 12-13, 1964

July 12	Hour ending											
	0100	**0200**	**0300**	**0400**	**0500**	**0600**	**0700**	**0800**	**0900**	**1000**	**1100**	**1200**
Brantford OWRC	0.00	0.00	0.00	0.11	0.08	0.00	0.01	0.02	0.05	0.05	0.00	0.01
Woodstock	0.00	0.00	0.05	0.01	0.03	0.03	0.01	0.05	0.06	0.09	0.05	0.04
London	0.07	0.00	0.01	0.03	0.07	0.01	0.06	0.16	0.06	0.02	0.00	0.02
Glen Allan	0.00	0.01	0.01	0.01	0.02	0.02	0.00	0.00	0.00	0.00	0.00	0.00
Fullarton	0.01	0.03	0.05	0.00	0.01	0.00	0.02	0.02	0.00	0.00	0.00	0.00
Delhi C.D.A.	0.00	0.00	0.10	0.07	0.00	0.00	0.05	0.03	0.08	0.02	0.00	0.01
Guelph W-1	0.00	0.00	0.05	0.05	0.00	0.04	0.04	0.04	0.09	0.06	0.00	0.08
July 12	**1300**	**1400**	**1500**	**1600**	**1700**	**1800**	**1900**	**2000**	**2100**	**2200**	**2300**	**2400**
Brantford OWRC	0.04	0.05	0.06	0.02	0.05	0.08	0.14	0.16	0.10	0.08	0.11	0.18
Woodstock	0.04	0.02	0.03	0.01	0.03	0.08	0.04	0.03	0.15	0.12	0.11	0.03
London	0.01	0.03	0.02	0.07	0.04	0.04	0.23	0.05	0.08	0.14	0.19	0.07
Glen Allan	0.00	0.00	0.00	0.00	0.01	0.01	0.01	0.09	0.06	0.08	0.17	0.15
Fullarton	0.00	0.00	0.00	0.00	0.00	0.00	0.00	0.00	0.07	0.01	0.05	0.26
Delhi C.D.A.	0.01	0.04	0.03	0.03	0.06	0.06	0.07	0.11	0.15	0.10	0.08	0.07
Guelph W-1	0.03	0.02	0.01	0.04	0.04	0.10	0.05	0.15	0.11	0.10	0.09	0.22
July 13	**0100**	**0200**	**0300**	**0400**	**0500**	**0600**	-	-	-	-	-	-
Brantford OWRC	0.46	0.27	0.07	0.05	-	-	-	-	-	-	-	-
Woodstock	0.18	0.30	0.20	0.02	-	-	-	-	-	-	-	-
London	0.09	0.70	-	-	-	-	-	-	-	-	-	-
Glen Allan	0.05	0.12	0.34	0.30	0.10	0.05	-	-	-	-	-	-
Fullarton	0.10	0.08	0.12	0.02	-	-	-	-	-	-	-	-
Delhi C.D.A.	0.39	0.33	0.04	-	-	-	-	-	-	-	-	-
Guelph W-1	0.34	0.22	0.11	-	-	-	-	-	-	-	-	-

Table 1.12 Mass-curve computations

Time (hour ending)	Cumulative rainfall (inch)							Seven-station rainfall (inch)	
	Bran tford	Woodst ock	London A	Glen Allan	Fullart on	Delhi C.D.A.	Guelph W-1	Total	Ave-rage
(1)	(2)	(3)	(4)	(5)	(6)	(7)	(8)	(9)	(10)
July 12									
0100	0.00	0.00	0.07	0.00	0.01	0.00	0.00	0.08	0.011
0200	0.01	0.00	0.14	0.01	0.04	0.00	0.00	0.12	0.017
0300	0.01	0.05	0.15	0.02	0.09	0.10	0.05	0.27	0.038
0400	0.12	0.06	0.18	0.03	0.09	0.17	0.10	0.28	0.040
0500	0.20	0.09	0.25	0.05	0.10	0.17	0.10	0.21	0.030
0600	0.20	0.12	0.26	0.08	0.12	0.17	0.14	0.13	0.018
0700	0.21	0.13	0.32	0.08	0.14	0.22	0.18	0.19	0.027
0800	0.25	0.18	0.48	0.08	0.16	0.25	0.22	0.32	0.045
0900	0.28	0.24	0.54	0.08	0.16	0.33	0.31	0.34	0.048
1000	0.33	0.33	0.56	0.08	0.16	0.35	0.37	0.24	0.034
1100	0.33	0.38	0.56	0.08	0.16	0.35	0.37	0.05	0.071
1200	0.34	0.42	0.58	0.08	0.16	0.36	0.45	0.16	0.0228
1300	0.38	0.46	0.59	0.08	0.16	0.37	0.48	0.13	0.018
1400	0.43	0.48	0.62	0.08	0.16	0.41	0.50	0.16	0.0228
1500	0.49	0.51	0.64	0.08	0.16	0.44	0.51	0.15	0.021
1600	0.51	0.52	0.71	0.08	0.16	0.47	0.55	0.17	0.024
1700	0.56	0.55	0.75	0.09	0.16	0.53	0.59	0.23	0.032
1800	0.64	0.63	0.79	0.10	0.16	0.59	0.69	0.37	0.0528
1900	0.78	0.67	1.02	0.11	0.16	0.66	0.74	0.44	0.062
2000	0.94	0.70	1.07	0.20	0.16	0.77	0.89	0.59	0.084
2100	1.04	0.85	1.15	0.26	0.23	0.92	1.00	0.72	0.1028
2200	1.12	0.97	1.29	0.34	0.24	1.02	1.12	0.63	0.09
2300	1.23	1.08	1.48	0.51	0.29	1.17	1.20	0.80	0.14
2400	1.41	1.11	1.55	0.66	0.55		1.29	0.98	0.14
July 13									
0100	1.87	1.29	1.64	0.71	0.65	1.56	1.85	1.61	0.23
0200	2.14	1.59	1.71	0.83	0.73	1.89	2.07	1.40	0.20
0300	2.21	1.79	1.79	1.17	0.85	1.93	2.18	0.88	0.125
0400	2.26	1.81	1.79	1.47	0.87	1.93	2.18	0.39	0.055
0500	2.26	1.81	1.79	1.57	0.87	1.93	2.18	0.10	0.014
0600	2.26	1.81	1.79	1.62	0.87	1.93	2.18	0.05	0.007

Table 1.13 Determination of 6-h, 12-h, and 24-h periods of maximum rainfall

Time (hour ending)	Seven-station rainfall (inch)	6-hour rainfall (inch)	12-hour rainfall (inch)	24-hour rainfall (inch)
July 12				
0100	0.08			
0200	0.12			
0300	0.27			
0400	0.28			
0500	0.21			
0600	0.13	1.09		
0700	0.19	1.20		
0800	0.32	1.40		
0900	0.34	1.47		
1000	0.24	1.43		
1100	0.05	1.27		
1200	0.16	1.30	2.39	
1300	0.13	1.24	2.44	
1400	0.16	1.08	2.48	
1500	0.15	0.89	2.36	
1600	0.17	0.82	2.25	
1700	0.23	1.00	2.27	
1800	0.37	1.21	2.51	
1900	0.44	1.66	2.76	
2000	0.59	2.05	2.93	
2100	0.72	2.62	3.31	
2200	0.63	3.08	3.76	
2300	0.80	3.65	4.45	
2400	0.98	4.26	5.27	7.76
July 13				
0100	1.61	5.33	6.75	9.29
0200	1.40	6.13	7.99	10.57
0300	0.88	6.29	8.72	11.18
0400	0.39	6.05	9.13	11.38
0500	0.10	5.35	9.00	11.29
0600	0.05	4.42	8.68	11.19

Table 1.14 Maximum rainfall values for 6-h, 12-h, and 24-h periods

Station	Max 6-h rainfall 2100 to 0300 h July 12-13 (inch)	Max 12-h rainfall 2100 to 0300 h July 12-13 (inch)	Max 24-h rainfall 2100 to 0300 h July 12-13 (inch)
Brantford OWRC	1.17	1.75	2.14
Woodstock	0.94	1.20	1.75
London	0.56	1.00	1.53
Glen Allan	0.91	1.39	1.44
Fullarton	0.62	0.71	0.78
Delhi C.D.A.	1.01	1.46	1.76
Guelph W-1	1.08	1.53	1.98

Table 1.15 Computation for average maximum rainfall (6-h period)

Sl. No.	Isohyetal line	Net area (mi^2)	Mean rainfall (inch)	Rainfall volume (inch -mi^2)	Average weighted rainfall (inch)
(1)	(2)	(3)	(4)	(5)	(6)
1	0.5	33	0.58	19.2	0.037
2	0.6	92	0.65	60.0	0.115
3	0.7	178	0.75	133.5	0.260
4	0.8	115	0.85	97.8	0.188
5	0.9	102	0.95	96.0	0.184
6	1.0				
		Sum		406.5	0.78
Average maximum rainfall = 406.5/520 = 0.78 inch					

Table 1.16 Computation for average maximum rainfall (12-h period)

Sl. No.	Isohyetal line	Net area (mi^2)	Mean rainfall (inch)	Rainfall volume (inch -mi^2)	Average weighted rainfall (inch)
(1)	(2)	(3)	(4)	(5)	(6)
1	0.9	41	0.95	39.0	0.075
2	1.0	130	1.05	136.2	0.261
3	1.1	154	1.15	187.0	0.357
4	1.2	155.5	1.25	194.0	0.372
5	1.3	39.5	1.35	53.0	0.102
6	1.4				
Sum				609.4	1.17
Average maximum rainfall = 609.4/520 = 0.78 inch					

Table 1.17 Computation for average maximum rainfall (6-h period)

Sl. No.	Isohyetal line	Net area (mi^2)	Mean rainfall (inch)	Rainfall volume	Average weighted rainfall (inch)
(1)	(2)	(3)	(4)	(5)	(6)
1	1.30	50.0	1.35	67.5	0.130
2	1.40	60.0	1.45	87.0	0.168
3	1.50	150.0	1.55	233.0	0.450
4	1.60	150.0	1.65	247.0	0.475
5	1.70	110.0	1.75	193.0	0.370
6	1.80				
Sum				827.5	1.59
Average maximum rainfall = 827.5/520 = 1.59 inch					

Table 1.18 Maximum depth-area calculations for the 6-h period

Sl. No.	Isohyets	Net area (mi^2)	Cumulative area (mi^2)	Average rainfall (inch)	Average weighted rainfall (inch)
(1)	(2)	(3)	(4)	(5)	(6)
1	>0.90	102	102.0	0.95	0.95
2	0.80	115	217.0	0.85	0.90
3	0.70	178	395.0	0.75	0.84
4	0.60	92	487.0	0.65	0.825
5	<.50	33	520.0	0.58	0.79

Table 1.19 Maximum depth-area calculations for the 12-h period

Sl. No.	Isohyets	Net area (mi^2)	Cumulative area (mi^2)	Average rainfall (inch)	Average weighted rainfall (inch)
(1)	(2)	(3)	(4)	(5)	(6)
1	>1.3	39.5	39.5	1.35	1.35
2	1.2	155.5	195.0	1.25	1.27
3	1.1	154.0	349.0	1.15	1.23
4	1.0	130.0	479.0	1.05	1.17
5	<0.9	41.0	520.0	0.95	1.102

Table 1.20 Maximum depth-area calculations for the 24-h period

Sl. No.	Isohyets	Net area (mi^2)	Cumulative area (mi^2)	Average rainfall (inch)	Average weighted rainfall (inch)
(1)	(2)	(3)	(4)	(5)	(6)
1	>1.70	110.0	110.0	1.75	1.75
2	1.60	150.0	260.0	1.65	1.69
3	1.50	150.0	410.0	1.55	1.64
4	1.40	60.0	470.0	1.45	1.62
5	<1.30				

1.4 Interception

The process of interception reduces rainfall by vegetation or other forms of surface cover. The vegetative cover and other surface objects retain a certain amount of rainfall that is either absorbed by the surface cover or eventually returned to the atmosphere through evaporation. A fraction of water reaching the ground is termed as throughfall. Thus, throughfall is that part of rainfall that reaches the ground by first passing through the vegetative cover. Interception losses are a function of storm characteristics, such as intensity, depth, and duration; wind velocity, direction, and duration; type, extent, and density of vegetative cover; and time of the year (season).

Interception is generally the first abstraction to occur during a storm. The frequently occurring light storms are abstracted substantially by the interception process. The annual amount of such an interception is taken approximately as 25% of the annual average rainfall. For moderate storms, interception losses vary widely, being greater during the growing season and smaller at other times of the year. Studies show a variation of interception losses from 7 to 36 percent of the total rainfall during the growing season and from 3 to 22 percent during the remainder of the year (Chow, 1964). For heavy storms, these losses usually amount to a small fraction of the total rainfall. For long-duration or infrequent storms, the contribution of interception to the total abstraction (or loss) is likely to be small. Studies dealing with floods usually ignore such losses for practical reasons.

The interception loss comprises two distinct elements (Horton, 1919). The first is the interception storage defined as the depth (or volume) of rainfall retained by the foliage against the forces of wind and gravity. The second is the evaporation loss from the foliage surface, which takes place throughout the duration of rain storm and afterwards. A combination of these processes can be expressed mathematically as (Chow, 1964):

$$L_i = S_i + KEt \tag{1.18}$$

where L_i is the interception loss (mm), S_i is the interception storage depth (mm), K is the ratio of evaporating foliage surface to its horizontal projection, E is the evaporation rate in mm/hour, and t is the storm duration in hours. S_i usually varies from 0.25 to 1.25 mm and its variation depends on the amount of rainfall as (Bras, 1990):

$$S_i = S_v(1 - e^{-P/S_v}) \tag{1.19}$$

where P is the total precipitation and S_v is the storage capacity of vegetation for the projected area of canopy.

1.5 Surface Detention and Depression Storage

In literature (for example, Ponce 1989), surface detention and depression storages are taken synonymous to each other. Surface depression refers to the fraction of the rainfall amount that is retained by unevenness due to surface puddles, ditches, and other natural or artificial depressions on the land surface. The water held in depressions as depression storage either evaporates or eventually contributes to soil moisture by infiltration. The spatial variability of storage in surface depressions precludes its precise determination. The surface detention is, however, primarily governed by the watershed's lag time that depends on the surface roughness, slope, and length of the watershed and eventually provides more time for infiltration to occur. While developing an SCS-CN-based infiltration model, Chen (1982) accounted for the surface detention component of the storage.

Intuitively, the milder the catchment's relief, the greater will be the effect of surface detention. Based on the field data, Viessman et al. (1977) concluded that depression storage was also inversely related to catchment slope. In general, an equivalent depth of depression storage can be estimated based on experience. For example, Gayle and Skaggs (1978) reported the estimated values to range from 0.6 to 7.5 mm of the average storage depth measured on three agricultural soils with different tillage practices. Hicks (1944) used depression storage depths of 5, 3.75, and 2.5 mm for sand, loam, and clay, respectively. Tholin and Keifer (1960) assumed 6.25 mm for pervious urban areas and 1.5 mm for paved areas. Onstad (1984) related the depression storage with surface roughness and its slope and gave an empirical relation. Where an accurate estimation is difficult, the depression storage amount can be lumped together with other more tractable hydrologic abstractions, such as interception or infiltration. According to Ponce (1989), the depression storage or surface detention can be accounted for using a peak-flow correction factor, similar to that in the SCS TR-55 graphical method.

Typically, the effect of depression storage varies in time and consequently with storm duration. At the beginning of a storm, the depression storage usually plays an active role in abstracting rainfall amounts. As time progresses, the depression storage volumes are eventually filled and any additional water contributes to runoff. This has led to the following conceptual model, expressed mathematically as

$$V_s = S_d(1 - e^{-\beta P_e}) \tag{1.20}$$

where V_s is the equivalent depth of depression storage (mm), P_e is the effective rainfall computed as the total rainfall amount minus the sum of interception loss and infiltration (mm), S_d is the depression storage (mm), and β is a constant (mm^{-1}). Linsley et al. (1982) recommended S_d to vary in the range 10-50 mm for most catchments. The value of β is estimated by assuming that for very small values of rainfall excess (P_e close to 0), essentially all the rainfall goes into depression storage ($dV_s/dP_e = 1$), leading to a definition of β as

$$\beta = \frac{1}{S_d} \tag{1.21}$$

1.6 Evaporation

Evaporation is the process by which the water on the land surface including that in surface depressions and water bodies, such as lakes and reservoirs, is converted to water vapour and returned to the atmosphere. It is expressed in terms of evaporation rate (mm/day) which is a function of several meteorological and environmental factors: solar radiation, sunshine hours, vapour pressure, humidity, air and water surface temperatures, wind velocity, and atmospheric pressure. Studies show that evaporation rates are high in arid and semiarid regions and low in humid regions. For example, the mean annual lake evaporation in the United States varies from 20 inches (508 mm) in the Northeast (Maine) and Northwest (Washington) to 86 inches (2184 mm) in the Southwest (California and Arizona) (Ponce, 1989). These high values call for a full evaluation of the potential reservoir evaporation for water resource planning in arid and semiarid regions. The computed evaporation rates determine, to a large extent, the feasibility of building surface water storage projects in the regions. Unlike other processes of the hydrologic cycle, lake evaporation cannot be measured directly and it has led to the development of several approaches for its determination. These are generally based on water budget, energy budget, mass transfer, or a combination thereof.

1.6.1 WATER BUDGET METHOD

The water budget method assumes that all relevant water-transport phases can be evaluated for a given time interval and expressed in terms of volumes or depth per unit area. The reservoir or lake evaporation is computed as

$$E = P + Q - O - I - \Delta V \qquad (1.22)$$

where E is the volume of water evaporated from the reservoir, P is the rainfall falling directly onto the reservoir, Q is the surface runoff into the reservoir, O is the outflow from the reservoir, I is the net volume of water infiltrated from the reservoir into the ground, and ΔV is the change in the stored volume. The time interval in water budget computations is generally taken as 1 week. Most terms in equation (1.22) can be evaluated directly. Rainfall is a measured quantity, and inflow and outflow can be obtained by integrating the flow records. The change in stored volume is determined by means of water stage recorders. The net infiltration can, however, be determined only indirectly, either by measuring soil permeability or monitoring changes in groundwater level in the neighbouring wells. The difficulty in measuring the net infiltration generally limits the water budget method to areas with little or no net infiltration. Despite these limitations, the water budget method has been found to work reliably under certain idealized conditions. The Lake Hefner (Oklahoma) studies show an accuracy of about 10% (USGS, 1974) in determination of the lake evaporation.

Example 1.7: Ignoring seepage losses, estimate the evaporation from a lake of 500 hectare for a month during which the inflow to the lake was 2.00 m^3/s, outflow was 1.00 m^3/s, and rainfall was 10 cm. At the end of the month, the water level of the lake dropped by 5 cm.

Solution: The following is given:
Monthly inflow = Q = 2 x 3600 x 24 x 30 m^3 = 51.84 x 10^5 m^3
Monthly outflow = O = 1 x 3600 x 24 x 30 m^3 = 25.92 x 10^5 m^3
Monthly rainfall = P = (10/100) x 500 x 10,000 m^3 = 5 x 10^5 m^3
Change in storage = ΔV = (5/100) x 500 x 10,000 m^3 = 2.5 x 10^5 m^3

From equation (1.22), the monthly evaporation E is computed as:
E = (51.84 + 5.0 –25.92 + 2.5) x 10^5 m^3 = 33.42 x 10^5 m^3

$$= \frac{33.42 \times 10^5}{500 \times 10^4} \times 100 \text{ cm} = 66.84 \text{ cm}$$

1.6.2 MASS TRANSFER METHOD

Evaporation rates depend on the temperature of the water surface and the prevailing atmospheric pressure. Higher water temperatures induce more vigorous molecular action and result in higher evaporation rates. However, a higher atmospheric pressure limits the movement of water molecules and results in lower evaporation rates. The overall effect of atmospheric pressure in reducing evaporation is small and can be ignored for practical reasons. Evaporation rates are usually computed using the Dalton law as

$$E = f(u)(e_s - e_a) \qquad (1.23)$$

where E is the evaporation rate, f(u) is a function of the horizontal wind speed, e_s is the saturation vapour pressure at the water surface temperature, and e_a is the vapour pressure of the overlying air. Based on equation (1.23), several equations referred to as mass-transfer equations have been developed. The commonly used Meyer equation is an example:

$$E = C(e_o - e_a)[1 + \frac{W}{10}] \tag{1.24}$$

where E is the evaporation rate (inch/month), C is a coefficient varying from 15 for small ponds to 11 for large lakes and reservoirs, e_o is the saturation vapour pressure at the mean monthly air temperature (inches of mercury), e_a is the vapour pressure of the air at the mean monthly air temperature (inches of mercury), and W is the mean monthly wind speed at 25 feet height (miles/hr).

Another version of the Meyer equation is given as (Meyer, 1944; Viessmann et al., 1977):

$$E = C(e_s - e_a)[1 + \frac{W}{16}] \tag{1.25}$$

where E is in cm per day, C is a coefficient that for daily data varies from 0.50 for small ponds to 0.36 for large lakes and reservoirs, e_s and e_a are in mb, and W is the daily wind speed at a height of 8 m in km per hour.

Example 1.8: Determine the lake evaporation for the month of June by Meyer's equation using the following observations:

Mean air temperature = 30 °C
Water temperature = 25 °C
Wind speed at 8 m above the ground = 15 km/hr
Relative humidity = 70%.
Saturated vapour pressure at 25 °C = 23.76 mm Hg
Saturated vapour pressure at 30 °C = 31.82 mm Hg

The coefficient of Meyer's formula for this lake can be taken equal to 0.36.

Solution: For the given data, the vapour pressure can be computed as 0.7 x 31.82 = 22.274 mm Hg. The lake evaporation E can be computed using equation (1.25) as:

$$E = C(e_s - e_a)[1 + \frac{W}{16}]$$

$$= 0.36 \times (23.76 - 22.274) \times \left(1 + \frac{15}{16}\right) = 1.037 \text{ cm/day}$$

= 1.037 x 30 = 31.11 cm/month

1.6.3 ENERGY BUDGET METHOD

In the energy budget method, the incoming energy is expressed as

$$Q_i = Q_s(1 - A') - Q_b + Q_a \tag{1.26}$$

where Q_i is the incoming energy; Q_s is the global radiation (short wave radiation from sun and sky); A' is the albedo, the reflectivity coefficient of a surface toward short wave radiation; Q_b is the long wave radiation loss by the water body; Q_a is the net energy advected into the water body by streams, rain, snow, and the like.

The energy expenditure, which must balance the incoming energy, is expressed as

$$Q_o = Q_h + Q_e + Q_t \tag{1.27}$$

where Q_o is the energy expenditure, Q_h is the sensible heat transfer from water body to the atmosphere by convection and conduction, Q_e is the energy expended in the evaporation process, and Q_t is the increase in energy stored in the water body. During condensation, Q_e is negative. All terms in equations (1.26) and (1.27) are given in calories per square centimetre per day or langleys per day.

The energy used in evaporation is converted into equivalent evaporation rate by the following formula:

$$Q_e = \rho \ell E \tag{1.28}$$

where ρ is the density of water (gm/cm^3); ℓ is the latent heat of vaporization, a function of temperature (calories per gram); and E is the evaporation rate (cm/day).

The terms Q_h and Q_e in equation (1.27) are difficult to determine directly. Bowen (1926) found their ratio to be more tractable and evaluated as

$$B' = \frac{Q_h}{Q_e} = \gamma \frac{T_s - T_a}{e_s - e_a} \frac{p}{1000} \tag{1.29}$$

where B' is Bowen's ratio; γ is a psychrometric constant equal to 0.66 mb/°C; T_s is the water surface temperature (°C); T_a is the overlying air temperature (°C); e_s is the saturation vapour pressure at the water surface temperature (mb); e_a is the vapour pressure of the overlying air (mb); and p is the atmospheric pressure (mb).

A balance of incoming energy (equation (1.26)) and energy expenditure (equation (1.27)) taking into account equations (1.28) and (1.29) leads to

$$E = \frac{Q_s(1 - A') - Q_b + Q_a - Q_t}{\rho \ell (1 + B')} \tag{1.30}$$

The quantities $Q_s(1\text{-}A')$ and Q_b can be measured with radiometers, instruments designed to measure radiation. The quantity Q_a can be determined by measuring volumes and temperatures of the water flowing into and out of the body, and Q_t is evaluated by periodic measurements of water temperatures. The study of Bruce and Roodgers (1962) illustrates the application of the energy budget method to Lake Ontario.

Example 1.9: Estimate the rate of lake evaporation by the energy budget method using the following data:

Air temperature	20 °C
Global radiation	580 cal/cm^2/day
Albedo	0.05
Long wave radiation loss by water body	60 cal/cm^2/day
Net energy advected into the water body	50 cal/cm^2/day
Increase in energy storage of the water body	27 cal/cm^2/day

Assume: Bowen ratio = 0.60, heat of vaporization at 20 °C = 586 cal/g, density of water = 0.998 gm/cm^3.

Solution: Given: Q_s = 580 cal/cm^2/day, A′ = 0.05, Q_b = 60 cal/cm^2/day, Q_a = 50 cal/cm^2/day, Q_t = 27 cal/cm^2/day, B′ = 0.60, ℓ = 586 cal/g, and ρ = 0.998 gm/cm^3. A substitution of these values in equation (1.30) yields the lake evaporation, E, as:

$$E = \frac{Q_s(1-A') - Q_b + Q_a - Q_t}{\rho\ell(1+B')}$$

$$= \frac{580(1-0.05)-60+50-27}{0.998 x 586 x (1+0.6)} = 0.55 \text{ cm/day}$$

1.6.4 COMBINATION METHOD

A combination method combines the energy-budget and mass-transfer methods. The most popular combination method for computing evaporation from free water surfaces was developed by Penman (1948). Another method developed by Priestley and Taylor (1972) is a simplified version of the Penman method.

(i) *Penman Method*

This method combines fundamental physical properties and empirical concepts based on standard meteorological observations. The physical principles include the energy-balance equation and the mass-transfer (or aerodynamic) equation. The method is expressed mathematically as (Ponce 1989; Singh, 1992):

$$E = \frac{\Delta E_n + \gamma E_a}{\Delta + \gamma} \tag{1.31}$$

where E is the evaporation rate (cm/day), E_n is the net evaporation rate (cm/day), E_a is the mass-transfer evaporation (cm/day) derived assuming the temperature of water surface to be equal to that of the overlying air, Δ is the slope of the curve of saturated vapour pressure against temperature (10^{-3} bars/°C), and γ is the psychrometric constant (= 0.66×10^{-3} bars/°C). Δ varies with air temperature, T_a, as follows (Linsley et al., 1982):

$$\Delta = (0.00815T_a + 0.8912)^7 \tag{1.32}$$

where Δ is in mb/°C and T_a is in °C. Equation (1.32) is applicable for $T_a > -25$ °C. Table 1.21 provides values of the Penman Δ/γ-ratio for various temperatures. The net radiation, E_n, can be computed from the net radiation, Q_n, as:

$$Q_n = \rho \upsilon E_n \tag{1.33}$$

where ρ is the density of water (gm/cm^3); υ is the latent heat of vaporization, depending on the temperature; and E is the rate of evaporation (cm/day).

The mass transfer evaporation rate E_a can, for example, be estimated from the following mass transfer evaporation equation given by Dunne and Leopold (1978):

$$E_a = (0.013 + 0.00016\,W)e_o\left(1 - \frac{RH}{100}\right) \tag{1.34}$$

where E_a is in cm/day, W is the wind velocity measured at 2-m above surface (km/day), e_o is the saturation vapour pressure at the overlying air temperature in millibars, and RH is the relative humidity (%).

Table 1.21 Values of Penman's Δ/γ ratio for various temperatures

Air temperature (°C)	Penman Δ/γ-ratio	Air temperature (°C)	Penman Δ/γ-ratio
0	0.68	25	2.86
5	0.93	30	3.69
10	1.25	35	4.73
15	1.66	40	6.00
20	2.19		

(ii) *Priestley-Taylor Formula*

Priestley and Taylor (1972) proposed that potential evaporation be taken as the radiation part of the Penman method (equation (1.31) with $E_a = 0$) modified by an empirical constant. The formula is expressed as:

$$E = \frac{1.26\Delta E_n}{\Delta + \gamma} \tag{1.35}$$

where E is the potential evaporation in cm/day, E_n is the net evaporation rate (cm/day), and γ and Δ are the same as described in equation (1.31).

Example 1.10: Estimate the rate of lake evaporation by the Penman method using the following meteorological data:

Air temperature	20 °C
Saturation vapour pressure at 20 °C	23.37 mm Hg
Net radiation	550 cal/cm^2/day
Wind speed at 2-m above the ground	200 km/day
Relative humidity	70%.

Assume: heat of vaporization at 20 °C = 586 cal/gm, density of water = 0.998 gm/cm^3, and Penman ratio at 20 °C = 2.19.

Solution: For using equation (1.34), e_o = 23.37 mm Hg, RH = 70%, and W = 200 km/day, derived from the given data. A substitution of these variables into equation (1.34) leads to the estimation of the mass transfer evaporation rate E_a as:

$$E_a = (0.013 + 0.00016\,W)e_o\left(1 - \frac{RH}{100}\right)$$

= (0.013 + 0.00016 x 200) x 23.37 x (1 – 70/100) = 0.315 mm Hg

The net radiation Q_n (= 550 cal/cm^2/day) can be converted to the net evaporation rate, E_n, from equation (1.33) as:

$E_n = Q_n / \rho \upsilon$ = 550/ (0.998 x 586) = 0.94 cm/day.

For computing the rate of lake evaporation, E, from equation (1.31), the Penman Δ/γ-ratio is given as 2.19. Thus,

$$E = \frac{\Delta E_n + \gamma E_a}{\Delta + \gamma} = \frac{(\Delta/\gamma)E_n + E_a}{1 + \Delta/\gamma} = \frac{2.19x0.94 + 0.316}{1 + 2.19} = 0.74 \text{ cm/day}$$

1.6.5 PAN EVAPORATION

Measurements of evaporation from water bodies are made most commonly using evaporation pans. The loss of water from the pan during a given time period represents evaporation. It provides a measurement of the integrated effect of the net radiation, wind, temperature, and humidity on the evaporation from an open surface. These measurements generally differ from the actual amount of lake evaporation. The ratio of pan-to-lake evaporation is an empirical constant referred to as the pan coefficient.

The basic pan-evaporation formula is written as

$$E = K_p E_p \tag{1.36}$$

where E is the potential evaporation, K_p is the pan coefficient, and E_p is the pan evaporation. Table 1.22 provides values of Class 'A' pan coefficient (or K_p) for several places in the United States of America.

Example 1.11: Compute the potential evaporation using the pan-evaporation formula. Given: pan coefficient = 0.7 and pan evaporation = 10 cm.

Table 1.22 Class 'A' pan coefficient

Location	Season	Years of record	K_p	Reference
Denver, Colo.	Annual	1915-16	0.67	Sleight (1917); 12 ft. pan
U.S.A.	Annual	1926-28	0.69	Rowher (1931)
Ft. Collins, Colo.	Apr.-Nov.	1926-27	0.70	Rowher (1934); 85 ft. reservoir
Utah	May-Oct.		0.67	White (1932); 12 ft. pan
U.S.A.	Annual		0.69	Hall (1934)
U.S.A.	Annual		0.70	Subcommittee (ASCE) (1934)
Texas	Annual	1939-47	0.68	U.S. Dept. of Interior (1952)
Florida	Annual	1940-46	0.81	U.S. Dept. of Interior (1952)
Oklahoma	Annual	1950-51	0.69	U.S. Dept. of Interior (1952)
Ariz., Nev.	Annual	1952-53	0.60	U.S. Dept. of Interior (1952)
California	Annual		0.77	Young (1947)
California	Annual		0.77	Young (1947); 12 ft. pan

Solution: Here, K_p = 0.7 and E_p= 10 cm. Therefore, from equation (1.36) potential evaporation is computed as:
Potential evaporation = K_pE_p = 0.7 x 10 = 7 cm

1.6.6 EVAPOTRANSPIRATION

Evapotranspiration is the process by which water on the land surface, in the soil, and on the vegetation surface is converted to the water vapour and returned to the atmosphere. It consists of evaporation from water, soil, vegetation, and other surfaces and includes transpiration by vegetation. Thus, evapotranspiration covers all the water converted to vapour and returned to the atmosphere and, therefore, is an important component of the long-term water balance.

In evaporation studies, the concept of potential evapotranspiration (PET) attributed to Thornthwaite (Thornthwaite et al., 1944) is widely used. PET is the amount of evapotranspiration that occurs under the ample availability of moisture at all times. PET is taken as an indicator of crop water requirement. Several equations have been developed for its computation, the formulae that are commonly used are briefly described below.

(i) *Thornthwaite Formula*

The Thornthwaite formula is the combination of the following equations expressed as

$$J = \sum_{N=1}^{12} I = \sum_{N=1}^{12} \left(\frac{T_N}{5}\right)^{1.514} \tag{1.37}$$

and

$$PET(0) = 1.62\left(\frac{10T}{J}\right)^{c} \tag{1.38}$$

where I is a heat index; T_N is the temperature of the Nth month (°C); N is an integer varying from 1 to 12; J is an annual temperature efficiency index, defined as the sum of 12 monthly values of heat index I; T is the mean monthly temperature (°C); PET(0) is the potential evapotranspiration at 0° latitude (cm/month); and c is an exponent determined as:

$$c = 0.000000675\,J^3 - 0.0000771\,J^2 + 0.01792\,J + 0.49239 \tag{1.39}$$

At latitudes other than 0°, PET is calculated as

$$PET = K\,(PET(0)) \tag{1.40}$$

where K is a constant for each month of the year, varying as a function of latitude.

Example 1.12: Estimate the potential evapotranspiration using the following data:

Wind velocity	10 km/hr
Mean monthly temperature	25 °C
Saturation vapour pressure	80 mb
Relative humidity	20%

Solution: From equation (1.37),

$$I = \left(\frac{25}{5}\right)^{1.514} = 11.435$$

Therefore, J = 12 x 11.435 = 137.221. From equation (1.39),

$$\begin{aligned} c &= 0.000000675\,J^3 - 0.0000771\,J^2 + 0.01792\,J + 0.49239 \\ &= 0.000000675 \times (137.221)^3 - 0.0000771 \times (137.221)^2 \\ &\quad + 0.01792 \times 137.221 + 0.49239 \\ &= 3.244 \end{aligned}$$

The potential evapotranspiration (PET) can be computed as:

$$PET = 1.62\left(\frac{10T}{J}\right)^c = 1.62 \times (10 \times 25/137.221)^{3.224} = 11.21 \text{ cm}$$

(ii) *Blaney-Criddle Formula*

The Blaney-Criddle Formula used for daily evapotranspiration is expressed in SI units as

$$U = p(0.46T + 8.13) \tag{1.41}$$

where U is the daily consumptive use factor (mm); p is the ratio of mean daytime hours for a given month to the total daytime hours in the year (%), a function of latitude; and T is the mean daily temperature for a given month (°C). For given actual insolation time, minimum relative humidity, and daytime wind speed, the consumptive use factor can be converted to evapotranspiration using standard graphical plots available elsewhere (Ponce, 1989; Singh, 1992; Singh and Singh, 2001).

The consumptive water requirement of a crop represents the amount of water required to meet its evaporation demands in the presence of ample supply of water. It is equal to the product of consumptive use factor, U, and an empirical consumptive use crop coefficient, k_c. The consumptive water requirements vary widely between climates having similar air temperatures and day lengths. Therefore, the effect of climate on crop water requirements is not fully described by the daily consumptive use factor, U. However, the crop coefficient, k_c, incorporates the effect of climate.

Example 1.13: Estimate the consumptive use factor for the reference crop using the Blaney-Criddle formula utilizing the following information:

Month of the year	March
Location	35 °N
Mean daily temperature	18 °C
Ratio of mean daytime hours in May to total daytime hours in the year	0.27

Solution: The consumptive use factor, U, can be computed from equation (1.41) as:
$U = p(0.46T + 8.13) = 0.27 \times (0.46 \times 18 + 8.13) = 4.43$ mm/day

(iii) *Jensen-Haise Method*

Jensen and Haise (1963) evaluated 3000 observations of evapotranspiration as determined by soil sampling procedures over a 35-year period. Based on the investigation, they suggested a formula to compute PET as:

$$PET = C_T(T - T_x)Q_s \tag{1.42}$$

where C_T is the temperature coefficient, T_x is the intercept of temperature axis, Q_s is the global radiation, and PET has the same dimension as of Q_s. C_T and T_x are considered as constants for an area. Jensen (1966) later defined C_T as

$$C_T = \frac{1}{C_1 + C_2 C_H} \tag{1.43}$$

$$C_H = \frac{50}{(e_2 - e_1)} \tag{1.44}$$

where e_2 and e_1 are the saturation vapour pressures (mb) at the mean maximum and mean minimum temperatures, respectively, for the warmest month of the year in an area; C_H is a nondimensional quantity; and C_2 = 13 °F or 7.6 °C. Jensen et al. (1970) defined C_1 = 68 °F - (3.6 °F x elevation in feet/1000 ft) and T_x = 27.5 °F – 0.25 (e_2 - e_1) °F/mb – [elevation (feet)]/1000 °F. For temperatures in °C, C_1 = 38 °C - (2 °C x elevation in m/305) and T_x = -2.5 – 0.14 (e_2 - e_1) °C/mb – elevation (m)/550. On the basis of relative humidity at noon, Stephens (1965) recommended the following values for C_T and T_x in °F:

Relative humidity at noon (%)	C_T	T_x (°F)
Greater than 60%	0.00820	23.2
50 – 60 %	0.00878	22.1
40 – 50 %	0.01067	21.1
Less than 40 %	0.014	26.4

(iv) Penman-Monteith Method

In the determination of PET, the Penman-Monteith method (Monteith, 1965) incorporates a wind function term based on the log-law wind profile and a leaf or canopy resistance term as:

$$PET = \frac{\Delta E_n + \rho_a c_p (e_s - e_a)/r_a}{\Delta + \gamma(1 + r_s / r_a)} \tag{1.45}$$

where E_n is the net evaporation rate (cm/day), ρ_a is the air density (gm/cm^3), r_a is the aerodynamic resistance to vapour and heat transfer (cm^{-1}day), r_s is the bulk surface resistance which is a function of the leaf area index (cm^{-1}day), and other terms are the same as described in equation (1.31). For further analytical details, reference can be made to ASCE (1996).

(v) *Hargreaves Method*

The Hargreaves method (Hargreaves and Samani, 1982; 1985) is suggested as a means for estimating grass reference PET (PET_o) in situations where data are limited and only maximum and minimum air temperature data are available. The Hargreaves equation is expressed as

$$PET_o = 0.0023\,(T_{max} - T_{min})^{0.5}\,(T_{mean} + 17.8)Q_{ex} \tag{1.46}$$

where T_{max} is the daily maximum temperature (°C), T_{min} is the daily minimum temperature (°C), and Q_{ex} is the extraterrestrial radiation (cm/day).

(vi) *van Bavel Method*

For computing PET, van Bavel (1968) suggested the following equation for well-watered grass (Linacre, 1969):

$$PET = \frac{0.167Q_n + 14x10^{-3} W_1 (T_{max} - T_{min})}{22 - 15x10^{-2}(T_{max} + T_{min})} \tag{1.47}$$

where PET is in mm/day, Q_n is the net radiation in cal-cm^{-2} day^{-1}, daily air temperatures are in °C, W_1 is the wind speed at 1 m in km/day.

1.7 Infiltration

The process of infiltration is a part of the process of soil moisture movement. The water seeps into the ground through soil pores. The infiltrated water is useful for plant growth, and irrigation demand actually arises when plants are unable to extract the water retained in the soil pores within the root zone depth. The water that percolates further down and meets the groundwater table becomes a part of the groundwater reservoir from which water can be extracted through wells for useful purposes such as domestic purposes and irrigation.

Infiltration refers to the water entry at the soil surface. Infiltration is distinguished from percolation in that the former takes place at the soil surface and the latter occurs below the surface and describes the downward movement of water. The amount of infiltration is described by the moisture content, θ, which is defined as

$$\theta = \frac{\text{Volume of water } (V_w)}{\text{Total volume } (V)} \tag{1.48}$$

where the total volume is the sum of the volume of air (V_a), water (V_w), and solids (V_s) as shown in Figure 1.18. Evidently, the water content represents the fractional amount of soil moisture which is available in the soil column.

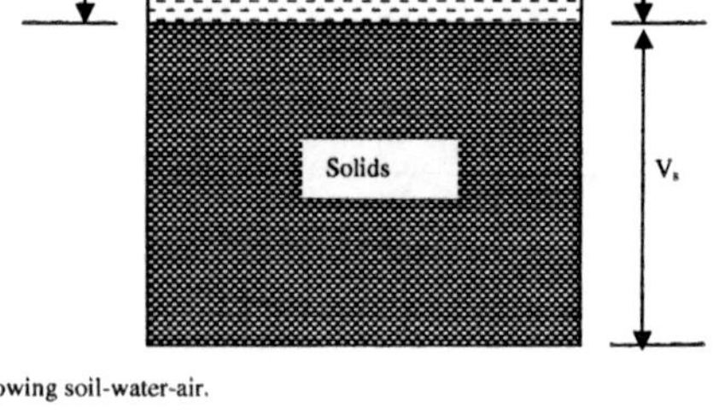

Figure 1.18. Schematic diagram showing soil-water-air.

There are two critical levels of moisture content. The first is the field capacity that is the amount of water which a soil will hold against gravitational forces. The field capacity is used to indicate the upper limit of the water available for plant growth, assuming that the amount in excess of field capacity drains away too quickly to be of any use. Thus, it represents a wet condition. The other is the extreme case of dry condition which occurs when the plant roots are unable to extract water from the soil pores. This leads to the permanent wilting of the plant and at this point, the plant fails to recover. Thus, the permanent wilting point is the lower limit of water available in the soil in the root zone. How water is actually held in the soil pores is of importance to understand the process of infiltration.

1.7.1 MECHANISM OF WATER RETENTION BY SOIL

The ability of soils to retain water is attributed to the presence of air-water interfaces similar to those in blotting papers or sponges. The surface tension acting at the air-water interface provides the mechanism for water retention. The difference between the hydrostatic pressure in a tube shown in Figure 1.19 on a level with the free water surface (Φ_2) and the hydrostatic pressure immediately below the interface (Φ_1) is equal to ρgh, where ρ is the water density, g is the gravitational acceleration, and h is the hydraulic head. Therefore,

$$\Phi_2 - \Phi_1 = \frac{\sigma(1 + \cos\alpha)}{r} \tag{1.49}$$

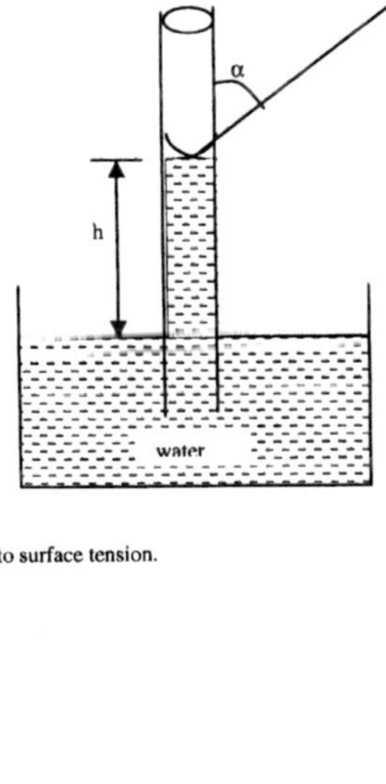

Figure 1.19. Capillary action due to surface tension.

where α is the contact angle, σ is the surface tension, and r is the radius of the tube. Equation (1.49) describes the lowering of the hydrostatic pressure across the curved air-water meniscus. The amount of lowering is commonly referred to as suction. The retention of water in soil particles can be thought of as a capillary phenomenon in which the soil particles are usually assumed to exhibit a contact angle $\alpha = 0$, leading to

$$\Phi_2 - \Phi_1 = \frac{2\sigma}{r} \tag{1.50}$$

Equation (1.50) shows that the suction apparently increases with a decrease in the pore diameter. In other words, water is more tightly held in narrow pores than in wide ones.

1.7.2 RETENTION CURVES

The forces of retention are the adhesion-cohesion forces resulting in concave air-water interfaces and the other kind of forces are adsorption forces arising due to ionic soil particles, as in clays. Combined together, these forces are known as matric forces because both result from the soil matrix. The existence of matric forces can be demonstrated by a tensiometer (Figure 1.20). A tensiometer is like a piezometer adapted for the measurement of negative pressure (suction). The water inside the porous cup of a tensiometer attains equilibrium with the water in the surrounding soil. The matric suction (h_m) is, thus, given by the position of the water level at the open end of the manometer.

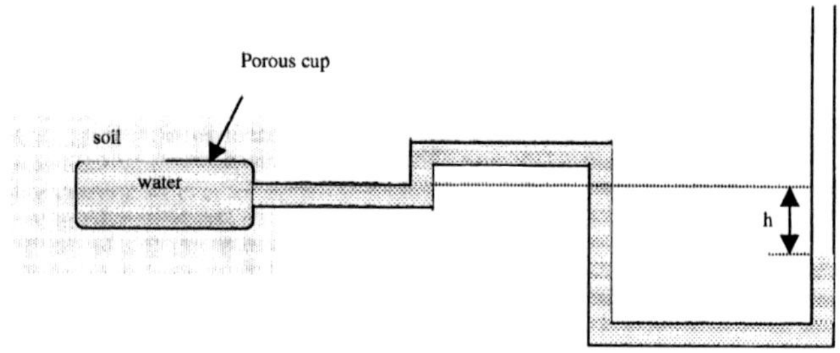

Figure 1.20. Tensiometer.

The matric suction is a function of soil water content. This can be realized from equation (1.50) implying that the greater the lowering in hydrostatic pressure (or suction), the smaller will be the radius of the water-filled pores or, in other words, the lower will be the resulting water content. If the suction is expressed in cm of water column and plotted against the water content, the resulting curve is called a retention or ψ-θ curve, where ψ refers to the suction head. This property signifies the water storing capacity of soils. Whether the water transmission through soil pores actually takes place depends on another important property known as hydraulic conductivity or permeability and is defined using Darcy's law.

1.7.3 DARCY'S LAW

Darcy's law of groundwater flow states that the flux (cross-sectionally averaged velocity of flow) (v) is directly proportional to the hydraulic gradient (i). Expressed mathematically,

$$v = k\,i \tag{1.51}$$

where k is the proportionality constant known as the hydraulic conductivity (LT^{-1}) of the soil. In unsaturated flow, k is a function of θ and is referred to as the capillary conductivity, K_h, which depends not only on the pore geometry but also on the water content, and is defined as the average velocity of flow under the unit hydraulic gradient. In equation (1.51), since v is a vector quantity, its direction is identified from the higher head to the lower head, identical to the flow of heat that flows from a higher temperature to a lower temperature. Physically, K_h refers to the openness of the soil-pores, or acts as an indicator of ease with which water can flow in a soil system. If K_h is plotted against θ on a graph sheet, the resulting curve depicts K_h-θ relation. Figures 1.21 and 1.22 respectively show typical ψ-θ (or retention curve) and K_h-θ relations for Yololight clay. As θ increases, the absolute value of ψ decreases and K_h increases and vice versa.

1.7.4 TRANSPORT OF SOIL MOISTURE

A mathematical expression for description of the soil-moisture transport in unsaturated soils can be derived from the Darcy law, the law of mass conservation, and the law of energy conservation. These are outlined below in partial differential form:

(1) *Darcy law*

For unsaturated flow where K_h depends on the degree of saturation, the Darcy law is expressed as

$$v = -K_h(\theta)\frac{\partial \Phi}{\partial z} \tag{1.52}$$

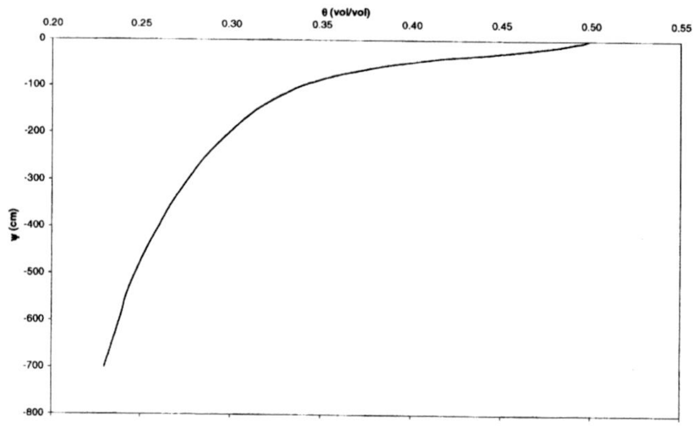

Figure 1.21. Soil moisture potential (ψ)- moisture content (θ) relation for Yololight clay.

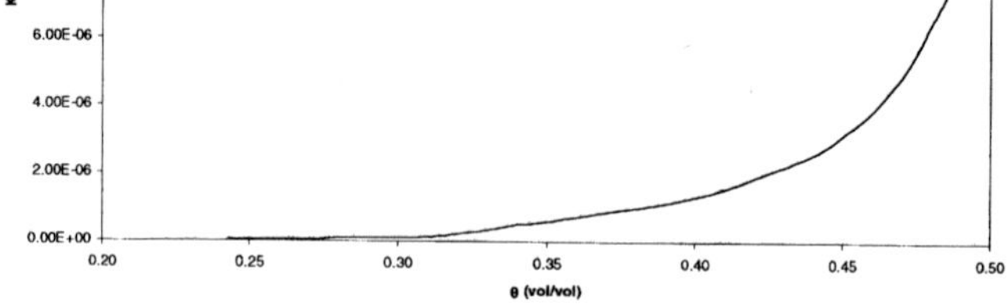

Figure 1.22. Hydraulic conductivity (K_h)-moisture content (θ) relation for Yololight clay.

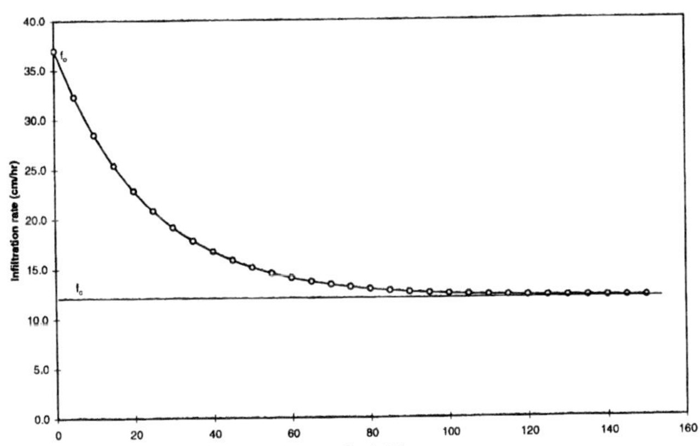

Figure 1.23. A typical infiltration decay curve for silty clay.

(2) *Continuity Equation*

$$\frac{\partial\theta}{\partial t}=\frac{\partial v}{\partial z} \tag{1.53}$$

where z is the height above some datum level.

(3) *Pressure or Energy Equation or Buckingham Equation*

$$H=\frac{\Phi}{\rho g}+z=\psi+z \tag{1.54}$$

where $\psi = \psi(\theta)$, H is the total potential head, Φ is the total pressure, ψ is the pressure head ($=\Phi/\rho g$), z and t are, respectively, the space and time coordinates. Combining equations (1.52) through (1.54), the result is the non-linear second order Fokker-Planck infiltration equation, the linear form of which is expressed as:

$$\frac{\partial\theta}{\partial t}+\frac{dK_h}{d\theta}\frac{\partial\theta}{\partial z}=D\frac{\partial^2\theta}{\partial z^2} \tag{1.55}$$

where D is the moisture diffusivity defined as

$$D = K_h \, d\psi/d\theta \tag{1.56}$$

which can be derived from relations (Feddes et al., 1988): $C(\theta) = d\theta/d$ and $D(\theta) = K_h(\theta)/C(\theta)$. A typical infiltration curve is shown in Figure 1.23. It is apparent that the infiltration rate decays with time monotonically and reaches a minimum constant value (f_c) after a sufficiently long time. An f-value at a time in the figure represents the infiltration capacity of the soil at that time. For example, f_c is called the infiltration capacity when the soil is completely saturated. It is the rate of water flow that percolates down to meet the groundwater table.

A practical implication of f_c occurring for a sufficiently long time due to prolonged irrigation in an area is to raise the groundwater table, leading sometimes to waterlogging and, consequently, to salinity and other environmental problems. The Mahi Right Bank Canal command area in Gujarat and some parts, viz., Indira Gandhi Canal command area in India are an example. The hydrologic implication of infiltration, f, can be described in terms of runoff as follows. After accounting for evaporation, if the effective rainfall intensity over a catchment $i \leq f$, the whole rainfall goes to infiltration and if $i > f$, $(i - f)$ will appear as direct surface runoff which may be of direct use in drinking, irrigation, and power generation.

1.7.5 MEASUREMENT OF INFILTRATION

Two types of equipments are in common use: (a) flooding type infiltrometer and (b) rainfall simulator.

a) *Flooding Type Infiltrometer*

This is a 60 cm long metal cylinder of 60 cm diameter and open at both ends. For conducting a field experiment, the cylinder is driven into the ground to a depth of 50 cm. A perforated disc is placed inside the cylinder to check the turbidity and clogging of soil pores at the soil surface. Water is filled up to 5 cm above the ground level and a pointer is set to mark the water level. As infiltration proceeds with time, the water is made up by a burette at a certain time interval to the tip of pointer. The added volume of water divided by the time interval and the x-sectional area of the cylinder yields the infiltration rate. The experiment continues till a reasonably steady rate is obtained. The experiment takes about 2-24 hours, depending on the type of the soil.

The major problem with such an instrument is that the infiltrated water spreads in the horizontal direction as shown by the dotted line in Figure 1.24 and the x-sectional area of the cylinder is, thus, not representative of the actual area of infiltration. To circumvent this problem, a ring infiltrometer, consisting of two concentric rings, is used. The cylinder is inserted into the ground and water is filled in both rings to a common level. The water in the outer ring serves as a water jacket and the infiltrating water from it restricts the water movement of the inner ring in the vertical direction, as shown in Figure 1.25. The volume of water added to the inner ring is used for computing the infiltration rate.

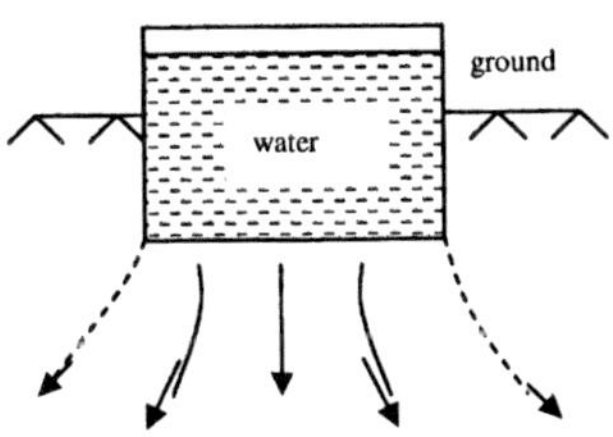

Figure 1.24. Simple infiltrometer.

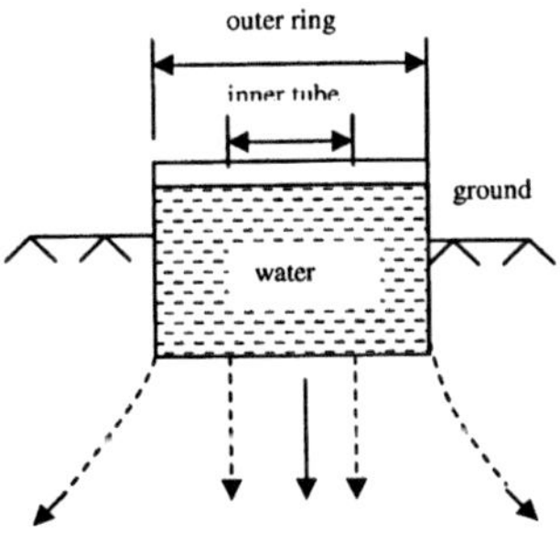

Figure 1.25. Ring infiltrometer.

The major disadvantages of flooding type infiltrometer are: (1) Driving of the cylinder into the ground disturbs the soil structure. (2) The infiltration rate, to some extent, depends on the size of the infiltrometer due to the border effect; those with the larger diameter yield less infiltration than the smaller ones. (3) The most important deficiency is that these do not simulate the effect of rainfall intensity, for which rainfall simulators are used.

b) *Rainfall Simulator*

A rainfall simulator is a series of nozzles on a 4m x 2m plot of land, fixed along the length, and a drainage arrangement at the border for collecting the drained water. The specifically designed nozzles sprinkle water at a height of 2 m. In a certain time interval Δt, the total rainfall P (= $i_o \Delta t$, where i_o is the uniform rainfall intensity) produces the runoff volume Q, and the infiltration rate f is computed as below:

$$f = \frac{P - Q}{A\Delta t} \tag{1.57}$$

where A is the plot area. The infiltration rates computed from such a simulator are usually less than those from flooding type infiltrometers due to the effect of rainfall intensity and water turbidity.

1.7.6 CONCEPTUAL INFILTRATION MODELS

Several models have been developed based on field measurements that describe the process of infiltration. Examples of some popular models that include empirical, semi-empirical, and theoretical ones, are listed in Table 1.23. Empirical relations have been developed based on field measurements that describe the infiltration process of a soil. The popular Horton (1933, 1938), modified Kostiakov, Philip, Green-Ampt, Singh-Yu, and Overton models are described below.

Table 1.23 Infiltration models

Sl. No.	Name of the model	Year	Sl. No.	Name of the model	Year
1.	Green and Ampt	1911	11.	Smith	1972
2.	Kostiakov	1932	12.	Dooge	1973
3.	Horton	1933	13.	Morel-Seytoux and Khanji	1975
4.	Philip	1957	14.	Parlange	1971
5.	Holtan	1961	15.	Collis-George	1977
6.	Overton	1964	16.	Smith and Parlange	1978
7.	Hydrograph model (Dunin, 1969)	1969	17.	Zhao	1981
8.	Modified Kostiakov	-	18.	HEC	1981
9.	Mein and Larson	1971	19.	Singh and Yu	1990
10.	Snyder	1971	20.	Mishra and Singh	2002

(i) *Horton Model*

The Horton infiltration model is based on the hypothesis that the reduction in the infiltration rate with time is directly proportional to the rate of infiltration. Horton (!933, 1938) showed it to be the result of action of energy of rain falling on the soil surface. Mathematically, the model is expressed as:

$$f = f_c + (f_o - f_c)e^{-\alpha t} \quad (1.58)$$

where f is the infiltration rate $[LT^{-1}]$ at time t, f_o is the initial infiltration rate $[LT^{-1}]$ at time t = 0, f_c is the minimum constant infiltration rate $[LT^{-1}]$, and α is the infiltration decay factor $[T^{-1}]$. The terms of equation (1.58) can be easily described from Figure 1.23. Both the constant minimum value of f_c and the rate of infiltration at time t = 0, f_o, can be determined directly from the figure and α can be determined from the plot of natural logarithmic values of the ratio $(f-f_c)/(f_o-f_c)$ on the ordinate (y-axis) and time on the abscissa (x-axis) as shown in Figure 1.26. The slope of the resulting straight line is the desired value of α. Its application is illustrated below.

Example 1.14: From field observations on silty clay in Narmada basin (India) the following infiltration rates were computed with time:

Time (min.)	0	5	10	15	20	25	30	35	40
Infiltration (cm/hr)	37.0	32.3	28.5	25.4	22.9	20.8	19.2	17.8	16.7

45	50	55	60	65	70	75	80	85	90	95
15.8	15.1	14.5	14.1	13.7	13.4	13.1	12.9	12.7	12.6	15.8

100	105	110	115	120	125	130	135	140	145
15.1	14.5	14.1	13.7	13.4	13.1	12.9	12.7	12.6	12.1

Find the Horton model parameters f_c, f_o, and α.

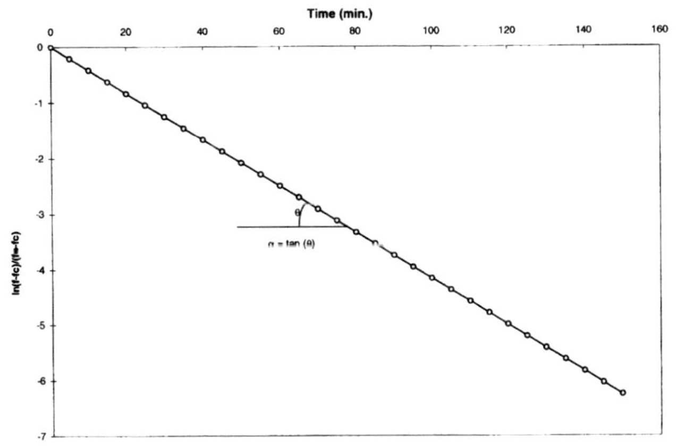

Figure 1.26. Determination of Horton model's decay factor.

Solution: The plot of the observed data is given in Figure 1.23. From the figure the values of f_c and f_o can be determined as 12 cm/hr and 37 cm/hr, respectively. For the computation of α, the logarithmic values of $(f-f_c)/(f_o-f_c)$ are plotted as a function of time, as shown in Figure 1.26. The tangent of the slope of the fitting line is the desired value of α, which is equal to 2.48 hr^{-1}.

(ii) *Modified Kostiakov Model*

The general form of the infiltration expression given by Kostiakov (1932) as cited by Rode (1965) is

$$F = a\,t^{b} \tag{1.59}$$

where a and b are constants (0<b<1). Equation (1.59) can be written for f as

$$f = \alpha(t)^{-\beta} \tag{1.60}$$

where α= ab and β=1-b. Equations (1.59) and (1.60) are applicable for $t \neq 0$. Equation (1.60) has primarily been used for irrigation applications.

Smith (1972) modified the Kostiakov method (equation (1.60)) by including the term f_c as

$$f = f_c + \alpha t^{-\beta} \tag{1.61}$$

where α and β are the same as described in equation (1.60). Equation (1.61) is similar to the Philip model.

(iii) *Philip Model*

The solution of equation (1.55) leads to Philip's (1957) two-term infiltration equation:

$$f = A'' + S_s t^{-1/2} \tag{1.62}$$

where f is the infiltration rate (LT^{-1}) at time t, A″ is a constant (LT^{-1}) and S_s (soil sorptivity) ($LT^{-1/2}$) are infiltration constants. S_s is derived from the ψ-θ relation of a soil and A″ corresponds to the minimum infiltration rate, f_c, derivable from field observations.

Equation (1.62) assumes an infinitely deep soil column so that water can percolate downward at a rate determined by the saturated hydraulic conductivity. In reality, the thickness of the soil column is finite. Therefore, the infiltration capacity is not necessarily limited by the saturated hydraulic conductivity of the soil but by the available space to store water in the soil column and by the rate of lateral saturated subsurface flow in the hillslope toward the stream (Schaake et al., 1996).

(iv) *Green-Ampt Model*

Green and Ampt (1911) presented a model based on the assumption that soil may be regarded as a bundle of tiny capillary tubes irregular in area, direction, and shape. Assuming a homogeneous, deep soil with uniform initial moisture content, and ponded surface, the Green-Ampt infiltration equation takes the form:

$$f = A\left[1 + \frac{B(H_c + H)}{F}\right] \tag{1.63}$$

where A and B are model parameters which depend on soil characteristics, H_c is the capillary potential at the wetting front [L], H is the head of water on the surface [L], and F is the cumulative infiltration. Smith and Parlange (1978) have shown that for initially ponded conditions (ponding time is zero) with rainfall rate approaching infinity and K varying slowly near saturation, equation (1.63) can be derived from equation (1.55). For fitting to infiltration data, equation (1.63) can be re-written as

$$f = A + C/F \tag{1.64}$$

where $C = AB(H_c + H)$. Parameters A and C can be estimated using observed data.

(v) *Singh-Yu Model*

The Singh-Yu (1990) model is based on two postulates: (1) The rate of infiltration in excess of the final infiltration rate at any time is directly proportional to the mth power of available storage space in the soil column at that time and (2) inversely proportional to the nth power of the cumulative infiltration up to that time. Expressed mathematically,

$$f(t) = f_c + \frac{a\,[S(t)]^m}{[S_o - S(t)]^n} \tag{1.65}$$

where f(t) is the infiltration rate $[LT^{-1}]$ at time t; S(t) is the available storage for water retention in a soil column at time t [L]; S_o is the potential storage space available for moisture retention in soil column [L]; a, m, and n are respectively the coefficient and exponents of the variables S(t) and $(S_o\text{-}S(t))$. The cumulative infiltration (F) is equal to $S_o - S(t)$. Singh and Yu suggested several relations for the variation of S(t) with time. This is a general model because it specializes into the models of Holtan, Horton, Kostiakov, modified Kostiakov, Overton, Green and Ampt, and Philip. The general model has five parameters namely, f_c, S_o, a, m, and n. The parameters f_c and S_o are determined from the available infiltration data and soil properties, for example, soil porosity and initial moisture content, and the other parameters a, m, and n are computed using a least squares approach.

Mishra et al. (1999) expressed equation (1.65) in terms of the SCS curve number (CN) as

$$f = f_c + a\,S_{abs}\left(\frac{CN_o}{100}\right)^n \frac{(100 - CN_t)^{n+1}}{(CN_t - CN_o)^n} \tag{1.66}$$

where CN_o is the CN at time t = 0, CN_t is the time varying CN, and S_{abs} is the maximum potential maximum retention described in Chapters 3 and 4.

(vi) *Overton Model*

The Overton (1964) infiltration equation is expressed as

$$f = f_c \sec^2[(a f_c)^{1/2}(t_c - t)] \tag{1.67}$$

where 'a' is a constant which varies widely with the soil moisture and t_c is a time parameter. From equation (1.67) it is possible to compute infiltration rate at any time during a storm, even when rainfall does not exceed the infiltration capacity or when there is a temporary interruption in rainfall.

1.7.7 INFILTRATION INDICES

To make hydrologic calculations easy and simple, there are two types of infiltration indices in common use: φ-index and W-index. Both the indices represent a constant infiltration capacity for the storm duration and are computed from the observed storm hyetograph and the corresponding direct runoff hydrograph. Steps for computation of the φ-index follow:

1. Subtract the baseflow, if any, from the available runoff hydrograph ordinates to determine the direct runoff hydrograph (DRH).
2. Compute the area under DRH which yields the total direct runoff volume.
3. Convert the direct runoff volume into the corresponding depth by dividing it by the area of the catchment.
4. Compute the total rainfall depth as follows. First prepare a table showing the time and rainfall intensity, multiply the rainfall intensity by their time interval, and then sum all the values to compute the total rainfall depth.
5. Compute the total infiltration which is the difference between the total rainfall depth (in step 4) and the computed DRH depth (in step 3).
6. Assume that the total effective duration (t_e) equals the storm duration. Here, the effective duration pertains to the effective period of DRH producing storm.
7. Compute the φ-index by dividing the total infiltration depth by t_e (computed in step 6).
8. Compare the rainfall intensities with the computed φ-index in step 7. If all rainfall intensities are greater than or equal to the φ-index, the assumed t_e is the desired value. If any of the values of the rainfall intensity in the whole storm duration is less than the φ-index, sum all those time intervals showing rainfall intensity less than or equal to the φ-index and revise t_e by subtracting them from the assumed t_e in step 5 and repeat steps 7 and 8.

Rainfall minus φ-index is the rainfall excess. If the rainfall intensity (= the rainfall depth/time interval) is less than or equal to the φ-index, the infiltration rate is equal to the rainfall intensity and if the rainfall intensity is greater than the φ-index, the difference between these two multiplied by the time interval yields the direct runoff depth.

Example 1.15: An isolated storm in a catchment produced a runoff of 3.5 cm. For the following time distribution of rainfall, estimate the φ-index and the distribution of rainfall excess.

Time (hr)	0	1	2	3	4	5	6
Rainfall intensity (mm/hr)	0	5.0	10.5	19.0	21.0	11.5	9.5

Solution: Here, the total direct runoff hydrograph (DRH) depth = 3.5 cm = 35.0 mm, and time interval of the hyetograph (Δt) = 1 hr. Total rainfall depth = (sum of rainfall intensities) x Δt = 76.5 x 1 = 76.5 mm. Thus, infiltration = total rainfall depth - DRH depth = 76.5 - 35.0 = 41.5 mm.

Assuming that the total effective duration (t_e) = 6 hrs, φ-index = total infiltration/t_e = 41.5/6 = 6.92 mm/hr, which is greater than the first hour rainfall intensity (i.e. 5 mm/hr). Therefore,
Total rainfall excess = (76.5 - 5.0) x Δt = 71.5 x 1 = 71.5 mm
Effective duration (t_e) = 6 - 1 = 5 hrs
Infiltration = 71.5 - 35.0 = 36.5 mm. Thus,
φ-index = 36.5/5 = 7.3 mm/hr
which is an appropriate value as all rainfall intensities (excluding that in the first hour) are greater than φ-index, implying that the last five-hour rainfall actually contributes to DRH. Therefore,
Rainfall excess = rainfall intensity - φ-index,
The time distribution of rainfall excess is given below. As a further check, the sum of rainfall excess = 35.0 x 1.0 = 35.0 mm = DRH.

Time (hr)	0	1	2	3	4	5	6
Rainfall excess (mm/hr)	0.0	0.0	3.2	11.7	13.7	4.2	2.2

It is noted that the φ-index includes all types of losses or abstractions, including the initial abstractions (I_a) such as surface detention, interception, and a part of infiltration at the beginning of infiltration. The W-index accounting for I_a is expressed as

$$W - \text{index} = \frac{(P - I_a) - Q}{t_e} \tag{1.68}$$

where P is the total rainfall (mm), I_a is the initial abstraction (mm), Q is the total direct runoff depth (mm), and t_e is in hr. I_a includes losses from interception, surface detention storage, evaporation, and infiltration.

Steps for the W-index computation are the same as those of the φ-index with the difference that the rainfall depth computed in step 4 excludes I_a from rainfall. The water intercepted through vegetation, surface detention storage, and a fraction of the infiltrated water get evaporated to the atmosphere and, therefore, do not contribute to direct runoff. In effect, these can be excluded from the total amount of rainfall in terms of I_a. Since I_a is usually difficult to measure, it is difficult to accurately estimate the W-index.

The φ-index depends on the soil type, vegetal cover, initial moisture condition, storm duration, rainfall intensity, and time of the year (or season, whether growing or dormant). An accurate estimate of the φ-index can be had from a detailed study of the catchment. Its employment for design flood estimation requires a determination of a critical storm and then the corresponding rainfall excess. As the flood peaks generally occur due to long duration storms, I_a is insignificant and can be assumed to be zero. The Central Water Commission, New Delhi (India), suggested an empirical relationship for daily runoff Q as:

$$Q = \beta i^{1.2} \tag{1.69}$$

and

$$\phi\text{ - index} = \frac{P - Q}{24} \tag{1.70}$$

where i is the rainfall intensity (cm/day) and β is a coefficient that depends on the soil type as shown in Table 1.24. Where no data are available, the φ-index can be assumed to be equal to 0.10 cm/hr for design purposes.

Table 1.24 β-values for different soil types

Sl. No.	Soil type	β
1.	Sandy soils and sandy loam	0.17 to 2.25
2.	Coastal alluvium and silty loam	0.25 to 0.34
3.	Red soils, clayey loam, and grey, brown alluvium	0.42
4.	Black-cotton and clayey soils	0.42 to 0.46
5.	Hilly soils	0.46 to 0.50

1.8 Runoff

Surface runoff refers to the water flowing on the surface of the land as overland sheet flow or channel flow or both. Water flows from higher elevations to lower elevations due to gravitational forces. Small streams eventually combine to form rivers that finally drain into the seas or oceans to complete the hydrologic cycle. Runoff consists of three components: surface flow, interflow, and groundwater flow. Surface runoff is produced by the effective rainfall and is equal to the total rainfall minus hydrologic abstractions. Surface flow is also known as direct runoff and produces large flows in rivers due to its relatively shorter time of concentration.

Interflow is subsurface flow that takes place in the unsaturated soil layers located beneath the ground surface. It consists of the slow lateral movement of water toward lower elevations and also includes a part of the rainfall abstracted by infiltration. Interflow eventually meets streams.

Groundwater flow takes place below the groundwater table in the form of saturated flow through alluvial deposits and other water-bearing formations located beneath the soil mantle. Groundwater flow includes the portion of the infiltrated amount of water that has reached the water table by percolation from the overlying soil. The groundwater flow is a continuous but slow process.

1.8.1 MODES OF RUNOFF GENERATION

Surface runoff is generated by a variety of surface and near-surface flow mechanisms, of which some of the most important are (Ponce and Hawkins, 1996): (a)

Hortonian overland flow, (b) saturation overland flow, (c) subsurface flow (d) throughflow processes, (e) direct channel interception, and (f) surface phenomena, such as crust development, hydrophobic soil layers, and frozen ground.

The Hortonian overland flow takes place when rainfall rate exceeds infiltration capacity, usually at the beginning of a storm (or season), when the soil profile is likely to be dry. The difference between the rate of rainfall and infiltration capacity is the rainfall excess that is converted to direct surface runoff. This mechanism is prominent in arid and semiarid regions.

The mechanism of saturation overland flow (Dunne and Black, 1970) occurs when the soil profile has become saturated, either from antecedent rainfall events or from a sufficient volume of rainfall within the same event. At this moment, any additional rainfall is converted to surface runoff. The saturation overland flow takes place during an infrequent storm or at the end of a particularly wet season, when the soil is likely to be already wet from prior rains. This mechanism dominates in humid regions.

Throughflow occurs in heavily vegetated areas with thick soil covers of less permeable layers, overlying relatively impermeable unweathered bedrock (Kirkby and Chorley, 1967). In a strict physical sense, throughflow is not direct runoff, for the flow takes place primarily as interflow, or lateral flow immediately below the ground surface. The relative quick response of the throughflow in time is of the same order as of surface runoff and is thus regarded as a mode of surface runoff generation.

The subsurface flow occurs in both the saturated and unsaturated zones underneath the ground surface. The percolating water meeting the groundwater table (saturated zone) vertically creates a local gradient to cause water movement in lateral or horizontal direction. The subsurface flow process primarily governed by the Darcy law takes much longer time to meet a stream and occurs as baseflow than does the direct surface runoff. The baseflow contribution is of significance in low flow studies.

The concept of partial-area runoff originated from the recognition that runoff estimates were improved by considering that only rainfall on a small or a fairly constant part of each drainage basin contributes to direct runoff (Kirkby and Chorley, 1967). Therefore, the concept of partial-area runoff can be interpreted as a combination of throughflow in the upper hillslopes and saturation overland flow in the lower hill slopes (Chorley, 1978; Branson et al. 1981; Hawkins, 1981).

The direct channel interception implies the runoff originating from the rainfall directly falling into channels. Such a mode of runoff generation can be important in watersheds of dense channel networks and certain humid bases, where the direct channel interception may be a significant source of streamflow generation (Hawkins, 1973).

Surface phenomena include processes of crust development, hydrophobic soil layers, and frozen ground, which make the soil surface impermeable and, thus, promote surface runoff. For example, a surface crust may develop following splash erosion in a denuded watershed, adversely affected by human activities or a natural hazard, such as fire. Under specific combinations of soil type and texture, the silt entrained by splash erosion may deposit on the surface and create a thin crust, which eventually reduces infiltration. Thus, any additional rainfall will be converted to surface runoff. This mode of runoff generation is typical of semiarid regions, where large amounts of surface runoff may take place, even though the underlying soil profile, below a relatively thin veneer, remains substantially dry (USDA, 1940; Le Bissonnais and Singer, 1993).

For a more complete treatment of the above processes, text books on hydrology, for example, the one by Singh (1992) among others can be referred to. The runoff concentration, an important characteristic of the runoff hydrograph, is described below.

1.8.2 RUNOFF CONCENTRATION

The runoff concentration refers to the concentration of the rainfall excess at the outlet of the watershed, leading to the generation of peak flow. It implies that the flow rate at the outlet will increase until rainfall from the entire catchment has had time to travel to the outlet and contributes to the flow at the outlet. At that time, the maximum or equilibrium flow rate is reached. If the rainfall duration is sufficiently long, the runoff from the entire watershed eventually concentrates at the watershed outlet. The time taken by a parcel of water to travel from the farthest point on the divide to the catchment outlet is referred to as the time of concentration. The equilibrium flow rate is computed as

$$Q_e = 2.78 i_e A \tag{1.71}$$

where Q_e is the equilibrium flow rate (l/s), i_e is the effective rainfall intensity (mm/hr), and A is the catchment area (ha). In general, the amount of surface runoff increases with the size of the watershed and consequently the surface flow increases. The available formulae relating peak flow to the drainage area can, in general, be expressed as:

$$Q_p = cA^n \tag{1.72}$$

where Q_p is the peak flow corresponding to the catchment area A, and c and n are parameters to be determined by regression analysis. The TR-55 method (Ponce, 1989), however, computes peak flow on a per unit area basis.

The runoff concentration relates to the travel time of the runoff through the watershed and, therefore, depends on the watershed characteristics. The watershed of a high drainage density will yield a quick and peaked runoff response, whereas a low drainage density characterizes a delayed runoff response. The role of catchment shape in catchment hydrologic response has not yet been clearly established although some attempts (Singh, 1996) have been made. If other parameters are equal, a high form ratio or compactness ratio close to 1 describes a catchment having a quick and peaked catchment response. Conversely, a low form ratio or a compactness ratio much larger than 1 describes a catchment with a delayed runoff response.

The process of runoff concentration can lead to three distinct types of catchment response (Ponce, 1989). The first type occurs when the effective rainfall duration is equal to the time of concentration. In this case, the runoff concentrates at the outlet or reaches the maximum (or equilibrium) rate after the elapsed time equal to the time of concentration. Rainfall stops at that time and subsequent flows at the outlet are no longer concentrated because the whole catchment does not contribute to the flow afterwards. Therefore, the flow gradually recedes and reduces to zero. Since it takes the concentration time for the farthest runoff parcels to travel to the outlet, the recession time is approximately equal to the concentration time, as shown in Figure 1.27a. Such a response is referred to as concentrated catchment flow.

The second type of catchment response occurs when the effective rainfall duration exceeds the concentration time. In this case, the runoff concentrates at the outlet or reaches the maximum (equilibrium) rate after an elapsed time equal to the concentration time. Since rainfall continues to occur, the whole catchment continues to contribute to the flow at the outlet, and subsequent flows remain concentrated and they all are equal to the equilibrium value. After the cessation of rainfall, the flow gradually recedes back to zero (Figure 1.27b). This type of response is referred to as super-concentrated catchment flow.

The third type of response occurs when the effective rainfall duration is shorter than the time of concentration. In this case, the flow at the outlet does not reach the equilibrium value. After the termination of rainfall, the flow recedes back to zero. The requirements of the volume conservation and the recession time to be equal to the concentration time leads to the idealized flattop response shown in Figure 1.27c. This type of response is referred to as sub-concentrated catchment flow.

In practice, the first two responses are typical of small watersheds whereas the third type is typical of midsize and large catchments. Small watersheds have small times of concentration and large catchments render large times of concentration.

1.8.3 TIME OF CONCENTRATION

Several empirical formulae for determining the time of concentration have been developed. Most of them relate the time of concentration to slope, length, rainfall, and roughness parameters (Papadakis and Kazan, 1987). The widely used Kirpich (1940) formula is expressed as

$$t_c = \frac{0.06628\, \ell^{0.77}}{Y^{0.385}} \tag{1.73}$$

where t_c is the time of concentration (hr), ℓ is the length of the principal watercourse from the outlet to divide (km), and Y is the slope between maximum and minimum elevations (m/m). Another formula by Haithaway (1945) is expressed as

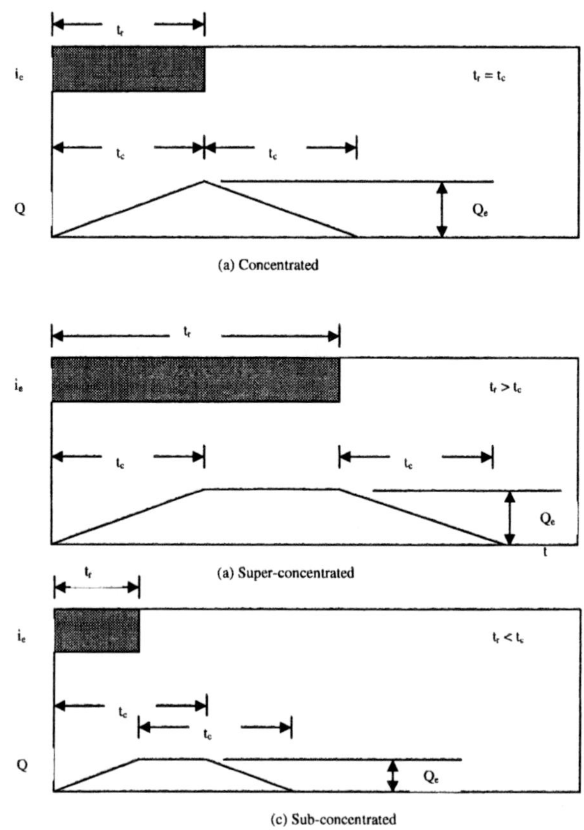

Figure 1.27. Types of catchment response (a) concentrated, (b) super-concentrated, (c) sub-concentrated. t_r time of rise, t_c = time of concentration.(Source: Ponce, 1989)

$$t_c = \frac{0.606\ (n\ell)^{0.467}}{Y^{0.234}} \tag{1.74}$$

where n is Manning's roughness coefficient and all other terms are the same as in equation (1.73).

For overland flow, the SCS TR-55 method computes the travel time (or time of concentration) using the formula expressed in metric units as (Ponce, 1989):

$$t_c = \frac{0.0288(n\ell)^{0.8}}{P_2^{0.5} Y^{0.4}} \tag{1.75}$$

where t_c is in hours, P_2 is the 2-year 24-hour rainfall depth in centimeters, Y is the slope in meters per meters, and the remaining terms are the same as described above.

It is also possible to compute the time of concentration using physically-based approaches, as for example, overland flow techniques using the kinematic wave approach (Singh, 1996).

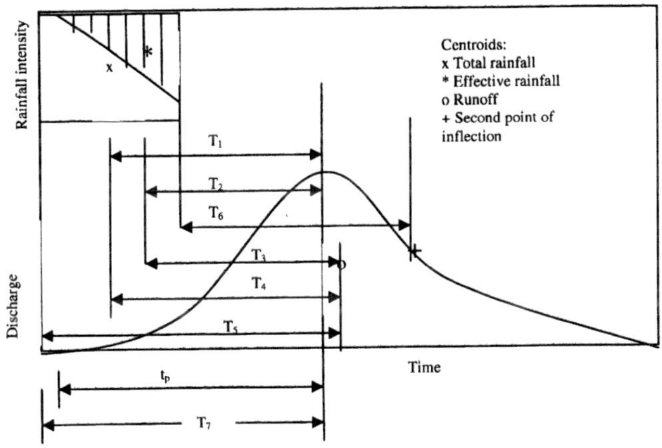

Figure 1.28. Alternative definitions of basin lag (Hall, 1984).

1.8.4 LAG TIME

The concept of catchment lag, basin lag, or lag time is central to rainfall-runoff modeling, unit hydrograph concept (described later) in particular. It is a measure of the time elapsed between the occurrence of unit rainfall and the occurrence of unit runoff.

Basin lag, embodying hydraulic length, catchment gradient, drainage pattern, drainage density, and other related factors, is an overall measure of the response time.

Hall (1984) suggested several definitions of the time lag, as shown in Figure 1.28. In this figure, lag times are measured from the beginning of the rainfall, beginning of runoff, centroid of the total rainfall, centroid of the effective rainfall, or the end of rainstorm to the time to peak discharge, time to the centroid of runoff, or time to the second point of inflection of the runoff hydrograph. Accordingly, the time lags are identified as T_1 through T_7. Among these, the T_2 lag defined as the time elapsed from the centroid of effective rainfall to the peak of runoff is most commonly used.

In practice, the catchment lag is determined from the catcment characteristics using the formula:

$$t_l = C\left(\frac{\ell L_c}{Y^{1/2}}\right)^N \tag{1.76}$$

where t_l is the time lag; ℓ is the catchment length measured along the main channel from outlet to catchment divide; L_c is the length measured along the main channel from outlet to a point located closest to the catchment centroid; Y is the measure of catchment slope; and C and N are empirical parameters. Here, ℓ describes catchment length, L_c is a measure of shape, and Y relates to relief.

Another formula after Snyder (1938) is widely used in the synthetic unit hydrograph analysis and is expressed as

$$t_l = C_t\,(\ell L_c)^{0.3} \tag{1.77}$$

where t_l is the catchment lag in hours, ℓ is the length along the main channel from outlet to divide, L_c is the length along the main channel from outlet to the point closest to the catchment centroid, and C_t is a coefficient dependent on the catchment gradient and the associated catchment storage. If ℓ and L_c are in kilometers, C_t ranges between 1.35 and 1.65 with a mean of 1.5 (Snyder, 1938).

The SCS method uses two methods for computing catchment lag: (a) the curve number method and (b) the velocity method. The curve number method is limited to catchments of areas ranging from 8 to 16 km^2 and expressed as (Ponce, 1989):

$$t_l = \frac{\ell^{0.8}(2540 - 22.86CN)^{0.7}}{14104CN^{0.7}Y^{0.5}} \tag{1.78}$$

where t_l is the catchment lag in hours, ℓ is the hydraulic length measured along the main channel from the outlet to the divide in meters, CN is the runoff curve number, and Y is the average catchment slope in meters per meter. The velocity method is considered appropriate for areas larger than 8 km^2 and curve numbers (discussed in the subsequent chapters in detail) varying in the range of 55-95. In this method, the time lag is taken equal to 60% of the time of concentration.

1.8.5 FLOW IN STREAM CHANNELS

The flow occurring in stream channels is known as streamflow. These channels are characterized by cross-sectional dimensions, cross-sectional shape, longitudinal slope, and boundary friction. Streamflow is determined using the empirical Manning or Chezy formula. The Manning formula is expressed as

$$v = \frac{1}{n} R^{2/3} s_o^{1/2} \tag{1.79}$$

where v is the mean flow velocity (m/s), R is the hydraulic radius (m), n is the Manning roughness coefficient, and s_o is the channel bed slope (m/m). In natural channels, n usually varies from 0.024 for large rivers with cross-section devoid of vegetation to as large as 0.079 for small streams with bed composed of large, angular boulders and banks consisting of exposed rocks, boulders, and trees. Typical values of n range between 0.03 and 0.05.

The Chezy formula is expressed as

$$v = C(Rs_o)^{1/2} \tag{1.80}$$

where C is the Chezy coefficient ($m^{1/2}/s$), and other terms are the same as for equation (1.79). The C-values calculated for the same conditions vary from 11 $m^{1/2}/s$ for large rivers to 79 $m^{1/2}/s$ for small rivers. Typical values of C lie in the range 40-70 $m^{1/2}/s$.

Flow rates can be grouped into low flow, average flow, and high flow. A low flow is typical of dry season, when streamflow is largely composed of baseflow originating from interflow and/or groundwater flow. A high flow occurs during wet seasons, when streamflow is primarily due to surface runoff. The average flow usually occurs in midseason and may have mixed contributions from surface runoff, interflow, and groundwater flow.

The study of low flows is needed for determining the minimum flow rates, below which a certain use will be impaired. Examples of such uses are irrigation water requirements, hydropower generation, river pollution control, and others. The average flows play an important role in the calculation of monthly and annual flow volumes for water resources planning (sizing of storage reservoirs) and management. The study of high flows is important for flood forecasting, flood plain zoning, flood warning, and flood protection works. The rivers during high flows tend to overflow the banks. In alluvial valleys, the land is apt to inundation.

1.8.6 RATING CURVE

The relation between river stage and the corresponding flow rate is referred to as a rating curve. The river stage represents the depth of flow above the river bed. The exact nature of the relationship is not readily known. The rating curves observed in laboratories and in field may be single-valued or double valued. For simplicity, the single-valued rating curves, known as steady-state rating curves, are usually preferred in hydrologic research and practice, and are expressed as

$$q = a(h - h_o)^b \tag{1.81}$$

where q is the discharge, h is the gauge height, h_o is the reference height, and 'a' and 'b' are constants to be determined from regression analysis. Double-valued rating curves are referred to as looped or hysteretic rating curves. The looped rating curves have been described by the non-dimensional hysteresis (ranging from 0 to 1), an index to represent the kind of wave behavior in natural open channels (Mishra and Singh, 1999). Flow rates in small catchments are primarily affected by infiltration that, in turn, also depends on the antecedent moisture condition of the soil, a brief description of which is in order.

1.8.7 ANTECEDENT MOISTURE

Surface runoff is directly related to the effective rainfall, and the effective rainfall is inversely related to the hydrologic abstractions including interception, surface detention, evaporation, evapotranspiration, and infiltration. During rainy periods, infiltration is the major abstraction. Actual infiltration rates and amounts vary widely, for they are heavily dependent on the initial soil moisture or antecedent moisture condition. Catchments with low initial soil moisture or dry catchments are not conducive to high runoff. Conversely, catchments with high initial soil moisture or wet catchments are likely to produce large quantities of runoff.

The concept of antecedent precipitation index (API) is generally used to define the antecedent moisture condition (AMC) of the watershed. The average level of moisture in a catchment varies daily. It is replenished by the current rainfall and subsequently depleted by evaporation and evapotranspiration. The assumption of a logarithmic depletion rate leads to a catchment's API for a day with no rain as

$$I_j = K\, I_{j-1} \tag{1.82}$$

where I_j is the index for day j, I_{j-1} is the index for day j-1, and K is a recession factor varying normally in the range (0.85, 0.98) (Linsley et al., 1982). If rain occurs on a day, the rainfall depth is added to the index. Parameters K and the initial value (0 day) of the index I are determined either from data or experience.

API is directly related to the amount of runoff. The greater the index, the greater will be the amount of runoff and vice versa. In practice, regression and other statistical tools are used to relate runoff to API. These empirically derived relations are strictly applicable to the situations similar to those for which they were derived.

Other measures of antecedent moisture have also been developed over the years. The Soil Conservation Service Curve Number (SCS-CN) method uses the concept of AMC grouped into three moisture levels: AMC I, a dry condition; AMC II an average condition; and AMC III a wet condition. The Illinois Urban Drainage Area Simulator (ILLUDAS) program of the Illinois State Water Survey (Terstriep and Stall,

1974) uses four levels of antecedent moisture described as bone dry, rather dry, rather wet, and saturated (Gray et al., 1982). As another example, the Streamflow Synthesis and Reservoir Regulation (SSARR) model (USAE, 1986) computes runoff volume based on a relationship linking runoff percent to a soil-moisture index (SMI), with rainfall as a third variable. The runoff percent is the ratio of runoff to rainfall, multiplied by 100.

1.9 Determination of Runoff Hydrograph

The description up to this point was focused on the elements of the rainfall-runoff process. In general, the objective of a hydrologist is to determine (a) the amount of runoff from a given amount of rainfall using the catchment characteristics, and (b) the time distribution of runoff at the outlet of the catchment, which is referred to as runoff hydrograph. In distributed hydrology, a watershed is divided in several hydrologic units. In practice, the sub-division is generally based on the size of tributaries joining the main river. The SCS-CN method is one of the most popular methods and can be used for computing the rainfall-excess hyetograph for each sub-catchment. Then, using the unit hydrograph technique, these rainfall-excess hyetographs can be convoluted to yield the runoff hydrographs at the outlet of these hydrologic units or sub-watersheds. These hydrographs are then routed through various channel segments treating the flows from intermediate tributaries as lateral inflows. Since this book is centered on the SCS-CN method, its use is demonstrated in detail in the forthcoming chapters. Here a brief account of the unit hydrograph and channel and reservoir routing techniques is provided.

1.9.1 UNIT HYDROGRAPH (UH)

The concept of unit hydrograph (UH) was developed by Sherman (1932). It is used for small to midsize catchments as a tool to develop a hydrograph for any given rainstorm. The term "unit" refers to a unit depth of rainfall-excess or direct surface runoff lasting for a "unit" increment of time. The unit depth of rainfall excess can be 1 mm, 1 cm, or 1 inch, and unit duration can be 1-hr, 2-hr, 6-hr, 12-hr, 24-hr, or any other suitable duration. For midsize catchments, unit durations are generally in the range 1-6 hrs.

The unit hydrograph is defined as the hydrograph produced by a unit depth of the rainfall-excess distributed uniformly in time and over the entire catchment and lasting for a specified unit duration. As an example, assume that a certain rainstorm produces 1 cm of rainfall-excess covering a catchment area of 25 km^2 over a period of 2 hrs. The resulting hydrograph at the outlet of the catchment is a 2 hr unit hydrograph.

The concept of unit hydrograph involves two crucial assumptions: linearity and superposition. Given a unit hydrograph, a hydrograph for a rainfall-excess depth other than unity can be obtained by simply multiplying the unit hydrograph ordinates by the indicated rainfall-excess depth. It follows the assumption of linearity. It is, however, possible only under the assumption that the time base remains constant regardless of the rainfall-excess depth.

The time base of all hydrographs obtained in this manner is equal to that of the unit hydrograph. Therefore, the procedure can be used to calculate hydrographs produced by a storm consisting of a series of rainfall-excess depths, each lagged in time one increment of the unit hydrograph duration. The summation of the corresponding ordinates of these hydrographs (superposition) allows the computation of the composite hydrograph. This procedure is known as the convolution of a unit hydrograph with a rainfall-excess hyetograph.

For a given catchment, a unit hydrograph can be determined either (a) directly, by using rainfall-runoff data for selected events, or (b) indirectly, by using a synthetic unit hydrograph formula. Though both the methods can be used for gauged catchments, the latter method is more suitable for ungauged catchments. These methods are described in detail in the text books, such as those by Ponce (1989), Bras (1990), and Singh (1988, 1992).

Limitations of the Unit Hydrograph Concept

According to Johnstone and Cross (1949), all the propositions of the unit hydrograph are empirical and cannot be proved mathematically. Nevertheless, a number of investigators support the view that the unit hydrograph theory is useful as a practical tool. The assumption of spatial invariance of the effective rainfall is hardly applicable to catchment areas greater than 5000 km^2. The nonuniform areal distribution of rainfall can cause variation in the hydrograph shape (Dickinson and Ayers, 1965). For example, if a high-intensity rainfall occurs over an area near the outlet, then the resulting hydrograph will exhibit a rapid rise, a sharp peak and rapid recession. On the other hand, if the same rainfall occurs over the upstream portion of the watershed, then the resulting hydrograph will exhibit a slow rise, lower and broader peak, and slow recession.

The effective rainfall does not occur uniformly, even for as short a duration as 5 minutes. However, the effect of temporal variation of rainfall intensity on the UH shape depends principally on the watershed size. For example, rainfall bursts lasting only a few minutes may produce well-defined peaks in the hydrograph from small watersheds (say, a few hectares in area), but may cause little change in the hydrograph shape from large watersheds. Only those changes that last for several hours will cause a distinguishable change in the hydrograph shape on large watersheds.

Thus, for the UH theory to be applicable, the time and space invariance assumptions should hold as nearly as possible. Rainfall from cyclonic storms is likely to cover large areas nearly uniformly, whereas convective storms cover only small areas uniformly. A factor that complicates the uniform runoff requirement is the spatial variation of infiltration characteristics on large drainage basins. Even where soil associations are shown to be of the same group, considerable differences in infiltration rates can exist. Such variations result in nonlinear conditions and hamper the concept of uniform runoff (or the UH principles of superposition and linearity). Thus, the UH theory should be used with caution.

Instantaneous Unit Hydrograph (IUH)

The difficulty arising from the dependence of the unit hydrograph on the duration D of the effective rainfall (ER) is circumvented by letting D be diminished indefinitely. The UH so obtained is called the instantaneous unit hydrograph (IUH). However, the volume of IUH remains the same as "unity" (say, 1 cm). Thus, the IUH is a hypothetical hydrograph and independent of time duration D. An IUH is a right-skewed hydrograph, that is, its peak is towards the left of the centroid with a long tail extending to the right.

The IUH can be derived using a conceptual linear reservoir represented by the continuity and storage equations, which, respectively, are:

$$I - O = \frac{dV}{dt} \tag{1.83}$$

$$V = KO \tag{1.84}$$

where I is the inflow to the reservoir, O is the outflow from the reservoir, V is the reservoir storage, t is the time, and K is the storage coefficient. Substitution of equation (1.84) into equation (1.83) leads to a first order partial differential equation which can be solved for O as:

$$O = e^{-t/K} \int e^{t/K} I dt \tag{1.85}$$

Replacing I(t) by Dirac delta function δ(t) in equation (1.85), the IUH {h(t)} can be obtained as

$$h(t) = \frac{1}{K} e^{-t/K} \tag{1.86}$$

In a series of papers, Nash (1957, 1958, 1959, 1960) developed a model based on a cascade of equal linear reservoirs for derivation of the IUH from a natural watershed. Kalinin and Milyukov (1958) independently derived an identical model for routing of flows in open channels. The finally derived IUH is expressed as

$$h(t) = \frac{1}{K\Gamma(n)} \left(\frac{t}{K}\right)^{n-1} e^{-t/K} \tag{1.87}$$

where n is the number of equal reservoirs in series and Γ is the Gamma function. Equation (1.87) that represents the Nash model is one of the most popular and frequently used models in applied hydrology.

S-Hydrograph (SH)

The S-hydrograph is used for the UH derivation and changing the time duration of UH. If the duration of ER is indefinitely long and its intensity is one unit per unit of time (say, 1 cm/h), then the resulting hydrograph is termed the summation hydrograph, S-hydrograph (SH). This hydrograph assumes a deformed S-shape and its ordinates ultimately approach the rate of ER in the limit or at the time of equilibrium. If the effective rainfall hyetograph (ERH) is broken down into pulses each of 1-hour duration and 1 cm/hr intensity, then the SH will be obtained by superposing the direct runoff hydrograph (DRH) of these pulses.

The IUH, UH, and SH are closely related to each other. For a detail mathematical treatment, the text books, for example, by Singh (1988, 1992) can be referred. The following examples, however, illustrate the derivation of the unit hydrograph using a simple S-hydrograph method and a simple UH application for computation of runoff hydrograph by convoluting the rainfall excess.

Example 1.16: A 1-hr unit hydrograph is given as a triangle whose base is 3 hours, peak is 2/3 cm/hr, and the times of rise and recession, respectively, are 1 h and 2 h. Compute a 2-hr unit hydrograph using the S-hydrograph method.

Solution: The computations are shown in Table 1.25. In this table, the first two cols. correspond to the time and unit hydrograph ordinates, respectively. The third col. gives the S-hydrograph (SH), which is the cumulative sum of the ordinates of col. 2. The fourth col. lags the SH of col. 3 by 2 hrs. The second last col. is the difference of the third and fourth cols. It represents the 2-hr hydrograph corresponding to 2 cm of rainfall-excess. (Note: The total volume of the unit hydrograph is equal to 1 cm.) Therefore, to derive a 2-hr unit hydrograph the ordinates of the second last col. are divided by 2 and the resulting values are given in the last col.

Table 1.25 Derivation of unit hydrograph

Time (hr)	1-hr UH (cm/hr)	SH (cm/hr)	SH lagged by 2 hours (cm/hr)	2-hr hydrograph (cm/hr)	2-hr UH (cm/hr)
0	0	0	0	0	0
1	2/3	2/3	0	2/3	1/3
2	1/3	1	0	1	½
3	0	1	2/3	1/3	1/6
4	0	1	1	0	0
5	0	1	1	0	0
6	0	1	1	0	0
7	0	1	1	0	0

Table 1.26 Convolution of rainfall-excess hyetograph

Time (hr)	Rainfall excess hyetograph (cm/hr)	1-hr UH (cm/hr)	Hydrograph due to 3 cm/hr rainfall-excess (cm/hr)	Hydrograph due to 6 cm/hr rainfall-excess (cm/hr)	Total hydrograph (cm/hr)
0	-	0	0	-	0
1	3	2/3	2	0	2
2	6	1/3	1	4	5
3		0	0	2	2
4		0	0	0	0

Example 1.17: Using the unit hydrograph of Example 1.16, convolute the rainfall-excess hyetograph consisting of the ordinates of 3 and 6 cm/hr lagged by 1 hr.

Solution: Computations for the convolution of the rainfall-excess hyetograph are given in Table 1.26. In this table, the first three cols. represent the time, rainfall-excess hyetograph, and 1-hr unit hydrograph, respectively. Note: The rainfall-excess is uniformly distributed in time (or time invariant). The volume of rainfall-excess in the 0-1 hr duration is 3 cm. Thus, a multiplication of UH of the third col. by 3 yields the hydrograph that will result from 3 cm/hr of rainfall excess and it is given in the fourth col. Similarly, the hydrograph due to 6 cm/hr (volume = 6 x 1 = 6 cm) can be computed. Since this rainfall lags by 1 hr, the resulting hydrograph is also lagged by 1 hr and it is given in the second last col. The sum of the values of the fourth and fifth cols. is given in the last col., which is the resulting hydrograph from the given rainfall-excess hyetograph.

1.9.2 CHANNEL AND RESERVOIR ROUTING

(a) *Channel Routing*

In a number of applications of engineering hydrology, it is necessary to compute the variation of flows in time and space. These applications involve reservoir design, design of flood-control structures, flood forecasting, and water resources planning and analysis. Routing is used to distribute flows in time and space. There are a number of techniques available for channel routing. The techniques applicable to channel routing are equally valid for reservoir routing. In brief, the techniques can be categorized into hydrologic, hydraulic, and semi-hydrologic methods.

Hydrologic methods utilize the continuity equation and a storage equation relating channel (or reservoir) storage to the inflow, outflow, or a combination thereof. On the other hand, hydraulic methods involve the solution of the combination of continuity and momentum equations. For a one-dimensional flow, these equations in partial differential form are popularly known as de Saint Venant equations. Due to their non-linearity, a closed form solution of these equations under general conditions is still not available. However, a number of simplified solutions are available in literature (Singh, 1996; Ponce, 1989; Mahmood and Yevjevich, 1975). A description of the most commonly used Muskingum-Cunge method of channel routing follows.

Muskingum-Cunge Method. The Muskingum-Cunge method is basically an improvement of the Muskingum (McCarthy, 1938) method that is a combination of the continuity equation (1.83) and the storage equation:

$$V = K[\varepsilon I + (1-\varepsilon)O] \tag{1.88}$$

where I is the inflow to the channel $[LT^{-3}]$, O is the outflow from the channel $[LT^{-3}]$, V is the channel storage $[L^3]$, t is the time [T], K is the storage coefficient [T], and ε is the weighting parameter varying in the range of (0, 0.5). If $\varepsilon = 0$, equation (1.88) describes the case of a single linear reservoir, in which the channel storage is directly proportional to the outflow (equation (1.84)). In the Muskingum method, parameters K and ε are determined from inflow and outflow data using a regression analysis.

Discretization of equation (1.83) in space (x)-time (t) plane (Figure 1.29) leads to

$$\frac{I_1 + I_2}{2} - \frac{O_1 + O_2}{2} = \frac{V_2 - V_1}{\Delta t} \tag{1.89}$$

where I_1 is the inflow at time level 1, I_2 is the inflow at time level 2, O_1 is the outflow at time level 1, O_2 is the outflow at time level 2, V_1 is the storage at time level 1, V_2 is the storage at time level 2, and Δt is the time interval. Equation (1.89) states that the average rate of inflow minus the average rate of outflow is equal to rate of change of storage in time interval Δt.

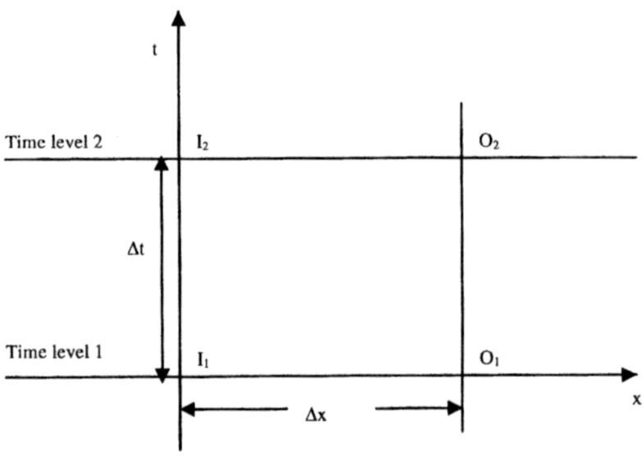

Figure 1.29. Discretization in space (x)-time (t) plane.

Equation (1.89) is expressed at time levels 1 and 2 as

$$V_1 = K[\varepsilon I_1 + (1-\varepsilon)O_1] \tag{1.90a}$$

$$V_2 = K[\varepsilon I_2 + (1-\varepsilon)O_2] \tag{1.90b}$$

A substitution of equations (1.90a) and (1.90b) in equation (1.89) and solving for O_2 yields

$$O_2 = d_o I_2 + d_1 I_1 + d_2 O_1 \tag{1.91}$$

where d_o, d_1, and d_2 are the routing coefficients defined in terms of Δt, K, and ε as

$$d_o = \frac{(\Delta t / K) - 2\varepsilon}{2(1-\varepsilon) + (\Delta t / K)} \tag{1.92a}$$

$$d_1 = \frac{(\Delta t / K) + 2\varepsilon}{2(1-\varepsilon) + (\Delta t / K)} \tag{1.92b}$$

$$d_2 = \frac{2(1-\varepsilon) - (\Delta t / K)}{2(1-\varepsilon) + (\Delta t / K)} \tag{1.92c}$$

Here, $d_o + d_1 + d_2 = 1$. Therefore, the routing coefficients can be interpreted as weighting coefficients.

In the Muskingum-Cunge method, parameters K and ε are defined, respectively, as

$$K = \frac{\Delta x}{c} \tag{1.93}$$

$$\varepsilon = \frac{1}{2}\left(1 - \frac{q_o}{s_o c \Delta x}\right) \tag{1.94}$$

where Δx is the length of the channel (Figure 1.29), c is the wave celerity, q_o is the reference discharge per unit width, and s_o is the channel bed slope. Equations (1.93) and (1.94) can be integrated (Ponce, 1989) using the definitions of Courant number and cell Reynolds number, which, respectively, are:

$$C = c\frac{\Delta t}{\Delta x} \tag{1.95}$$

$$D = \frac{q_o}{s_o c \Delta x} \tag{1.96}$$

Coupling of equation (1.95) with equation (1.93) yields an expression for parameter K:

$$K = \Delta t / C \qquad \text{or} \qquad C = \Delta t / K \tag{1.97}$$

Similarly, a coupling of equation (1.96) with equation (1.94) yields an expression for ε as

$$\varepsilon = \frac{1}{2}(1 - D) \tag{1.98}$$

Substitution of equations (1.97) and (1.98) into equations (1.92a) through (1.92c) and a little manipulation yield

$$d_o = \frac{-1+C+D}{1+C+D} \tag{1.99a}$$

$$d_1 = \frac{1+C-D}{1+C+D} \tag{1.99b}$$

$$d_2 = \frac{1-C+D}{1+C+D} \tag{1.99c}$$

For practical applications, the wave celerity can be computed from $c = bv_o$ or from a rating curve as: $c = (1/B)dq/dh$, where b is the rating exponent (equation (1.81)), v_o is the uniform flow velocity, B is the top width of the channel, q is the discharge, and h is the depth of flow. Theoretically, both the formulae are the same. These are further discussed in Chapter 8.

Example 1.18: Route the following hydrograph using the Muskingum-Cunge method:

Time (h)	0	1	2	3	4	5	6	7	8	9	10
Discharge (m^3/s)	0	200	400	600	800	1000	800	600	400	200	0

Assume: channel slope = 0.000868, channel cross-sectional area = 400 m^2, top width at the peak flow = 100 m, rating exponent = 1.6, channel reach length = 14.4 km, and time interval = 1 h.

Solution: From the given inflow hydrograph peak discharge = 1000 m^3/s. For the given cross-sectional area (= 400 m^2), mean flow velocity v_o = 1000/400 = 2.5 m/s. Therefore, wave celerity c = 1.6 x 2.5 = 4 m/s. The reference flow per unit width (based on peak discharge) can be computed as: q_o = 1000/100 = 10 m^2/s. From equation (1.95) the Courant number can be calculated as

$$C = c\frac{\Delta t}{\Delta x} = 4 \text{ x } 1 \text{ x } 3600/(14.4 \text{ x } 1000) = 1$$

Using equation (1.96), the cell Reynolds number can be computed as

$$D = \frac{q_o}{s_o c \Delta x} = 10/(0.000868 \text{ x } 4 \text{ x } 14.4 \text{ x } 1000) = 0.2$$

The routing coefficients can be computed from equations (1.99a) through (1.99c) as:

$$d_o = \frac{-1+C+D}{1+C+D} = \frac{-1+1+0.2}{1+1+0.2} = 0.091$$

$$d_1 = \frac{1+C-D}{1+C+D} = \frac{1+1-0.2}{1+1+0.2} = 0.818$$

$$d_2 = \frac{1-C+D}{1+C+D} = \frac{1-1+0.2}{1+1+0.2} = 0.091$$

It is noted that the sum of these coefficients is equal to 1, which is a check for the accuracy of the computed routing coefficients. The routing calculations are shown in Table 1.27.

Table 1.27 Channel routing using the Muskingum-Cunge method (Example 1.18)

Time (h)	Inflow (I) (m^3/s)	Partial flows (m^3/s)			Outflow (O) (m^3/s)
		d_oI_2	d_1I_1	d_2O_1	
0	0	-	-	-	0.0
1	200	18.2	0.0	0.0	18.20
2	400	36.4	163.6	1.66	201.66
3	600	54.6	327.2	18.35	400.15
4	800	72.8	490.8	36.41	600.01
5	1000	91.0	654.4	54.60	800.00
6	800	72.8	818.0	72.80	963.60
7	600	54.6	654.4	87.69	796.69
8	400	36.4	490.8	72.50	599.70
9	200	18.2	327.2	54.57	399.97
10	0	0.0	163.6	36.40	200.00
11	0	0.0	0.0	18.20	18.20
12	0	0.0	0.0	1.66	1.66
13	0	0.0	0.0	0.16	0.16

(b) *Reservoir Routing*

When a flood wave passes through a full reservoir, it experiences attenuation of its peak, elongation of its peak time, and lengthening of its time base. This is the effect of a reservoir on a flood hydrograph. For better understanding, consider an inflow hydrograph and the resulting hydrograph, as shown in Figure 1.30. The inflow hydrograph is at the upstream of the reservoir and is not influenced by the reservoir. The reservoir modifies this inflow hydrograph in the form shown as the outflow hydrograph in the figure. It is seen that the peak of the outflow hydrograph is smaller than the peak of the inflow hydrograph, the peak of the outflow hydrograph occurs after the peak of the inflow hydrograph, the time base of the outflow hydrograph is longer than the inflow hydrograph, and the volume of flow is the same for both inflow and outflow hydrographs. The time difference between the peak of inflow and that of outflow is called lag time, or lag. The attenuation and lag are two important parameters of flood routing. These effects are similar to that in channel routing. For example, in Example 1.18, the lag is of 1 hour and attenuation is 1000 – 963.60 = 36.40 m^3/s. In channel routing, the storage of the channel actually acts as a reservoir that modifies the inflow hydrograph. The difference, however, lies in the fact that a reservoir has a well-defined stage (or storage)-outflow relation due to a regular shape of the reservoir spillway, whereas this relation is more conceptual for channels. The following example illustrates the reservoir routing using the most commonly used storage indication method.

Storage Indication Method. The storage indication method is also frequently referred to as the modified Puls method. This method employs equation (1.89), which can be expressed as

$$V_2 - V_1 = \left(\frac{I_1 + I_2}{2}\right)\Delta t - \left(\frac{O_1 + O_2}{2}\right)\Delta t \tag{1.100}$$

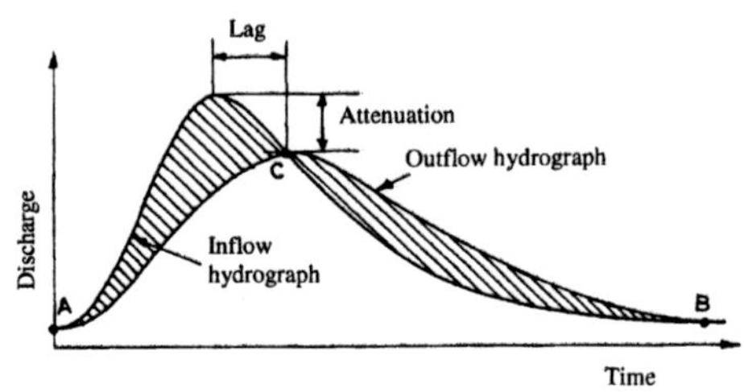

Figure 1.30. Routing of an inflow hydrograph through a reservoir

In a reservoir, the outflow O may be the sum of two components: controlled, O_c, and uncontrolled, O_s. Thus,

$$O = O_c + O_s \tag{1.101}$$

Combination of equations (1.100) and (1.101) and then separation of the unknown quantities from the known ones and rearrangement of equation (1.100) yield

$$(I_1 + I_2) - (O_{c_1} + O_{c_2}) + \left(\frac{2V_1}{\Delta t} - O_{s_1}\right) = \frac{2V_2}{\Delta t} + O_{s_2} \tag{1.102}$$

In equation (1.102), the left-hand side is known and the right-hand side is unknown. The inflow hydrograph is known. The controlled discharge may pass through turbines, outlet works, or over the spillway and, therefore, it is known. The remaining terms on the left side, V_1 and O_1 are assumed known at the start of routing, and are determined for each subsequent step. The right side of equation (1.102) is known as storage indication quantity and can be written as

$$\frac{2V}{\Delta t} + O = f(y) \tag{1.103}$$

where y is the water surface elevation. In order to utilize equation (1.102), elevation-storage and elevation-discharge relationships must be specified. For convenience and simplicity, assume that O_c is negligible so that O can be taken to imply O_s and can be determined using a weir formula:

$$O = CLh^{3/2} \tag{1.104}$$

where C is the weir coefficient, L is the length of the spillway crest, and h is the depth of water above the spillway crest or stage.

Thus, equation (1.102) can then be written as

$$(I_1 + I_2) + \left(\frac{2V_1}{\Delta t} - O_1 \right) = \frac{2V_2}{\Delta t} + O_2 \tag{1.105}$$

The elevation-discharge relationship of equation (1.104) is utilized to determine outflow, given the depth of water over the spillway. From topographic information, elevation y corresponding to h can be obtained. Then, the curve for either y versus O or h versus O can be constructed.

The elevation-storage curve can be constructed from the topographic map. The depth of water above spillway h multiplied by the reservoir surface area can yield the (temporary) storage of the reservoir. The reservoir surface area is derived from the topographic map. Therefore, the temporary storage V for any h can be determined approximately as

$$V = \frac{h}{2}(A_1 + A_2) \tag{1.106}$$

where A_1 represents the surface area of the reservoir when h = 0, and A_2 is the surface area when the depth of flow is h. Then, curves for either y versus V or h versus V can be constructed. The obtained curves of elevation-discharge and elevation-storage can be easily combined to form a storage-discharge curve.

An important aspect of both channel and reservoir routing is the selection of time interval Δt. Its value should neither be too large nor too small. If it is too long and exceeds the travel time through the reservoir, then the crest segment of outflow containing the peak discharge may pass through the reservoir in between time intervals and, therefore, may not be computed. On the other hand, if Δt is too short, then it takes longer to perform flood routing. Linsley et al. (1982) recommended Δt to be one-third to one-half the travel time through the reservoir. As a thumb rule, Δt is approximately 20 to 40% of the time of rise of the inflow hydrograph.

It is noted that subtracting 2O from $[2V/\Delta t + O]$ yields $[2V/\Delta t - O]$, which appears on the left-hand side of equation (1.105). Before routing is performed, the storage indication curves of $[2V/\Delta t \pm O]$ are thus constructed. This can be accomplished as follows. Select an initial outflow discharge O. Estimate the corresponding elevation using the elevation-discharge curve. Determine the corresponding storage from the elevation-storage curve. Then, determine $[2V/\Delta t + O]$ and plot against O.

The steps for reservoir routing using storage indication method are as follows:

(i) At the beginning of routing, inflow (I_1 and I_2), storage V_1, and outflow O_1 are known from initial reservoir conditions. Compute the left side of equation (1.105), which equals $(2V_2/\Delta t) + O_2$.

(ii) Determine O_2 for $[2V_2/\Delta t + O_2]$ specified in step (i). This value of O_2 is the instantaneous outflow at the end of Δt.

(iii) Determine $[2V_2/\Delta t - O_2]$ by subtracting twice the value of O_2 of step (ii). This gives the value of $[2V_1/\Delta t - O_1]$ for the next time interval.

(iv) Add the two inflows I_1 and I_2 and the value in step (iii). This gives $[2V_2/\Delta t + O_2]$ for the next time interval.

(v) Determine O_2 against $[2V_2/\Delta t - O_2]$ of step (iv) from the above described storage indication curve.

(vi) Determine $[2V_2/\Delta t - O_2]$. This gives the value of $[2V_1/\Delta t - O_1]$ for the next time interval.

(vii) Go back to step (iv) and repeat the procedure.

The major shortcoming of this method is that the outflow begins at the same time as does inflow, implying that the inflow passes through the reservoir instantaneously regardless of its length. In addition to this, it is often difficult to choose an appropriate value of Δt. Its selection may lead to negative outflows whenever $\Delta t > 2V_2/O_2$ or $O_2/2 > V_2/\Delta t$. This also means that the rising hydrograph is distorted. This problem can be circumvented by plotting the storage indication curve on a log-log paper and comparing the plot with the line of equal values (Singh, 1992). If the plotted value lies above the line of equal values, the constructed storage indication curve must be abandoned and a new value of Δt selected. Furthermore, negative outflows can usually be avoided by taking Δt less than the time of travel.

Table 1.28 Storage and discharge characteristics of a dam

Elevation (m)	Discharge (m^3/s)	Storage (x10^8 m^3)	Elevation (m)	Discharge (m^3/s)	Storage (x10^8 m^3)
125	0.02	0.21	320	185.0	5.70
130	0.13	0.90	340	230.0	6.00
140	0.45	1.59	350	256.0	6.15
160	2.7	2.40	355	268.0	6.19
180	7.5	3.03	360	280.0	6.30
200	15.5	3.54	365	293.0	6.39
220	30.0	3.97	370	306.0	6.43
240	50.0	4.54	375	321.0	6.49
260	75.0	4.75	390	360.0	6.70
280	105.0	5.10	400	392.0	6.85
300	142.0	5.40	420	456.0	7.15

Table 1.29 Inflow hydrograph

Time (h)	Discharge (m^3/s)	Time (h)	Discharge (m^3/s)	Time (h)	Discharge (m^3/s)	Time (h)	Discharge (m^3/s)
0	20	60	730	120	330	180	22
6	60	66	780	126	280	186	21
12	220	72	870	132	200	192	20
18	350	78	770	138	160	198	19
24	380	84	700	144	130	204	18
30	430	90	670	150	110	210	17
36	480	96	630	156	90	216	17
42	530	102	550	162	60	222	17
48	610	108	490	168	50		
54	670	114	360	174	30		

Example 1.19: The storage and discharge characteristics of a reservoir are given in Table 1.28. Consider the routing period as 6 hours. The dam has an unregulated

opening 5 m high with its elevation at 125 m and a spillway-crest elevation of 355 m. Discharges for pool elevations above 355 m include both unregulated and spillway discharges. Route the inflow hydrograph given in Table 1.29 by the storage indication method. Determine the attenuation and lag.

Solution: The reservoir elevation-discharge-storage characteristics are plotted in Figure 1.31. The quantities $(2V/\Delta t - O)$ and $(2V/\Delta t + O)$ are computed and plotted in Figure 1.32. It is seen from this figure that both the curves are close to each other. It is because O is much smaller than $(2V/\Delta t)$. The inflow hydrograph is routed following the steps outlined before, and calculations are given in Table 1.30. The inflow and outflow hydrographs are plotted in Figure 1.33. From Table 1.30 and Figure 1.33,

Outflow peak discharge $\approx 430\ m^3/s$

Reduction in peak flow $\approx 870 - 430 = 440\ m^3/s$

Lag between inflow and outflow peaks = 108 − 68.5 = 39.5 hrs

1.10 Scope of the SCS-CN Concept in Hydrology

The existing SCS-CN method has been widely used for computing the direct runoff volume from a given rainfall amount. Apart from such an application, the method has also been applied to long-term hydrologic simulation but only to a limited extent. In this book, the SCS-CN method is shown to be viable in dealing with the problems involving the computation of cumulative infiltration and the direct runoff amount, determination of infiltration rates, short-term and long-term hydrologic simulation, and simulation of pollutant transport.

1.10.1 COMPUTATION OF INFILTRATION AND DSRO VOLUMES

The determination of infiltration or runoff volumes is vital to watershed management activities. For computing infiltration and runoff using the SCS-CN method, it is necessary to have the background knowledge of its parameter: potential maximum retention, S, or curve number, CN. The computational accuracy increases with the accuracy of S. Recent studies have shown that the existing SCS-CN method does not account for the static portion of infiltration and the antecedent moisture. The fundamental proportionality concept of the existing SCS-CN method has been modified to account for static infiltration and antecedent moisture as presented in Chapter 4.

1.10.2 COMPUTATION OF INFILTRATION RATES

The problem of computing infiltration rates has been recognized and has long been discussed in hydrologic literature. Although the SCS-CN method is construed as an infiltration model, it has not yet been employed in field for computing infiltration rates. The results of a preliminary investigation by Gray et al. (1982) were also not very encouraging. In this book, while discussing the physical significance of the SCS-CN parameter S in Chapter 4, an SCS-CN-based infiltration model is developed and its applications are presented in Chapter 6.

Table 1.30 Reservoir routing by storage indication method

Time (h)	I (m^3/s)	I_1+I_2	O (m^3/s)	$2V/\Delta t-O$ (m^3/s)	$2V/\Delta t+O$ (m^3/s)	Reservoir elevation (m)
0	20	80	268.00	57093.34	57173.34	355.00
6	60	280	255.50	56662.35	56942.35	349.81
12	220	570	251.25	56439.84	57009.84	348.17
18	350	730	252.49	56504.85	57234.85	348.65
24	380	810	256.96	56720.93	57530.93	350.40
30	430	910	265.24	57000.43	57910.43	353.85
36	480	1010	271.43	57367.57	58377.57	356.43
42	530	1140	277.12	57823.31	58963.31	358.80
48	610	1280	285.37	58392.55	59672.55	362.07
54	670	1400	299.43	59073.68	60473.68	367.47
60	730	1510	321.27	59831.14	61341.14	375.10
66	780	1650	338.32	60665.58	62314.48	381.66
72	870	1640	357.46	61559.55	63239.55	389.02
78	770	1470	377.93	62483.69	63953.69	395.60
84	700	1370	394.04	63165.67	64535.67	400.63
90	670	1300	407.12	63721.44	65021.44	404.72
96	630	1180	418.06	64185.32	65365.32	408.14
102	550	1040	425.80	64513.71	65553.69	410.56
108	490	850	430.04	64693.60	65543.56	411.89
114	360	690	429.81	64683.93	65373.93	411.82
120	330	610	425.99	64521.94	65131.94	410.62
126	280	480	420.54	64290.84	64770.84	408.92
132	200	360	412.41	63946.01	64306.01	406.38
138	160	290	401.94	63502.12	63792.12	403.11
144	130	240	390.37	63011.37	63251.37	399.49
150	110	200	378.19	62494.98	62694.98	395.68
156	90	150	365.66	61963.64	62113.64	391.77
162	60	110	353.51	61406.61	61516.61	387.50
168	50	80	341.77	60833.06	60913.06	382.99
174	30	52	329.91	60253.25	60305.25	378.43
180	22	43	316.93	59671.39	59714.39	373.64
186	21	41	300.70	59112.99	59153.99	367.96
192	20	39	288.30	58577.38	58616.38	363.19
198	19	37	280.04	58056.28	58093.28	360.02
204	18	35	273.66	57545.95	57580.95	357.36
210	17	34	266.65	57047.66	57081.66	354.44
216	17	34	253.81	56574.03	56608.03	349.16
222	17		245.11			345.81

1.10.3 TIME-DISTRIBUTED EVENT-BASED HYDROLOGIC SIMULATION

Flood studies often require a determination of the peak flow rate and the time of concentration. The lack of short-term real time data can be attributed to the development of the SCS-CN method for computing runoff volumes rather than rates. The coupling of the SCS-CN method-based infiltration model with a routing mechanism makes it possible to compute runoff rates at discrete times during a storm. The development of the methodology along with applications is described in Chapter 6.

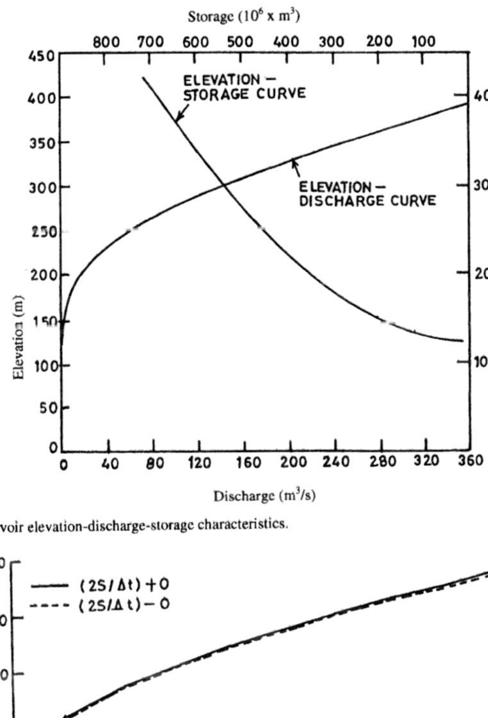

Figure 1.31. Reservoir elevation-discharge-storage characteristics.

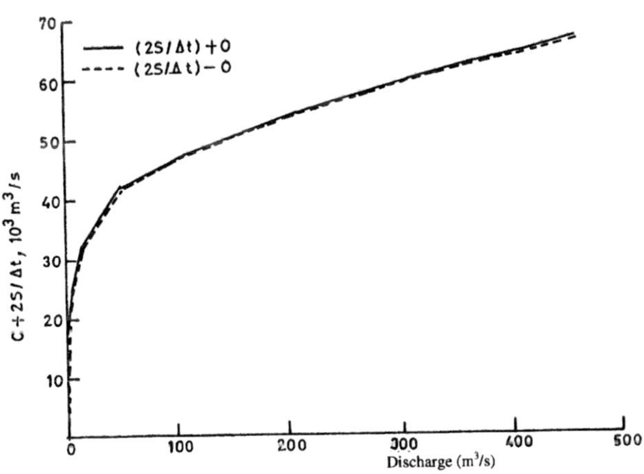

Figure 1.32. (2S/Δt)+O versus O curve

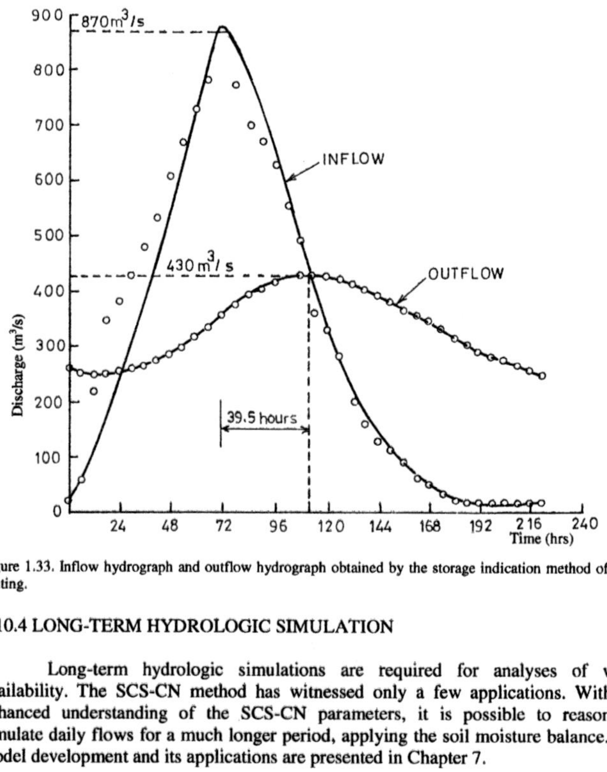

Figure 1.33. Inflow hydrograph and outflow hydrograph obtained by the storage indication method of flood routing.

1.10.4 LONG-TERM HYDROLOGIC SIMULATION

Long-term hydrologic simulations are required for analyses of water availability. The SCS-CN method has witnessed only a few applications. With the enhanced understanding of the SCS-CN parameters, it is possible to reasonably simulate daily flows for a much longer period, applying the soil moisture balance. The model development and its applications are presented in Chapter 7.

1.10.5 TRANSPORT OF URBAN POLLUTANTS

Storm water runoff affected by urban transportation activities and associated infrastructure transports significant pollutant loads of dissolved, colloidal, and suspended solids in a complex heterogeneous mixture that includes heavy metals, inorganic, and organic compounds (Hamilton and Harrison, 1991). It often contains elevated levels of heavy metals and particles (Sansalone and Buchberger, 1997; Sansalone et al., 1998; Li et al., 1999). Heavy metals, such as lead, cadmium, copper, nickel, zinc, and chromium can pose acute and chronic threats to receiving water bodies and soils. They often exceed surface water discharge criteria on an event basis for both dissolved and particulate-bound fractions (Foster and Charlesworth, 1996; Sansalone and Buchberger, 1997; Sansalone et al., 1998; Sansalone and Tribouillard, 1999). Unlike organic compounds, heavy metals are not degraded in the environment. Therefore, it is desirable to have a prior knowledge of the pollutant load for necessary treatment. To this end, the fundamental proportionality concept of the SCS-CN method is extended for computing the mass of heavy metals and for metal partitioning during rainfall, river flow, and snowmelt on urban highways and is discussed in Chapter 8.

1.10.6 SEDIMENT YIELD

Sediment has been characterized as the largest carrier of pollutants to receiving water bodies. Much of the sediment carried by streams originates in upland areas. Many methods have been developed for estimation of erosion and sediment yield from agricultural, urban and rural watersheds. One of the most popular methods is the empirically derived Universal Soil Loss Equation (USLE). A close examination of this equation suggests that the SCS-CN methodology can be employed to derive a sediment yield equation which is similar to USLE. This is presented in Chapter 9.

1.11 Organization of the Book

Chapter 1 describes the basic elements of the Soil Conservation Service Curve Number (SCS-CN) methodology. Chapter 2 presents the existing version of the method along with its applications. Chapter 3 provides a detailed analytical treatment of the method. Chapters 4 and 5 discuss the physical significance of the SCS-CN parameter, the potential maximum retention or curve number using the volumetric principles and the dynamical concept of soil physics, respectively. Chapters 6 and 7 present the short-term and long-term hydrologic simulation, respectively. Chapter 8 extends the SCS-CN application to transport of urban pollutants and Chapter 9 to modelling of sediment yield.

CHAPTER 2

SCS-CN METHOD

The Soil Conservation Service Curve Number (SCS-CN) method was developed in 1954 and is documented in Section 4 of the National Engineering Handbook (NEH-4) published by the Soil Conservation Service (now called the Natural Resources Conservation Service), U.S. Department of Agriculture in 1956. The document has since been revised in 1964, 1965, 1971, 1972, 1985, and 1993. The SCS-CN method is the result of exhaustive field investigations carried out during the late 1930s and early 1940s and the works of several early investigators, including Mockus (1949), Sherman (1949), Andrews (1954), and Ogrosky (1956). The passage of Watershed Protection and Flood Prevention Act (Public Law 83-566) in August 1954 led to the recognition of the method at the Federal level and the method has since witnessed myriad applications all over the world. It is one of the most popular methods for computing the volume of surface runoff for a given rainfall event from small agricultural, forest, and urban watersheds. The method is simple, easy to understand and apply, stable, and useful for ungauged watersheds. The primary reason for its wide applicability and acceptability lies in the fact that it accounts for most runoff producing watershed characteristics: soil type, land use/treatment, surface condition, and antecedent moisture condition. This chapter describes the existing SCS-CN method, the concept of curve number and factors affecting it, the procedure for its application, sensitivity of its parameters, and its advantages and limitations.

2.1 Historical Background

2.1.1 EXPERIMENTAL WATERSHEDS AND INFILTRATION STUDIES

In mid 1930's, an acute need for hydrologic data for design of conservation practices was felt and eventually, the Soil Conservation Service (SCS) was established under the United States Department of Agriculture (USDA). The major objectives of SCS were to set up demonstration conservation projects and evaluate the design and construction of soil and water conservation practices. To that end, several experimental watersheds were set up at different locations for collecting data on rainfall, runoff, and associated factors. According to the Flood Control Act of 1936 (Public Law 74-738), USDA carried out surveys and investigations for installing measures for retarding flows from watersheds, which is a classical hydrologic problem. It eventually led to the evaluation of the effect of watershed treatment and/or conservation measures on the rainfall-runoff process. The data collected from experimental watersheds were, however, found to be scant and covering only a marginal fraction of the conditions affecting the rainfall-runoff process in watersheds (Andrews, 1954). Therefore, a need for collecting data for carrying out infiltration studies was felt.

Using the sprinkler-type infiltrometer, thousands of infiltration tests on field plots of 6 feet wide and multiples of 12 feet long were carried out during the late 1930's and early 1940's. For economic reasons, another FA infiltrometer (Rallison and Miller, 1982) was devised for plots of 12 x 30 inches and it was used extensively. Using these infiltration data, a rational method for estimating runoff under various cover conditions was developed. For that purpose, SCS hired three private consultants, W. W. Horner, R. E. Horton, and R. K. Sherman. Horton (1933) characterized the infiltration capacity curves and Horner (1940) concentrated on the development of infiltration capacity from small watershed data. Their studies resulted in the development of a series of rainfall retention rate curves and rainfall-excess and time-of-excess curves for computing runoff volume from field plots. This method, however, required time-distributed rainfall data, and therefore, its application was severely restricted in many areas.

2.1.2 DEVELOPMENT OF RAINFALL-RUNOFF METHODS

Rainfall-runoff methods developed during the early 1940's utilized infiltration data for computing the runoff amount. Andrews (1954) grouped the infiltration data gathered from Texas, Oklahoma, Arkansas, and Louisiana, using soil texture as a soil characteristic. He eventually developed a graphical rainfall-runoff method taking into account the soil texture, type and amount of cover, and conservation practices, combined into what is referred to as soil-cover complex or soil-vegetation-land use (SVL) complex (Miller and Cronshey, 1989). Subsequent to the development of this method and others, it was recognized that there was a need for a method that could generally be used nationally and was also applicable to ungauged watersheds. To that end, according to Rallison and Miller (1982), the methods given by Mockus (1949) and Andrews (1954) were transformed and generalized to yield the existing SCS-CN method. Both the models of Mockus and Zoch (1934, 1936) can be described to be a combination of the Horton model and the SCS-CN method and this is discussed in Chapter 3.

2.2 SCS-CN Method

The SCS-CN method is based on the water balance equation and two fundamental hypotheses. The first hypothesis equates the ratio of the actual amount of direct surface runoff (Q) to the total rainfall (P) (or maximum potential surface runoff) to the ratio of the amount of actual infiltration (F) to the amount of the potential maximum retention (S). The second hypothesis relates the initial abstraction (I_a) to the potential maximum retention. Thus, the SCS-CN method consists of

(a) water balance equation:

$$P = I_a + F + Q \qquad (2.1)$$

(b) proportional equality hypothesis:

$$\frac{Q}{P - I_a} = \frac{F}{S} \qquad (2.2)$$

and (c) I_a-S hypothesis:

$$I_a = \lambda S \tag{2.3}$$

where P = total rainfall; I_a = initial abstraction; F = cumulative infiltration excluding I_a; Q = direct runoff; and S = potential maximum retention or infiltration, also described as the potential postinitial abstraction retention (McCuen, 2002). All quantities in equations (2.1) through (2.3) are in depth or volumetric units.

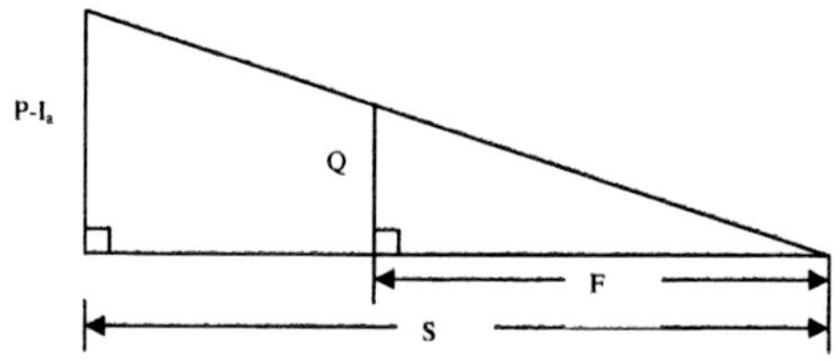

Figure 2.1. Proportionality concept.

The fundamental hypothesis (equation (2.2)) is primarily a proportionality concept, as shown in Figure 2.1. Apparently, as $Q \rightarrow (P-I_a)$, $F \rightarrow S$. This proportionality enables partitioning (or dividing) $(P-I_a)$ into two surface water (Q) and subsurface water (F) for given watershed characteristics or S. This partitioning, however, undermines the saturated overland flow or source area concept that allows runoff generation from only saturated or wet portions of the watershed. Consequently, the statistical theory (Moore and Clarke, 1981; 1982; 1983; Moore, 1983; 1985) based on the runoff production from only saturated (independent or interacting) storage elements is negated. According to the SCS-CN method, the extent of runoff contribution of a storage element depends on its capacity or, alternatively, the magnitude of S and, therefore, the whole watershed should contribute to runoff, if S is taken to be a definite quantity. Thus, the ratio of the wet and total areas describing the contributing portion should be equal to one.

Parameter S of the SCS-CN method depends on the soil type, land use, hydrologic condition, and antecedent moisture condition (AMC). The initial abstraction accounts for the short-term losses, such as interception, surface storage, and infiltration. Parameter λ is frequently viewed as a regional parameter dependent on geologic and climatic factors (Bosznay, 1989; Ramasastri and Seth, 1985). The existing SCS-CN method assumes λ to be equal to 0.2 for practical applications. Many other studies carried out in the United States and other countries (SCD, 1972; Springer et al., 1980; Cazier and Hawkins, 1984; Ramasastri and Seth, 1985; Bosznay, 1989) report λ to vary in the range of (0, 0.3).

The second hypothesis of the SCS-CN method (equation (2.3)) linearly relates the initial abstraction to the potential maximum retention. It is based on the results of Figure 2.2 (SCS, 1971) depicting the plot between I_a and S. The data for S and I_a were derived from rainfall-runoff records of watersheds less than 10 acres in area. The S values were derived from rainfall-runoff plots prepared for determining CN for AMC II, discussed in the next section. It is apparent from Figure 2.2 that more than 50% of the data points lie within the limits $0.095 \leq \lambda \leq 0.38$. Errors in S were largely attributed to the computation of the average rainfall of the watershed. The I_a-values were computed by accumulating the rainfall amount from the beginning to the time of start of runoff. The large scatter in the data points in Figure 2.2 was attributed to the errors in the estimates of I_a due to (SCS, 1971): (i) the difficulty in determining the actual time of the start of rainfall because of storm travel and lack of instrumentation, (ii) the difficulty in determining the time of the start of runoff largely due to time lag in runoff from the watershed, and (iii) impossible determination of the amount of interception losses prior to runoff and its delayed contribution to runoff. As originally hypothesized (SCS, 1971), parameter S includes I_a (see Appendix A). For this condition, equation (2.3) can be re-written as (Chen, 1982):

$$I_a = \frac{\lambda}{1-\lambda} S \tag{2.4}$$

For $\lambda = 0.2$ (Figure 2.2), equation (2.4) is recast as $I_a = 0.25$ S. Such an analysis is further elaborated and clarified in Chapter 4 which presents the generalized and other variations of the SCS-CN model derived using the volumetric principles.

Combining equations (2.1) and (2.2), the popular form of the SCS-CN method is obtained:

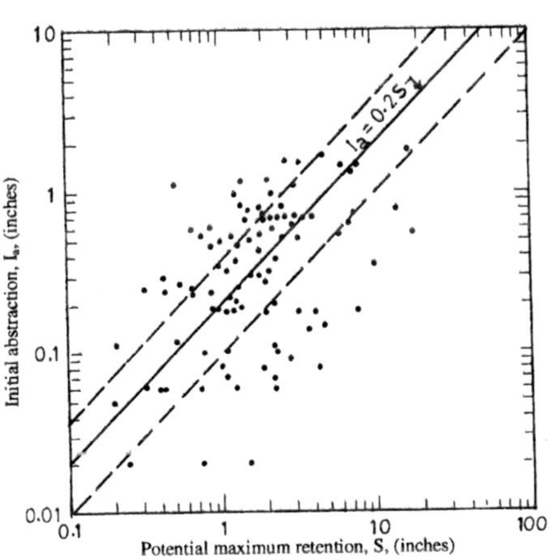

Figure 2.2. Relation between initial abstraction, I_a, and potential maximum retention, S. (Data source: SCS, 1971).

$$Q = \frac{(P - I_a)^2}{P - I_a + S} \tag{2.5}$$

Equation (2.5) is valid for $P \geq I_a$; Q = 0 otherwise. For $\lambda = 0.2$, equation (2.5) can be re-written as

$$Q = \frac{(P - 0.2S)^2}{P + 0.8S} \tag{2.6}$$

Thus, the existing SCS-CN method (equation (2.6)) is a one-parameter model for computing surface runoff from daily storm rainfall, for the method was originally developed using daily rainfall-runoff data of annual extreme flows (Rallison and Cronshey, 1979). Mockus (1964) [In: Rallison, 1980] described the physical significance of parameter S of equation (2.6) as follows: " ... S is that constant and is the maximum difference of (P-Q) that can occur for the given storm and watershed conditions. S is limited by either the rate of infiltration at the soil surface or the amount of water storage available in the soil profile, whichever gives the smaller S value. Since infiltration rates at the soil surface are strongly affected by the rainfall impact, they are strongly affected by the rainfall intensity." This description, however, compares the magnitude of infiltration rate with the volume of water retention in the soil, which is unwarranted. It is discussed in detail in Chapter 4.

Since parameter S (equation (2.6)) can vary in the range of $0 \leq S \leq \infty$, it is mapped into a dimensionless curve number (CN), varying in a more appealing range $0 \leq CN \leq 100$, as follows:

$$S = \frac{1000}{CN} - 10 \tag{2.7}$$

The underlying difference between S and CN is that the former is a dimensional quantity [L] whereas the latter is a non-dimensional quantity. Although CN theoretically varies from 0 to 100, the practical design values validated by experience lie in the range (40, 98) (Van Mullem, 1989).

2.3 Factors Affecting CN

Major watershed characteristics that affect the SCS-CN parameter S or curve number, CN, are soil type, type of vegetation cover, land use/treatment, hydrologic condition, antecedent moisture condition, and climate of the watershed. The combination of soil type, vegetation cover, and land use/treatment is referred to as soil-vegetation-land use (SVL) complex (Miller and Cronshey, 1989). These characteristics primarily affect the infiltration potential of a watershed. For a given rainfall amount, the magnitude of Q depends on S or the infiltration potential. It is elaborated in subsequent chapters. NEH-4 presents CN-values for several typical vegetation cover or land use/treatment and its combinations, as shown in Table 2.1. Apparently, the SVL combinations can broadly be classified into urban, agricultural, and woods and forest. A description of these factors affecting CN or infiltration follows.

2.3.1 SOIL TYPE

Soils are broadly classified as sand, silt, and clay on the basis of the grain size which decreases in their order, as shown in Figure 2.3. The size of grains affects the size of pores that, in turn, affects the surface tension (or suction) or the ψ-θ relation (Chapter 1) and, therefore, affect infiltration. Other major factors in this category include soil texture, structure, hydraulic conductivity, and initial moisture content that affect K_h-θ and ψ-θ relations. A loose conductive sandy soil will exhibit larger infiltration rate than the tightly packed clay. A soil with good under-drainage system will exhibit larger infiltration than in the otherwise situation. A layered soil with different textural properties and the order of their arrangement will also affect the overall infiltration. A dry soil will show larger infiltration than will a wet soil.

The Soil Conservation Service identified four hydrologic groups of soils A, B, C, and D, based on their infiltration and transmission rates. Infiltration is defined as the rate at which the water enters the soil at its surface and is thus controlled by surface conditions. On the other hand, the transmission is defined as the rate at which water moves in the soil and is controlled by the soil horizon. The former refers to the infiltration capacity of the soil defined by various formulae whereas the latter refers to the hydraulic conductivity of the soil (Chapter 1).

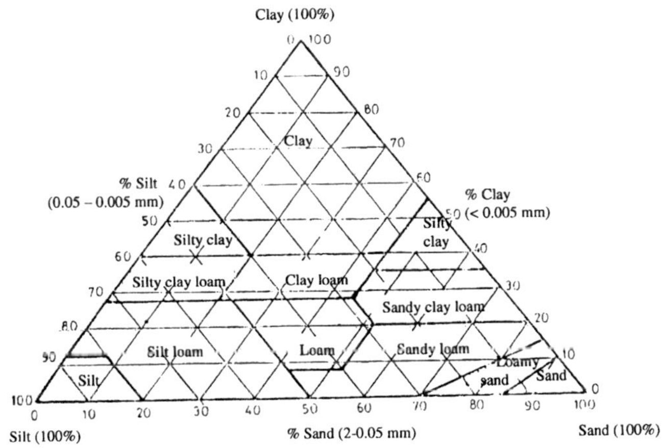

Figure 2.3. Soil textural classification of the U.S. Public Road Administration.

Table 2.1 Runoff curve numbers for hydrologic cover complexes (Antecedent moisture condition II and $I_a = 0.2\ S$)

Sl. No.	Land use Description/Treatment	Hydrologic Condition /% impervious area	Hydrologic Soil Groups			
			A	B	C	D
	Urban					
1	[1]Residential:					
	Average lot size- 1/8 acre or less	65*	77	85	90	92
	1/4 acre	38*	61	75	83	87
	1/3 acre	30*	57	72	81	86
	1/2 acre	25*	54	70	80	85
	1 acre	20*	51	68	79	84
	2 acre	12*	46	65	77	82
2	Paved parking lots, roofs, driveways, etc.[2] (excluding right-of-way)		98	98	98	98
3	Streets and roads :					
	Paved with curbs and storm sewers[2] (excluding right-of-way)		98	98	98	98
	Paved, open ditches (including right-of-way)		82	89	92	93
	Gravel (including right-of-way)		76	85	89	91
	Dirt (including right-of-way)		72	82	87	89
4	Western desert areas:					
	Natural desert landscaping (pervious areas only)[3]		63	77	85	88
	Artificial desert landscaping (impervious weed barrier, desert shrub with 1- to 2-inch sand or gravel mulch and basin borders)		96	96	96	96
5	Urban districts :					
	Commercial and business areas	85	89	92	94	95
	Industrial districts	72	81	88	91	93
6	Developing areas :					
	Newly graded areas (pervious areas only, no vegetation)[4]		77	86	91	94
	Idle lands[5]					
7	Open spaces, lawns, parks, golf courses, cemeteries, etc.					
	Grass cover on 75% or more of the area	Good	39	61	74	80
	Grass cover on 50% to 75% of the area	Fair	49	69	79	84
	Agricultural					
	Cultivated lands :					
8	Fallow :					
	Bare soil Straight row	----	77	86	91	94
	Crop residue cover	Poor	76	85	90	93
		Good	74	83	88	90
9	Row crops :					
	Straight row	Poor	72	81	88	91
	Straight row	Good	67	78	85	89
	Crop residue cover Straight row	Poor	71	80	87	90
	Crop residue cover Straight row	Good	64	75	82	85
	Contoured	Poor	70	79	84	88
	Contoured	Good	65	75	82	86
	Crop residue cover Contoured	Poor	69	78	83	87
	Crop residue cover Contoured	Good	64	74	81	85
	Contoured & terraced	Poor	66	74	80	82

Table 2.1 Continued

Sl. No.	Land use Description/Treatment	Hydrologic Condition /% impervious area	Hydrologic Soil Groups			
			A	B	C	D
	Contoured & terraced	Good	62	71	78	81
	Crop residue cover Contoured & terraced	Poor	65	73	79	81
	Crop residue cover Contoured & terraced	Good	61	70	77	80
10	Small grain :					
	Straight row	Poor	65	76	84	88
	Straight row	Good	63	75	83	87
	Crop residue cover Straight row	Poor	64	75	83	86
	Crop residue cover Straight row	Good	60	72	80	84
	Contoured	Poor	63	74	82	85
	Contoured	Good	61	73	81	84
	Crop residue cover Contoured	Poor	62	73	81	84
	Crop residuc cover Contoured	Good	60	72	80	83
	Contoured & terraced	Poor	61	72	79	82
	Contoured & terraced	Good	59	70	78	81
	Crop residue cover Contoured & terraced	Poor	60	71	78	81
	Crop residue cover Contoured & terraced	Good	58	69	77	80
11	Close-seeded legumes[6] or rotation meadow :					
	Straight row	Poor	66	77	85	89
	Straight row	Good	58	72	81	85
	Contoured	Poor	64	75	83	85
	Contoured	Good	55	69	78	83
	Contoured & terraced	Poor	63	73	80	83
	Contoured & terraced	Good	51	67	76	80
	Uncultivated lands :					
12	Pasture or range :	Poor	68	79	86	89
		Fair	49	69	79	84
		Good	39	61	74	80
	Contoured	Poor	47	67	81	88
	Contoured	Fair	25	59	75	83
	Contoured	Good	6	35	70	79
13	Meadow- continuous grass, protected from grazing, and generally mowed for hay	Good	30	58	71	78
	Brush-brush weed grass mixture with brush being the major element	Poor	48	67	77	83
		Fair	35	56	70	77
		Good	30	48	65	73
14	Farmsteads- buildings, lanes, driveways, and surrounding lots	----	59	74	82	86
	Woods and forests					
	Humid rangelands or agricultural uncultivated lands :					
15	Woods or forest land	Poor	45	66	77	83
		Fair	36	60	73	79
		Good	25	55	70	77
16	Woods-grass combination (orchard or tree farm)	Poor	57	73	82	86
		Fair	43	65	76	82
		Good	32	58	72	79
	Arid and Semiarid rangelands[7] :					
17	Herbaceous	Poor		80	87	93

Table 2.1 Continued

Sl. No.	Land use Description/Treatment	Hydrologic Condition /% impervious area	Hydrologic Soil Groups			
			A	B	C	D
		Fair		71	81	89
		Good		62	74	85
18	Oak-aspen	Poor		66	74	79
		Fair		48	57	63
		Good		30	41	48
19	Pinyon-juniper	Poor		75	85	89
		Fair		58	73	80
		Good		41	61	71
20	Sagebrush with grass understory	Poor		67	80	85
		Fair		51	63	70
		Good		35	47	55
21	Desert shrub	Poor	63	77	85	88
		Fair	55	72	81	86
		Good	49	68	79	84

Notations:

[1]Curve numbers are computed assuming the runoff from the house and driveway is directed towards the street with a minimum of roof water directed to lawns where additional infiltration could occur.

[2]In some warmer climates of the country, a curve number of 95 may be used.

[3]Composite CNs should be computed from Figure 2.10 (Ponce, 1989) based on % impervious area (CN=98) and the pervious area CN. The pervious area CNs are taken to be equivalent to desert shrub in poor hydrologic condition.

[4]Composite CNs for the design of temporary measures during grading and construction should be computed from Figure 2.10 (Ponce, 1989), based on the degree of development (impervious area percentage) and the CNs for the newly graded pervious areas.

[5]Compute CN from Figure 2.10.

[6]Close-drilled or broadcast.

[7]Curve numbers for Group A have been developed only for desert shrub.

[*]The remaining pervious areas (lawns) are considered to be in good pasture condition.

Group A. The soils falling in Group A exhibit high infiltration rates even when they are thoroughly wetted, high rate of water transmission, and low runoff potential. Such soils include primarily deep, well to excessively drained sands or gravels.

Group B. These soils have moderate infiltration rates when thoroughly wetted and consist primarily of moderately deep to deep, moderately well to well drained soils with fine, moderately fine to moderately coarse textures, for example, shallow loess and sandy loam. These soils exhibit moderate rates of water transmission.

Group C. Soils in this group have low infiltration rates when thoroughly wetted. These soils primarily contain a layer that impedes downward movement of water. Such soils are of moderately fine to fine texture as, for example, clay loams, shallow sandy loam, and soils low in organic content. These soils exhibit a slow rate of water transmission.

Group D. The soils of this group exhibit very low rates of infiltration when they are thoroughly wetted. Such soils are primarily clay soils of high swelling potential, soils with a permanent high water table, soils with a claypan or clay layer at or near the surface, and shallow soils over nearly impervious material. These soils exhibit a very slow rate of water transmission.

At this point, it is appropriate to understand the basis of the above classification. A thoroughly wetted soil refers to the situation when the soil pores are completely exhausted for retention. It implies that the dynamic (or capillary) portion of infiltration, as for example, the component that is in excess of the static (or gravitational) infiltration, becomes insignificant when the soil is fully saturated. To further elaborate, in an infiltration process, the static portion of infiltration is defined by the infiltration below the line describing the minimum infiltration rate (f_c), and the dynamic portion by the decaying portion of infiltration. Thus, after a soil is fully wetted or saturated, the infiltration due to the existence of the minimum infiltration rate dominates. Therefore, the major criterion that governs the above soil classification is the magnitude of the minimum infiltration rate. Sandy soils exhibit the highest f_c, whereas clay soils exhibit the lowest f_c. The minimum infiltration rate can also be described by the permeability or transmissibility of a soil. It points to the capability of soil to transmit water through soil pores. It underscores the underlying difference between water retention that refers to the holding of water in soil pores, not movement, and transmission that refers to the water movement through soil pores. The water retention is governed by the soil suction (ψ) characteristic and the water transmission by the hydraulic conductivity (K_h) (Chapter 1). These characteristics vary with the moisture content. This aspect is further discussed in Chapters 4 and 5.

NEH-4 provides a list of more than 4000 soils of the United States of America. These are classified into the above hydrologic groups. It is worth mentioning that the original classification of these soils was based on the rainfall-runoff data from small agricultural watersheds or infiltrometer plots. However, most of the soils presented in the recent versions of the NEH-4 are based on the judgment and experience of soil scientists. This classification assumes that (a) soil surfaces are bare, (b) the maximum swelling has taken place, and (c) rainfall rates exceed infiltration rates. Thus, the soils that are similar in depth, organic-matter content, structure, and the degree of swelling when saturated will respond in an essentially similar fashion during a storm of excessively high rainfall intensities. The classification based on the minimum infiltration rates is given in Table 2.2 (McCuen, 1982). Thus, the soil type of a watershed significantly affects the runoff potential of the watershed. It increases as the soil type changes from Group A to Group D and vice versa.

Table 2.2 Description of hydrologic groups

Hydrologic Soil Group	Minimum infiltration rate (inch/hr)
A	0.30-0.45
B	0.15-0.30
C	0.05-0.15
D	0-0.05

2.3.2 LAND USE

The land use characterizes the uppermost surface of the soil system and has a definite bearing on infiltration. It describes the watershed cover and includes every kind of vegetation, litter and mulch, and fallow as well as nonagricultural uses, such as water surfaces, roads, roofs, etc. A forest soil, rich in organic matter, allows greater infiltration than a paved one in urban areas. On an agricultural land or a land surface with loose soil whose particles are easily detached by the impact of rainfall, infiltration is affected by the process of rearrangement of these particles in the upper layers such that the pores are clogged and lead to reduction in the infiltration rate. A grassy or vegetated land will help reduce such a clogging and allow more infiltration. Land treatment applies mainly to agricultural land uses and includes mechanical practices, such as contouring and terracing and management practices, such as grazing control or rotation of crops. Classes refer to the combinations of use and treatment actually to be found on watersheds. The land use and treatment classes listed in Table 2.1 can be broadly classified into urban, cultivated land, and woods and forest.

(i) *Urban Lands*

Urban lands refer to the areas of low or insignificant permeability. Table 2.1 lists curve numbers for residential; paved parking lots; streets and roads; western desert areas; commercial and industrial areas; developing areas; and open spaces, including lawns, parks, etc. Paved areas are assigned the curve number equal to 98. Most of these classifications are based on the amount of percent imperviousness of the area. The remaining pervious areas are assumed to be in a good pasture condition. Thus, the amount of imperviousness plays an important role in the rainfall-runoff process. The larger the impervious area, the higher will be the runoff potential of the watershed and vice versa.

The above described impervious areas include only directly connected impervious areas (DCIA) (SCS, 1986). If runoff from an impervious area passes over a pervious area as sheet flow prior to being discharged, that impervious area is not considered directly connected, rather it is considered as part of the pervious area because it undergoes infiltration while passing over a pervious area as sheet flow. It is noted that infiltration occurs in the same manner as if the runoff originated from the pervious area. There may be significant differences between DCIA and the total impervious area of the watershed. For example, based on the calibration of the observed rainfall/runoff data in a residential watershed during storm events, Pandit and Regan (1997) found that while the total impervious area was 61%, DCIA was only 40% of the watershed area. This difference was mainly because portions of roofs and the driveway discharged water over lawns as sheet flow and were, therefore, not directly connected.

According to Schueler (1987), the runoff potential measured in terms of the runoff factor, C (= Q/P), is related with the percent imperviousness, I, as:

$$C = 0.05 + 0.9\,(I/100) \tag{2.8}$$

Equation (2.8) was derived with a coefficient of determination, r^2, equal to 0.71 using the data of 16 sites of the United States National Urban Runoff Program (NURP) (EPA, 1983). Thus, for I = 0 implying a complete pervious watershed, C = 0.05 and C is equal to 0.95 for I = 100 implying completely impervious area. C = 0.95 for a complete impervious area infers that there exists a 5% loss of water due to evaporation. The term percent imperviousness is frequently used for computing annual runoff from annual rainfall and, in turn, for computing annual amount of pollutant loads, which is discussed in Chapter 8. The available formulae for determination of annual runoff are listed in

Table 2.3.

In urban areas, the percent imperviousness is largely governed by the population density; the larger the population density, the larger will be the percent imperviousness. Stankowski (1974) correlated I with the population density, PD_d, as

$$I = 9.6\,PD_d^{(0.573-0.0391\,\log_{10} PD_d)} \tag{2.9}$$

Table 2.3 Formulae for computing annual runoff

Sl. No.	Formula	Definition of variables	Reference
1	$Q_a/P_a = 0.15(1 - I/100) + 0.90\,(I/100)$	Q_a = annual runoff (cm) P_a = annual rainfall (cm) I = percent imperviousness	Heany et al. (1976),Schueler (1987)
2	$Q_a/P_a = 0.15 + 0.75\,(I/100) - 5.234(V_s)^{0.5957}/P_a$ $V_s = 0.635(1 - I/100) + 0.1588\,(I/100)$	Q_a = annual runoff (inches) P_a = annual rainfall (inches) V_s = depression storage (inches) I = percent imperviousness	Heany et al. (1976)
3	$Q_a/P_a = 0.15 + 0.75\,(I/100) - 7.6297(V_s)^{0.5957}/P_a$ $V_s = 0.635(1 - I/100) + 0.1588\,(I/100)$	Q_a = annual runoff (cm) P_a = annual rainfall (cm) V_s = depression storage (cm) I = percent imperviousness	Pandit and Gopalakrishnan (1996)
4	$Q_a/P_a = 0.045 + 0.81\,(I/100)$	Q_a = annual runoff (cm) P_a = annual rainfall (cm) I = percent imperviousness	Schueler (1987)

where PD_d is in persons per acre. In metric units, equation (2.9) is expressed as (Pandit and Gopalakrishnan, 1996):

$$I = 9.6\,PD_d^{(0.573-0.0391\,\log_{10} PD_d - 0.398)} \tag{2.10}$$

where PD_d is in persons per hectare.

Consistent with the above description, Table 2.1 also reflects an increase in the runoff potential or curve number with the percent imperviousness of the area. It is emphasized that most typical combinations of the urban cover complexes are included in the NEH-4 list. The size of a residential area is shown to have a great impact on the runoff potential. As the size of the watershed increases, the curve number decreases or runoff potential decreases. It is because of the increase in the infiltration amount with the increase in the size of watershed. For larger watersheds, water takes more time to reach the outlet, rendering more time for water to infiltrate than in small watersheds. The right-of-way, indicated in Table 2.1, is relatively more pervious than the impervious areas and allows more infiltration than paved areas. Therefore, the runoff potential will increase with the reduction in the right-of-way and vice versa.

(ii) *Cultivated Land*

It is apparent from Table 2.1 that agricultural watersheds can also be classified as cultivated and uncultivated. The agricultural land uses are classified as fallow land, row crops, small grain crops, close-seeded legumes or rotation meadow, pasture or range, and meadow. Fallow refers to bare agricultural land use treatment having the highest runoff potential. A row crop implies any field crop (for example, maize, sorghum, soybean, sugar beat, tomato, tulip, etc.) planted in rows far apart so that most of the soil surface is exposed to the rainfall impact throughout the growing season. In the areas experiencing snowfall, growing seasons occur when soils are not frozen and there is no snow on the ground. At the planting time, it is tantamount to fallow land and may be so again after harvest. Row crops can be planted either in straight rows or on the contour and they can be in poor or good rotation.

Straight row fields are farmed in straight rows either up and down the hill or across the slope. Where land slopes are less than 2%, farming across the land slope in straight rows is equivalent to contouring. Contoured fields are farmed as nearly as possible along the contour. Contouring delays runoff to increase infiltration. As an example, from Table 2.1, row crops on soil A in poor hydrologic condition exhibit a CN-value of 72 whereas a CN-value of 70 on the same soil in the same hydrologic condition but on contoured rows.

Rotation refers to a planned sequence of crops to maintain soil fertility or reduce erosion or provide an annual moisture supply to a particular crop. Rotation can be poor, fair, or good depending on the allowance for density of vegetative cover in the rotation. Poor rotations are generally one-crop land uses, such as continuous corn (maize) or continuous wheat or combinations of row crops, small grains, and fallow. On the other hand, good rotations generally contain alfalfa or other close-seeded legume or grass to improve tilth and increase infiltration. Another terraced class of land use refers to the systems containing open-end level or graded terraces, grassed waterway outlets, and contour furrows between terraces. In furrows and terraces, a low infiltration land use, if replaced by grassed waterways, will increase opportunities for infiltration.

Small grain crops (for example, wheat, oat, barley, flax, etc.) are planted in rows close enough so that the soil surface is not exposed except during planting and shortly thereafter. Close-seeded legumes or rotation meadow (for example, alfalfa, sweet-clover, timothy, etc. and combinations) are either planted in close rows or broadcast. This cover may remain for more than a year to allow year-round protection to the soil from rainfall impact. Meadows represent fields on which grass is continuously grown. These are protected from grazing and are generally mowed for hay. If a wet meadow is drained, its soil group classification as well as its land use and treatment class may change. Contour furrows on native pasture or range last longer than on cultivated land. Their life span depends on the type of soil, intensity of grazing, and the density of cover. The CN-values in Table 2.1 are based on the data from contour grassland watersheds in central and southern Great Plains. Terraces are seldom used for grasslands. In the event of their use, the construction methods expose bare soils such that the terraced grassland acts hydrologically like a terraced cropland for the next 2-3 years. In other words, the runoff potential of the terraced grassland is the same as of the terraced cropland in the coming few years.

(iii) *Woods and Forests*

Woods and forests may fall under humid rangelands that are tantamount to uncultivated agricultural watersheds and arid and semiarid rangelands. Woods are usually small isolated grooves of trees raised for farm or ranch use. Based on the cover effectiveness (not on timber production), woods can be evaluated as shown in Table

2.4. The hydrologic condition, described in the following section, is estimated visually. Forests, on the other hand, generally cover a large part of the watershed. In humid forest regions of the United States, soil group, humus type, and humus depth are the major factors affecting CN. The undecomposed leaves or needles, twigs, bark, and other vegetative debris on the forest floor form litter, from which humus is derived. Humus is the organic layer immediately below the litter layer. Thus, litter protects humus from oxidation. Humus may consist of mull, an intimate mixture of organic matter and mineral soil or of mor which is practically pure organic matter indistinguishable from the original material lying on the forest floor. Humus increases with the age of forest and because of its porous nature, it increases infiltration. Good management practices refer to proper use, protection and improvement of humus content for increasing infiltration, whereas poor management practices allow burning, over-cutting or overgrazing, and thereby, reduce infiltration. Humus content is evaluated in terms of the degree of compaction: compact, moderately compact, and loose or friable. If mulls are firm and mors are felty, it is compact; otherwise, it is friable. The moderately compact situation represents the transition stage. Frost in compact humus acts like concrete and inhibits infiltration and in loose humus, the frost is granular and favors infiltration. Since frost and humus type are closely related, the effect of frost was not separately accounted for.

Table 2.4 Classification of woods (Source: SCS, 1971)

Sl. No.	Vegetation Condition	Hydrologic Condition
1	Heavily grazed or regularly burned. Litter, small trees, and brush are destroyed.	Poor
2	Grazed but not burned. Some litter exists, but these woods not protected.	Fair
3	Protected from grazing and litter and shrubs cover the soil.	Good

In the forest-range regions (arid and semiarid) of the western United States, soil group, cover type, and cover density are the major factors affecting CN. Figures 2.4 and 2.5 depict the relation between these factors and CN for soil-cover complexes. These are based on the U.S. Forest Service publication "Handbook on Methods of Hydrologic Analysis". The rangelands of arid and semiarid regions have been classified as herbaceous, oak-aspen, pinyon-juniper, sagebrush with grass understory, and desert shrub.

Herbaceous. It represents the mixture of grass, weeds, and low growing brush, with brush being the minor element.

Oak-Aspen. It represents the mountain brush mixture of oak brush, aspen, mountain mahogani, bitter brush, maple, and other brush.

Pinyon-Juniper. It contains pinyon, juniper, or both with grass understory.

Sagebrush. It represents the sage with an understory of grass.

Desert Shrub. The major plants in this class include saltbrush, greasewood, creosotebush, blackbrush, bursage, palo verde, mesquite, and cactus.

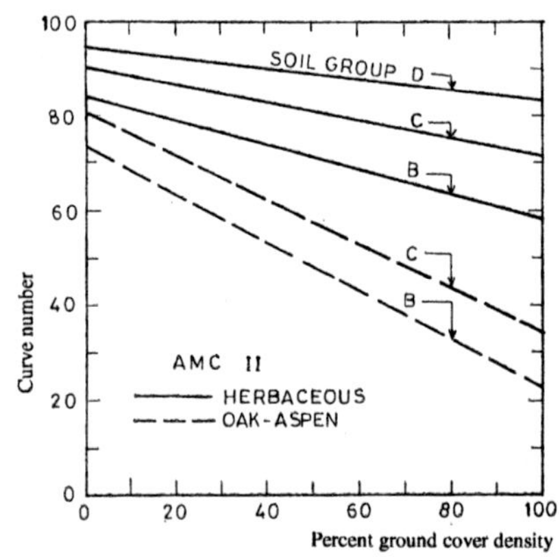

Figure 2.4. Curve number estimation for herbaceous and oak-aspen forests.

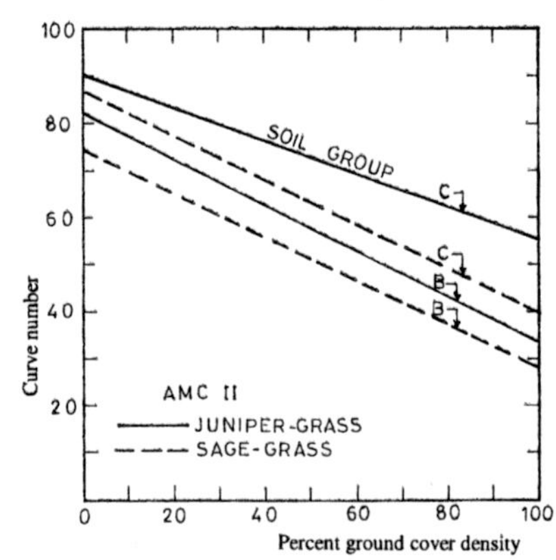

Figure 2.5. Curve number estimation for juniper-grass and sage-grass forests.

For a given soil type and hydrologic condition, Table 2.1 shows that the herbaceous forests exhibit the highest runoff potential or curve number and oak-aspen forests do the lowest runoff potential. The pinyon-juniper forests yield curve numbers less than the herbaceous but greater than others. The desert shrub falls next to the pinyon-juniper and the sagebrush with grass understory next to the desert shrub in respect of the runoff potential. It is noted that the hydrologic condition of forests is defined on the basis of the extent of humus content that enhances infiltration. However, the hydrologic conditions of pastures are defined in terms of the extent of grazing and of woods in terms of either the extent of grazing or the burning of litter, as shown in Table 2.4.

Besides the above cover complexes, SCS has also developed curve numbers for soil-cover complexes of Puerto Rico, a typical watershed of Contra Costa County (California), and Hawaii. These complexes include fallow, grass, coffee, minor crops (garden or truck crops), tropical kudzu, sugarcane, scrub, irrigated pasture, range (annual grass), and urban areas.

2.3.3 HYDROLOGIC CONDITION

The hydrologic condition of an agricultural watershed is defined in terms of the percent area of grass-cover. The larger the area of grass cover in a watershed, the lesser will be the runoff potential of the watershed and more will be infiltration. Such a situation describes the watershed to be in a good hydrologic condition. It is good because it favors the protection of watershed from erosion for soil conservation purposes. Similarly, a watershed having lesser acreage of grass cover can be defined to be in a poor hydrologic condition. Alternatively, a good hydrologic condition allows more infiltration than does a poor hydrologic condition. Thus, the hydrologic condition of a forest area also represents its runoff-producing potential. The curve number will be the highest for poor, average for fair, and the lowest for good condition, leading to categorizing the hydrologic condition into three groups: good, fair, and poor, depending on the areal extent of grasslands or native pasture or range, as shown in Table 2.5. These conditions are based on cover effectiveness. It is emphasized that the cover effectiveness does not imply forage protection. The percent area covered and the intensity of grazing are estimated visually. Grazing on dry soils results in lowering of infiltration rates due to the compaction of the soil by hooves. Its effect may carry over for a year or more even without further grazing. Meadows representing fields on which grass is continuously grown are protected from grazing. Drained meadows (with low water table) have little or no surface runoff except during rainfalls of high intensity. Undrained meadows (with high water table) may be as much wet as water surfaces for which CN is equal to 100.

Table 2.5 Classification of native pasture or range (Source: SCS, 1971)

Sl. No.	Vegetation condition	Hydrologic Condition
1	Heavily grazed and no mulch or plant cover on less than 1/2 of the area	Poor
2	Not heavily grazed and plant cover on less than 1/2 to 3/4 of the area	Fair
3	Lightly grazed and plant cover on more than 3/4 of the area	Good

A determination of CN for forest areas for various hydrologic conditions is primarily guided by the U.S. Forest Service (USFS) (1959). SCS (1985) has also briefly described it. The class of hydrologic condition is defined on the scale of 1 to 6; the higher number refers to the lower runoff potential and vice versa. The relation between classes and humus type and depth is shown in Figure 2.6. If the depth of litter is less than 1/2 inch, the humus is considered unprotected and the hydrologic condition class is reduced by 0.5.

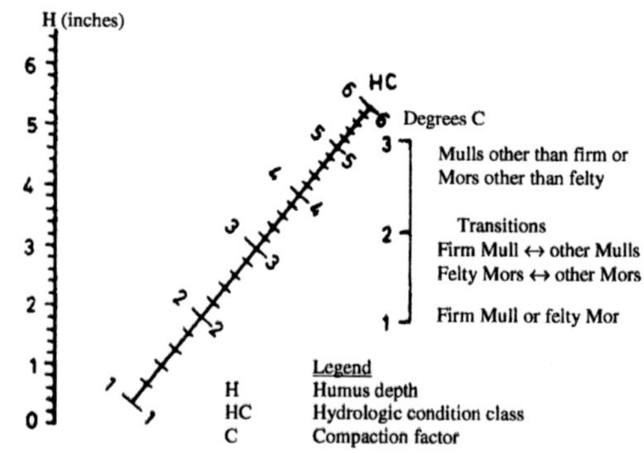

Figure 2.6. Nomograph for determination of the present hydrologic condition class of forest and woodland.

2.3.4 AGRICULTURAL MANAGEMENT PRACTICES

Agricultural management systems involve different types of tillage, vegetation, and surface cover. Freebairn et al. (1989) illustrated the effects of tillage practices (moldboard plough, chisel plough, and no till) on infiltration. Such practices primarily alter the soil porosity that is defined as

$$n = \frac{V_v}{V} = \frac{e}{1+e} = 1 - \frac{BD}{PD} \tag{2.11}$$

where V_v is the volume of voids in a soil profile, V is the total volume of the soil profile, 'BD' is the bulk density, 'PD' is the particle density, and 'e' is the void ratio defined as

$$e = \frac{V_v}{V_s} \tag{2.12}$$

where V_s is the volume of solids. The terms V_v, V_s, and V have been described in Figure 1.18 (Chapter 1).

Brakensiek and Rawls (1988) reported that moldboard plough increases soil porosity from 10-20%, depending on the soil texture and, in turn, increases infiltration rates over non-tilled soils. It is shown (Rawls, 1983) that an increase in organic matter in the soil lowers bulk density or increases porosity, and hence increases infiltration and, in turn, decreases the runoff potential. Rawls (1983) found the bare soil to decrease the hydraulic conductivity, K_h, between planting and midseason and then K_h was found to stay stable because of crusting. The residue maintains a high steady state rate until harvest while the canopy and canopy-residue combination increase K_h. Thus, the residue cover as such increases the soil porosity, n, and hence reduces the runoff potential. The combined effect of these factors on the bulk density (or n) can be interpreted quantitatively from the bulk density (BD) as (Rawls et al., 1992):

$$BD = 1.51 + 0.0025\,(pS) - 0.0013\,(pS)\,(pOM) - 0.0006\,(pC)(pOM) - 0.0048(pC)(CEC) \tag{2.13a}$$

or alternatively, from porosity derived using equations (2.13a) and (2.11) as:

$$n = 1 - (1/PD)\,[1.51 + 0.0025\,(pS) - 0.0013\,(pS)\,(pOM) - 0.0006\,(pC)(pOM) - 0.0048(pC)(CEC)] \tag{2.13b}$$

where BD is the bulk density of less than 2 mm material (gm/cm^3), pS is the percent sand, pC is the percent clay, pOM is the percent organic matter (= 1.7 times the percent organic carbon), and CEC is the ratio of cation-exchange capacity of clay to the percent clay. CEC ranges from 0.1-0.9.

2.3.5 ANTECEDENT MOISTURE CONDITION

As also described in Chapter 1, the antecedent moisture condition (AMC) refers to the wetness of the soil surface or the amount of moisture available in the soil profile, or alternatively the degree of saturation before the start of the storm. In the event that the soil is fully saturated, the whole amount of rainfall will directly convert to runoff without infiltration losses and if the soil is fully dry, it is possible that the whole rainfall amount is absorbed by the soil, leading to no surface runoff. Thus, the antecedent moisture condition affects the process of rainfall-runoff significantly. Laboratory infiltration data (Mein and Larson, 1971) exhibit a significant impact of initial water content of the soil on infiltration rates. For completeness, a brief description of the antecedent moisture condition is in order (Ponce and Hawkins, 1996).

A conceptual model like the existing SCS-CN method working with the mean values leaves room for some variability. Experience (SCS, 1985; Hjelmfelt, 1991) indicates that a set of curve numbers can exist for a given watershed. Ponce and Hawkins (1996) summarized the likely sources to lie in the spatial and temporal variability of rainfall, quality of measured rainfall-runoff data, and the variability of antecedent rainfall and the associated soil moisture amount. The last source of variability was recognized very early leading to the development of the concept of antecedent moisture condition. This concept is also referred to as the antecedent runoff condition (ARC), implying a departure of emphasis from soil moisture to runoff (SCS, 1986). In the plot of rainfall-runoff data, the scatter of the data points is usually interpreted as the measure of natural variability of soil moisture in the associated rainfall-runoff relation. There are three concepts generally used in hydrologic literature to identify the antecedent moisture condition (AMC) of the soil. These are the antecedent precipitation index (API), antecedent baseflow index (ABFI), and the soil-moisture-index (SMI).

API is based on the amount of antecedent rainfall. The term antecedent varies from previous 5 to 30 days (d). However, there is no explicit guideline available to vary the soil moisture with the antecedent rainfall of certain duration. The National Engineering Handbook (SCS, 1971) uses the antecedent 5-d rainfall as API for AMC (Table 2.6) and it is generally used in practice. AMC is categorized into three levels: AMC I, AMC II, and AMC III. AMC I refers to the dry condition of a soil, AMC II to the normal or average, and AMC III to the wet situation of the watershed. Thus, the CN corresponding to AMC I refers to the dry CN or the lowest runoff potential; the CN corresponding to AMC III refers to the wet CN or the highest runoff potential, and the CN corresponding to AMC II stands for the average CN or the average runoff potential. In other words, higher the antecedent moisture or rainfall amount, higher the CN and higher the runoff potential of the watershed and vice versa. Table 2.7 is used for converting CN of AMC II to CNs of AMC I and AMC III. AMC is further discussed in Chapter 3.

Table 2.6 Antecedent soil moisture conditions (AMC)

AMC	Total 5-day antecedent rainfall (cm)	
	Dormant season	Growing season
I	Less than 1.3	Less than 3.6
II	1.3 to 2.8	3.6 to 5.3
III	More than 2.8	More than 5.3

The above concept of AMC can also be visualized in terms of the soil wetness, as described under the classification of soils. A thoroughly wet soil yields the minimum infiltration rate and, in turn, the maximum runoff rate. Thus, these AMCs also refer to the levels of soil wetness. AMC I stands for dry condition, implying the soil is not wet at all, allowing the maximum rate of infiltration or the minimum rate of runoff. On the other hand, AMC III stands for almost completely wet situation of the soil, yielding the minimum infiltration rate and, in turn, producing the maximum runoff rate. AMC II refers to the normal situation that falls in the middle of these two extreme conditions.

The above dry, wet, and normal situations are statistically found to correspond, respectively, to 90%, 10%, and 50% cumulative probability of exceedance of runoff depth for a given rainfall (Hjelmfelt et al., 1982). Alternatively, AMC II represents the central tendency, whereas AMC I and AMC III account for dispersion in the data. Depending on the amount of antecedent 5-d rainfall, AMC is varied according to Table 2.6. The cases of 3-d antecedent rainfall have also been reported (for example, Mishra et al., 1998). SCS (1971) also recommended for developing AMC criteria for individual watersheds using rainfall-runoff data. The advantage of the existing approach of API is that it is simple, easy to grasp, and easy to apply in field.

Table 2.7 Curve numbers for three antecedent moisture conditions

AMC II	AMC I	AMC III	AMC II	AMC I	AMC III
100	100	100	60	40	78
99	97	100	59	39	77
98	94	99	58	38	76
97	91	99	57	37	75
96	89	99	56	36	75
95	87	98	55	35	74
94	85	98	54	34	73
93	83	98	53	33	72
92	81	97	52	32	71
91	80	97	51	31	70
90	78	96	50	31	70
89	76	96	49	30	69
88	75	95	48	29	68
87	73	95	47	28	67
86	72	94	46	27	66
85	70	94	45	26	65
84	68	93	44	25	64
83	67	93	43	25	63
82	66	92	42	24	62
81	64	92	41	23	61
80	63	91	40	22	60
79	62	91	39	21	59
78	60	90	38	21	58
77	59	89	37	20	57
76	58	89	36	19	56
75	57	88	35	18	55
74	55	88	34	18	54
73	54	87	33	17	53
72	53	86	32	16	52
71	52	86	31	16	51
70	51	85	30	15	50
69	50	84			
68	48	84	25	12	43
67	47	83	20	9	37
66	46	82	15	6	30
65	45	82	10	4	22
64	44	81	5	2	13
63	43	80	0	0	0
62	42	79			
61	41	78			

The concept of ABFI is based on the amount of antecedent baseflow, which, of course, is seldom used in practice. It, however, appears to be a better index than API, for it eliminates the problem of the selection of the antecedent duration. However, since baseflows are largely governed by the ground water flow, there exists the problem of separating baseflows from total runoff. Using baseflow as an index, a relation developed for a watershed is depicted in Figure 2.7 (SCS, 1971).

The concept of SMI is generally used for long-term hydrologic simulation, where soil moisture needs to be accounted for water balance reasons. Such simulations utilize evapotranspiration, and thus incorporate other climatic factors, such as daily temperature, solar radiation, etc. (Chapter 1). Such studies include those of Williams and Laseur (1976), Hawkins (1978), Soni and Mishra (1985), Mishra et al. (1998), among others.

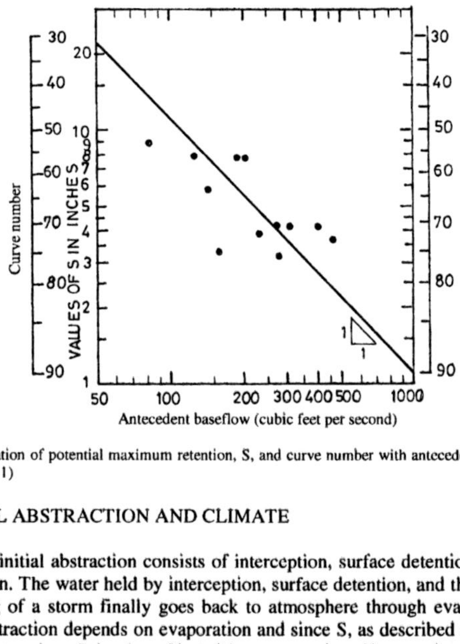

Figure 2.7. Variation of potential maximum retention, S, and curve number with antecedent baseflow. (Data source: SCS, 1971)

2.3.6 INITIAL ABSTRACTION AND CLIMATE

The initial abstraction consists of interception, surface detention, evaporation, and infiltration. The water held by interception, surface detention, and the infiltration at the beginning of a storm finally goes back to atmosphere through evaporation. Thus, the initial abstraction depends on evaporation and since S, as described above, includes the initial abstraction, S is also affected by evaporation. Evaporation (Chapter 1) is primarily governed by meteorological factors, such as radiation, temperature, humidity, wind, sun-shine hours etc., which describe the climate. Thus, the effect of the climatic condition of the watershed is accounted for by the existing SCS-CN method in terms of the initial abstraction. The higher is the amount of initial abstraction, the lower will be the runoff for a given rainfall amount in a watershed and vice versa. Thus, the initial abstraction reduces the runoff potential of the watershed and, in turn, the curve number. The temperature also affects the viscosity of water, which, in turn, affects, to some extent, the surface tension or the ψ-θ relation (Chapter 1). Thus, the magnitude of temperature may have a bearing on infiltration rate. The impact of initial abstraction on the curve number is also discussed in Chapter 4.

2.3.7 RAINFALL INTENSITY AND DURATION AND TURBIDITY

As described in Chapter 1, for a given rainfall amount, the rain duration affects rainfall intensity which influences infiltration (Mein and Larson, 1971). To further visualise the rainfall-runoff process rainfall is assumed as the product of rainfall intensity and storm duration. For a given rainfall amount, the rainfall intensity and time duration are inversely related; the greater the rainfall intensity, the lesser will be the time duration and vice versa. A greater intensity rainfall of lesser storm duration will render lesser time for rain water to stay over the land surface, leading to a lesser amount of infiltration and consequently, a greater amount of direct runoff. The reverse will also hold. As described above, the NEH-4 tables exhibit higher curve numbers for straight row ploughings than those for contoured ones, for the former yield lesser time for rain water to infiltrate than do the latter. Similarly, a longer duration rainfall will result in a greater amount of infiltration for a given rainfall amount than otherwise. Since S in NEH-4 does not account for the time duration of rainfall separately, its impact appears to have been integrated with the curve number. This aspect is further investigated in Chapter 4.

In reality, a high intensity rainfall or raindrop breaks down the soil structure to make soil fines move into the soil surface or near-surface pores, leading to the formation of crust that impedes infiltration. The crust formation actually decreases the effective soil depth responsible for infiltration and also the soil porosity, and thus, decreases S or increases CN. It is for this reason that a fallow land exposed to rain produces a higher runoff for a given rainfall amount than does the unexposed or covered land. Materials covering the surface (such as residues) protect the soil surface. A similar description can be advanced for a good or poor rotation. A good rotation implying the following crop of denser cover than the previous one protects the soil from direct hits of raindrops and allows more infiltration or lesser runoff and, in turn, protects the soil from erosion. A good rotation also protects the humus content to enhance the soil fertility and, in turn, protects from the land to become barren or fallow. A poor rotation does not protect the soil. The impact of rainfall intensity on the curve number is further discussed in Chapter 4.

The term turbidity refers to impurities of water that affect infiltration by the process of clogging of soil pores and consequently, affecting the soil conductivity or ease with which water is transmitted into the soil. The contaminated water with dissolved minerals, such as salts, affects the soil structure and consequently, infiltration. As shown in equation (2.13b), the porosity is greatly influenced by dissolved minerals and organic matters.

2.4 Determination of Curve Number

In most cases, the curve numbers were developed using daily rainfall-runoff records corresponding to the maximum annual flows derived from gauged watersheds (SCS, 1985) for which information on their soils, cover, and hydrologic conditions was available (SCS, 1972). Rainfall (P)-runoff (Q) data were plotted on the arithmetic paper having a grid of plotted curve numbers, as shown in Figure 2.8. The curve number that represented the watershed was taken as the median curve number. Thus, the developed curve numbers represented the averages of median site values for soil groups, cover, and hydrologic conditions. The upper enveloping curve was taken to correspond to AMC III and the lower curve to AMC I. The average was later extended to imply the average soil moisture condition (Miller and Cronshey, 1989). It is worth mentioning that not all soils, cover types, and hydrologic conditions were represented by rainfall-runoff data, rather these were interpolated to complete the information contained in NEH-4 (SCS, 1971).

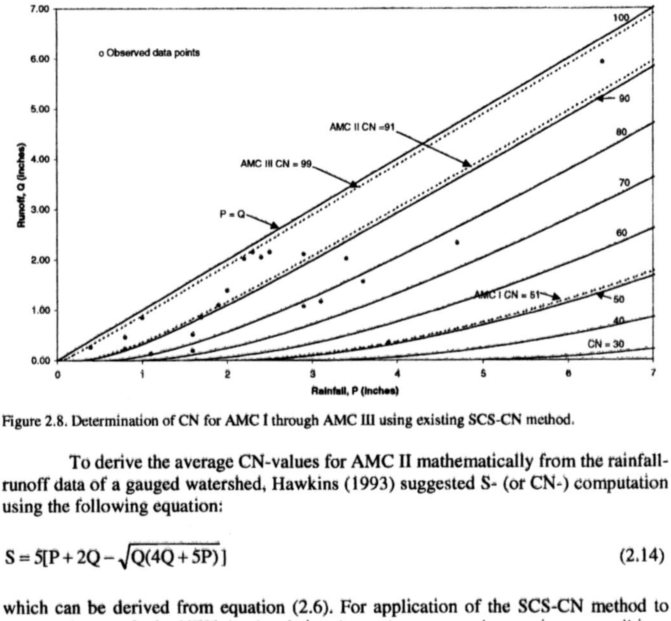

Figure 2.8. Determination of CN for AMC I through AMC III using existing SCS-CN method.

To derive the average CN-values for AMC II mathematically from the rainfall-runoff data of a gauged watershed, Hawkins (1993) suggested S- (or CN-) computation using the following equation:

$$S = 5[P + 2Q - \sqrt{Q(4Q + 5P)}] \tag{2.14}$$

which can be derived from equation (2.6). For application of the SCS-CN method to ungauged watersheds, NEH-4 related the above three antecedent moisture conditions with the amount of antecedent 5-d rainfall and the crop season (Table 2.6).

For determining CN for the present hydrologic condition of a forest area, sample plots are located in the area and soil type, litter depth, humus type and depth are determined from shallow soil wells that are dug in the plots. Using this information, hydrologic condition is determined from Figure 2.6 to obtain the corresponding CN from Figure 2.9.

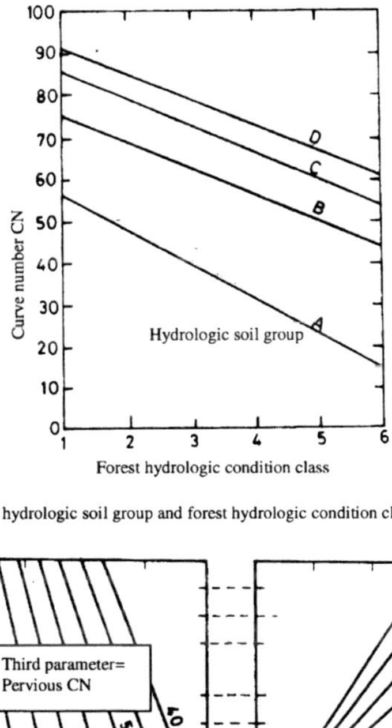

Figure 2.9. Curve numbers by hydrologic soil group and forest hydrologic condition classes.

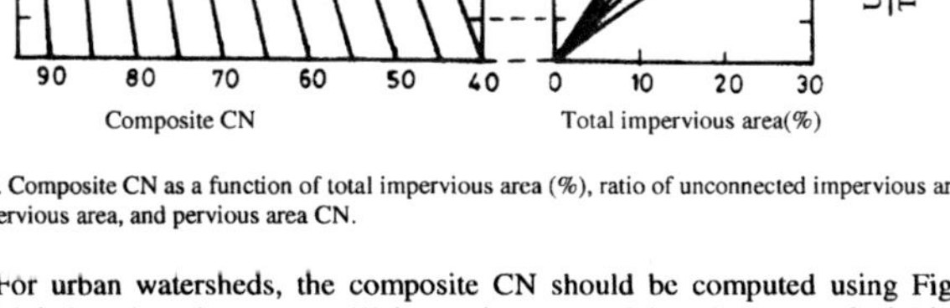

Figure 2.10. Composite CN as a function of total impervious area (%), ratio of unconnected impervious area to total impervious area, and pervious area CN.

For urban watersheds, the composite CN should be computed using Figure 2.10, which is based on the percent (%) impervious area of the urban watershed. Figure 2.10 should also be used to compute composite CNs for design of temporary measures during grading and construction. In such cases, the percent impervious area refers to the degree of development.

2.4.1 DEVELOPMENT OF CN FOR COMPLEXES

Since the NEH-4 table (Table 2.1) forms the key to the application of the existing SCS-CN method, it is necessary to mention a little about the development of this table or on the assignment of CN to the above described complexes. To this end, according to Mockus (SCS, 1971), data were searched for the watersheds in single cover complexes (one soil group and one cover). For most of the listed complexes, watersheds were available. As described above, an average CN for each watershed was obtained using rainfall-runoff data corresponding to annual extreme flows (storms). The selected watersheds were generally less than 1 square mile in size, the number of watersheds for a complex varied, and the storms were of 1 day or less duration. The CN values of the watersheds in the same complex were averaged, all CN-values for a cover were plotted as shown in Figure 2.8, a curve for each cover was drawn with greater weight given to the CN based on the data from more than one watershed, and each curve was extended as far as necessary to provide CN for ungauged watersheds. Except for arbitrary complexes (in Table 2.1:- Sl. No. 3 for "gravel and dirt"; and Sl. No. 14), the CN-values for all agricultural watersheds were derived in this fashion. For agricultural watersheds of complexes not listed in Table 2.1, CN is taken equal to 100, which is equivalent to a water body. For arbitrary complexes, the proportions of different covers were estimated and CN was computed from the previously derived CN.

2.4.2 RATIONALE OF CURVE NUMBER

In an attempt to justify the rationale for developing individual curve numbers, Mockus (1964) explained: "The CN associated with the soil-cover complexes are median values, roughly representing average conditions of a watershed. We took the average condition to mean average soil moisture condition because we had to ignore rainfall intensity." Since the sample variability in CN can be due to infiltration, evapotranspiration, soil moisture, lag time, rainfall intensity, temperature, etc., the antecedent moisture condition (AMC) was supposedly used to represent this variability.

2.5 Use of NEH-4 Tables for SCS-CN Application

For field applications, the procedure for using Tables 2.1, 2.2, 2.6, and 2.7 is as follows:

a) Subdivide large watersheds of different soils and land uses into subwatersheds or hydrological units. Each hydrological unit represents a drainage basin of a tributary joining the major river.
b) Identify the kind of land use, class of treatment, and the type of soils in the hydrological unit of the watershed. A soil can be classified according to Table 2.2 that is based on the minimum infiltration rates of soils.
c) Determine the percent acreage of each of the identified classes.
d) Read CN-values for each of the classes from Table 2.1.
e) Multiply the CN-values by the respective percent areal coverage by each class to compute the weighted CN. This CN corresponds to AMC II.

f) Sum the antecedent 5-d rainfall amounts and determine AMC from Table 2.6.
g) Revise the weighted CN value of AMC II (step e) according to the identified AMC using Table 2.7.
h) Use the revised CN-value in equation (2.6) for computing the direct runoff volume from the watershed for a given amount of rainfall.

The above procedure is illustrated by the following examples.

Example 2.1: Determine the direct runoff for an average storm rainfall depth of 4.3 inches which occurred on a watershed having a cover of good pasture, soil type C, and antecedent moisture condition, AMC II.

Solution: In this example, 'good pasture' watershed implies that the watershed cover is of pasture and it is in good hydrologic condition. This SVL class is defined at Sl. No. 12 of Table 2.1. The curve number for soil type C for this SVL is equal to 74, which is for AMC II. For computing the direct runoff Q, the potential maximum retention, S, is determined as:

$$S = \frac{1000}{CN} - 10 = 1000/74 - 10 = 3.51 \text{ inch.}$$

Compute the amount of initial abstraction, I_a, as

$I_a = 0.2 \times S = 0.2 \times 3.51 = 0.70$

which is less than the amount of rainfall, P = 4.3 inches. If P were less than I_a, Q would be equal to zero. Since $P > I_a$, Q is computed from equation (2.6) as

$$Q = \frac{(P - 0.2S)^2}{P + 0.8S} = \frac{(4.3 - 0.2 \times 3.51)^2}{4.3 + 0.8 \times 3.51} = 1.82 \text{ inches}.$$

Example 2.2: Taking the same watershed as of Example 2.1, compute the amount of direct runoff for AMC I through III for the same rainfall amount and compare them.

Solution: For the given CN = 74 for AMC II, the CN values corresponding to AMC I and III (Table 2.7) are 55 and 88, respectively. Therefore, the potential maximum retention, S, for AMC I and AMC III, respectively, are:

S = 1000/55 – 10 = 8.18 inches

and

S = 1000/88 – 10 = 1.36 inches

Thus, the magnitudes of direct runoff for AMC I and AMC III are, respectively, computed using equation (2.6) as

$$Q = \frac{(4.3 - 0.2 \times 8.18)^2}{4.3 + 0.8 \times 8.18} = 0.65 \text{ inch}$$

and

$$Q = \frac{(4.3 - 0.2 \times 1.36)^2}{4.3 + 0.8 \times 1.36} = 3.01 \text{ inches}$$

The former is 35.7 % (= 0.65 x 100/1.82) and the latter is 165.4 % (= 3.01 x 100/1.82) of the runoff amount for AMC II, viz., 1.82 inches.

Example 2.3: A watershed of soil group B is of 630 acres. Its 400 acres of area is under 'row crop, contoured, good rotation' and the remaining 230 acres is occupied by 'rotation meadow, contoured, and good rotation'. Compute the direct runoff for the rainfall amount of 5.1 inches when the watershed is in AMC II.

Solution: This example can be solved in two ways (a) by weighted-Q method and (b) by weighted-CN method.

(a) *Weighted-Q Method*

The computations are performed in Table 2.8. It is noted that good rotation implies good hydrologic condition. Good rotation infers the rotation of crops on agricultural fields in such a way that enhances infiltration. The computation of runoff is shown in Table 2.8. Col. 1 of this table shows the given SVL complex; col. 2 the area in acres; col. 3 the % area covered by each complex; col. 4 the CN values for each complex derived from Table 2.1; col. 5 the potential maximum retention, S, computed using equation (2.7); col. 6 the computed runoff using equation (2.6), and col. 7 computes the weighted-Q as the product of the Q (col. 6) and % area (col. 3) divided by 100. The sum of the weighted Qs represents the total direct runoff of the watershed for the given rainfall P = 5.1 inches.

Table 2.8 Computation of runoff using the weighted-Q method

SVL Complex	Area (acres)	% area	CN (for AMC II)	S (inches)	Q (inches)	Weighted Q (inches)
1	2	3	4	5	6	7
Row crop, contoured, good rotation	400	63.49	75	3.33	2.53	1.61
Rotation meadow, contoured, and good rotation	230	36.51	58	7.24	1.22	0.45
Sum	630					2.06

Table 2.9 Computation of runoff using the weighted-CN method

SVL Complex	Area (acres)	% area	CN (for AMC II)	Weighted CN	S (inches)	Q (inches)
1	2	3	4	5	6	7
Row crop, contoured, good rotation	400	63.49	75	48	---	----
Rotation meadow, contoured, and good rotation	230	36.51	58	21		
Sum	630			69	4.49	2.03

(b) *Weighted-CN Method*

The computations for the direct runoff are shown in Table 2.9. In the weighted-CN method, the CN computed in col. 4 is areally weighted by multiplying the CN-value with the % area (col. 2) and divided by 100. The computed weighted CNs are summed up to obtain the corresponding S (col. 6). Using S = 4.49 inches, Q is computed as equal to 2.03 inches. It is noted that the computed Q (= 2.06 inches) by the weighted-Q method compares fairly well with the present Q = 2.03 inches. The small discrepancy is, however, attributed to the rounding off errors in computations.

It is worth mentioning here that the weighted-Q method is superior to the weighted-CN method, for the former is more rational than the latter for water balance reasons. However, the former method requires more computational effort than the latter. The weighted-CN is easier to work with the watershed having many complexes or with a series of storms. However, when there is a large difference in CN for various complexes in a watershed, the computed direct runoff by the weighted-CN method significantly deviates from that by the weighted-Q method, as shown in the following example.

Example 2.4: For an urban watershed with 20 acres of impervious area and 175 acres of lawn classed as good pasture on a B soil, compute the direct runoff, Q, values for rainfall values of 1, 2, 4, 8, 16, 32 inches. Based on the computed Q-values, compare the weighted–Q and weighted–CN methods. Assume CN = 100 for impervious area.

Solution: For the SVL complex defined by 'lawn, good pasture, and soil group B', CN = 61 (Table 2.1). As in Example 2.3, compute the Q values for the rainfall values of 1, 2, 4, 8, 16, 32 inches using the weighted Q and CN methods, as shown in Tables 2.10 and 2.11, respectively.

For comparison, the Q values computed by both the above methods (weighted-Q and weighted-CN methods) are listed in Table 2.12 along with the percent deviation of Qs computed using the weighted-CN method from those computed using the weighted-Q method. It is seen from this table that the weighted-Q method is preferable when the rainfall amount is small, for the % deviation is quite large. On the other hand, for high rainfalls, the weighted-CN computes Q values that deviate from those computed by weighted-Q method in the range from 7.85 to 8.67 %, which is an acceptable range.

Table 2.10 Computation of runoff using the weighted-Q method

SVL Complex	Area (acre)	% area	CN	S (in)	Weighted-Q for rainfalls (inches):					
					1	2	4	8	16	32
Impervious area	20	10.26	100	0	0.10	0.21	0.41	0.82	1.64	3.28
Lawn, good pasture	175	89.74	61	6.39	0	0.07	0.81	3.45	10.26	25.43
Sum	195				0.10	0.28	1.22	4.27	11.90	28.71

Table 2.11 Computation of runoff using the weighted-CN method

SVL Complex	Area	% area	CN	^{a}CN	S (in)	Direct runoff for rainfalls (inches):					
						1	2	4	8	16	32
Impervious area	20	10.26	100	10							
Lawn, good pasture	175	89.74	61	55							
Sum	195			65	5.38	0	0.14	1.03	3.90	10.97	26.3

Area in acres.

[a]Weighted CN

Table 2.12 Comparison of the weighted-Q and –CN methods

Storm rainfall →	1	2	4	8	16	32
Weighted-Q method	0.12	0.28	1.22	4.27	11.90	28.71
Weighted-CN method	0	0.14	1.03	3.90	10.97	26.34
% deviation →	100	50	15.57	8.67	7.85	8.28

Example 2.5: Determine the runoff amount from a series of rainfalls of 1, 2, 4, 8 16, and 32 inches that occurred over W-1 watershed located in Waco, Texas, U.S.A. The major soil type of this watershed is Houston Black Clay or equivalent and the acreage of land use and treatment are as below:

Land use and treatment	% area
Row crop, straight row, poor rotation	58
Small grain, straight, poor rotation	25
Pasture (including hay), fair condition	15
Farmsteads and roads	2

Solution: The Houston Black Clay soil can be characterized by the hydrologic soil group D. CNs for AMC II for various land uses and treatments are derived from Table 2.1 and weighted CN is determined as below:

Land use and treatment	% area	CN	Weighted CN
Row crop, straight row, poor rotation	58	91	89.2≈89
Small grain, straight, poor rotation	25	88	
Pasture (including hay), fair condition	15	84	
Farmsteads and roads	2	94	

Corresponding to the weighted CN = 89 that holds for AMC II, the respective CN values for AMC I and AMC III are derived from Table 2.7 as 76 and 96. The computations are shown in Table 2.13. In this table, col. 1 indicates the event number, cols. 2-4 describe the date, col. 5 shows the observed rainfall (inches) of the day, col. 6 the antecedent 5-d rainfall (inches), col. 7 the observed runoff (inches) of the day, col. 8 AMC, col. 9 the corresponding curve number, col. 10 the computed runoff using the curve number of col. 9, col. 11 the errors between the observed and computed runoff amounts, and col. 12 the square of the error for computing the root mean square error. It is noted that within a storm, for example, the storm of 1940 occurred on Nov. 22-25, AMC is computed by adding the previous rainfall amounts. On day 22, the antecedent 5-d rainfall (P_5) was equal to 0.18 inches, it was 4.92 inches on November 23 computed by adding 0.18 to the previous day rainfall, and so on. It assumes negligible rainfall contribution during the beginning days of the antecedent 5-days or, in other words, the antecedent rainfall amount of November 23 corresponds to the 6-d antecedent rainfall instead of 5-d one. Based on these P_5-values, AMC is determined from Table 2.6 and the resulting CN from Table 2.7. Using these curve numbers, the direct runoff values are computed for each rainfall amount using equation (2.6). It is apparent from Table 2.13 that the errors between observed and computed runoff values range from –1.3 to 0.56 inches and the root mean square error is 2.04 inches.

Example 2.6: Taking the Example 2.5 watershed as a gauged watershed, derive CN values for AMC I through AMC III for this watershed.

Solution: The CN-values from given P-Q data sets are computed using equation (2.14). The computed values from P-Q data of Example 2.5 are shown in Table 2.14. It is apparent from this table that CN ranges between 51 and 99. The median CN value is equal to 91. Thus, the CN values for AMC I, AMC II, and AMC III are 51, 92, and 99, respectively. A graphical plot of the data points on the SCS-CN grids is shown in Figure 2.8. Here, the exclusion of event 6 (yielding the minimum CN = 51) from the P-Q data sets yields a median CN value equal to 92, and CN = 74 for AMC I. The NEH-4 table (Table 2.7) yields CN values equal to 76, 89, and 96 for AMC I, AMC II, and AMC III, respectively, which are close to the computed values, provided event 6 is excluded from the data set.

Table 2.13 Computation of direct runoff for a series of rainfalls (Example 2.5)

Event	Year	Month	Day	Rainfall (inch)	P_5 (inch)	Obs. Runoff (inch)	AMC	CN	Comp. Runoff (inch)	Error (inch)	$(\text{Error})^2$
1	1940	Nov.	22	4.74	0.18	2.32	I	76	2.32	0.00	0.00
2			23	2.2	4.92	2.02	III	96	1.77	0.25	0.06
3			24	2.03	6.94	1.39	III	96	1.61	-0.22	0.05
4			25	0.38	8.97	0.26	III	96	0.13	0.13	0.02
5	1941	June	10	2.39	1.38	2.05	III	96	1.96	0.09	0.01
6	1942	Sept.	7	3.89	0.22	0.35	I	76	1.65	-1.30	1.69
7			8	3.36	4.11	2.02	III	96	2.91	-0.89	0.79
8			9	0.78	7.47	0.46	III	96	0.44	0.02	0.00
9	1943	June	5	1.58	0.09	0.51	I	76	0.22	0.29	0.08
10	1944	April	29	3.63	0	1.56	I	76	1.45	0.11	0.01
11			30	2.64	3.63	2.15	III	96	2.21	-0.06	0.00
12		May	1	6.37	6.27	5.92	III	96	5.90	0.02	0.00
13			2	1.1	12.64	0.13	III	96	0.73	-0.60	0.36
14	1945	March	2	0.77	0.41	0.23	I	76	0.00	0.23	0.05
15			3	2.5	1.18	2.15	III	96	2.07	0.08	0.01
16	1946	May	12	2.9	1.08	2.11	III	96	2.46	-0.35	0.12
17			13	0.95	3.98	0.84	III	96	0.59	0.25	0.06
18	1947	March	18	1.74	0	0.85	I	76	0.29	0.56	0.31
19	1948	April	25	3.1	0.5	1.17	I	76	1.08	0.09	0.01
20	1949	July	4	2.86	0.03	1.07	I	76	0.92	0.15	0.02
21	1950	Feb.	12	1.94	1.08	1.09	III	96	1.52	-0.43	0.18
22	1951	June	16	1.64	1.28	0.19	II	89	0.74	-0.55	0.30
Sum											4.15
Root Mean Square Error (inches)											2.04

Table 2.14 CN computation for a gauged watershed (Example 2.6)

Event	Year	Month	Day	Rainfall (inch)	Runoff (o) (inch)	S	CN
1	1940	Nov.	22	4.74	2.32	3.16	76
2			23	2.2	2.02	0.16	98
3			24	2.03	1.39	0.69	94
4			25	0.38	0.26	0.13	99
5	1941	June	10	2.39	2.05	0.31	97
6	1942	Sept.	7	3.89	0.35	9.44	51
7			8	3.36	2.02	1.55	87
8			9	0.78	0.46	0.38	96
9	1943	June	5	1.58	0.51	1.74	85
10	1944	April	29	3.63	1.56	2.91	77
11			30	2.64	2.15	0.47	96
12		May	1	6.37	5.92	0.39	96
13			2	1.1	0.13	2.38	81
14	1945	March	2	0.77	0.23	0.91	92
15			3	2.5	2.15	0.32	97
16	1946	May	12	2.9	2.11	0.81	92
17			13	0.95	0.84	0.1	99
18	1947	March	18	1.74	0.85	1.16	90
19	1948	April	25	3.1	1.17	2.9	77
20	1949	July	4	2.86	1.07	2.71	79
21	1950	Feb.	12	1.94	1.09	1.03	91
22	1951	June	16	1.64	0.19	3.58	74
Median CN value							91

Data: Example (2.4)

2.6 Sensitivity Analysis

In dealing with uncertainty in the estimation of Q, it is often desired to know as to how sensitive the computed Q is to changes in the actual values of λ, P, and CN which cannot be measured with certainty. Sensitivity analysis, thus, presents a viable method to evaluate, without repetitious computations, the proper weight and consideration that must be taken when each of the influencing factors is evaluated. In what follows is the sensitivity analysis using first-order analysis proposed by Singh and Yu (1990) based on Taylor series expansion and the conventional analysis by Chen (1982).

2.6.1 FIRST-ORDER SENSITIVITY ANALYSIS

To illustrate the procedure, equation (2.5) is coupled with equations (2.3) and (2.7) as

$$Q = \frac{\left[P - \lambda\left(\frac{1000}{CN} - 10\right)\right]^2}{P + (1-\lambda)\left(\frac{1000}{CN} - 10\right)} \tag{2.15}$$

Thus, equation (2.15) can be expressed as

$$Q = g\,(P, \lambda, CN) \tag{2.16}$$

where g is some function. If the true values of P, λ, and CN differ from their corresponding nominal (average) values, $\overline{P}$, $\overline{\lambda}$, and $\overline{CN}$, where

$$\overline{Q} = g(\overline{P}, \overline{\lambda}, \overline{CN}) \tag{2.17}$$

then the effect of this discrepancy on Q can be evaluated by expanding the function g(·) around the point defined by $P = \overline{P}$, $\lambda = \overline{\lambda}$, $CN = \overline{CN}$.

For purposes of simplicity, let x_1 = P, x_2 = λ, and x_3 = CN. Then the Taylor series expansion of equation (2.16) yields

$$Q = g(\overline{P}, \overline{\lambda}, \overline{CN}) + \sum_{i=1}^{3}\left[\left(\frac{\partial g}{\partial x_i}\right)_{-}\right](x_i - \overline{x}_i) + \frac{1}{2}\sum_{i=1}^{3}\sum_{j=1}^{3}\left[\left(\frac{\partial^2 g}{\partial x_i \partial x_j}\right)_{-}\right](x_i - \overline{x}_i)(x_j - \overline{x}_j) + \text{higher-order terms} \tag{2.18}$$

where $[(\partial g/\partial x_i)_{-}]$ = the first-order partial derivative of the function g with respect to x_i evaluated at the point of expansion $(\overline{x}_1, \overline{x}_2, \overline{x}_3)$; and $(\partial^2 g/\partial x_i \partial x_j)_{-}$ = the second-order partial derivative of g with respect to x_i and x_j evaluated at the point of expansion.

If the second- and higher-order terms are neglected, the resulting first-order expression for the error in Q is

$$Q - \overline{Q} = E = \sum_{i=1}^{3}\left(\frac{\partial g}{\partial x_i}\right)_{-}(x_i - \overline{x}_i) \tag{2.19}$$

The variance of this error, S_Q^2, is

$$S_Q^2 = E[(Q-\overline{Q})^2] \tag{2.20a}$$

or

$$S_Q^2 = E\left\{\sum_{i=1}^{3}\left[\left(\frac{\partial g}{\partial x_i}\right)_{-}(x_i - \overline{x}_i)\right]^2\right\} \tag{2.20b}$$

where E is expectation operator. For simplicity, it is assumed that the variables x_i are independent.

$$S_Q^2 = \sum_{i=1}^{3}\left(\frac{\partial g}{\partial x_i}\right)_{-}^2 S_{x_i}^2 \tag{2.21a}$$

or

$$S_Q^2 = \left(\frac{\partial g}{\partial P}\right)_{-}^2 S_P^2 + \left(\frac{\partial g}{\partial \lambda}\right)_{-}^2 S_\lambda^2 + \left(\frac{\partial g}{\partial CN}\right)_{-}^2 S_{CN}^2 \tag{2.21b}$$

where $S_{x_i}^2$ = the variance of x_i. From equation (2.15), the partial derivatives ($\partial g/\partial x_i$ = $\partial Q/\partial x_i$) are

$$\frac{\partial Q}{\partial P} = \frac{\left[P-\lambda\left(\frac{1000}{CN}-10\right)\right]\left[P+(2-\lambda)\left(\frac{1000}{CN}-10\right)\right]}{\left[P+(1-\lambda)\left(\frac{1000}{CN}-10\right)\right]^2} \tag{2.22a}$$

$$\frac{\partial Q}{\partial \lambda} = \frac{-\left(\frac{1000}{CN}-10\right)\left[P-\lambda\left(\frac{1000}{CN}-10\right)\right]\left[P+(2-\lambda)\left(\frac{1000}{CN}-10\right)\right]}{\left[P+(1-\lambda)\left(\frac{1000}{CN}-10\right)\right]^2} \tag{2.22b}$$

$$\frac{\partial Q}{\partial CN} = \frac{\frac{1000}{CN^2}\left[P-\lambda\left(\frac{1000}{CN}-10\right)\right]\left[(1+\lambda)P+\lambda(1-\lambda)\left(\frac{1000}{CN}-10\right)\right]}{\left[P+(1-\lambda)\left(\frac{1000}{CN}-10\right)\right]^2} \tag{2.22c}$$

By substituting the values of these partial derivatives in equations (2.21a) and (2.21b), the error in Q can be computed. If the error in Q is computed due to error in only one variable, say P, then

$$S_Q^2 = \mathrm{var}(Q) = \left(\frac{\partial g}{\partial P}\right)_{\bar{P}}^2 S_P^2 = \frac{\left[P-\lambda\left(\frac{1000}{CN}-10\right)\right]^2\left[P+(2-\lambda)\left(\frac{1000}{CN}-10\right)\right]^2}{\left[P+(1-\lambda)\left(\frac{1000}{CN}-10\right)\right]^4} S_P^2 \quad (2.23a)$$

The value of S_Q^2 is the standard error of estimate of Q. If error in Q is calculated due to error in λ only, then

$$S_Q^2 = \mathrm{var}(Q) = \left(\frac{\partial g}{\partial \lambda}\right)_{\bar{\lambda}}^2 S_\lambda^2 =$$

$$\frac{\left(\frac{1000}{CN}-10\right)^2\left[P-\lambda\left(\frac{1000}{CN}-10\right)\right]^2\left[P+(2-\lambda)\left(\frac{1000}{CN}-10\right)\right]^2}{\left[P+(1-\lambda)\left(\frac{1000}{CN}-10\right)\right]^4} S_\lambda^2 \quad (2.23b)$$

If only CN is to be considered, then

$$S_Q^2 = \mathrm{var}(Q) = \left(\frac{\partial g}{\partial CN}\right)_{\overline{CN}}^2 S_{CN}^2 =$$

$$\frac{\frac{10^6}{CN^4}\left[P-\lambda\left(\frac{1000}{CN}-10\right)\right]^2\left[(1+\lambda)P+\lambda(1-\lambda)\left(\frac{1000}{CN}-10\right)\right]^2}{\left[P+(1-\lambda)\left(\frac{1000}{CN}-10\right)\right]^4} S_{CN}^2 \quad (2.23c)$$

The total variance of Q can also be expressed as a function of the coefficients of variation of P, λ, and CN as a sum of the right-hand sides of equations (2.23a) through (2.23c) as:

$$S_Q^2 = \frac{\left[P - \lambda\left(\frac{1000}{CN} - 10\right)\right]^2}{\left[P + (1-\lambda)\left(\frac{1000}{CN} - 10\right)\right]^4} \left\{ \begin{array}{l} \left[P + (2-\lambda)\left(\frac{1000}{CN} - 10\right)\right]^2 S_P^2 + \\ \left(\frac{1000}{CN} - 10\right)^2 \left[\begin{array}{l}(1+\lambda)P + \\ \lambda(1-\lambda)\left(\frac{1000}{CN} - 10\right)\end{array}\right]^2 S_\lambda^2 \\ + \frac{10^6}{CN^4} \left[\begin{array}{l}(1+\lambda)P + \\ \lambda(1-\lambda)\left(\frac{1000}{CN} - 10\right)\end{array}\right]^2 S_{CN}^2 \end{array} \right\} \tag{2.24}$$

It is seen from equation (2.24) that the variance of Q not only depends on the variance of each parameter but also on the values of the parameters themselves.

Example 2.7: Compute errors in runoff estimation due to a +10% deviation from the mean values of λ, P, and CN, which are 0.2, 5 inches, and 70, respectively.

Solution: Given: $\overline{\lambda} = 0.2$, $\overline{P} = 5$ inches, and $\overline{CN} = 70$. Since the measured values are 10% higher than the mean values, λ = 0.2 x 1.1 = 0.22, P = 5 x 1.1 = 5.5 inches, and CN = 70 x 1.1 = 77. Furthermore, $S_P^2 = S_\lambda^2 = S_{CN}^2 = 0.1 \times 0.1 = 0.01$. Substitution of these values in equation (2.24) yields

$S_Q^2 = 0.0062 \times [117.0 \times 0.01 + 465.43 \times 0.01 + 1.48 \times 0.01] = 0.0362$

The root mean square of S_Q^2 exhibits the error in runoff from its mean value and is computed as equal to 0.19. It implies that the computed runoff will be 19% higher than its desired value.

2.6.2 CONVENTIONAL ANALYSIS

The conventional analysis is based on the work of Chen (1982), who performed the sensitivity analysis in terms of the runoff factor, C (= Q/P), which exhibits the runoff producing potential of a watershed.

Coupling of equation (2.5) with equations (2.3) and (2.7) along with C = Q/P yields

$$C = \frac{\left[1 - \frac{\lambda}{P}\left(\frac{1000}{CN} - 10\right)\right]^2}{1 + \frac{(1-\lambda)}{P}\left(\frac{1000}{CN} - 10\right)} \tag{2.25}$$

which holds for $P \geq \lambda$ [1000/CN – 10]. Thus, it is evident from equation (2.25) that C is a function of λ, P, and CN. Therefore, the total derivative of C can be given as

$$dC = \frac{\partial C}{\partial \lambda} d\lambda + \frac{\partial C}{\partial P} dP + \frac{\partial C}{\partial CN} dCN \tag{2.26}$$

where ∂C/∂λ, ∂C/∂P, ∂C/∂CN are the partial derivatives of C with respect to λ, P, and CN, respectively. The total differential, dC, corresponding to the increments dλ, dP, and dCN can be physically interpreted as the total variation of C due to the variations of λ, P, and CN at any point in the (λ, P, CN) domain. The variations of C with respect to the variable under consideration can be derived from equation (2.25).

A more useful form of equation (2.26) follows:

$$\frac{dC}{C} = \left(\frac{\partial C}{\partial \lambda}\frac{\lambda}{C}\right)\frac{d\lambda}{\lambda} + \left(\frac{\partial C}{\partial P}\frac{P}{C}\right)\frac{dP}{P} + \left(\frac{\partial C}{\partial CN}\frac{CN}{C}\right)\frac{dCN}{CN} \tag{2.27}$$

where (∂C/∂λ)(λ/C), (∂C/∂P)(P/C), (∂C/∂CN)(CN/C) are referred to as ratios of the error in the runoff coefficient (dC/C) to the error in the initial abstraction ratio (dλ/λ), to the error in the storm rainfall (dP/P), and to the error in the runoff curve number (dCN/CN), respectively. Expressing CN in terms of the initial abstraction I_a by combining equations (2.3) and (2.7), all individual right-hand side components of equation (2.27) can be derived as:

$$\frac{\partial C}{\partial \lambda}\frac{\lambda}{C} = -\frac{\frac{I_a}{P}\left[1 + \frac{2(1-\lambda)}{\lambda}\frac{I_a}{P} - \frac{2-\lambda}{\lambda}\left(\frac{I_a}{P}\right)^2\right]}{\left(1-\frac{I_a}{P}\right)^2\left(1+\frac{1-\lambda}{\lambda}\frac{I_a}{P}\right)} \tag{2.28}$$

$$\frac{\partial C}{\partial P}\frac{P}{C} = \frac{\frac{1}{\lambda}\frac{I_a}{P}\left[1+\lambda+(1-\lambda)\frac{I_a}{P}\right]}{\left(1-\frac{I_a}{P}\right)\left(1+\frac{1-\lambda}{\lambda}\frac{I_a}{P}\right)} \tag{2.29}$$

$$\frac{\partial C}{\partial CN}\frac{CN}{C} = \frac{\left(\frac{I_a}{\lambda}+10\right)\left[\frac{1+\lambda}{P} + \frac{1-\lambda}{\lambda}\left(\frac{I_a}{P}\right)^2\right]}{\left(1-\frac{I_a}{P}\right)\left(1+\frac{1-\lambda}{\lambda}\frac{I_a}{P}\right)} \tag{2.30}$$

Equations (2.28) through (2.30) are plotted for a given λ, say $\lambda = 0.25$, as shown in Figures 2.11-2.13, respectively. These figures illustrate the variation or the sensitivity of C to the above three variables.

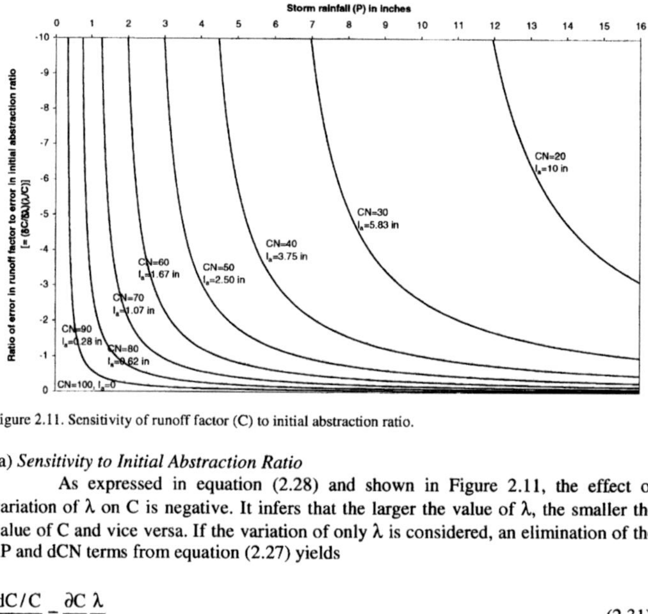

Figure 2.11. Sensitivity of runoff factor (C) to initial abstraction ratio.

(a) *Sensitivity to Initial Abstraction Ratio*

As expressed in equation (2.28) and shown in Figure 2.11, the effect of variation of λ on C is negative. It infers that the larger the value of λ, the smaller the value of C and vice versa. If the variation of only λ is considered, an elimination of the dP and dCN terms from equation (2.27) yields

$$\frac{dC/C}{d\lambda/\lambda} = \frac{\partial C}{\partial \lambda}\frac{\lambda}{C} \tag{2.31}$$

The left-hand side of equation (2.31) represents the ratio of the error in C to the error in λ. Here, it is noted that this ratio approaches infinity as P approaches I_a. It is because C approaches zero when P approaches I_a. It is apparent from Figure 2.11 that the smaller is the value of CN, the larger is the effect of variation of λ on C. On the other hand, when CN and P are large, the corresponding C-value is also large, but insensitive to the variation of λ.

(b) *Sensitivity to Storm Rainfall*

Similar to the above analysis, if the variation of only P is considered ignoring the impact of λ and CN, equation (2.27) reduces to

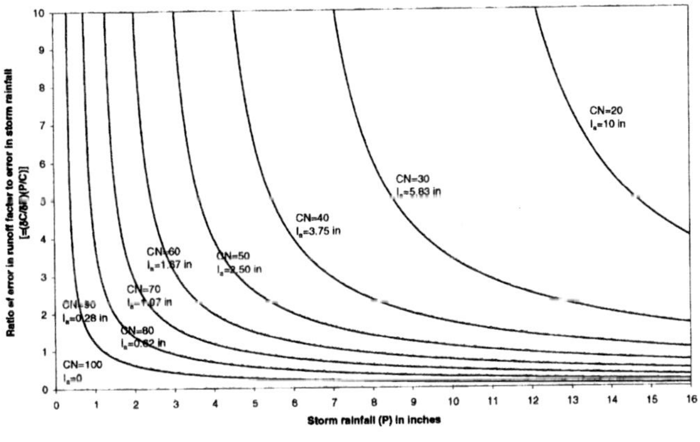

Figure 2.12. Sensitivity of runoff factor (C) to storm rainfall (P).

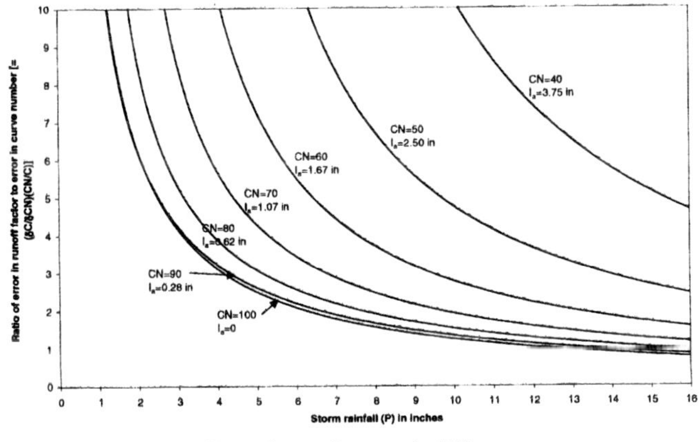

Figure 2.13. Sensitivity of runoff factor (C) to runoff curve number (CN).

$$\frac{dC/C}{dP/P} = \frac{\partial C}{\partial P}\frac{P}{C} \tag{2.32}$$

Analogous to the preceding analysis, the left-hand side of equation (2.32) represents the ratio of the error in C to the error in P. It is again emphasized that for the same reason described in the previous analysis, the ratio of equation (2.32) approaches infinity as P approaches I_a. It is apparent from Figure 2.12 that smaller values of CN show a greater effect of the variation of P on C and vice versa.

(c) *Sensitivity to Runoff Curve Number*

Consideration of the variation of only CN reduces equation (2.27) to

$$\frac{dC/C}{dCN/CN} = \frac{\partial C}{\partial CN}\frac{CN}{C} \tag{2.33}$$

The left-hand side of equation (2.33), thus, represents the ratio of the error in C to the error in CN. This ratio is also infinity at $P = I_a$. Depicting the plot of this ratio against P and CN for $\lambda = 0.25$, Figure 2.13 shows the same trends of the results as shown in Figures 2.11 and 2.12. It follows that the smaller is the value of CN, the larger is the effect of the variation of CN on C, and vice versa.

It is quite possible that two or all of the above three factors vary simultaneously from the conditions assumed in the above individual cases. For such combined conditions, the corresponding two or all of equations (2.28) through (2.30) can be substituted into equation (2.27) for complete expression of the total effect of the two or three variations. If greater accuracy than two significant numbers is not required, Figures 2.11 through 2.13 may be used to estimate individual effects and then added. It is emphasized that Figures 2.11 through 2.13 hold for $\lambda = 0.25$. To determine the sensitivity of C to any value of λ other than 0.25, one must use equations (2.27) through (2.30). To use equation (2.27), it is noted that the positive or negative input errors in λ, P, and CN will reflect themselves in the signs of $d\lambda$, dP, and dCN, and the resulting positive or negative output error will be indicated by the sign of dC.

From the above sensitivity of C to λ, P, and CN, it can be concluded that errors in CN have much more serious consequences than errors of similar magnitude in λ or P. It is consistent with Hawkins (1975) who derived the same conclusion for a specific simplified situation. However, there is a marginal disagreement between his iso-error contours and that which is computed from equations (2.29) and (2.30) for λ = 0.20. Chen (1982) found that Hawkins' iso-error contours agree with the results obtained from equation (2.30) for $\lambda = 0.2$ multiplied by dCN/CN = 0.1 (equation (2.33)), except for a portion (about P > 9 inches) which, according to the Hawkins analysis, was attributed to an error source from P rather than CN. Furthermore, a small portion of the Hawkins iso-error contours attributed to a negative error source from CN can hardly be accounted for by equations (2.30) and (2.33). The fact that the values of dC/C computed from equation (2.30) are always greater than those from equation (2.29) casts doubt on the accuracy of the results of Hawkins.

An important conclusion drawn from the above analysis is that the smaller CN-values make C (or Q) more sensitive to changes (or errors) in the actual values of λ, P, and CN. In an investigation of the rainfall-runoff data of five watersheds of the United States, Hjelmfelt (1980) concluded the same. He also questioned the validity of the second SCS-CN hypothesis (equation (2.3)), especially for taking λ equal to 0.2 for arid west. It was due to the fact that the runoff frequency distribution transformed from the rainfall frequency distribution for the Sonoita Creek (Arizona) watershed with the smallest CN failed to fit the measured data using the existing SCS-CN method.

(d) *Sensitivity to Variation of AMC*

According to Chen (1982), more than 10% of variation in CN may arise solely from differences in antecedent moisture conditions (AMC) or in extreme situations, or from changing land conditions over time. Therefore, the sensitivity of C to AMC is analyzed.

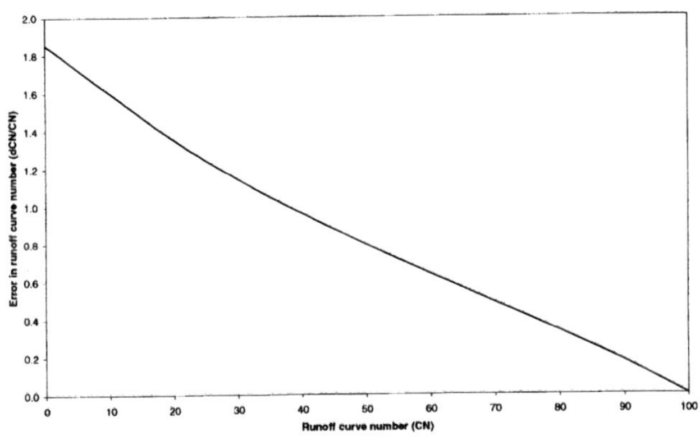

Figure 2.14. Possible maximum error in runoff curve number (CN) due to variations in antecedent moisture condition (AMC).

The Soil Conservation Service related the variation of CN with three levels of AMC, viz., AMC I, AMC II, and AMC III, as shown in Table 2.7. Mathematically, these relations are expressed as (Ponce and Hawkins, 1996):

$$CN_I = \frac{CN_{II}}{2.281 - 0.01281CN_{II}} \tag{2.34}$$

and

$$CN_{III} = \frac{CN_{II}}{0.427 + 0.00573CN_{II}} \tag{2.35}$$

where subscripts I, II, and III refer to AMC I, AMC, II, and AMC III, respectively. The derivation of equations (2.34) and (2.35) is described in Chapter 3. These are, however, used for the purpose of sensitivity analysis in what follows.

Assume that the CN-variation (dCN) in a basin results entirely from the differences in AMC. If CN_{II} is the true CN, the possible maximum error in the curve number (dCN/CN) is assumed to be as follows:

$$\frac{dCN}{CN} = \frac{CN_{III} - CN_{I}}{CN_{II}}\bigg|_{CN_{II}=CN} \tag{2.36}$$

which is valid for any CN. Substitution of equations (2.34) and (2.35) into equation (2.36) leads to

$$\frac{dCN}{CN} = \frac{0.001854CN\left(\dfrac{1000}{CN} - 10\right)}{(0.427 + 0.00573CN)(2.281 - 0.01281CN)} \tag{2.37}$$

For illustration, equation (2.37) is plotted in Figure 2.14 and some of the corresponding values of CN, dCN, and dCN/CN are listed in Table 2.15 for comparison. It is, however, noted that in the interpretation of the results of Figure 2.14 and Table 2.15, equation (2.37) represents only the error in CN from the assumption of either AMC I or AMC III when, in fact, AMC II is the correct condition. The error could be larger than that expressed by equation (2.37) as, for example, AMC III conditions were assumed when AMC I conditions were appropriate.

Table 2.15 Variations and possible maximum errors in CN due to changes in AMC from assumed true CN_{II}

CN	dCN	dCN/CN	CN	dCN	dCN/CN
100	0.0	0.000	40	38.3	0.959
90	15.7	0.174	30	34.3	1.142
80	26.7	0.333	20	27.1	1.353
70	34.0	0.485	10	16.0	1.600
60	38.2	0.636	0	0.0	1.854
50	39.6	0.792			

It is interesting to note from Table 2.15 that the possible maximum error in CN (= dCN/CN) due to changes in AMC from the assumed true CN_{II} increases from zero at CN = 100 to about 1.854 at CN = 0 although the values of dCN are zero at both extreme values of CN and about the maximum midway between them. Consequently, this rising trend of dCN/CN values aggravates the susceptibility of dC/C (equation (2.33)) to the variations of CN. Thus, from equation (2.33) the possible maximum error in C (= dC/C) due to AMC can be obtained by multiplying the expression of ($\partial C/\partial CN$)

(CN/C) (equation (2.30)) to that of dCN/CN (equation (2.37)). The dC/C values so obtained again vary with λ, P, and CN, and are similar to a family of curves drawn for various CN-values, as shown in Figure 2.13, except that each curve must be multiplied by the corresponding constant dCN/CN listed in Table 2.15.

An examination of Table 2.1 reveals that the CN values for most of the listed field soil cover complexes are higher than 40. This is equivalent to having dCN/CN equal to less than one (Table 2.15). It is encouraging in a sense that because the multiplier, dCN/CN, being less than unity in magnitude in equation (2.33) will in effect lessen the corresponding possible maximum error in C (= dC/C) for various P and CN values within the ranges of interest (i.e. P < 16 inches and CN > 40), as depicted in Figure 2.13. However, according to Chen (1982), the actually computed dC/C values derived from equation (2.33) have indicated that the extent of favorable influence on dC/C by dCN/CN due to changes in AMC is much smaller than anticipated. In most cases within the ranges of interest, the dC/C values so computed are found to be 10 percent or more, increase rapidly beyond the tolerable limits with decreasing CN values, and break down without bound at $P = I_a$. Therefore, a misjudgment in AMC or a drastic change in AMC over a short period of time may cause a serious error in the direct runoff estimate. It is also illustrated in the following examples.

Example 2.8: Compute errors in runoff estimation due to a 10% error in each of the variables, λ, P, and CN values for the base values of these variables as λ equal to 0.1, 0.2, 0.3; P equal to 1, 5, and 10 inches; and CN equal to 50, 75, and 95.

Solution: Table 2.16 shows computations of errors in runoff estimation due to 10% error in each of variables, the initial abstraction coefficient (λ), storm rainfall (P), and runoff curve number (CN), for given base values of λ, P, and CN. In this table, Sl. Nos. 1-3 describe the values of λ, P, and CN, respectively. Sl. No. 4 shows the values of the initial abstraction (I_a), below which runoff (Q) will be equal to zero. Sl. Nos. 5 and 6 show computations of the runoff factor C (= Q/P) determined from equation (2.25) and corresponding Q, respectively. The ratios of the errors given by equations (2.28)-(2.30) are shown at Sl. Nos. 7-9, respectively. For given values of λ, P, and CN and the computed C (Sl. No. 5), the errors in C (= $\partial C/C$) due to 10% error in the given base values of λ, P, and CN are shown at Sl. Nos. 10-12, respectively. A sum of these errors gives the total value of $\partial C/C$, as shown at Sl. No. 13. Thus, dC (equation (2.26)) at Sl. No. 14 is computed multiplying the total value of $\partial C/C$ (Sl. No. 13) by C (Sl. No. 5). The erroneous Q (Sl. No. 15) computed as: Erroneous Q = (C + dC)P and the resulting % error in Q at Sl. No. 16 is determined as: Percent error in Q = (erroneous Q – Q) x 100/Q.

Example 2.9: Evaluate the sensitivity of the runoff estimation to λ, P, and CN measurements using data of Example 2.8 and find the most sensitive variable.

Table 2.16 Computation of percent errors due to errors in λ, P, and CN (Example 2.7)

Sl. No.	Variable/error	Computations								
1	λ	0.1								
2	P (inch)	5			10			15		
3	CN	50	75	95	50	75	95	50	75	95
4	I_a (inch)	1.00	0.33	0.05	1.00	0.33	0.05	1.00	0.33	0.05
5	C	0.06	0.22	0.66	0.08	0.23	0.67	0.09	0.24	0.67
6	Q (inch)	0.32	1.09	3.32	0.81	2.34	6.71	1.31	3.59	10.11
7	$(\partial C/\partial \lambda)(\lambda/C)$	-0.43	-0.10	-0.01	-0.17	-0.04	-0.01	-0.10	-0.03	0.00
8	$(\partial C/\partial P)(P/C)$	1.14	0.52	0.11	0.70	0.30	0.06	0.52	0.21	0.04
9	$(\partial C/\partial CN)(CN/C)$	5.18	2.32	2.15	2.34	1.27	1.11	1.52	0.88	0.75
10	$\partial C/C$ due to λ	-0.04	-0.01	0.00	-0.02	0.00	0.00	-0.01	0.00	0.00
11	$\partial C/C$ due to P	0.11	0.05	0.01	0.07	0.03	0.01	0.05	0.02	0.00
12	$\partial C/C$ due to CN	0.52	0.23	0.21	0.23	0.13	0.11	0.15	0.09	0.08
13	Total $\partial C/C$	0.59	0.27	0.22	0.29	0.15	0.12	0.19	0.11	0.08
14	DC	0.04	0.06	0.15	0.02	0.04	0.08	0.02	0.03	0.05
15	Erroneous Q (inch)	0.51	1.39	4.07	1.04	2.69	7.50	1.56	3.97	10.90
16	% error in Q	58.93	27.38	22.44	28.65	15.30	11.64	19.35	10.69	7.86

Table 2.16 Continued.

Sl. No.	Variable/error	Computations								
1	λ	0.2								
2	P (inch)	5			10			15		
3	CN	50	75	95	50	75	95	50	75	95
4	I_a (inch)	2.00	0.67	0.11	2.00	0.67	0.11	2.00	0.67	0.11
5	C	0.04	0.20	0.67	0.07	0.24	0.69	0.08	0.25	0.69
6	Q (inch)	0.20	1.02	3.37	0.71	2.38	6.89	1.25	3.74	10.41
7	$(\partial C/\partial\lambda)(\lambda/C)$	-1.18	-0.22	-0.02	-0.39	-0.09	-0.01	-0.22	-0.06	-0.01
8	$(\partial C/\partial P)(P/C)$	1.95	0.66	0.12	0.94	0.35	0.06	0.66	0.24	0.04
9	$(\partial C/\partial CN)(CN/C)$	11.28	3.12	2.40	3.89	1.55	1.23	2.27	1.04	0.83
10	$\partial C/C$ due to λ	-0.12	-0.02	0.00	-0.04	-0.01	0.00	-0.02	-0.01	0.00
11	$\partial C/C$ due to P	0.19	0.07	0.01	0.09	0.04	0.01	0.07	0.02	0.00
12	$\partial C/C$ due to CN	1.13	0.31	0.24	0.39	0.16	0.12	0.23	0.10	0.08
13	Total $\partial C/C$	1.21	0.36	0.25	0.44	0.18	0.13	0.27	0.12	0.09
14	DC	0.05	0.07	0.17	0.03	0.04	0.09	0.02	0.03	0.06
15	Erroneous Q (inch)	0.44	1.39	4.21	1.03	2.81	7.77	1.59	4.19	11.30
16	% error in Q	120.51	35.56	24.95	44.44	18.17	12.80	27.09	12.30	8.61

Table 2.16 Continued.

Sl. No.	Variable/error	Computations								
1	λ	0.3								
2	P (inch)	5			10			15		
3	CN	50	75	95	50	75	95	50	75	95
4	I_a (inch)	3.00	1.00	0.16	3.00	1.00	0.16	3.00	1.00	0.16
5	C	0.02	0.19	0.69	0.06	0.24	0.71	0.08	0.26	0.72
6	Q (inch)	0.10	0.96	3.43	0.61	2.43	7.08	1.20	3.92	10.73
7	$(\partial C/\partial \lambda)(\lambda/C)$	-2.75	-0.36	-0.04	-0.68	-0.14	-0.02	-0.36	-0.09	-0.01
8	$(\partial C/\partial P)(P/C)$	3.58	0.82	0.13	1.27	0.41	0.07	0.82	0.28	0.05
9	$(\partial C/\partial CN)(CN/C)$	22.92	4.02	2.66	5.71	1.84	1.35	3.07	1.20	0.90
10	$\partial C/C$ due to λ	-0.28	-0.04	0.00	-0.07	-0.01	0.00	-0.04	-0.01	0.00
11	$\partial C/C$ due to **P**	0.36	0.08	0.01	0.13	0.04	0.01	0.08	0.03	0.00
12	$\partial C/C$ due to CN	2.29	0.40	0.27	0.57	0.18	0.13	0.31	0.12	0.09
13	Total $\partial C/C$	2.38	0.45	0.28	0.63	0.21	0.14	0.35	0.14	0.09
14	DC	0.05	0.09	0.19	0.04	0.05	0.10	0.03	0.04	0.07
15	Erroneous Q (inch)	0.34	1.39	4.37	1.00	2.94	8.07	1.62	4.47	11.74
16	% error in Q	237.50	44.70	27.54	63.03	21.12	13.98	35.23	13.92	9.37

Solution: From Table 2.16 is seen that as λ increases from 0.1 to 0.3, errors in the runoff (Q) computation also increase. For a given λ, the percent error in Q decreases as P increases. For given λ and P, these errors decrease with the increase in CN. The error due to only λ is negative and in a way is compensatory to others due to P and CN. For λ greater than or equal to 0.2, the absolute errors due to P and λ are of the same order and their sum is negligible, considering the errors due to CN, which are quite high. Thus, CN is the most sensitive variable and needs to be determined most carefully, especially when both P and CN are low. For $P \geq 5$ inches and $CN \geq 75$, The maximum possible error will be less than or equal to approximately 45% for $\lambda \leq 0.3$.

2.7 Advantages and Limitations of the SCS-CN Method

The SCS-CN method has several advantages over other methods. It is a simple conceptual method for estimation of the direct runoff amount from a storm rainfall amount, and is well supported by empirical data. The method relies on only one parameter, the curve number CN, which is a function of the major runoff-producing watershed characteristics: Four hydrologic soil groups; the land use and treatment classes, including agricultural, range forest, and urban (SCS, 1986); and the hydrologic surface condition of native pasture, such as poor, fair, and good. It is fairly well documented for its inputs (soil, land use/treatment, surface condition, and antecedent moisture condition), its features are readily grasped, well established, and accepted for use in the United States and other countries.

Mockus (1964) noted several problems associated with the SCS-CN method. For example, it does not contain any expression for time and ignores the impact of rainfall intensity and its temporal distribution. As indicated by Cowan (1957) (In: Rallison and Miller, 1982), time was not incorporated in the method because (a) sufficiently reliable data were not available to describe infiltration rates for a wide range of SVL complexes and (b) there was no reliable method available for distributing rainfall in time. Rallison and Miller (1982) described several other limitations which were of concern, for example, the availability of reliable data in varying geographic and environmental conditions, reproducibility of the runoff amount by the method even beyond the enveloping SVL curves, and so on.

As envisaged by Ponce and Hawkins (1996), Rallison and Miller (1982) among others, the SCS-CN method has yet to be established for its credibility and acceptance. Being an agency methodology, the method has been isolated from the rigors of peer review. The information supplied in NEH-4 is less than complete (Rallison and Cronshey, 1979). There is a lack of clear guidance on how to vary antecedent condition, especially for lower curve numbers and/or rainfall amounts, for the curve number exhibits sensitivity to the antecedent condition (Hawkins, 1975; Bondelid et al., 1982; Ponce, 1989). Since the method was originally developed for agricultural sites, it performs best on these watersheds, fairly on range sites, and poorly in application to forest sites (Hawkins, 1984, 1993). There is no explicit provision for spatial scale effects. Simanton et al. (1973) showed that CN for areas less than 227 ha in southeastern Arizona tended to decrease with increasing watershed size, exhibiting a significant role of channel transmission losses. Recently, Grove et al. (1998) and Moglen (2000) compared lumped and distributed approaches of the SCS-CN method and found the resulting runoff magnitudes to be significantly different for low rainfall magnitudes. Although Mockus (in: SCS, 1971) indicated the applicability of the method to even large catchments, the lack of clear guidance still exists if it can reliably be applied to small, mid, or large size catchments. Ponce and Hawkins (1996), however, cautioned against the use of the method to watersheds larger than 250 sq. km. The following discussion (i) presents Moglen's approach that extends SCS-CN application to distributed watershed modeling and (ii) modifies the Moglen method incorporating the time-area concept.

2.8 SCS-CN Application to Distributed Watershed Modeling

With the advancement of high-speed computers, distributed watershed modeling has become much popular in the last decade. Before describing the extended application of the SCS-CN method, it is appropriate to provide a brief account of the availability of watershed characteristics at a high spatial resolution to highlight the viability of its application to distributed watershed modeling.

2.8.1 AVAILABILITY OF DATA

(a) *Topography*

Spatially distributed topographical details available in the form of a set of elevation values at a regular interval (grid) are referred to as a digital elevation model (DEM). Each grid element or pixel takes on a single value representative of the elevation of the extent of that element. The U.S. Geological Survey (USGS) provides such data at 90 m resolution for all locations in the continental United States (USGS, 2000). The USGS also provides 30 m resolution DEMs for many locations within the United States. Knowing elevation and applying basic rules allowing water to flow in the direction of the downhill gradient, it is possible to infer flow directions, flow lengths, slopes, drainage area, and watersheds (O'Callaghan and Mark, 1984; Jenson and Domingue, 1988; Tarboton et al., 1991). Using a Geographic Information System (GIS), the process of determining from a DEM any of these quantities is largely an automated process (Moglen and Casey, 1998).

(b) *Land Use*

In digital format, land use is often described by a series of polygons (vectors) over a region, with each polygon indicating a region of assumed homogeneous land use. There are several schemes available to describe land use according to the kind of application. The most common procedure (Anderson et al., 1976) divides the land use into broad categories, such as urban, agricultural, forest, and water/wetlands. Such digital representations of land use at a relatively coarse resolution (about 200 m) covering the entire United States are available with the USGS (USGS, 2000). Other descriptions of land use may be found from state or regional organizations, such as state planning offices or departments of environment.

(c) *Soil Type*

As described earlier in this chapter, hydrologic soils are categorized into A, B, C, and D groups. Digital representations of these data can be obtained directly from the Natural Resources Conservation Service (NRCS, 2000) at several resolutions: SSURGO, STATSGO, and NATSGO. SSURGO data correspond to the highest available resolution, digitized at scales ranging from 1:12,000 to 1:31,680, which is quite appropriate and adequate for small watersheds of the size of a few square miles. STATSGO data are discretized at 1:250,000 scale and these are useful for analyzing watersheds at the multi-county to state scale. NATSGO data describe variations in soil type from a multi-state to regional scale and are not appropriate for the applications being discussed here.

For application of the SCS-CN method, the land use and soil type are combined to determine a curve number. This procedure is generally the same as described earlier in this chapter, viz., employment of the NEH-4 table (Table 2.1) that associates a curve number with a given land use and soil type. Using GIS, the process of inferring the spatial distribution of curve number, given the land use and soil type is a process that is readily automated (Ragan, 1991; Moglen and Casey, 1998).

2.8.2 MOGLEN METHOD

Following the work of Grove et al. (1999), Moglen (2000) suggested two methods based on the existing SCS-CN method (equation (2.5) or equation (2.6)) to account for the spatial and orientation effects in distributed watershed modeling. These methods are named as α and β methods; the former is a specific form of the latter.

(i) *α Method*

The α method allows an infinite infiltration downstream as follows. The runoff produced by equation (2.5) or equation (2.6) for an arbitrarily chosen pixel will naturally proceed downhill, eventually finding its way to a location of the concentrated flow (which is termed as swale or channel by the Soil Conservation Service). From the perspective of the downhill pixel, runoff and rainfall are the same: they are both sources of an input volume of water and will contribute to infiltration in this pixel. To reflect this perspective, Moglen modified equations (2.5) and (2.6), respectively, as:

$$R_d = \frac{[(\Sigma R_u + P) - I_a]^2}{\Sigma R_u + P + S_d - I_a} \tag{2.38}$$

and

$$R_d = \frac{[(\Sigma R_u + P) - 0.2S_d]^2}{\Sigma R_u + P + 0.8S_d} \tag{2.39}$$

where R_d is the runoff leaving the downstream pixel (in units of pixel-inches); ΣR_u is the summation of the runoff from all immediately upstream pixels (in pixel-inches); and S_d is the storage of the downstream pixel (in pixel-inches). This new unit of "pixel-inch" is introduced because of the runoff accumulation nature of equation (2.38) or equation (2.39). Pixel-inches are converted back to inches after runoff from all pixels within the watershed has been determined. The result is an areal average depth of runoff analogous to the SCS-CN method's Q at each pixel throughout the watershed. Most importantly, the areal average runoff depth at the outlet of the watershed reflects the net runoff produced by the watershed allowing for both spatially varied and oriented curve numbers.

Example 2.10: For a rainfall amount of 4.8 inches, determine the corresponding runoff using the Moglen method for the watershed consisting of three sub-watersheds of curve numbers 56, 78, and 60, arranged as shown in Figure 2.15.

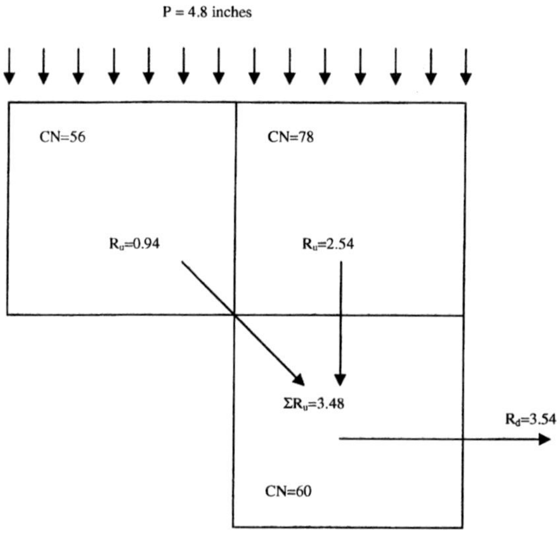

Figure 2.15. Schematic diagram illustrating runoff aggregation and accounting structure associated with the application of the Moglen concept. R_u, ΣR_u, and R_d are in units of pixel-inches.

Solution: Figure 2.15 illustrates the use of equations (2.38) and (2.39) in distributed watershed modeling. In this figure, a DEM has been used to determine that two pixels flow into the downstream pixel in question. In this case, $\Sigma R_u = 0.94 + 2.54 = 3.48$ pixel-inches. This upstream runoff amount of 3.48 pixel-inches is added to the rainfall at the downstream pixel (P = 4.8 pixel-inches) to produce a net input of 3.48 + 4.8 = 8.28 pixel-inches of water. If the curve number is 60 at this downstream pixel, resulting in a storage space, S_d, of 6.67 pixel-inches, then the downstream runoff, R_d, is 3.54 pixel-inches. If there are no other inputs to the three upstream pixels, then the area associated with the downstream pixel, A_d, is 3 pixels (e.g., the number of pixels draining into downstream pixel plus the downstream pixel itself). The areal average runoff at the downstream pixel for this example is:

$$Q_\alpha = \frac{R_d}{A_d} = \frac{1}{3} x \frac{[(3.48 + 4.8) - 0.2 x 6.67]^2}{3.48 + 4.8 + 0.8 x 6.67} = 3.54/3 = 1.18 \text{ inches}$$

It is noted that symbol Q is used to denote runoff in inches and symbol R is used to indicate runoff in pixel-inches. Furthermore, subscript α indicates the determination of Q using equation (2.39).

Example 2.11: Consider a case where the runoff paths are exclusively parallel such that the flow is accumulating in a strictly linear fashion (one-dimensional flow) moving downstream as illustrated in Figure 2.16. A 100-pixel path is divided into two 50-pixel sections and subjected to a rainfall, P = 3.0 inches. In "Case A," the upper 50 pixels are grassed (CN = 78) and the lower 50 pixels are paved (CN = 98). In 'Case B," the order is reversed with the paved surface upstream of the grassed surface. Determine the net runoff values for both the cases using the Moglen and the existing SCS-CN methods. Also analyze (a) the variation of direct runoff with the number of pixels and (b) the effect of watershed orientation on the runoff.

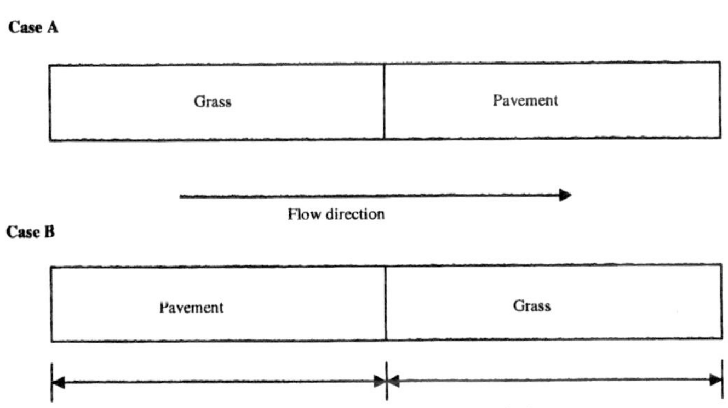

Figure 2.16. Two cases of one-dimensional flow: 1. In Case A, the grassed area is upstream of the paved area, 2. In case B, the paved area is upstream of the grassed area.

Solution: Figures 2.17a and 2.17b present the runoff production of both the systems as a function of position along the flow path. In both cases, the incremental runoff contributed by a single pixel is high for the paved surface as evidenced by the steep slope of the curves in Figure 2.17a. In Case A, the grassed surface produces a relatively little incremental runoff as evidenced by the shallow positive slope in Figure 2.17a. In Case B, the grassed surface actually serves as a small net sink to runoff production indicated by the shallow negative slope in Figure 2.17a. At the downstream end of both systems, i.e., after 100 pixels of one-dimensional flow, Case A produces runoff equal to 149.9 pixel-inches. On the other hand, Case B produces runoff of 121.5 pixel-inches, which is less than that in Case A. Despite the fact that both the systems have identical make-up, the difference in runoff is attributed to the orientation of the two sub-watersheds.

To obtain the net runoff for Cases A and B, divide the runoff computed in terms of pixel-inches by A_d = 100 pixels for each system to yield as for Case A: $Q_{\alpha A}$ = (149.9 pixel-inches)/(100 pixels) = 1.50 inches. Similarly, the net runoff can be computed for Case B as: $Q_{\alpha B}$ = (121.5 pixel-inches)/(100 pixels) = 1.22 inches.

In the traditional (lumped or weighted-CN) SCS-CN method, average CN = (50 x 78 + 50 x 98)/100 = 88. Therefore, the parameter potential maximum retention, S, is computed from equation (2.7) as: S = 1000/88 – 10 = 1.36 inches. Using equation (2.6), Q is determined as: $Q = (3-0.2 \times 1.36)^2/(3 + 0.8 \times 1.36) = 1.81$ inches.

To derive the runoff using the weighted-runoff method, Q = 1.13 inches from grassed area and Q = 2.77 inches for the paved area. Since orientation does not matter in the existing SCS-CN approach, the runoff for either case is a weighted average over 100 pixels evenly divided between the two runoff surfaces producing Q = {(50 pixels) x (1.13 inches) + (50 pixels) x (2.77 inches)}/(100 pixels) = 1.94 inches. As also illustrated in Example 2.3, the runoff from the weighted-Q method is greater than the runoff from the weighted-CN method, which is greater than $Q_{\alpha A}$, and $Q_{\alpha A} > Q_{\alpha B}$.

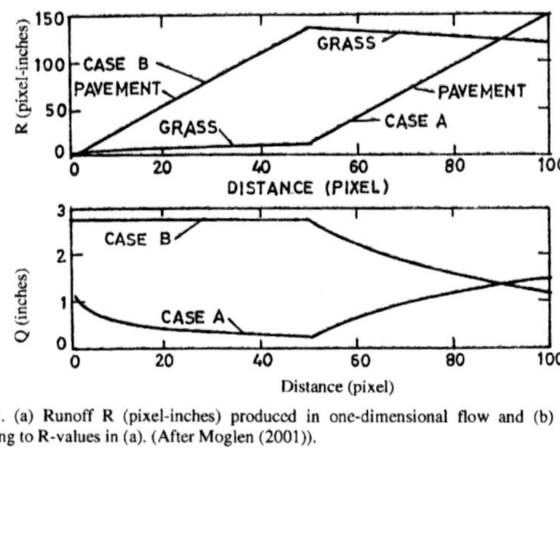

Figure 2.17. (a) Runoff R (pixel-inches) produced in one-dimensional flow and (b) runoff Q (inches) corresponding to R-values in (a). (After Moglen (2001)).

(ii) *β Method*

In the β method, the occurrence of infiltration is allowed only up to a certain distance in a watershed, following the concept of the SCS TR-55 model (SCS, 1986). For runoff generation and routing, the TR-55 model identifies three fundamentally different flow regimes within the watershed. Over the first 100-300 feet from the watershed divide to the channel, overland flow is assumed to dominate. Immediately downstream of this regime "shallow concentrated flow" dominates. Once the channel is encountered, open channel flow conditions as characterized by Manning's friction (equation (1.79)) are dominant.

Such a description of the progression of flow aggregation leads to the modification of the α method. Equation (2.38) or equation (2.39) accounts for the possibility of runoff generated upstream infiltrating later somewhere downstream. Such downstream infiltration can physically occur only in the region of overland flow. Once the flow has reached a swale or channel, it is concentrated and infiltration is assumed minimal because saturated conditions have been encountered. For pixels in this concentrated flow region, it is appropriate to employ the original runoff generation model (equation (2.5) or equation (2.6)). Together, equations (2.6) and (2.39) amount to perhaps more realistic model allowing for downhill infiltration only upstream of swales or channels and reverting to the classic SCS-CN equation where swales or channels are present.

Thus, in this method, the runoff generated from the watershed depends on the length, ℓ, of the watershed to the onset of the channel flow. The larger the length ℓ, the lesser will be the runoff produced, and vice versa. If $\ell = 0$, the runoff (= Q_β for $\ell = 0$) is equal to the weighted runoff. On the other hand, Q_β for $\ell = \infty$ is equal to Q_α. A typical variation of Q_β with overland flow length for the Rock Creek watershed is shown in Figure 2.18. Apparently, as $\ell \rightarrow \infty$, $Q_\beta \rightarrow Q_\alpha$.

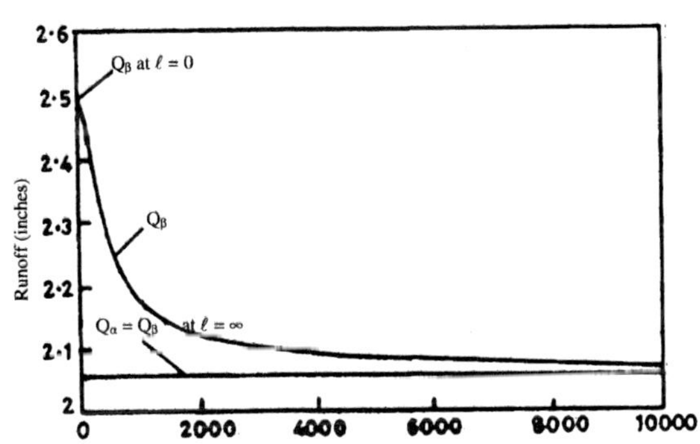

Figure 2.18. Predicted runoff Q_β from the Rock Creek watershed as a function of overland flow length. (After Moglen (2001)).

2.8.3 ADVANTAGES AND LIMITATIONS OF THE MOGLEN METHOD

The major advantage of the Moglen method is that it is simple, easy to understand and apply. It can be appreciated from the view point that generalizing a lumped model into a spatially distributed form is not a straightforward task (Finnerty et al., 1997). The disadvantage, however, is that the method produces the unrealistic negative runoff under certain circumstances. Following Moglen (2000), it can be illustrated as below.

Depending on the accumulated runoff (ΣR_u), the incremental runoff production ($R_d - \Sigma R_u = \Delta R$) for the grassed surface in Case A may be positive or it may be negative in Case B, as shown in Figure 2.17. This suggests that a CN = 78 as used here for grass will result in a limiting condition where the runoff volume is no longer increasing as it moves from an upstream pixel to a downstream pixel. Under these conditions,

$$R_d = \Sigma R_u \tag{2.40}$$

A substitution of equation (2.40) into equation (2.39) yields

$$R_d = \frac{(P - 0.2S_d)^2}{1.2S_d - P} \tag{2.41}$$

Though the numerator of equation (2.41) is physically meaningful according to the existing SCS-CN method (equation (2.6)), the denominator limits P to $1.2S_d$ for R_d to be positive. This is in contrast with the general notion that rainfall (P) grows unbounded (Ponce and Hawkins, 1996). This aspect is further discussed in Chapter 3. The Moglen method assumes a watershed to be of infinite length, which is also in contrast with the reality. In addition to these, while routing the flow from upstream pixel to the downstream pixel, the rainfall on the downstream pixel is forced to wait for the upstream flow to join it and thus, violates the widely used time-area concept. These limitations are eliminated in the modified version of the method.

2.8.4 MODIFIED MOGLEN METHOD

In the modified Moglen approach, the concept of time-area incorporating infiltration during runoff travel is considered, as shown in Figure 2.19. In this figure, the rainfall of amount P has occurred on a watershed of area A during the time interval Δt. Thus, the average rainfall intensity, i, is equal to $P/\Delta t$. The watershed is divided into two parts of areas A_1 and A_2 such that $A_1 = A_2$. These notations are used for distinguishing the runoff generation from these watersheds at different times. It is, however, worth emphasizing that these sub-watersheds may be of the same or different curve numbers. Assuming that each watershed is of length Δx and the flow wave celerity (discussed in detail in chapter 8) is equal to $\Delta x/\Delta t$. The total runoff generated from the watershed, which is equal to Q, can be computed as follows.

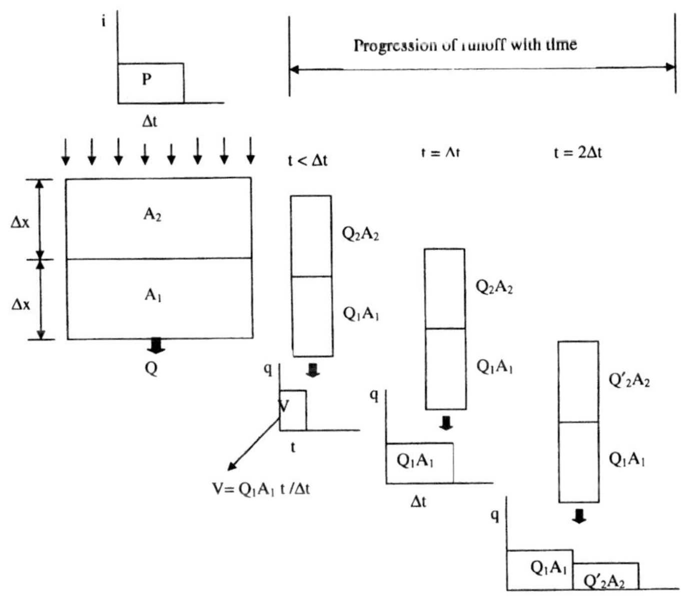

Figure 2.19.Time-area concept incorporating infiltration for the modified Moglen method.

Figure 2.19 also shows the progression of runoff with time. At time $t < \Delta t$, the runoff volume (V) produced at the outlet in time t will be equal to the product of (Q_1A_1) and $(t/\Delta t)$. Thus, the runoff depth at the outlet in time t will be equal to V/A and the runoff rate, q, will be equal to V/t. Similarly, at time $t = \Delta t$, $V = Q_1A_1$, which will be generated by the area A_1. During this period, sub-watershed A_2 will produce the runoff volume equal to Q_2A_2 at the outlet of A_2 (or at the inlet of A_1). The resulting runoff depth Q during Δt will be equal to Q_1A_1/A. During the period $\Delta t < t \leq 2\Delta t$, the runoff volume Q_2A_2 will pass through the process of infiltration and will, therefore, convert to runoff volume equal to Q'_2A_2 at the end of the $2\Delta t$ period. Thus, the total runoff volume generated at the outlet of A_1 is equal to $(Q_1A_1 + Q'_2A_2)$ and the corresponding runoff depth $Q = (Q_1A_1 + Q'_2A_2)/A$. The resulting hydrographs for these times are shown in Figure 2.19.

In addition to the above, while computing runoff from a sub-watershed, the SCS-CN parameter S is varied according to the volumetric concept that is discussed in greater detail in Chapter 4, and the initial abstraction, I_a, is accounted for only once on a sub-watershed. To this end, the computation of Q'_2 employs $P = Q_2$ and $I_a = 0$ in equation (2.6), in which S is taken equal to the initial S-value (S-value before the start of the rainstorm) minus the infiltration amount, F, computed from the water balance equation (2.1). This procedure is illustrated in the following example.

Example 2.12: Using the modified Moglen approach, determine the weighted runoff generated on a watershed of 120 acres for a rainfall amount of 4 inches that occurred in 2 hrs. Assume CN of the watershed is equal to 70 and the total time of travel for this watershed is equal to 4 hrs. Also compute runoff rates and compare the resulting runoff depth with that derived from (a) the traditional (lumped) SCS-CN approach and (b) the Moglen method.

Solution: Since the time of travel for the given watershed is equal to 4 hrs, the total area of the watershed can be divided into two parts of equal time of travel (= 2 hrs), which is the time duration of the rainstorm. Since the watershed is homogeneous in character, the area of each part can also be taken equal to half of the total area. Thus, the configuration of the watershed is exactly the same as shown in Figure 2.19. The runoff computations are shown in Table 2.17. In this table, the computations are performed for quantities S, Q, and F for given P and CN for the sub-watersheds 1 and 2. Time steps 1 and 2 are in multiple of 2 hrs. Thus, the computations are consistent with the time-area concept.

Table 2.17 Runoff computation using the modified Moglen approach (Example 2.12)

Time step	Variables	Sub-watershed 2	Sub-watershed 1	Runoff volume (acre-inch)	Runoff rate, q (acre-inch/hr)
1	S	4.29	4.29		
	P	4.00	4.00		
	Q	1.33	1.33	60 x 1.33	39.90
	F	1.81	1.81		
2	S	2.48	2.48		
	P	0.00	1.33		
	Q	0.00	0.47	60 x 0.47	14.01
	F	0.00	0.86		
Sum				60 x 1.80	
Weighted runoff, Q = 1.80 x 60/120 = 0.90 inch					

At the first time step, S is computed from equation (2.7) for a given CN = 70. It is equal to 4.29 inches. Using equation (2.6), compute Q which is equal to 1.33 inches for both sub-watersheds and convert it to runoff volume as shown in the second last column of the table. This is the computed runoff volume from sub-watershed 1 appearing at its outlet. Compute the total infiltration, F, using the water balance equation (2.1) as: F = 4.0 – 0.2 x 4.29 - 1.33 = 1.81 inches.

For computation of time step 2, the computed infiltration, F, is subtracted from S of the time step 1 (= 4.29 inches). It equals 2.48 inches. Since the initial abstraction, I_a, is already accounted for in the computations for the time step 1, $I_a = 0$ for this and other time steps. Using equation (2.6), computations are performed for this time step as follows:

For sub-watershed 2, Q will be equal to zero because the rainfall amount in this time step is equal to zero, resulting into F = 0. However, at the end of 2 hrs, the runoff amount of 1.33 inches from sub-watershed 2 will reach the sub-watershed 1 to pass through the process of infiltration during 2-4 hrs. Again, since the rainfall amount in this time step is equal to zero, the depth of water to experience infiltration on sub-watershed 1 is equal to 1.33 inches, which is indicated in Table 2.17 at time step 2 against P under sub-watershed 1. Compute runoff from equation (2.6) taking $I_a = 0$ as:

$$Q = \frac{P^2}{P+S} = \frac{(1.33)^2}{1.33+2.47} = 0.47 \text{ inch}$$

and

$$F = P - Q = 1.33 - 0.47 = 0.86 \text{ inch}$$

Computations are stopped at this time step because the whole runoff corresponding to the given rainfall has reached the outlet of the watershed. It is noted that the runoff coming out of sub-watershed 1 is the resulting runoff in the time interval.

The addition of runoff volumes shown in the last column of Table 2.17 is equal to 60 x 1.33 + 60 x 0.47 = 60 x 1.80 acre-inches. Therefore, the weighted runoff corresponding to 4 inches of rainfall can be computed as: Q = 60 x 1.80 acre-inches/120 acres = 0.90 inch. The resulting runoff hydrograph can be derived by dividing the runoff volume shown in the second last column of Table 2.17 by the time interval (= 2 hrs). The computed values are shown in the last column of this table.

The SCS-CN method (whether weighted-runoff or weighted-CN) yields a runoff depth equal to 1.33 inches, which is about 32% higher than that computed using the modified Moglen method. According to the Moglen method, ΣR_u = 1.33 pixel-inches, P = 4.0 inches. Using S = 4.29 inches, R_d can be computed from equation (2.39) as:

$$R_d = \frac{[(\Sigma R_u + P) - 0.2S_d]^2}{\Sigma R_u + P + 0.8S_d} = \frac{[(1.33+4.0) - 0.2x4.29]^2}{(1.33+4.0) + 0.8x4.29} = 2.28 \text{ pixel} - \text{inches}$$

Therefore, the runoff depth Q = 2.28/2 = 1.14 inches, which is about 21% higher than that from the modified Moglen method.

Example 2.13: Using the modified Moglen method, compute the weighted runoff for three configurations of a watershed shown in Figure 2.20 and compare with that computed using the traditional (weighted-CN) SCS-CN method. Assume rainfall as 4 inches and the areas of two sub-divisions in the figure are equal.

Solution: For configuration (a) (Figure 2.20), the runoff corresponding to P = 4 inches and CN = 70 can be computed as equal to 1.33 inches. For configurations (b) and (c) of Figure 2.20, the computations are shown in Table 2.18. The runoff computed for configurations (a)-(c) are 0.67, 0.53, and 0.97 inches, respectively. Configuration (c) yields a higher runoff than configuration (b) and thus, shows the impact of orientation. According to the traditional (weighted-CN) SCS-CN method, the weighted CN = 70 and thus, S = 4.29 inches. Therefore, Q from equation (2.6) can be computed as equal to 1.33 inches, which is higher than the runoff value computed using the modified Moglen method for any configuration.

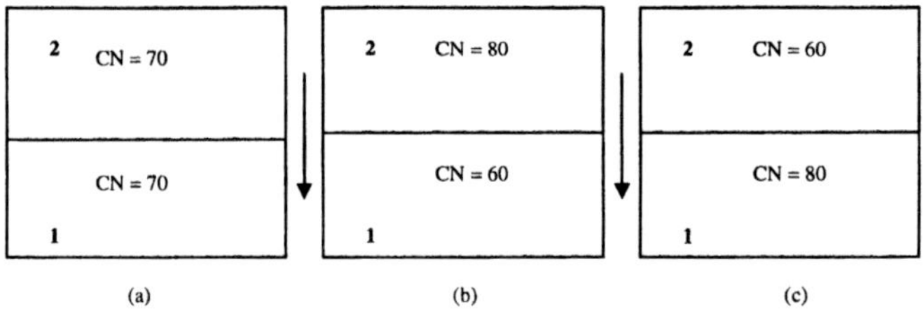

Figure 2.20. Three watersheds of unit area. Arrows indicate flow direction and nos. 1 and 2 indicate sub-watershed number.

Table 2.18 Runoff computation for Example 2.13

Time step	Variable	Configuration (a)		Configuration (b)		Configuration (c)	
		2*	1*	2*	1*	2*	1*
1	S	4.29	4.29	2.50	6.67	6.67	2.50
	P	2.00	2.00	2.00	2.00	2.00	2.00
	Q	0.24	0.24	0.56	0.06	0.06	0.56
	F	0.90	0.90	0.94	0.61	0.61	0.94
2	S	3.38	3.38	1.56	6.06	6.06	1.56
	P	2.00	2.24	2.00	2.56	2.00	2.06
	Q	0.74	0.89	1.12	0.76	0.50	1.17
	F	1.26	1.35	0.88	1.80	1.50	0.89
3	S	2.13	2.04	0.69	4.26	4.56	0.67
	P	0.00	0.74	0.00	1.12	0.00	0.50
	Q	0.00	0.20	0.00	0.23	0.00	0.21
	F	0.00	0.54	0.00	0.89	0.00	0.29
Sum (pixel-inches)			1.33		1.06		1.94
Weighted runoff (inch)			0.67		0.53		0.97

Note: superscript '*' denotes the number of the sub-watershed.

Example 2.14: Taking rainfall equal to 4 inches and CN = 70, analyze the impact of size and shape of the watershed shown in Figure 2.21.

Solution: In Figure 2.21, watershed in configuration (a) is 120 acres and the area of each of the other watersheds of configurations (b) and (c) is 90 acres. The curve number of all the segments of each watershed is the same, implying that all the watersheds are of the same soil-vegetation-land use complex or in the same hydrologic regime. Configuration (b) is a representation of a watershed narrow upstream and wide downstream and configuration (c) is the reverse of it. Thus, the computed runoff values will exhibit the impact of the size and shape of the watershed. The runoff computations are shown in Table 2.19.

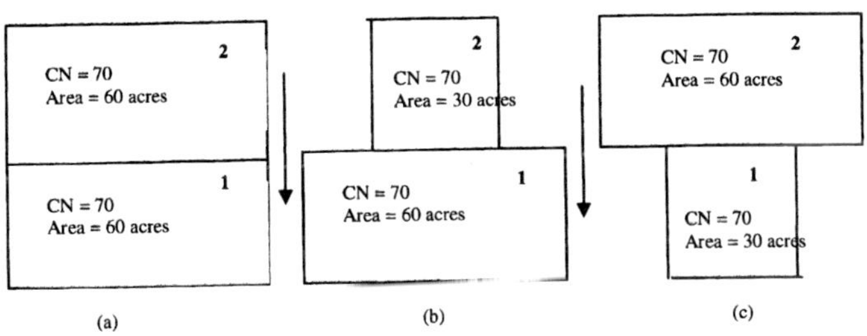

Figure 2.21. Three watersheds of (a) 120, (b) 90, and (c) 90 acres. Arrows indicate flow direction and nos. 1 and 2 indicate sub-watershed number.

In Table 2.19, the procedure for the runoff computation is generally the same as described in Example 2.13. It, however, differs a little for configurations (b) and (c) because of the varying areas of the sub-watersheds. In configuration (b), P-values for sub-watershed 1 at time steps 2, 3, and so on are computed as: P = rainfall of the current time step + the runoff of the previous time step generated by sub-watershed 2 x (area of sub-watershed 2/area of sub-watershed 1). For example, at time step 2, the P-value (shown by shaded area in Table 2.19) for sub-watershed 1 is computed as: P = 2.0 + 0.24 x 30/60 = 2.12.

Table 2.19 Runoff computation for the configurations of Example 2.14

Time step	Variable	Configuration (a)			Configuration (b)			Configuration (c)		
		2	1	Vol.	2	1	Vol.	2	1	Vol.
1	S	4.29	4.29		4.29	4.29		4.29	4.29	
	P	2.00	2.00		2.00	2.00		2.00	2.00	
	Q	0.24	0.24	14.44	0.24	0.24	14.44	0.24	0.24	7.22
	F	0.90	0.90		0.90	0.90		0.90	0.90	
2	S	3.38	3.38		3.38	3.38		3.38	3.38	
	P	2.00	2.24		2.00	2.12		2.00	2.48	
	Q	0.74	0.89	53.56	0.74	0.82	49.01	0.74	1.05	31.49
	F	1.26	1.35		1.26	1.30		1.26	1.43	
3	S	2.13	2.04		2.13	2.08		2.13	1.95	
	P	0.00	0.74		0.00	0.37		0.00	1.49	
	Q	0.00	0.20	11.92	0.00	0.06	3.38	0.00	0.64	19.27
	F	0.00	0.54		0.00	0.32		0.00	0.84	
Sum				79.91			66.82			57.98
Weighted runoff (inch)				0.67			0.74			0.64

Note: Vol. denotes runoff volume passing through sub-watershed 1.

Since the runoff finally passes through sub-watershed 1, the corresponding runoff volume is computed by multiplying the runoff depth with the area of sub-watershed 1. For example, at time step 2, the runoff depth is equal to 0.82 corresponding to P equal to 2.12. The corresponding runoff volume will be equal to 0.82 x 60 = 49.20 acre-inches. Here, the negligible discrepancy in decimal places is attributed to rounding-off errors. Thus, the weighted runoff values for configurations (a)-(c) are 0.67, 0.74, and 0.64 inches, respectively. Thus, the impact of the size depends on the shape of the watershed. Configuration (b) yields higher runoff than configuration (a) and vice versa holds for configuration (c). However, the watershed narrow upstream and wide downstream (configuration (b)) yields a higher runoff than the watershed wide upstream and narrow downstream (configuration (c)). It is noted that higher runoff-values imply a higher average curve number for a watershed and vice versa. Furthermore, in this case, the effect of watershed size on runoff is not that prominent as the shape of the watershed. It is because configurations (b) and (c) yield, respectively, 11.5% higher and 3.3% lower runoff than that in configuration (a). On the other hand, the runoff in configuration (b) is lowered by 14.8% with respect to the runoff due to configuration (a).

Example 2.15: For the configuration (a) watershed of Example 2.14, compute the weighted runoff from a rainfall hyetograph showing 1 inch of rainfall in the first 2 hrs and 3 inches of rainfall in the next 2 hrs. Assume the time of travel of each segment of example sub-watersheds is equal to 2 hrs. Also analyze the impact of the time distribution of rainfall if it is reversed.

Table 2.20 Computation of the weighted runoff for varying time distribution of rainfall (Example 2.15)

Time step	Variable	Sub-watersheds of Case A		Sub-watersheds of Case B	
		2	1	2	1
1	S	4.29	4.29	4.29	4.29
	P	1.00	1.00	3.00	3.00
	Q	0.00	0.00	0.71	0.71
	F	0.14	0.14	1.43	1.43
2	S	4.15	4.15	2.86	2.86
	P	3.00	3.00	1.00	1.71
	Q	1.26	1.26	0.26	0.64
	F	1.74	1.74	0.74	1.07
3	S	2.41	2.41	2.12	1.79
	P	0.00	1.26	0.00	0.26
	Q	0.00	0.43	0.00	0.03
	F	0.00	0.83	0.00	0.23
Sum of Q-values			1.70		1.39
Weighted runoff (inch)			0.85		0.70

Solution: The runoff computations are shown in Table 2.20. In case A (Table 2.20), the rainfall of 1-inch precedes the 3-inch rainfall and case B corresponds to the reverse situation. Thus, case A represents the case of a rising rainfall hyetograph and case B refers to the receding hyetograph. It is apparent from the table that case A yields higher runoff (= 0.85 inch) than the runoff (= 0.70 inch) in case B. Important to note is that both these runoff values are higher than the runoff (= 0.67 inch) due to a uniform rainfall (see Example 2.14). Though the total rainfall in these cases is equal to 4 inches, the runoff values differ significantly. Thus, the rainfall distribution affects significantly the weighted runoff.

Example 2.16: For the watershed of Example 2.15, compute the weighted runoff for 2 inches of rainfall occurring on sub-watershed 1 in a certain time duration and 6 inches on sub-watershed 2 in the same time duration. Analyze the impact of the spatial variation of rainfall on runoff if sub-watershed 1 experiences 6 inches of rainfall and sub-watershed 2 does 2 inches. Assume that the example watershed is divided into two equal segments.

Solution: The computations are shown in Table 2.21. In case A, sub-watershed 1 experiences 6 inches of rainfall and sub-watershed 2 does 2 inches of rainfall in a certain time duration. Case B is reverse of case A. Since the watershed is divided into two equal parts, the rainfall of a watershed is also divided into two equal parts of equal time interval, for consistency with the time-area concept. Case A yields the weighted runoff equal to 0.52 inch and case B yields 1.14 inches of runoff. Since the lower segment of the watershed does not pass through infiltration, case B with high rainfall on the lower sub-watershed (sub-watershed 1) yields higher runoff than case A. It is quite close to the physical reality. Furthermore, the average rainfall of the watershed is equal to 4 inches (= 6 x 0.5 + 2 x 0.5). As shown in Example 2.15, the runoff corresponding to P = 4 inches and CN = 70 is equal to 0.67 inch. The spatial impact, however, is to reduce the runoff to 0.52 inch in case A and increase the runoff to 1.14 inches in case B. Thus, the computed runoff greatly depends on the spatial distribution of rainfall.

Table 2.21 Computation of the weighted runoff for varying spatial distribution of rainfall (Example 2.16)

Time step	Variable	Sub-watersheds of Case A		Sub-watersheds of Case B	
		2	1	2	1
1	S	4.29	4.29	4.29	4.29
	P	3.00	1.00	1.00	3.00
	Q	0.71	0.00	0.00	0.71
	F	1.43	0.14	0.14	1.43
2	S	2.86	4.15	4.15	2.86
	P	3.00	1.71	1.00	3.00
	Q	1.54	0.50	0.19	1.54
	F	1.46	1.21	0.81	1.46
3	S	1.39	2.93	3.34	1.39
	P	0.00	1.54	0.00	0.19
	Q	0.00	0.53	0.00	0.02
	F	0.00	1.01	0.00	0.17
Sum of Q-values			1.03		2.28
Weighted runoff (inch)			0.52		1.14

2.8.5 FEATURES OF THE MODIFIED MOGLEN METHOD

The modified method basically considers two types of the watershed orientation: parallel and serial. The existing SCS-CN method assumes the parallel orientation. In the parallel configuration, each element (pixel or sub-watershed) of the watershed is arranged in a parallel orientation, such that the flow from each pixel directly reaches the outlet of the watershed. Alternatively, according to Moglen's β method, pixels in the watershed length ℓ are assumed to be in serial orientation, implying that the runoff from the upstream pixels passes through the process of infiltration on the downstream pixels. Afterwards, the flow concentrates or meets the channel, where infiltration is insignificant. The categorization of pixels into parallel and serial orientation is, however, more convincing in a way that it can be readily grasped and understood. For example, consider a typical open book configuration of a watershed, as shown in Figure 2.22. In this figure, sub-watersheds (1 and 2) and (3 and 4) are in series and sub-watersheds (1 and 3) and (2 and 4) are in parallel configuration.

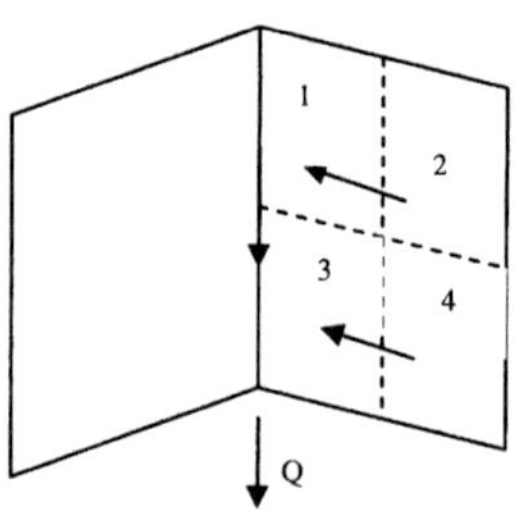

Figure 2.22. Open book configuration of a typical watershed, showing parallel and serial orientations of sub-watersheds. Sub-watersheds (1 and 2) and (3 and 4) are in series and sub-watersheds (1 and 3) and (2 and 4) are in parallel orientation.

In the serial configuration of sub-watersheds, it is important to understand the variation of the weighted runoff with the number of sub-divisions or discretization. To this end, hypothetical homogeneous watersheds of curve numbers 70, 80, and 90 experiencing rainfalls of 4 and 10 inches were selected and runoff values were computed, as follows.

If the watershed is discretized, for example, into 3 parts (or pixels), then the rainfall of (say) 10 inches was also divided into three equal parts such that the rainfall intensity does not change. Here, for application of the time-area concept, one assumes that the time of travel in each sub-division is equal to the time duration of the total 10-inch rainfall divided by the number of sub-divisions. This problem of discretization is similar to the selection of an appropriate time interval Δt in rainfall-runoff modeling (Singh, 1992). Further computations for the curve number equal to 70 are shown in Table 2.22. The computed weighted runoff is equal to 4.97 inches.

Similar computations were performed for each rainfall and CN-values. The variation of the computed runoff values is depicted in Figure 2.23. It is seen that the runoff magnitude decreases with the increase in discretization (or number of sub-divisions) of the watershed. The trend, however, is to attain a constant runoff value with the increase in the number of sub-divisions. It is because the SCS-CN parameter S goes on reducing with the level of discretization. After a certain time that depends on the amount of rainfall and the curve number, parameter S approaches zero, leading to no infiltration. At this stage, the modified Moglen method mimics the existing time-area concept that is generally used for no infiltration condition. Here, it is noted that the number of sub-divisions is restricted to less than or equal to the total rainfall divided by the initial abstraction to avoid negative infiltration. For example, the total number of sub-divisions for 10-inch rainfall on the watershed with a CN value equal to 70 will be computed as follows. For CN = 70, S = 4.29 inches. Therefore, $I_a = 0.2S = 0.2 \times 4.29 = 0.86$ inch. Therefore, the maximum number of sub-divisions (or pixels) < 10/0.86 = 11.63. Thus, the maximum number of sub-divisions can be equal to 11.

Table 2.22 Computation of the weighted runoff for P = 10 inches and CN = 70

Time step	Variable	Sub-watershed 3	Sub-watershed 2	Sub-watershed 1	Runoff (pixel-inches)
1	S	4.29	4.29	4.29	
	P	3.33	3.33	3.33	
	Q	0.91	0.91	0.91	0.91
	F	1.57	1.57	1.57	
2	S	2.72	2.72	2.72	
	P	3.33	4.24	4.24	
	Q	1.84	2.58	2.58	2.58
	F	1.50	1.66	1.66	
3	S	1.22	1.06	1.06	
	P	3.33	5.17	5.92	
	Q	2.44	4.29	5.02	5.02
	F	0.89	0.88	0.90	
4	S	0.33	0.18	0.16	
	P	0.00	2.44	4.29	
	Q	0.00	2.27	4.13	4.13
	F	0.00	0.17	0.16	
5	S	0.33	0.01	0.01	
	P	0.00	0.00	2.27	
	Q	0.00	0.00	2.27	2.27
	F	0.00	0.00	0.01	
Sum					14.91
Weighted runoff = 14.91/3 = 4.97 inches					

2.8.6 ADVANTAGES AND LIMITATIONS OF MODIFIED MOGLEN METHOD

The modified Moglen method is based on the popular time-area concept and thus, to a certain extent, accounts for the physical process of watershed routing. It is, therefore, more rational than the Moglen method. The existing time-area concept is generally applied to route the rainfall excess hyetograph through the watershed. Thus, the modified Moglen method, in a way, extends the time-area concept for infiltration. As shown through the examples, it is easy to ascertain the impact of temporal and spatial distribution of rainfall on runoff in addition to the examination of the impact of urbanization (or land use changes or changes in soil-vegetation-land use complex), shape, size, and orientation of the watershed on the resulting runoff. The computations for simple networks can be easily performed in the form of a spread-sheet. A simple computer coding of the procedure will help analyze the complex networks of large watersheds of any complex. The method, however, assumes a constant wave celerity irrespective of the watershed characteristics. It can be circumvented by first determining a suitable time-area diagram using the available tools to perform the analysis. Besides, there remains to ascertain the level of discretization of a watershed. To this end, the criterion suggested based on the initial abstraction can be useful.

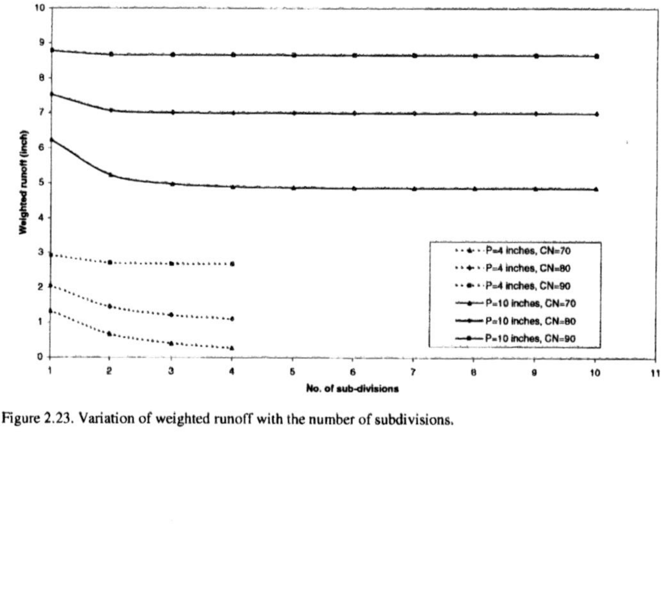

Figure 2.23. Variation of weighted runoff with the number of subdivisions.

CHAPTER 3

ANALYTICAL DERIVATION OF THE SCS-CN METHOD

Since the inception of the SCS-CN method, the issues such as the rational derivation of the method (equation (2.5)), the rationale of initial abstraction, the analytics of the S-CN mapping relation (equation (2.7)), and the CN-AMC relations have been of major concern. These relations appear to be mysterious in a sense that no analytical or physical explanation of their development is yet available in the literature. Thus, the objective of this chapter is to revisit the existing SCS-CN method from an analytical perspective and explore the fundamental proportionality concept (equation (2.2)). The general notion that the SCS-CN method is a generalization of the Mockus (1949) method (Rallison and Miller, 1982) is proved analytically. Alternate analytical means are proposed to derive the existing SCS-CN method and these are shown to be an improvement over the existing derivations. The description of its functional behaviour leads to the development of criteria useful for field applications. The empirical S-CN relationship is investigated for its analytical derivation. The relations linking CN with AMC are also proposed and discussed. A few case studies are presented to support the analytical derivations and finally, a brief investigation is made for using the SCS-CN concept as an alternative to the power law widely used as a surrogate to the popular Manning's equation described in Chapter 1.

The chapter begins with a brief description of the early rainfall runoff methods which led to the development of the SCS-CN method as pointed out in Chapter 2. The Mockus method is derived analytically by coupling the SCS-CN proportionality hypothesis (equation (2.2)) with the popular Horton (1933) model, and then it is generalised to yield the existing SCS-CN method.

3.1 Early Rainfall-Runoff Methods

Rainfall-runoff methods developed during the early 1940's utilized infiltration data for computing the runoff amount. For example, Andrews (1954) grouped the infiltration data gathered from Texas, Oklahoma, Arkansas, and Louisiana using soil texture as a soil characteristic and developed a graphical rainfall-runoff method taking into account the soil texture and type and amount of cover, and conservation practices, combined into what is referred to as soil-cover complex or soil-vegetation-land use (SVL) complex (Chapter 2).

Sherman (1949) was probably the first to propose plotting direct runoff versus storm rainfall. Using this concept, Mockus (1949) presented a rainfall-runoff model expressed as

$$Q = P_e[1 - 10^{-b P_e}] \tag{3.1}$$

where P_e is the storm rainfall amount excluding the initial abstraction (inches), Q is the direct runoff (inches), and b is an index related to the watershed and storm characteristics as:

$$b = \frac{0.0374(10)^{0.229P_5}(LU)^{1.061}}{\kappa^{1.990}T_s^{1.333}(10)^{2.271(K_h/T_s)}} \tag{3.2}$$

where P_5 is the 5-day antecedent rainfall (inches); LU is the land cover practice index; κ is the seasonal index, a function of the time of the year and temperature (°F); T_s is the storm duration (hr); and K_h is the soil index (inches/hr). Since the model parameter 'b' is estimated using meteorological data and physical watershed characteristics, the Mockus method can also be used for ungauged watersheds. Based on the results of application, the method is more suitable for large storms than for small storms, for short storms than for long storms, and for mixed-cover watersheds rather than single-cover watersheds. Here, a large watershed implies a watershed larger than several hundred acres. It is important to note that equation (3.1) is also valid for runoff rates and, therefore, parallels the model proposed by Zoch (1934, 1936), as shown below.

The Zoch model for the rising hydrograph can be expressed as

$$q = i_o(1 - e^{-t/c}) \tag{3.3}$$

where q is the runoff rate (inch/hr), i_o is the uniform rainfall intensity (inch/hr), t is the time, and c is described as the runoff coefficient. For comparing with the Mockus method, equation (3.1) is recast in exponential form as:

$$Q = P_e[1 - e^{-BP_e}] \tag{3.4}$$

where B = b ln(10). A comparison of equation (3.3) with equation (3.4) yields

$$\frac{Q}{P_e} = \frac{q}{i_o} \tag{3.5}$$

and

$$P_e = \frac{1}{cB}t \tag{3.6}$$

Equation (3.6) implies that the cumulative rainfall grows linearly with time or uniform rainfall intensity, i_o, is constant during a storm. However, during a typical isolated storm, the rainfall intensity first generally grows with time, attains a peak, and then finally reduces to zero. From equation (3.6) a relation between the uniform rainfall intensity, i_o, B, and c can be derived as

$$i_o = 1/(cB) \tag{3.7}$$

The assumption of constant uniform rainfall intensity is reasonable in the sense that it conforms with the infiltration derived from field measurements. Such an assumption also helps signify parameter 'b' of the Mockus model, as shown later. The following section derives analytically the Mockus and Zoch models from the combination of the Horton model and the SCS-CN proportionality concept (equation (2.2)).

3.2 Analytical Derivation of the Mockus and Other Methods

The Mockus method and the Andrews method, which lead to the existing SCS-CN method, compute the runoff volume or rate or indirectly compute the infiltration volume or rate using the water balance equation. Therefore, as also pointed out by Ponce and Hawkins (1996) among others, the SCS-CN method can also be construed as an infiltration loss model. Further analysis also shows analytically that the SCS-CN method is an infiltration model. To this end, using the SCS-CN concept of proportionality (equation (2.2) for $I_a = 0$), the Mockus method is derived from the Horton model.

3.2.1 DERIVATION OF MOCKUS METHOD

The Horton infiltration model (Chapter 1) is expressed as

$$f = f_c + (f_o - f_c)e^{-\alpha t} \tag{1.58}$$

where f is the infiltration rate [LT^{-1}] at time t, f_o is the initial infiltration rate [LT^{-1}] at time t = 0, α is the decay constant [T^{-1}], and f_c is the final infiltration rate [LT^{-1}]. It is noted that equation (1.58) is valid for the time 't' past ponding. An integration of equation (1.58) leads to the cumulative infiltration F at time t as

$$F = \frac{f_o - f_c}{\alpha}(1 - e^{-\alpha t}) \tag{3.8}$$

where F excludes f_c (or static or gravitational) component of infiltration. Thus, as $t \to \infty$, $F \to (f_o - f_c)/\alpha$. From equation (2.2) as $Q \to (P - I_a)$, $F \to S$, which is valid for time t approaching infinity. Therefore,

$$S = \frac{f_o - f_c}{\alpha} \tag{3.9}$$

From common experience on infiltration tests $f_o = i_o$, where i_o is the uniform rainfall intensity at time t (time past ponding) equal to zero. Its substitution into equation (3.9) yields

$$f_o - f_c = i_o - f_c = i_e = \alpha S \tag{3.10}$$

where i_e is the effective uniform rainfall intensity. Equation (3.10) describes the relationship among four parameters f_o, f_c, α, and S and defines the Horton parameter α equal to the ratio of the effective uniform rainfall intensity, i_e, to the potential maximum retention, S. This implies that α increases as i_e increases and decreases as S increases or CN decreases, and vice versa. Thus, α depends on the magnitude of the rainfall intensity and soil type, land use, hydrologic condition, antecedent moisture that affect S. Further substitution of equation (3.9) into equation (3.8) yields

$$\frac{F}{S} = (1 - e^{-\alpha t}) \tag{3.11}$$

Coupling equation (3.11) with equation (2.2) (for I_a = 0) (the SCS-CN hypothesis) yields

$$\frac{Q}{P} = (1 - e^{-\alpha t}) \tag{3.12}$$

An assumption of rainfall P growing linearly with time t leads to

$$P = i_e t \tag{3.13}$$

where P excludes static infiltration. Equation (3.13) asserts the general notion that P grows unbounded (Ponce and Hawkins, 1996). Substitution of equation (3.10) into equation (3.13) yields

$$\frac{P}{S} = \alpha t \tag{3.14}$$

Substituting equation (3.14) into equation (3.12) leads to

$$Q = P(1 - e^{-P/S}) \tag{3.15}$$

which is equivalent to equation (3.4) for B = 1/S and I_a = 0. Replacing P by P_e (= P-I_a) in equation (3.15) yields the Mockus equation (equation (3.1)) for S = 1/[b ln(10)]. For P to be the total rainfall for derivation of the Mockus method, f_c should be equal to zero. Furthermore, coupling of equation (3.14) with equation (3.13) yields the same parametric relation as described by equation (3.10).

Equation (3.15) (the Mockus method) can also be derived using a first-order linear hypothesis for the variation of S with time or rainfall as

$$\frac{dS_t}{dt} = -\alpha S_t \tag{3.16a}$$

or

$$\frac{dS_t}{dP} = -BS_t \tag{3.16b}$$

whose solution is

$$S_t = S_o e^{-\alpha t} \tag{3.17a}$$

or

$$S_t = S_o e^{-BP} \tag{3.17b}$$

for $S_t = S_o$ (or S in equation (3.15)) at t = 0 or P = 0. It implies that for a given antecedent moisture condition (AMC), $S = S_o$ represents the maximum possible amount of infiltration or the potential maximum space available for moisture retention. Replacing S_t by $(S_o - F)$ and B by 1/S, equation (3.17a) or equation (3.17b) can be recast for F as

$$\frac{F}{S_o} = 1 - e^{-\alpha t} = 1 - e^{-P/S} \tag{3.18}$$

Coupling equation (3.18) with equation (2.2) for I_a = 0 leads to equation (3.15), which represents the alternative form of the Mockus method. It is worth emphasizing that equation (3.17a) or equation (3.17b) describes the variation of S with time or rainfall. As time or rainfall increases, S_t decreases exponentially. For antecedent dry, wet, or normal condition, S_o corresponds to dry, wet, or normal condition of the soil before the start of the storm. Thus, $S_o = S$.

3.2.2 DERIVATION OF ZOCH MODEL

The Zoch model (equation (3.3)) can be analytically derived assuming that the basic SCS-CN proportionality concept also holds for rainfall-runoff rates as below:

$$\frac{Q}{P} = \frac{F}{S} = \frac{q}{i_o} \tag{3.19}$$

which is equivalent to equation (3.5). Combination of equations (3.19) and (2.2) for I_a and f_c equal to zero leads to a relation among runoff rate (q), rainfall intensity (i_o), and the average storage capacity (s_{av}) as:

$$q = \frac{i_o^2}{i_o + s_{av}} \tag{3.20}$$

which is similar to the equation proposed by Yu (1998).

For a given storm duration, equation (3.20) computes the storm-averaged runoff rate corresponding to the storm-averaged rainfall intensity, and thus, it does not compute the actual instantaneous runoff rate corresponding to the instantaneous rainfall intensity. The average storage capacity, s_{av}, refers to the ratio of S to the storm duration T_s, and thus, it is the storm-average of the potential maximum storage capacity (rate). Therefore, it can be defined as the maximum possible storm-average rate of infiltration. In other words, s_{av} is analogous to the potential maximum phi-index (Chapter 1) for a storm.

Similar to the derivation of the Mockus method, parametric relations can be derived assuming that

$$\frac{P}{S} = \alpha t = t / c \tag{3.21}$$

which also implies that the cumulative rainfall grows linearly with time. Substitution of equation (3.21) into equation (3.15) yields equation (3.3), which is the Zoch model. From equation (3.21) α can be derived as

$$\alpha = 1/c \tag{3.22}$$

Coupling of equation (3.22) with equation (3.10) (for $f_c = 0$) leads to

$$S = c\, i_o \tag{3.23}$$

Equation (3.23) misleadingly describes c as the runoff coefficient because it multiplies the rainfall intensity in the equation. However, c in equation (3.23) is a dimensional quantity bearing the dimension of time, unlike the runoff coefficient which is non-dimensional. Furthermore, S does not represent the runoff amount, rather it is the potential maximum retention, a storage property of the soil profile (see Chapter 4). In this context, c can be better interpreted as storage coefficient rather than runoff coefficient, similar to storage equation (1.84).

3.2.3 DERIVATION OF DEPRESSION AND INTERCEPTION STORAGE MODELS

In Chapter 1, the depression storage model (equation (1.20)) is expressed as:

$$\frac{V_s}{S_d} = 1 - e^{-\beta P_e} \tag{3.24}$$

Similar to the fundamental proportionality concept (equation (2.2)) of the SCS-CN method, the following proportionality hypothesis can be assumed to hold:

$$\frac{Q}{P_e} = \frac{V_s}{S_d} \tag{3.25}$$

which states that the ratio of the actual runoff (Q) to the potential runoff (P or P_e) is equal to the ratio of the actual depth of depression storage (V_s) to the potential maximum depression storage (S_d). Coupling of equation (3.25) with equation (3.4) yields a relation analogous to equation (3.24), which is the depression storage model, for $\beta = 1/S_d$. Similarly, the interception model (equation (1.19)) can also be derived.

3.3 Derivation of the SCS-CN Method

3.3.1 GENERALIZATION OF THE MOCKUS METHOD

The SCS-CN method is an approximation or generalization of the Mockus equation (equation (3.1)). To this end, equation (3.4), which is a variant of the Mockus equation, is approximated to yield the SCS-CN method as shown below.

An expansion of equation (3.4) leads to

$$\frac{Q}{P} = 1 - \left[1 - BP + \frac{(BP)^2}{2!} - \frac{(BP)^3}{3!} ...... + (-1)^N \frac{(BP)^N}{N!}\right] \tag{3.26}$$

where N is an integer. For BP < 1, equation (3.26) can be reasonably approximated as

$$\frac{Q}{P} = 1 - [1 - BP + (BP)^2 - (BP)^3 ...... + (-1)^N (BP)^N] \tag{3.27}$$

or

$$\frac{Q}{P} = BP[1 - (BP) + (BP)^2 - (BP)^3 ...... + (-1)^N (BP)^N] \tag{3.28}$$

Equation (3.28) can be written alternatively as

$$\frac{Q}{P} = \frac{BP}{1 + BP} \tag{3.29}$$

which is also valid for BP < 1. For B = 1/S, as above, and its substitution into equation (3.29) leads to

$$\frac{Q}{P}=\frac{P}{S+P} \tag{3.30}$$

Thus, it is seen that the existing SCS-CN method is the generalization of the Mockus method. Replacement of P by P_e (= $P-I_a$) leads to equation (2.5), which is the existing SCS-CN equation.

Fundamental to the analytical derivation of equation (3.30) is the assumption that BP < 1, involved in the transformation of equation (3.26) to equation (3.27) and equation (3.28) to equation (3.29). To check the validity of this assumption, the ratio of the right side of equation (3.27) to that of equation (3.26), designated as $RATIO^1$ and the ratio of the right side of equation (3.29) to that of equation (3.28), designated as $RATIO^2$, were computed as shown in Table 3.1. For a given value of BP, $RATIO^1$ and $RATIO^2$ are the ratios of the sum of 33 terms (N = 33) on the right-hand side of equations (3.27) and (3.26), (3.29) and (3.28), respectively. For exact transformation from equation (3.26) to equation (3.27) and from equation (3.28) to equation (3.29), respectively, these ratios should be equal to 1 for any value of BP. Evidently, the former transformation is valid for BP < 0.9801 with approximately ±20% deviation from the results of equation (3.26). With ±10% deviation, it is valid for BP < 0.2645 and within the BP range (0.9410-0.9717). Similarly, the second transformation is valid for BP < 0.9143 and BP < 0.9319 with respective deviations of approximately +5% and +10% from the results of equation (3.28).

Table 3.1 Verification of the assumption, BP < 1, used in the derivation of existing SCS-CN method [equation (3.30)]

BP	$RATIO^1$	$RATIO^2$	BP	$RATIO^1$	$RATIO^2$
.0001	.9999500	1.0000000	.8376	.8058624	1.0024227
.0010	.9995006	1.0000000	.9143	.8385266	1.0499078
.0100	.9950578	1.0000000	.9319	.8733593	1.0999873
.0199	.9902765	1.0000000	.9410	.9019374	1.1448031
.1000	.9553029	1.0000000	.9664	1.0500433	1.4552857
.1134	.9500372	1.0000000	.9717	1.1003277	1.6045860
.2645	.9000184	1.0000000	.9801	1.2004531	2.0197347

Note: $RATIO^1$ is the ratio of the right hand terms of equation (3.27) and equation (3.26) and $RATIO^2$ is the ratio of those of equations (3.29) and (3.28).

3.3.2 STATISTICAL DERIVATION OF THE SCS-CN METHOD

The probability-based derivation of the SCS-CN method attempted by Schaake et al. (1996) and Yu (1998, 2000) is provided.

(a) *Schaake et al. Derivation*

Schaake et al. (1996) assumed exponential distributions for the variation of infiltration and rainfall, respectively, as

$$g(i_c)=\frac{1}{I_c}e^{\left(-\frac{i_c}{I_c}\right)} \tag{3.31}$$

and

$$g(p) = \frac{1}{P} e^{(-\frac{p}{P})} \quad (3.32)$$

where g(.) stands for the probability density function, i_c is the point infiltration capacity during a rain storm, I_c is the spatially averaged infiltration capacity, p is the storm total rainfall at a point, and P is the spatially averaged storm total rainfall. Assume that at a point during a storm the total amount of water that would infiltrate would be equal to the rainfall amount at that point if the rainfall were less than or equal to the infiltration capacity. This neglects the possibility that instantaneous rainfall rates during the storm might exceed the instantaneous infiltration capacity. Accordingly, the point amount of surface runoff, q_s, will be

$$q_s = \begin{cases} p - i_c & \text{if } p > i_c \\ 0 & \text{otherwise} \end{cases} \quad (3.33)$$

If 'p' and 'i_c' are the spatially distributed random variables, q_s will be a random variable, leading to the derivation of spatially averaged surface runoff Q as:

$$Q = \int_0^{\infty} q_s g(q_s) dq_s \quad (3.34)$$

where $g(q_s)$ is the derived distribution (in frequency domain) of point surface runoff. The density function for $g(q_s)$ depends on the joint probability distribution of p and i_c. Since p and i_c apply to the entire storm period, it is reasonable to assume that the spatial distribution of p is independent of the initial state of the catchment and, therefore, of i_c. Thus,

$$g(p, i_c) = g(p) g(i_c) \quad (3.35)$$

Here, it is worth emphasizing that although p is taken as independent of i_c, there is strong evidence of i_c being dependent on p (Cook, 1946; Moldenhauer et al., 1960; Murai and Iwasaki, 1975; Hawkins, 1982; Flanagan et al., 1988; and Yu et al., 1997). Thus, both i_c and q (or Q) are not independent random variables.

The cumulative distribution function, G, for q_s is

$$G(q_s^* \le q_s) = \iint_{R(q_s^* \le q_s)} g(p) g(i_c) \, di_c \, dp \quad (3.36)$$

$$= \int_0^{\infty} \int_0^{q_s} g(p) g(i_c) \, di_c \, dp \quad (3.37)$$

where $R(q_s^* \leq q_s)$ is the region of (p, i_c) plane where q_s^* is less than or equal to q_s. Substitution of equations (3.31) and (3.32) into equation (3.37) and then integration yield

$$G(q_s) = 1 - \frac{P}{P + I_c} e^{(-\frac{q_s}{P})} \tag{3.38}$$

The density function for q_s has a finite probability that $q_s = 0$, expressed as:

$$G(q_s = 0) = \frac{I_c}{P + I_c} \tag{3.39}$$

The density function is

$$g(q_s) = \frac{dG(q_s)}{dq_s} = \frac{P}{P + I_c} e^{(-\frac{q_s}{P})} \tag{3.40}$$

It follows that the average value of q_s (= Q) is

$$Q = \frac{P^2}{P + I_c} \tag{3.41}$$

This equation is analogous to the SCS-CN method (equation (2.5) for $I_a = 0$).

(b) *Yu (1998) Derivation*

Yu (1998) derived the SCS-CN method assuming that (a) the spatial variation of infiltration capacity has an exponential distribution and (b) the temporal variation of rainfall rate also follows an exponential distribution. These postulates are expressed, respectively, as:

$$G_f(f) = 1 - e^{-f/\bar{f}} \tag{3.42}$$

and

$$G_i(i) = 1 - e^{-i/\bar{i}} \tag{3.43}$$

where G_f (f) and G_i (i) are the cumulative distribution functions of the infiltration capacity, f, and rainfall rate, i, respectively; $\bar{i}$ is the average rainfall rate; and $\bar{f}$ is the average infiltration capacity across the watershed when the entire watershed generates runoff. G_i can be interpreted as a fraction of the storm duration when the rainfall rate is less than or equal to $\bar{i}$. Similarly, G_f can be interpreted as the fraction of the watershed where the infiltration capacity is less than or equal to $\bar{f}$.

The actual infiltration rate, f_a, is related to rainfall rate, i, as

$$f_a = \begin{cases} i & \text{if } i<f \\ f & \text{if } i \ge f \end{cases} \tag{3.44}$$

The actual infiltration rate is limited by either rainfall rate i or infiltration capacity f, whichever is smaller. It follows that the rainfall-excess, q, at this point is

$$q = \begin{cases} 0 & \text{if } i<f \\ i-f & \text{if } i \ge f \end{cases} \tag{3.45}$$

For a given rainfall rate, the rainfall-excess is, therefore, regarded as a function of the infiltration capacity that varies in space and the rainfall rate that varies in time. Although this rainfall-excess is spatially variable because of the spatial variability in infiltration capacity, it can be integrated over the watershed to yield an average rainfall-excess rate for the watershed, $\bar{q}$, defined as

$$\bar{q} = \int_0^\infty q \, dG_f(f) \tag{3.46a}$$

or

$$\bar{q} = \int_0^i (i-f) \, dG_f(f), \text{ since } q = 0 \text{ when } f > i \tag{3.46b}$$

This watershed-wide rainfall-excess rate can be considered as a function of the rainfall rate and the spatial distribution of the infiltration capacity. Integration of equation (3.46b) yields

$$\bar{q} = \int_0^i G_f(f) \, df \tag{3.47}$$

The fraction or the partial area of the watershed that contributes to runoff, r, is given by

$$r = \int_0^i dG_f(f) \tag{3.48a}$$

or

$$r = G_f(i) \tag{3.48b}$$

It can be seen from equations (3.47) and (3.48) that as rainfall rate increases, both the watershed-wide average rainfall-excess rate and the area of the watershed from which runoff occurs will increase.

When an exponential distribution of the infiltration capacity is assumed, the rate of rainfall-excess for the watershed as a function of the rainfall rate is given by combining equations (3.47) and (3.42) as:

$$\bar{q} = i - \bar{f}(1 - e^{-i/\bar{f}}) \tag{3.49}$$

Therefore, the apparent infiltration rate for the watershed is

$$i - \bar{q} = \bar{f}(1 - e^{-i/\bar{f}}) \tag{3.50}$$

This infiltration rate increases as the rainfall rate, i, increases.

The rate of rainfall-excess can be weighted by the frequency distribution of the rainfall rate to yield an average watershed-wide rainfall-excess rate, $\overline{Q}$, of

$$\overline{Q} = \int_0^\infty \bar{q}\, dG_i(i) \tag{3.51a}$$

$$\overline{Q} = \frac{1}{\bar{i}} \int_0^\infty [i - \bar{f}(1 - e^{-i/\bar{f}})e^{-i/\bar{i}}\, di \tag{3.51b}$$

$$\overline{Q} = \bar{i} - \bar{f} + \frac{\bar{f}^2}{\bar{i} + \bar{f}} \tag{3.51c}$$

$$\overline{Q} = \frac{\bar{i}^2}{\bar{i} + \bar{f}} \tag{3.51d}$$

Equation (3.51d) is similar to equation (3.20), leading to $q = \overline{Q}$, $i_o = \bar{i}$, and $\bar{f} = s_{av}$. Here, $\bar{f}$ is similar to the φ-index (Chapter 1) (Istanbulluoglu, 2000), except for the difference that the former is the space-averaged infiltration and the latter is the time-averaged infiltration.

For a particular rainstorm event, multiplying equation (3.51d) by the storm duration, T_s, yields the total rainfall-excess, Q, for the event as

$$Q = \overline{Q}T_s = \frac{(\bar{i}T_s)^2}{\bar{i}T_s + \bar{f}T_s} = \frac{P_e^2}{P_e + S} \tag{3.52}$$

The total rainfall-excess, Q ($= \overline{Q}T_s$), is the same as storm runoff for individual events; $\bar{i}T_s$ is the same as P_e if $\bar{i}$ refers to the average rainfall rate after runoff has begun; and $\bar{f}T_s$ can be interpreted as the potential maximum retention, S, or equivalently the CN, because $\bar{f}T_s$ is the maximum amount of infiltration that could have occurred during the runoff event. Equation (3.52), therefore, has a functional form identical to that of equation (2.5), the SCS-CN method.

(c) *Yu (2000) Derivation*

Assume X and Y to be exponentially distributed random variables with means E(X) and E(Y), respectively. Now, define a new random variable Z such that

$$Z = \begin{cases} X - Y & \text{if } X > Y \\ 0 & \text{otherwise} \end{cases} \tag{3.53}$$

It can be shown that if X and Y are independent, then the mean of Z, E(Z) is given by

$$E(Z) = \frac{E^2(X)}{E(X) + E(Y)} \tag{3.54}$$

Equation (3.54) is similar to SCS-CN equation (2.5) for $I_a = 0$. Schaake et al. (1996) considered X as rainfall depth and Y as storage capacity, apparently consistent with the storage concept of Moore (1985), and accordingly, variable Z corresponds to the direct runoff or rainfall-excess amount (depth). Yu (2000) defined the storage capacity as the maximum amount of water holding capacity of a soil at a given location. Here, the spatial variation of the rainfall depth was ignored for the reason of validity of the assumption of uniform depth over small or plot scale watersheds, for which the SCS-CN method was originally developed. However, the apparent basic problem with this derivation too is that infiltration rates are not independent of rainfall intensity and thus, leads to an ambiguity in the SCS-CN derivation.

3.3.3 SCS-CN DERIVATION FROM THE FIRST-ORDER HYPOTHESIS

Similar to the derivation of the SCS-CN method from the generalization of the Mockus method, it is possible to show the variation of S with P from equation (3.17b) exhibiting the first-order exponential variation with rainfall as:

$$\frac{S_t}{S_o} = e^{-BP} \tag{3.17b}$$

An expansion of the right-hand side of equation (3.17b) analogous to equation (3.26) yields

$$\frac{S_t}{S_o} = \left[1 - BP + \frac{(BP)^2}{2!} - \frac{(BP)^3}{3!} ...... + (-1)^N \frac{(BP)^N}{N!}\right] \tag{3.55}$$

where N is an integer. Similar to the transformation of equation (3.26) to equation (3.27), equation (3.55) can be approximated as

$$\frac{S_t}{S_o} = [1 - BP + (BP)^2 - (BP)^3 ...... + (-1)^N (BP)^N] \tag{3.56}$$

which is valid for BP < 1. Equation (3.56) can also be written as

$$\frac{S_t}{S_o} = \frac{1}{1 + BP} \tag{3.57}$$

which is also valid for BP < 1. For $B = 1/S_o$, equation (3.57) can be re-written as

$$S_t = \frac{S_o^2}{P + S_o} \tag{3.58}$$

which is the equation describing the variation of S_t with P. It is again noted that $S_o = S$, as above. Substitution of $S_t = S_o - F$ in equation (3.58) yields

$$\frac{F}{S_o} = \frac{P}{S_o + P} \tag{3.59}$$

Coupling equation (3.59) with the proportionality hypothesis (equation (2.2) for $I_a = 0$) or with the water balance equation with $I_a = 0$ yields equation (3.30), which is the existing SCS-CN method for $I_a = 0$.

3.3.4 DERIVATION OF SCS-CN PROPORTIONAL EQUALITY

It is further possible to analytically derive the basic proportionality concept of the existing SCS-CN method using the Horton or the first-order storage hypothesis and water balance equation. Coupling of equation (3.11) with equation (3.14) with the assumption of the cumulative rainfall growing linearly with time yields

$$\frac{F}{S} = (1 - e^{-P/S}) \tag{3.60}$$

An approximation of the right-hand side of equation (3.60), as above, leads to

$$\frac{F}{S} = \frac{P}{S + P} \tag{3.61}$$

which is equivalent to equation (3.59), derived from the first-order hypothesis. Substitution of equation (3.61) into the water balance equation (equation (2.1)) yields

$$\frac{Q}{P} = \frac{P}{S + P} \tag{3.62}$$

which is the same as equation (3.30). A comparison of equation (3.62) with equation (3.61) yields

$$\frac{Q}{P} = \frac{F}{S} \tag{3.63}$$

which is the basic proportional equality expressed by equation (2.2) for $I_a = 0$. Thus, the basic SCS-CN hypothesis can be derived from the Horton equation or the first-order storage hypothesis and the water balance equation with the assumption that the cumulative rainfall grows linearly with time.

3.3.5 NON-LINEAR DERIVATION OF SCS-CN METHOD

Based on the second-order non-linear hypothesis for storage variation with time directly yields the SCS-CN method. The hypothesis is expressed as:

$$\frac{dS_t}{dt} = -\gamma S_t^2 \tag{3.64}$$

where γ is a proportionality coefficient. Integration of equation (3.64) yields

$$-\frac{1}{S_t} = -\gamma t + D \tag{3.65}$$

where D is an integration constant, which can be derived for the condition that at t = 0, $S_t = S_o$ (or S). Therefore,

$$D = -\frac{1}{S_o} \tag{3.66}$$

Substituting equation (3.66) into equation (3.65) leads to

$$\frac{S_t}{S_o} = \frac{1}{1 + \gamma S_o t} \tag{3.67}$$

Analogous to equation (3.21), it is assumed that the following holds:

$$\frac{P}{S_o} = \gamma S_o t \tag{3.68}$$

which yields

$$i_e = \gamma S_o^2 \qquad (3.69)$$

where i_e is the effective uniform rainfall intensity $[LT^{-1}]$. Equation (3.69) describes γ as the ratio of i_e to the square of S_o, with a dimension $[L^{-1}T^{-1}]$. Further coupling of equation (3.69) with equation (3.10) yields

$$\gamma = \frac{\alpha}{S_o} \qquad (3.70)$$

which expresses γ as the ratio of the Horton decay parameter to the potential maximum retention, S_o (= S).

Substitution of equation (3.68) into equation (3.67) leads to

$$\frac{S_t}{S_o} = \frac{S_o}{P + S_o} \qquad (3.71)$$

which is the same as equation (3.58), which when coupled with the water balance equation (equation (2.1)) along with $S_t = (S_o - F)$ leads to the SCS-CN equation (3.30), as described above.

The above non-linear approach also leads to inclusion of the evapotranspiration directly in the SCS-CN method. The derivation is based on the following hypothesis of storage variation with time:

$$\frac{dS_t}{dt} = -\gamma S_t^2 \left(1 - \frac{E_t}{E_p}\right) \qquad (3.72)$$

where γ is a proportionality coefficient, E_t is the rate of evapotranspiration at time t, and E_p is the potential maximum rate of evapotranspiration. Integration of equation (3.72) yields

$$-\frac{1}{S_t} = -\gamma\left(1 - \frac{\overline{E}}{E_p}\right)t + D \qquad (3.73)$$

where $\overline{E}$ is the average rate of evapotranspiration during the storm and D is an integration constant, which can be derived for the condition that at t = 0, $S_t = S_o$ (or S). Therefore,

$$D = -\frac{1}{S_o} \tag{3.74}$$

Substituting equation (3.74) into equation (3.73) leads to

$$\frac{S_t}{S_o} = \frac{1}{1+\gamma\left(1-\frac{\overline{E}}{E_p}\right)S_o t} \tag{3.75}$$

Similar to equation (3.68), it is assumed that the following relation holds:

$$\frac{P}{S_o} = \gamma\left(1-\frac{\overline{E}}{E_p}\right)S_o t \tag{3.76}$$

Equation (3.76) yields

$$i_e = \gamma\left(1-\frac{\overline{E}}{E_p}\right)S_o^2 \tag{3.77}$$

where i_e is the effective uniform rainfall intensity $[LT^{-1}]$ and γ has the dimensions $[L^{-1}T^{-1}]$. Combination of equation (3.77) with equation (3.10) yields

$$\gamma = \frac{\alpha}{S_o\left(1-\frac{\overline{E}}{E_p}\right)} \tag{3.78}$$

which expresses the dependence of γ on the evapotranspiration. The usual substitution of equation (3.76) into equation (3.75) leads to equation (3.71), and, in turn, the SCS-CN method. It is worth emphasizing here that since evapotranspiration losses are negligible during a storm, such a derivation is useful for long duration storms, such as a day or longer.

3.3.6 SCS-CN DERIVATION INCLUDING INITIAL ABSTRACTION

The above derivation of the Mockus method or the SCS CN method from the Horton equation (equation (1.58)) assumed the time to ponding, t_p, equal to zero. Alternatively, these methods are valid for the time past ponding. It is, however, possible to incorporate t_p in the Horton model as

$$f - f_c = (f_o - f_c)\,e^{-\alpha(t-t_p)} \tag{3.79}$$

Integration of equation (3.79) leads to the cumulative infiltration F at time t as

$$F = \frac{f_o - f_c}{\alpha}[1 - e^{-\alpha(t-t_p)}] \tag{3.80}$$

where F excludes the static infiltration. From equation (3.10) $i_e = f_o\text{-}f_c = \alpha S$ and its substitution into equation (3.80) yields

$$\frac{F}{S} = [1 - e^{-P_e/S}] \tag{3.81}$$

where P_e is described as

$$P_e = P - F_c - I_a = \alpha(t\text{-}t_p)S \tag{3.82}$$

where F_c = static (or gravitational) infiltration (= $f_c t$).

Substitution of $i_e = \alpha S$ in equation (3.82) leads to an expression for I_a as

$$I_a = i_e\, t_p \tag{3.83}$$

which describes I_a as the product of the effective uniform rainfall intensity and the time to ponding. In the context of infiltration, such a computation of I_a is reasonable and consistent with that occurring in field. It, however, ignores water losses due to surface detention, evaporation from vegetative cover, and so on. An approximation of equation (3.81), as for the above SCS-CN derivation from the Mockus method, yields

$$\frac{F}{S} = \frac{P_e}{P_e + S} \tag{3.84}$$

Coupling of equation (3.84) with the water balance equation (equation (2.1)) for deriving the Q/P_e ratio and comparing the resulting expression with equation (3.84) yields the proportional equality described by equation (2.2).

Alternatively, the derivation of Q/P_e ratio yields equation (2.5):

$$Q = \frac{(P - I_a)^2}{P - I_a + S} \tag{2.5}$$

which is the existing SCS-CN equation for $F_c = 0$. Thus, it is proved that the existing SCS-CN method is the generalization or approximation of the Mockus method, which holds for P excluding both I_a and F_c. The important inference derived from such a derivation is that the formulation of the existing SCS-CN method for I_a has a sound justification within the framework of infiltration phenomena observed in the field and is not a mere replacement of P by P_e in equation (3.30). Since the original concept of the SCS-CN method (SCS, 1971) assumes I_a as a part of S for deriving $\lambda = 0.2$, on which the NEH-4 curve number tables are based, the corresponding analytical treatment of Chen (1982) is provided in Appendix A. Here, it is noted that $\lambda = 0.25$ for S including I_a (equation (2.4)).

3.3.7 DEVELOPMENT OF AN INITIAL ABSTRACTION MODEL

The initial abstraction includes interception, surface depression, infiltration, and evaporation. Using either the generalized approach or the second-order non-linear approach, the interception (equation (1.18)) and surface depression (equation 1.20)) models can be transformed, respectively, to

$$L_i = \frac{PS_v}{P + S_v} + KEt \tag{3.85}$$

and

$$V_s = \frac{P_e^{'} S_d}{P_e^{'} + S_d} \tag{3.86}$$

where L_i is the interception loss, P is the total rainfall, $P_e^{'}$ is the effective rainfall (= P-L_i), F is the cumulative infiltration, K is the ratio of vegetal surface area to its projected area, E is the evaporation rate from plant surfaces during storm, t is the rainfall duration (= T_s), S_v is the storage capacity of vegetation, V_s is the depression storage, and S_d is the storage capacity of surface depressions. The initial abstraction due to infiltration can be computed from equation (3.83).

Thus, a complete initial abstraction model based on the SCS-CN concept can be derived as

$$I_a = \frac{PS_v}{P + S_v} + KEt + \frac{(P - L_i - F)S_d}{P - L_i - F + S_d} + i_e t_p \tag{3.87}$$

where (P-L_i-F) is the rainfall-excess and cumulative infiltration F is given as:

$$F = \frac{(P - L_i)S}{P - L_i + S} \tag{3.88}$$

Thus, the process of initial abstraction is modelled as follows. First, the evaporation and interception demands are met and then the remaining water fulfils the surface depression and infiltration demands. A simplified version of this model can be given as below.

Let $S_v + S_d = S'$. Assuming S' varies with time according to the second-order law described in Section 3.3.5. At time t = 0, $S' = S'_0$. Then, at any time t, I_a excluding evaporation (= KEt) and infiltration (= $i_e t_p$) can be given as

$$I_a = S'_o - S'_t = \frac{PS'_o}{P + S'_o} \tag{3.89}$$

Coupling of equation (3.89) with equation (2.5) leads to

$$Q = \frac{P^4}{(P + S'_o)[P^2 + (P + S'_o)S]} \tag{3.90}$$

which is another simplified form of the SCS-CN model (equation (2.5)). It is noted that equation (3.90) ignores evaporation and the initial (time to ponding) infiltration losses. For the existing SCS-CN method with $\lambda = 0.2$, S'_o can be derived from the coupling of equation (3.89) with equation (2.3) as

$$S'_o = \frac{0.2SP}{P - 0.2S} \tag{3.91}$$

For any other value of λ, S'_o can be given as:

$$S'_o = \frac{\lambda SP}{P - \lambda S} \tag{3.92}$$

If there exists a possibility to determine S'_o by some other means, then λ can alternatively be determined from equation (3.92) as

$$\lambda = \frac{PS'_o}{S(P + S'_o)} \tag{3.93}$$

Equation (3.93), however, excludes evaporation and initial infiltration losses. Therefore, a general expression for computing λ can be given as

$$\lambda = \frac{PS'_o}{S(P + S'_o)} + \frac{KEt}{S} + \frac{i_e t_p}{S} \tag{3.94}$$

Equation (3.94) exhibits the dependence of λ on meteorological factors (Chapter 2) besides the factors governing the soil-vegetation-land use complex.

3.4 Implications of Generalization of the Mockus Method

3.4.1 MODIFICATION OF THE SCS-CN METHOD

Neglecting the third and higher order terms in equation (3.26) and doing a little algebraic manipulation lead to

$$\frac{Q}{P}=\frac{P}{S+0.5P} \tag{3.95}$$

Introducing I_a in equation (3.95), S/P can be derived as below:

$$\frac{S}{P}=\frac{[4\lambda+2C-\lambda C]-\sqrt{C[C(2-\lambda)^2+16\lambda]}}{4\lambda^2} \tag{3.96}$$

where C is the runoff factor (= Q/P). From equation (3.96) it can be shown that S/P always assumes a real value. Evidently, as $\lambda \to 0$, S/P $\to$ 0/0 or S is indeterminate for a definite P; and as $C \to 0$, S/P $\to 1/\lambda$ or $I_a \to P$. Since Q = 0 for $P \le I_a$, therefore, $\lambda S \to P$ leads to two concepts: (i) Given S, λ varies such that $\lambda S \to P$; in such a case, λ is dependent on P. (ii) Given λ, S varies such that the aforementioned inference holds; in this case, S becomes a function of P. The normal practice, however, is to follow the second concept, consistent with equation (3.58). For C = 0, the maximum S, S_{max}, = P/λ; as an example, for $\lambda = 0.2$, $S_{max} = 5P$. As $C \to 1$, S/P $\to$ 0.39 for $\lambda = 0.2$. It can be shown that S will approach P if either $I_a \to 0$ or $C \to 0.4143$ if $\lambda = 0.2$ or $\lambda \to 0.59$ if $C \to 0$.

3.4.2 GENERAL FORM OF THE SCS-CN MODEL

In general, equation (3.95) can be written as

$$Q=\frac{P^2}{S+aP} \tag{3.97}$$

where 'a' is the coefficient replacing 0.5 in equation (3.81). The first derivative of equation (3.97) with respect to P leads to

$$\frac{\partial Q}{\partial P}=\frac{\frac{P}{S}\left[2+a\frac{P}{S}\right]}{\left[1+a\frac{P}{S}\right]^2} \tag{3.98}$$

Here again, if $\partial Q/\partial P \to 0$, either $P \to 0$ or $S \to \infty$ and if $\partial Q/\partial P = 1$,

$$a = \frac{1}{2} - \frac{S}{P} + \sqrt{\frac{1}{4} + \frac{S}{P}} \tag{3.99}$$

where the "-" sign before the square root sign in equation (3.99) is omitted. a = 1/2 (for the modified method) at P/S = 0.83; as S/P → 0, a → 1 or the general form reduces to the existing SCS-CN method. For ∂Q/∂P = 0, either P = 0 or S = ∞. The variation of 'a' with S/P is shown in Figure 3.1.

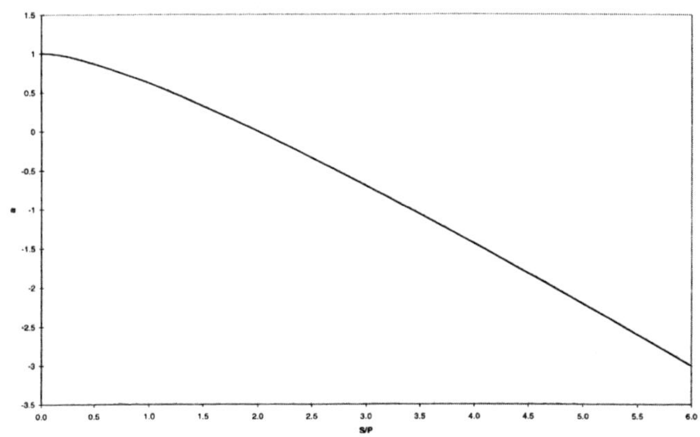

Figure 3.1. Variation of 'a' with S/P for ∂Q/∂P=1.

3.5 Characteristics of the SCS-CN and Mockus Methods

3.5.1 MOCKUS METHOD

Since the derivation of the SCS-CN, Mockus, and Zoch methods exclude the static portion of infiltration f_c, they all underestimate infiltration. For application, other major characteristics of the Mockus method can be derived from equation (3.15). For Q ≤ P, $e^{-P/S} \leq 1$ or (-P/S) ≤ 0 implying that S ≤ ∞ and P ≥ 0. Here, S (or b) can have a range (0, ∞). As Q → 0, either P → 0, thus, F → 0 or S → ∞, and thus, F → P. Introduction of I_a in P implies that the former condition is valid when P ≥ I_a. As P → S, F → P/e (here, e = 2.72) or F → 0.37 P; and Q will approach P when S → 0 for P is a finite quantity. The usually defined condition, F → S, is possible only when the following condition is satisfied:

$$(P/S)\, e^{-P/S} = 1 \tag{3.100}$$

which is indeterminate implying that F never approaches or equals S. The maximum possible F is equal to P/e and occurs at P = S.

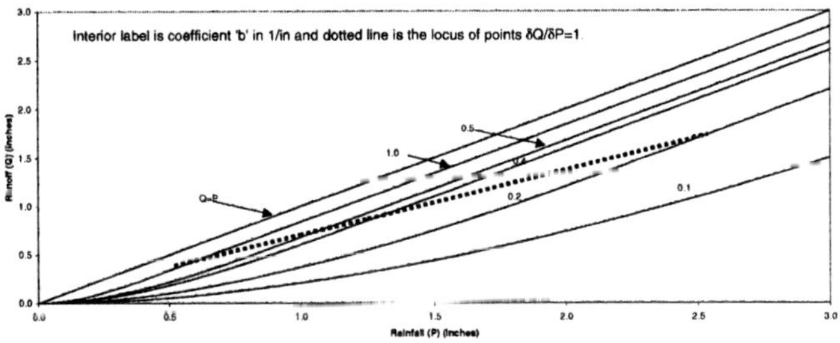

Figure 3.2. Useful part (below dotted line) of Mockus method

In addition to the above, the Mockus equation, as shown in Figure 3.2, depicts a double curvature above the dotted line indicating $\partial Q/\partial P$ even greater than 1, which is not possible. It can also be shown mathematically as follows.

Differentiation of equation (3.4) with respect to P_e for B = 1/S yields

$$\frac{\partial Q}{\partial P} = 1 + (P_e / S - 1) e^{-P_e/S} \tag{3.101}$$

The condition that the left-hand side of equation (3.101) be less than or equal to 1 yields $(P_e/S) \leq 1$. It implies that the Mockus method is applicable only for the rainfall amount less than or equal to S. For P greater than S, the Mockus method will yield runoff rates greater than the rainfall intensity, which is not possible. Alternatively, for application of the Mockus method, it can be shown that the time duration (t) should be less than or equal to the inverse of the Horton parameter α:

$$t \leq \frac{1}{\alpha} \tag{3.102}$$

Thus, the Mockus method is restricted in its application. Similarly, the applicability of the SCS-CN method can also be ascertained.

3.5.2 SCS-CN METHOD

The first derivative of equation (3.30) with respect to P is

$$\frac{\partial Q}{\partial P} = \frac{(P/S)[2 + P/S]}{[1 + P/S]^2} \tag{3.103}$$

Since the direct runoff rate ($\partial Q/\partial t$) does not exceed the corresponding rainfall intensity ($\partial P/\partial t$), $\partial Q/\partial P$ can range between 0 and 1, as also suggested in the variable-source area concept (for example, Steenhuis et al., 1995), which, however, is not supported by the SCS-CN concept (Chapter 2). Evidently, from equation (3.103), $\partial Q/\partial P \geq 0$ for all values of $P/S \geq 0$; $\partial Q/\partial P = 0$ at $P = 0$. For the upper limit, it can be shown that as $\partial Q/\partial P \rightarrow 1.0$, $S \rightarrow 0$. Similar inferences can be drawn if P is replaced by P_e in equation (3.103); for $\partial Q/\partial P \geq 0$, $P \geq I_a$ since S is a non-negative quantity. Another condition that $\partial Q/\partial P \leq 1$ always holds for S is a non-negative value. Thus, the existing SCS-CN method is more stable than the Mockus method.

Table 3.2 Comparison of existing SCS-CN method and modified method with the Mockus method

BP	RATIO^3	RATIO^4	BP	RATIO^3	RATIO^4
.010	.995058	1.000008	5.0	.838986	1.438262
.100	.955303	1.000793	6.0	.859273	1.503727
.200	.919443	1.003028	7.0	.875799	1.556975
.300	.890376	1.006512	8.0	.889187	1.600537
.400	.866641	1.011082	9.0	.900111	1.636566
.600	.831138	1.022940	10.0	.09132	1.666742
.900	.798214	1.045935	20.0	.952381	1.818182
1.000	.90988	1.054651	100.0	.990099	1.960784
1.788	.770164	1133695	1000.0	.999001	1.996008

Note: RATIO^3 is the ratio of the right-hand side terms of equation (3.30) existing SCS-CN method) and those of equation (3.4) (Mockus method); and RATIO^4 is the ratio of those of equations (3.95) (modified method) and (3.4).

3.5.3 NUMERICAL COMPARISON OF METHODS

To compare the SCS-CN method (equation (3.30)) and the modified method (equation (3.95)) with the Mockus method (equation (3.1) or equation (3.4)), the two ratios, designated as RATIO^3 and RATIO^4 are computed. RATIO^3 is the ratio of the right-hand side terms of equation (3.30) to those of equation (3.4), and RATIO^4 is the ratio of the right side terms of equation (3.95) to those of equation (3.4), as shown in Table 3.2. For given BP (= P/S), the existing SCS-CN method underestimates and the modified version overestimates Q, respectively, as compared with the Mockus method. Within ±5% deviation, the results due to the existing SCS-CN method are in agreement with those of the Mockus method for BP < 0.1 and BP > 20 as well as with the results due to the modified method for BP < 1. The SCS-CN method deviates most (about - 1.69 %) from the Mockus method at BP = 1.788 whereas the modified method deviates increasingly from the Mockus method with increasing BP; at BP = 1.0, the +5% deviation from the results of the Mockus method is apparent. These results also hold for P excluding I_a.

Table 3.3 Comparison of existing and modified methods with Mockus method using WS-1 data ($I_a = 0$)

Year	P	Q	C	Mockus Method					Existing SCS-CN		Modified SCS-CN	
				BP	B	b	S	CN	Q	RATIO[3]	Q	RATIO[4]
	mm	mm			mm^{-1}	mm^{-1}	mm		mm		mm	
1964	27.90	14.50	.52	.73	.0263	.0114	38.04	86.97	11.80	.8141	14.97	1.0325
1965	76.50	38.10	.50	.69	.0090	.0039	110.99	69.59	31.21	.8192	39.21	1.0292
1966	25.10	9.10	.36	.45	.0179	.0078	55.74	82.00	7.79	.8564	9.23	1.0137
1967	153.20	106.90	.70	1.20	.0078	.0034	128.03	66.49	83.46	.7807	114.70	1.0729
1968	22.90	5.30	.23	.26	.0115	.0050	86.99	74.49	4.77	.9004	5.33	1.0051
1969	57.20	5.10	.09	.09	.0016	.0007	612.49	29.31	4.89	.9580	5.10	1.0007
1970	61.70	22.90	.37	.46	.0075	.0033	133.01	65.63	19.55	.8538	23.23	1.0145
1971	43.20	30.70	.71	1.24	.0287	.0125	34.84	87.94	23.92	.7790	33.07	1.0772
1972	42.40	15.70	.37	.46	.0109	.0047	91.68	73.48	13.41	.8540	15.93	1.0144
1973	57.20	8.40	.15	.16	.0028	.0012	360.15	41.36	7.84	.9333	8.42	1.0019
1974	13.20	2.00	.15	.16	.0124	.0054	80.34	75.97	1.86	.9314	2.00	1.0021
1975	40.90	8.10	.20	.22	.0054	.0023	185.32	57.82	7.39	.9129	8.13	1.0037
1976	9.10	0.30	.03	.03	.0037	.0016	271.46	48.34	.30	.9839	.30	1.0001
1977	14.20	7.40	.52	.74	.0519	.0225	19.29	92.94	6.02	.8138	7.64	1.0327
1978	37.80	8.40	.22	.25	.0066	.0029	150.41	62.81	7.59	.9038	8.44	1.0047

Table 3.4 Comparison of existing and modified methods with Mockus method using WS-2 data ($I_a = 0$)

Year	P	Q	C	Mockus Method					Existing SCS-CN		Modified SCS-CN	
				BP	B	b	S	CN	Q	RATIO[3]	Q	RATIO[4]
	mm	mm			mm^{-1}	mm^{-1}	mm		mm		mm	
1964	30.00	14.70	.49	.67	.0224	.0097	44.55	85.08	12.07	.8212	15.11	1.0281
1965	74.70	36.80	.49	.68	.0091	.0039	110.09	69.76	30.20	.8206	37.85	1.0284
1966	26.40	10.20	.39	.49	.0185	.0080	54.06	82.45	8.66	.8492	10.36	1.0159
1967	145.00	95.50	.66	1.07	.0074	.0032	134.91	65.31	75.11	.7865	101.37	1.0614
1968	24.60	7.10	.29	.34	.0138	.0060	72.24	77.86	6.25	.8802	7.16	1.0082
1969	56.60	4.30	.08	.08	.0014	.0006	716.34	26.18	4.14	.9639	4.30	1.0005
1970	48.80	17.80	.36	.45	.0093	.0040	107.55	70.25	15.23	.8557	18.05	1.0139
1971	36.80	20.60	.56	.82	.0223	.0097	44.85	84.99	16.59	.8051	21.41	1.0393
1972	37.10	10.70	.29	.34	.0092	.0040	109.04	69.97	9.42	.8802	10.79	1.0082
1973	109.00	16.50	.15	.16	.0015	.0007	664.07	27.67	15.37	.9314	16.53	1.0021
1974	28.40	2.50	.09	.09	.0032	.0014	308.21	45.18	2.40	.9585	2.50	1.0007
1975	42.20	7.60	.18	.20	.0047	.0020	212.52	54.45	6.99	.9199	7.62	1.0030
1976	14.50	.50	.03	.04	.0024	.0011	413.21	38.07	.49	.9831	.50	1.0001
1977	45.20	16.80	.37	.46	.0103	.0045	97.27	72.31	14.34	.8536	17.04	1.0146
1978	28.40	5.10	.18	.20	.0070	.0030	143.48	63.90	4.69	.9201	5.12	1.0030

A comparative analysis of the above methods is also performed using the data of two watersheds WS-1 and WS-2 located in Treynor, Iowa (Hjelmfelt et al., 1982). The $I_a = 0$ assumption is invoked in the analysis. The results of calculation are given in Tables 3.3 and 3.4. Apparently, BP ranges as (0.03-1.24) within which the runoff computed by the modified method deviates from that by the Mockus method by +7.29% whereas the deviation when the SCS method was applied was -22.1%.

3.5.4 MODELS PERFORMANCE ON FIELD DATA

a) *Application 1*

The performance of the modified model (equation (3.95)) and the modified model in general form (equation (3.97)) is compared with the performance of other existing models, viz. the existing SCS-CN method, Mockus method, and the model developed by Fogel and Duckstein (1970) and expressed as:

$$Q - c'(P\ I_a) \tag{3.104}$$

where c' is the model parameter and other terms are the same as described above. To that end, the annual extreme rainfall-runoff events of watersheds WS-1 and WS-2, given in Tables 3.3 and 3.4, respectively, and a number of rainfall-runoff events, including the longest duration events of several days, of the Ramganga and Hemavati catchments of India are considered. A brief description of the latter catchments follows.

The Upper Ramganga catchment (Figure 3.3) lies in the foothills of Himalayas in the Uttaranchal State of India. River Ramganga, originating at Diwali Khel, is a major tributary of River Ganga. It emerges out of the hills at Kalagarh (District Almora) where a major multi-purpose Ramganga dam is situated. The river traverses approximately 158 km before it joins the reservoir and then continues its journey in the downstream plains for 370 km before joining River Ganga at Farrukhabad (Uttar Pradesh). During its travel up to Ramganga dam, the river is joined by several main tributaries: Ganges, Binoo, Khatraun, Nair, Badangad, Mandal, Helgad, and Sona Nadi. Its catchment (area = 3134 sq. km) lies between elevation 262 and 2926 m above the mean sea level and it is considerably below the perpetual snow line of the Himalayas. About 50% of the drainage basin is covered with forest, 30% is under cultivation on terraced fields, and the remaining 20% is urban/barren land. The Ramganga valley experiences approximately an annual rainfall of 1550 mm. The raingauge network consists of Ranikhet, Chaukhatia, Naula, Marchulla, Lansdowne and Kalagarh besides the other existing stations. Streamflow record of the Ramganga River, including river stages, instantaneous as well as monthly, are available at Kalagarh since 1958.

The Hemavati catchment (area = 600 sq. km) up to Sakleshpur is a sub-catchment of River Cauvery (Figure 3.4). It lies between $12^0 55'$ and $13^0 11'$ North latitudes and $75^0 29'$ and $75^0 51'$ East longitudes in the southwest part of Karnataka State (India). The upper part of the catchment is hilly with an average elevation of 1240 m above the mean sea level and the lower part forms a plain terrain with an average elevation of 890 m. It traverses a total length of about 55.13 km up to Sakleshpur. Its topography can be broadly divided into three parts: (i) low land (valley lands), (ii) semi-hilly (gently sloping lands), and (iii) hilly (hill ranges of steep to moderate slopes). The average rainfall of the basin for the period 1942-71 is 2972 mm. The daily rainfall-runoff data from June 1, 1974 to December 31, 1979 were available for use.

For convenience, the SCS-CN method with varying λ is designated as Model 1, the existing SCS-CN method with $\lambda = 0.2$ as Model 2, the modified method as Model 3, the modified method in general form as Model 4, the Mockus method as Model 5, and the method by Fogel and Duckstein (1970) as Model 6. The parameters of all these models are estimated using the above rainfall-runoff data by employing the Marquardt constrained least square approach (Appendix B). The model performance on these watersheds is evaluated in terms of the root mean square (RMS) error computed as:

$$RMS = \frac{1}{N} \sum_{j=1}^{N} \sqrt{(Q_{obs} - Q_{comp})_j^2} \tag{3.105}$$

Figure 3.3. Ramganga catchment.

where Q_{obs} is the observed storm runoff (mm), Q_{comp} is the computed runoff (mm), N is the total number of rainfall-runoff events, and 'j' is an integer that varies from 1 to N. The higher the RMS value, the poorer is the performance of the model and the lower

RMS value shows a better performance of the model. The RMS value equal to zero indicates a perfect fit.

The resulting root mean square error values are given in Table 3.5 and the optimized values of parameters of the six models are given in Table 3.6. It is apparent from these tables that the modified method (Model 3) performs better than Model 2 (existing SCS-CN method with $\lambda = 0.2$) in all applications and Model 2 (SCS-CN method with varying λ) in three applications, except for the application on the Ramganga data. Model 5 (the Mockus method) performs poorer than Models 1 through 4 except in one application of Model 2 on watershed WS-2 where the Mockus method performs better. Model 6 performs better than Model 5 (the Mockus method) on two watersheds, Ramganga and Hemavati, and poorer on WS-1 and WS-2. The derived general form of the model performs the best of all and thus supports the validity of the implied generality of the developed model.

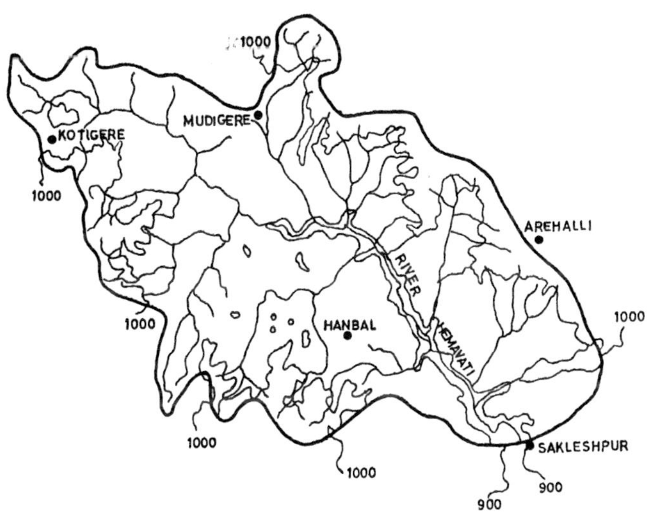

Figure 3.4. River Hemavati up to Sakleshpur.

Table 3.5 Root mean square of the error (mm) between the observed and computed runoff

Catchment	Model 1	Model 2	Model 3	Model 4	Model 5	Model 6
WS-1	31.747	32.188	31.129	29.975	32.214	37.094
WS-2	46.465	48.386	45.300	44.939	47.866	48.610
Ramganga	93.724	100.185	94.380	93.674	125.332	95.747
Hemavati	522.400	521.969	503.602	490.912	597.694	542.885

Note: Model 1:- SCS-CN with varying λ [$Q/(P-I_a) = (P-I_a)/(S + P-I_a)$]
Model 2:- Existing SCS-CN with λ=0.2 [$Q/(P-I_a) = (P-I_a)/(S + P-I_a)$]
Model 3:- Modified model [$Q/(P-I_a)= P/(S + 0.5(P-I_a)$]
Model 4:- Modified model in general form [$Q/(P-I_a)=(P-I_a)/(S+a(P-I_a))$]
Model 5:- Mockus Model [$Q/(P-I_a) = 1 - 10^{-bP}$]
Model 6:- Fogel and Duckstein(1970) model [$Q = c'(P-I_a)$]

Table 3.6 Optimized parameters of various rainfall-runoff models

Catchment	Location	No. of events	Model 1		Model 2	Model 3	
			CN	λ	CN	CN	λ
WS-1	USA	15	77.54	0.0417	82.52	61.85	0.0000
WS-2	USA	15	62.29	0.0000	77.09	55.48	0.0000
Ramganga	India	35	29.56	0.0044	47.49	24.71	0.0000
Hemavati	India	54	45.37	0.2072	44.42	18.02	0.0000

Table 3.6 Continued.

Catchment	Model 4			Model 5		Model 6		
	CN	λ	a	CN	λ	CN	λ	c′
WS-1	66.45	0.0000	0.6081	41.68	0.0000	68.29	0.2098	1.0000
WS-2	50.46	0.0000	0.1187	32.54	0.0000	70.65	0.1712	0.5549
Ramganga	35.30	0.0244	1.3850	42.92	0.3157	56.03	0.2817	0.3496
Hemavati	17.80	0.0000	0.4633	54.62	1.0000	39.15	0.7303	1.0000

b) *Application 2*

A summary of the storm rainfall-runoff data used in this study is given in Table 3.7. The catchments are the sub-watersheds of the Luni basin falling in the arid region of India. These watersheds vary significantly in size, ranging from 83.1 to 3050 sq. km, and character. Using physical characteristics of the watersheds, viz., soil type and land use, and NEH-4 tables, areally-weighted curve numbers were computed for each watershed and these are shown in Table 3.7. For this purpose, remote sensing data of IRS-1A LISS II were used to prepare land use, and soil maps were prepared from irrigation atlas. After converting to digital form using a geographical information system (GIS), these two maps were superimposed to classify different land use and soil types. The resulting map was re-classified using the CN values obtained from NEH-4 tables and a CN map was prepared. From this CN map, the average CN values for each sub-watershed were derived and these are presented in Table 3.7.

Using the above RMS error criterion (equation (3.105)), model parameters were computed using the same Marquardt algorithm. In all applications, the initial

Table 3.7 Computation of CN and RMS error for existing and modified SCS-CN methods

Sl. No.	River	Gauging site	Catchment area (sq. km)	No. of rainfall-runoff events	Parameter Estimation: Existing SCS-CN method		Parameter Estimation: Modified SCS-CN method			Root mean square (RMS) error (mm): Existing SCS-CN method	Root mean square (RMS) error (mm): Modified SCS-CN method
					CN^a	CN^b	λ	CN^a	CN^c		
1	Sukri	Sojat Road	358.6	39	42.23	89.13	0.0000	17.86	80.4	53.678	42.199
2	Sukri	Sojat	626.0	37	40.01	85.66	0.0021	16.33	82.2	24.527	14.339
3	Sukri	Sarangwan	316.9	37	43.25	79.75	0.0000	17.09	81.3	63.714	49.030
4	Sukri	Sheopura	1285.0	36	36.21	81.46	0.0000	12.08	87.1	22.581	12.788
5	Guhia	Jhupelav	349.2	32	47.27	87.58	0.0000	20.56	77.1	67.405	59.523
6	Modiya	Hariamali	177.5	38	43.69	61.25	0.0000	15.58	83.1	67.325	56.862
7	Lilri	Dhaneri	463.3	42	38.96	86.03	0.0000	16.10	82.5	36.151	23.646
8	Phupheria	Dataredan	373.3	35	39.73	80.80	0.0000	15.88	82.8	26.713	11.420
9	Radia	Kohat	251.0	37	37.92	84.79	0.0000	13.42	85.6	29.100	14.388
10	Guria	Sandia	264.0	40	43.87	78.31	0.0000	19.37	78.6	34.796	26.692
11	Guhia Bala	Artia	209.0	36	39.92	78.81	0.0000	14.12	84.8	23.177	12.452
12	Guhia	Karniyali	3050.0	37	33.96	80.78	0.0000	7.58	92.1	34.883	24.568
13	Guhia	Singari	2837.6	37	41.04	80.59	0.0007	16.69	81.8	19.455	8.417
14	Guhia	Sabalpura	83.1	41	42.67	78.31	0.0000	13.87	85.1	44.006	34.725

Notations: Superscripts 'a' and 'c' stand for CN computed from rainfall-runoff data using equations (2.7) and (3.166), respectively, and superscript 'b' stand for CN computed from NEH-4 tables.

estimate of parameter CN of both the models was taken as 70 and the additional parameter of the modified SCS-CN model, λ, was taken as 0.2 (a standard value for the existing version of the SCS-CN method). CN ranged between 0 and 100 and λ was assumed to vary in the range (0, 1). The computed parameter-values are shown in Table 3.7. From the table it is apparent that CN of the existing SCS-CN method varies from 36.21 to 47.27 in all applications. It is noted that the computed curve numbers that largely depend on the antecedent moisture condition correspond to the values fitting the above events. Since the basic hypothesis of the modified SCS-CN model differs from the existing SCS-CN method for which $\lambda = 0.2$, the computed CN values of both the models in an application will be different from each other. The modified model yields CN values ranging from 7.58 to 19.37 in these applications. Except for two watersheds, watersheds at Sl. Nos. 2 and 13 in Table 3.7, for which λ is equal to 0.0021 and 0.0007, respectively, the value of λ of the modified model is equal to zero. The fitting of data of two sample watersheds is depicted in Figures 3.5 and 3.6. It is apparent that the existing SCS-CN method attempts to fit only high rainfall-runoff data whereas the modified method fits the whole data reasonably well.

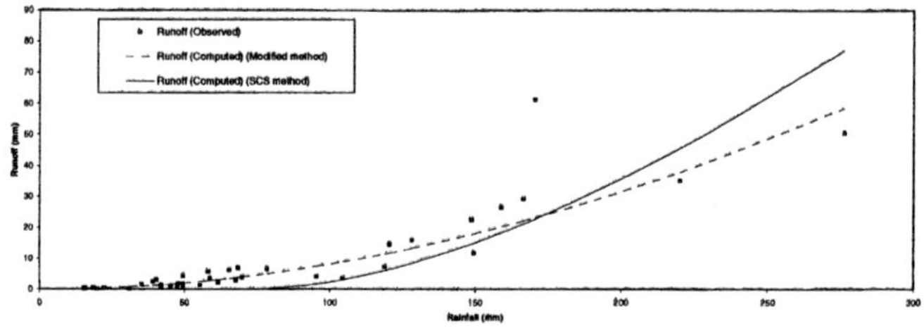

Figure 3.5. Fitting of the existing and modified SCS-CN methods to data of watershed 1 (Table 3.7).

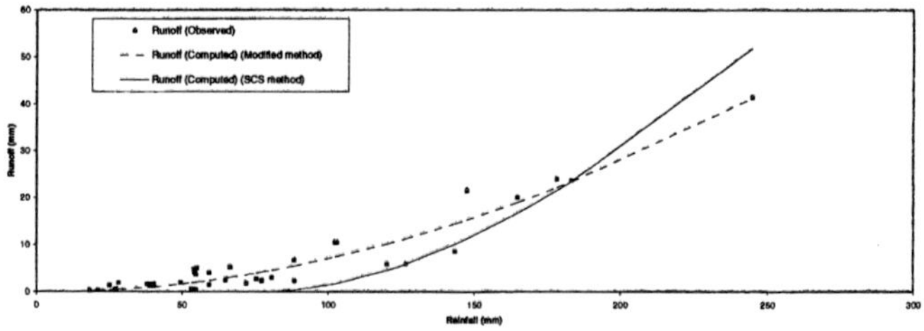

Figure 3.6. Fitting of the existing and modified SCS-CN methods to data of watershed 2 (Table 3.7).

The CN-values derived from rainfall-runoff data using equation (2.7) and NEH-4 tables are shown in Table 3.7 by superscripts 'a' and 'b', respectively. These data are plotted in Figures 3.7 and 3.8. It is seen that except for one data point, all others exhibit the existence of a correlation between the two CN values derived from different sources. It is worth noting that the far-deviating data point belonged to the watershed dominated by forest, on which the existing SCS-CN method performs poorly (Ponce and Hawkins, 1996; Mishra and Singh, 1999a), as also shown in the forthcoming section. Figures 3.7 and 3.8 exhibit a possibility for the adjustment of CN-values derived from NEH-4 tables utilizing only a limited number of watershed characteristics. On the other hand, CN values derived from rainfall-runoff data exhibit the integrated response of all runoff producing characteristics of the watershed.

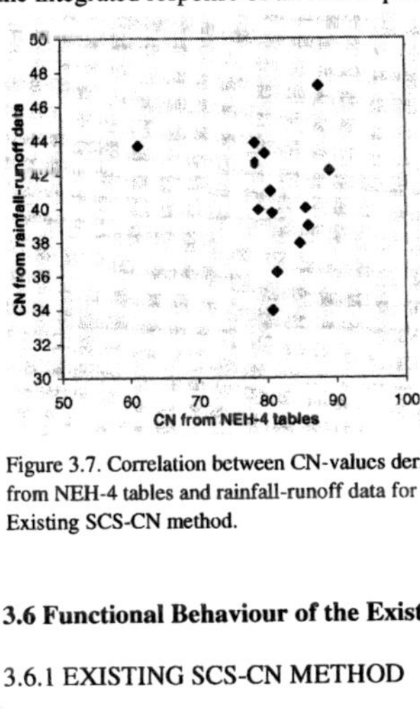

Figure 3.7. Correlation between CN-values derived from NEH-4 tables and rainfall-runoff data for Existing SCS-CN method.

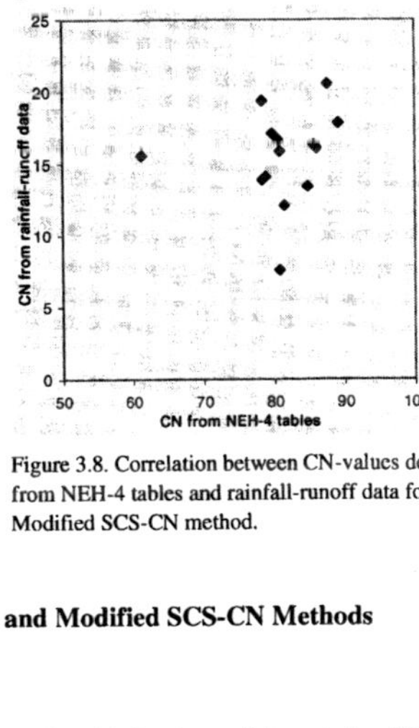

Figure 3.8. Correlation between CN-values derived from NEH-4 tables and rainfall-runoff data for Modified SCS-CN method.

3.6 Functional Behaviour of the Existing and Modified SCS-CN Methods

3.6.1 EXISTING SCS-CN METHOD

As part of the description of the functional behaviour of the existing SCS-CN method, a brief insight into the method can be obained by considering two extreme cases. For an immediate ponding upon rainfall, an extreme case, $I_a = 0$. For this condition, equation (2.5) reduces to

$$\frac{Q}{P} = [1 + S/P]^{-1} \tag{3.106}$$

If S and P are equal, then Q = 0.5P, i.e., the volume of surface runoff is half of the rainfall amount. For a specific case, $I_a = S$, equation (2.5) reduces to

$$Q = (P - S)^2 / P = P(1 - S / P)^2 \tag{3.107}$$

If S = P, Q = 0, i.e., the entire rainfall is absorbed by the soil and consequently, no surface runoff volume is produced.

For a complete description of the functional behavior of the method, equation (2.5) is coupled with equation (2.3) to yield

$$\lambda^2 (S / P)^2 - [2\lambda + (1 - \lambda)C](S / P) + 1 - C = 0 \tag{3.108}$$

which solves for S/P as

$$\frac{S}{P} = \frac{[2\lambda + C(1 - \lambda)] \pm \sqrt{C[C(1 - \lambda)^2 + 4\lambda]}}{2\lambda^2} \tag{3.109}$$

where C = Q/P = runoff factor. Since the term in the square root of the numerator of equation (3.109) is always greater than or equal to zero and is always less than the first bracketed term in the numerator, S/P will always be a non-negative quantity. For S/P = 0, $4\lambda^2(1\text{-}C) = 0$, therefore, either $\lambda = 0$ or C = 1. At $\lambda = 0$, S/P becomes indeterminate (0/0), interpreted as ∞ (Chen, 1982) or P tending to infinity; for C = 1, S/P = 0 (either S = 0 or $\lambda = \infty$ or $P = \infty$) if only '-' sign, as normally used, is retained in place of $\pm$ sign in the equation. Since P is a definite quantity, $\lambda = 0$ leads to describing $S = \infty$, and C = 1 to S = 0. These deliberations can be further clarified with the aid of equations (2.1)-(2.3), as follows.

Under the state of complete saturation the basic SCS-CN hypothesis (equation (2.2)) fails to describe Q, for both F and S are equal to zero; rather Q is described by the water balance equation (equation (2.1)). Since S = 0 and therefore F = 0, $I_a = 0$ for $\lambda = 0.2$ and Q = P from equation (2.1). However, for the condition $I_a \neq 0$, $\lambda = \infty$ from equation (2.3) and from equation (2.1) Q = P-I_a. Thus, Q or runoff factor C is significantly governed by I_a, which is consistent with the notion that I_a governs S (McCuen, 1982) that, in turn, governs Q (equation (2.5) with $I_a = 0.2S$). On the other hand, no runoff (or Q = 0) condition can possibly occur if P = 0, $I_a \geq P$, or $S = \infty$. If P = 0, then each term of equation (2.1) should be equal to zero. In such a situation, the left-hand side of equation (2.2) equals 0/0 and the right-hand side is equal to zero for S > 0; it is indeterminate (= 0/0) if S = 0. Under the condition $I_a \geq P$, the entire rainfall is initially abstracted. Therefore, both F and Q are taken as equal to zero to restrict the violation of water balance (equation (2.1)). These conditions ($I_a \geq P$ and F = Q = 0), however, do not limit the validity of equation (2.2), for both its left- and right-hand sides are equal to zero for S > 0. Finally, if $S = \infty$ (for example, an infinitely deep sandy soil), then, according to the existing SCS-CN method, F = Q = 0 because $I_a = \infty$ for $\lambda = 0.2$. However, if $I_a \neq 0$, then for $S = \infty$, $\lambda = 0$ (equation (2.3)) and Q = 0 from equation (2.2) and consequently, F = P-I_a from equation (2.1). Thus, the condition $\lambda = 0$ does not necessarily describe $I_a = 0$. However, if $I_a = 0$, then $\lambda = 0$, as shown in what follows.

The above explanation confirms that the functional behavior of the SCS-CN method greatly depends on both λ and I_a. It, however, does not explicitly describe the range of λ variation and a more appealing reason for retaining only '-' sign in place of mathematically valid $\pm$ sign. These can be described using I_a as a descriptor of S as follows.

Multiplication of equation (3.109) by λ yields

$$I_a^* = \frac{[2\lambda + C(1-\lambda)] - \sqrt{C[C(1-\lambda)^2 + 4\lambda]}}{2\lambda} \tag{3.110}$$

where $I_a^* = I_a/P$. Equation (3.110) can be utilized for describing the behavior of the existing SCS-CN method in C-I_a^*-λ space, for the reason that the variation of I_a^* has a range of $0 \le I_a^* \le 1$ and if $I_a^* > 1$, C = 0. For the condition $0 \le I_a^* \le 1$, equation (3.110) yields $\lambda \ge 0$ and $0 \le C \le 1$ and is valid for only '-' sign before the square root in the numerator of equation (3.110) instead of '$\pm$' sign. The solution of equation (3.110) for λ yields

$$\lambda = \frac{C I_a^*}{(1 - I_a^*)(1 - I_a^* - C)} \tag{3.111}$$

and $\lambda = 0$. Equation (3.111) can also be directly derived from equation (2.5). In equation (3.111), if $I_a^* \to 1$ or $(I_a^* + C) \to 1$, $\lambda \to \infty$. Equation (3.111) yields prohibitive negative values of λ [$= -1/(1-I_a^*)$] for C approaching 1. Thus, for λ to be a non-negative value, the following should hold:

$$I_a^* + C \le 1 \tag{3.112}$$

Thus, when $I_a^* \to 1$, $C \to 0$, and thus, both the numerator and denominator of equation (3.111) approach 0, leading to λ approaching 0/0, which can be taken as equal to ∞ (Chen, 1982). Equation (3.111) can, alternatively, be written as:

$$Q + I_a \le P \tag{3.113}$$

Coupling of equation (3.113) with equation (2.1) yields $F \ge 0$ and combination of equation (2.3) with equation (3.113) leads to

$$S \le \frac{P - Q}{\lambda} \tag{3.114}$$

It implies that $S \leq 5(P-Q)$ for $\lambda = 0.2$. For practical application of the existing SCS-CN method, equation (3.111) is further explained in the C- I_a^* -λ space graphically (Figure 3.9), in what follows.

Depicting the variation of λ with C and I_a^*, Figure 3.9 shows that λ varies in the range $(0, \infty)$; for a given value of I_a^*, λ increases with C and reaches ∞ as $(C+I_a^*)$ approaches 1; for a given value of C, λ increases with I_a^*; as $I_a^* \rightarrow 0$, $\lambda \rightarrow 0$, as described above, and from equation (2.3) $S \rightarrow 0/0$ (taken as equal to ∞). Furthermore, as $\lambda \rightarrow 0$, all curves including $I_a^* > 0$ converge. It implies that λ can assume a value even when I_a is not necessarily equal to zero, as described earlier. It is also seen from this figure that as $C \rightarrow 0$, $\lambda \rightarrow 0$ and consequently, $S \rightarrow \infty$. Thus, the employment of equation (2.3) with $\lambda = 0.2$ (existing SCS-CN method) will yield I_a much larger than zero. For this reason, the existing SCS-CN method performs poorly on very low runoff producing (or low C-values) lands, such as sandy soils and forest lands. Similarly, when λ increases sharply and approaches infinity for a given I_a^* (or $C = 1-I_a^*$) (Figure 3.9), the SCS-CN method may perform poorly.

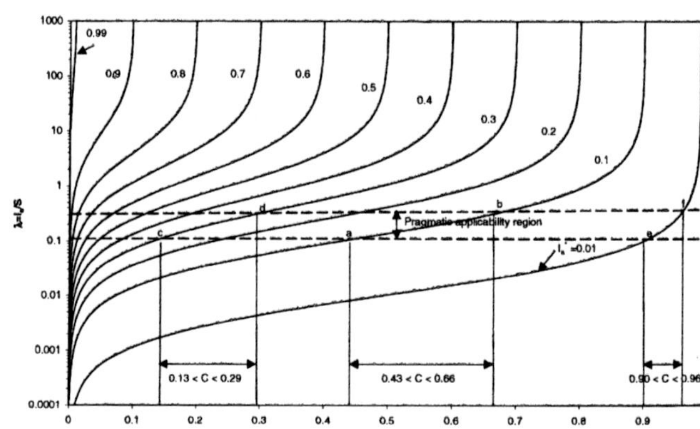

Figure 3.9. Variation of initial abstraction coefficient, λ, with runoff factor, C, and nondimensional initial abstraction, I_a^*.

Since λ is taken to vary in the range (0.1, 0.3) in field applications (SCD, 1972), a region of practical applicability is shown by thick dashed-lines in Figure 3.9. Apparently, as C increases and I_a^* decreases, the applicability region in terms of the runoff-generation potential widens or the range of C-values increases to a certain extent and then, decreases as $C \rightarrow 1$ and $I_a^* \rightarrow 0$. As an example, for $I_a^* = 0.1$, the region lying between 'a' and 'b' in the figure describes applicability of the SCS-CN method to watersheds showing C-values in the range (0.43, 0.66). On the other hand, for $I_a^* = 0.3$, the applicability region between 'c' and 'd' narrows to C-values ranging from 0.13-0.29. Similarly, for $I_a^* = 0.01$, the region between 'e' and 'f' shrinks to C-values ranging from 0.90-0.96. Mathematically, the maximum C-range (0.39, 0.62) can be described for $I_a^* = 0.113$. It implies that the existing SCS-CN method has widest applicability on those watersheds exhibiting C-values in the approximate range of (0.4, 0.6) and the initial abstraction amount of the order of 10% of the total rainfall.

Thus, the applicability of the SCS-CN method largely depends on the C-I_a^* values. Based on these values, it can be tacitly inferred that the extreme left end of Figure 3.9 refers to those watersheds exhibiting very high I_a^*-values, for example, forest watersheds as described above, for which the C-values are generally low and the I_a^*-values are high. Similarly, the extreme right end refers to those watersheds showing very low I_a^*-values and very high C-values, which generally holds for urban watersheds. Thus, it is easy to ascribe the middle portion of the figure to agricultural watersheds, for which the SCS-CN method was originally devised, and it works best on these watersheds (Ponce and Hawkins, 1996). Thus, given a large set of P-Q-I_a data, it is possible to further sub-divide the applicability region for various SVL complexes.

Here, it is worth emphasizing that the SCS-CN method with $\lambda = 0.2$ is likely to perform well on urban watersheds (SCS, 1986) because of their very low S-values, yielding realistically insignificant initial abstraction values and consequently, exhibiting insignificant bearing on the computed Q-values. It follows that, for applicability of the existing SCS-CN method, C is not limited to 0.6, as above. Thus, $C \geq 0.4$ and from equation (3.114) $S \leq (10/3)(P-Q)$ for $\lambda = 0.3$ form the bounds for applicability of the method. Coupling of these conditions yields a broad criterion expressed as: $S \leq 2P$, which is close to the criterion: $S < 2.17\ P$ (Hjelmfelt, 1982) derived statistically for 90% probability, and it is recommended for field use by Hawkins et al. (1985). It implies that the existing SCS-CN method is applicable if S is less than or equal to twice the total rainfall amount. In terms of CN (equation (2.7)), the criterion is

$$CN \geq \frac{1000}{10+2P} \tag{3.115}$$

As a text example, if P = 1 inch, $CN \geq 83$, for which $Q \geq 0.23$ inch. Similarly, if P = 10 inches, $CN \geq 33$ and, in turn, $Q \geq 2.29$ inches. Thus, the applicability bounds in Figure 3.9 are: $\lambda \leq 0.30$, $I_a \leq 0.35P$, and $C \geq 0.23$. It is noted that the condition $S \leq 2P$ allows the SCS-CN method to work even when $S < P$, where the Mockus method fails because of equation (3.101) yielding the unrealistic values of $\partial Q/\partial P$ greater than one in this region (Figure 3.2). Since equation (3.115) holds for AMC I in contrast with the recommendation of Hawkins et al. (1985) for AMC II, this criterion is further revised for AMC II in Chapter 4, and this revision leads to a modification of the bounds.

3.6.2 MODIFIED SCS-CN METHOD

The modified version of the existing SCS-CN method is expressed as

$$Q = \frac{(P - I_a)^2}{0.5(P - I_a) + S} \tag{3.116}$$

which is applicable for $S \geq 0.5(P-I_a)$, for Q to be less than or equal to $(P-I_a)$. Equation (2.7) is assumed to hold for S-CN conversion. For a brief insight into the method, two specific cases can be considered. First, for $I_a = 0$, equation (3.116) reduces to

$$Q/P = (0.5 + S/P)^{-1} \tag{3.117}$$

If S = P, Q = (2/3) P, implying that the volume of the surface runoff is equal to the two-third of the rainfall amount. For another case, if $I_a = S$, then

$$Q = \frac{2(P - S)^2}{P + S} \tag{3.118}$$

If S = P, Q = 0, i.e. the entire rainfall is absorbed by the soil and consequently, no surface runoff volume is produced. Furthermore, if $I_a = P$, Q = 0 from equation (3.116).

For a complete description of the behavior of the modified SCS-CN method, equations (3.116) and (3.118) are combined to lead to equation (3.96). Keeping $I_a^* = I_a/P$, equation (3.96) can be re-written as

$$I_a^* = \frac{[4\lambda + C(2 - \lambda)] \pm \sqrt{C[C(2 - \lambda)^2 + 16\lambda]}}{4\lambda} \tag{3.119}$$

Similar to the above analysis for the behavior of the existing SCS-CN method, the behavior of equation (3.119) can be described for the following four conditions of I_a^*.

(a) *Condition 1:* $I_a^* = 0$

For the condition $I_a^* = 0$, the result of the sum of two terms before and after the ± sign in the numerator of equation (3.119) should be equal to zero. Since these terms are always greater than or equal to zero for any real value of C and λ, the numerator of equation (3.119) can be equal to zero only for '-' sign. Therefore, equation (3.119) can be written as

$$I_a^* = \frac{[4\lambda + C(2-\lambda)] - \sqrt{C[C(2-\lambda)^2 + 16\lambda]}}{4\lambda} \tag{3.120}$$

The condition $I_a^* = 0$ also leads to

$$8\lambda^2[2 - C] = 0 \tag{3.121}$$

Since $C \le 1$ by experience, $\lambda = 0$.

(b) *Condition 2:* $I_a^* > 0$

The employment of the condition $I_a^* > 0$ leads to

$$8\lambda^2[2 - C] > 0 \tag{3.122}$$

Again, since $C \le 1$, $\lambda > 0$.

(c) *Condition 3:* $I_a^* \le 1$

The application of the condition $I_a^* \le 1$ leads to the following condition:

$$16\lambda C \ge 0 \tag{3.123}$$

which implies that both λ and C are greater than or equal to zero.

(d) *Condition 4:* $I_a^* > 1$

Similar to the above, the condition $I_a^* > 1$ leads to

$$16\lambda C \le 0 \tag{3.124}$$

Since C is a non-negative quantity and is equal to zero when $I_a^* > 1$. Therefore, $\lambda \le 0/0$, which is indeterminate, but interpreted as infinity as described above. Thus, for the condition $I_a^* > 1$, $\lambda \le \infty$.

To summarize, the condition $0 \le (I_a/P) \le 1$ yields that $\lambda \ge 0$ because $C \ge 0$ and the condition $I_a^* > 1$ yields $\lambda \le \infty$. Thus, λ can assume any value ranging from zero to infinity. This condition also warrants the use of '-' sign before the square root in the numerator of equation (3.119). For further elaboration, equation (3.120) can be solved for λ as

$$\lambda = \frac{2CI_a^*}{(1 - I_a^*)(2 - C)} \tag{3.125}$$

and $\lambda = 0$. In equation (3.125), if $I_a^* \to 1$, $\lambda \to \infty$ as $C \le 1$. For $C = 1$, $\lambda = 2I_a^*/(1-I_a^*)$ and if $C = 0$, $\lambda = 0$. Thus, λ can vary between zero and infinity. The variation of λ with C and I_a^* is depicted in Figure 3.10.

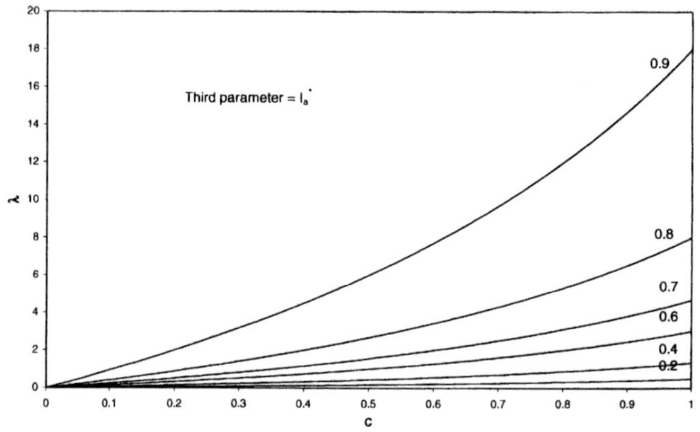

Figure 3.10. Behaviour of the modified SCS-CN method in C-I_a^*-λ spectrum.

3.7 Significance of the Proportional Equality

It is apparent from the above derivation of the SCS-CN method from the Mockus method, the Zoch method, or the first-order storage hypothesis that the proportional equality (equation (2.2)) is the result of the generalization of the first-order infiltration process coupled with the water balance equation with the assumption that the cumulative rainfall grows linearly with time. Thus, this proportional equality is an improvement over the exponentially decaying infiltration process, for the severe limitation of the Mockus method exhibiting runoff (rainfall-excess) rate to exceed the rainfall intensity, which is impossible. It is further noted that the Mockus method assumes the exponential decay of the infiltration rate with time. Since the Horton model is an empirical method, the SCS-CN method can also be taken as an empirical method. Besides, derivation of the SCS-CN method from the second-order storage hypothesis leads to its categorization as a conceptual method. However, the applicability of the SCS-CN method to most hydrological conditions serves as a motivation to explore the physical basis of the method. The following discussion explores this proportional equality using the concept of soil porosity.

3.7.1 SOIL POROSITY

The space occupied by a soil column can be divided into three main parts: volume of solids, water, and air, as shown in Figure 1.18. In this figure, V is the total volume, V_w is the volume of water, and V_a is the volume of air for a unit surface area. The sum of the volume of air and the volume of water represents the volume of voids V_v. Expressed mathematically,

$$V_v = V_a + V_w \tag{3.126}$$

In volumetric terms, the water or moisture content (θ) is defined as

$$\theta = \frac{V_w}{V} \tag{3.127}$$

and porosity (n) is defined as

$$n = \frac{V_v}{V} = \frac{e}{1+e} \tag{2.11}$$

where e is the void ratio defined as

$$e = \frac{V_v}{V_s} \tag{2.12}$$

where V_s is the volume of solids (Figure 1.18). The moisture content, θ, can also be expressed as

$$\theta = n\, S_r \tag{3.128}$$

where S_r is the degree of saturation, varying from 0 to 1, or alternatively,

$$S_r = \frac{\theta}{n} = \frac{V_w}{V_v} \tag{3.129}$$

3.7.2 PROPORTIONAL EQUALITY

In the basic SCS-CN hypothesis (equation (2.2) for I_a = 0), F represents the infiltrated amount of water (= V_w) and S is equal to the maximum possible amount of infiltration equal to the maximum (P-Q) difference which, in turn, is equal to the volume of voids V_v. Therefore, F/S represents the degree of saturation of the soil (equation (3.129)). Expressed mathematically,

$$\frac{F}{S} = S_r \tag{3.130}$$

The left-hand side of equation (3.130) represents the runoff factor C which has a range of (0, 1) as:

$$C = \frac{Q}{P} \tag{3.131}$$

From the source area concept (Hawkins, 1982) for daily P-Q data, C is equal to the ratio of the wet and total areas of the catchment designated by A_w and A_T, respectively. Expressed mathematically,

$$C = \frac{A_W}{A_T} \tag{3.132}$$

It is, however, emphasized that the SCS-CN methodology does not support the source area concept, as described in Chapter 2. Thus, substitution of equations (3.130) and (3.131) into equation (2.2) for $I_a = 0$ leads to

$$C = S_r \tag{3.133}$$

which is the equation that integrates surface flow and subsurface flow linearly on 1:1 scale, implying that the SCS-CN method is not only a runoff method but also an infiltration method (Aron et al., 1977; Chen, 1982; Gray et al., 1982; Ponce and Hawkins, 1996; Mishra, 1998; Mishra and Singh, 2002). Thus, the SCS-CN method can also be categorised as an infiltration model. Using this concept of proportional equality, it is possible to signify the curve number, CN.

3.7.3 SIGNIFICANCE OF CN

Equation (2.7) of the SCS-CN method defines CN and its link with S that represents the maximum possible amount of infiltration. CN can be signified from equation (3.61), which is re-written for convenience:

$$\frac{F}{S} = \frac{P}{P+S} \tag{3.61}$$

Equation (3.61) describes the variation of the degree of saturation (equation (3.130)) with rainfall P. This equation leads to a derivation of the S-CN mapping relation (equation (2.7)) as follows.

The ratio F/S varies from 0 to 1. To map it on a scale of 0-100, it is necessary to multiply equation (3.61) by 100 leading to

$$100\frac{F}{S} = \frac{100P}{P+S} \tag{3.134}$$

Defining the left-hand side of equation (3.134) as CN leads to

$$CN = \frac{100P}{P+S} = \frac{100}{1+S/P} \tag{3.135}$$

which describes the variation of CN with P for a given S. Assuming P = 10 inches leads to equation (2.7) which can be recast as:

$$CN = \frac{1000}{S+10} \tag{3.136}$$

Equation (3.136) represents equation (2.7), the S-CN mapping relation. Thus,

$$CN = 100\,\frac{F}{S} \tag{3.137}$$

Equation (3.137) defines CN as the percent degree of saturation of the watershed due to a 10-inch rainfall amount. It is worth noting that the direct use of CN in the proportionality hypothesis (equation (2.2)) for computing Q is restricted, because CN, by definition, corresponds to the 10-inch base-rainfall, not to the actual. Therefore, equation (2.5) with $I_a = 0$ should be resorted to computation of Q for a given rainfall amount. It can also be asserted as follows.

In terms of CN, the runoff factor C (= Q/P) can be defined from equation (2.5) (for $I_a = 0$) as

$$C = \frac{1}{1+\frac{10}{P}\left(\frac{100}{CN}-1\right)} \tag{3.138}$$

To describe C physically it is necessary to explain the bracketed portion in the denominator of equation (3.138) in terms of the volumetric elements of the soil (Figure 1.18) as

$$\frac{100}{CN} - 1 = \frac{100}{100V_W / V_v} - 1 = \frac{V_v - V_W}{V_W} = \frac{V_a}{V_w} \tag{3.139}$$

where V_v is the void space, V_w is the available moisture due to 10-inch rainfall, and V_a is the space available for water retention after 10-inch rainfall. An actual rainfall P greater than 10 inch would result in higher V_w and, consequently, lesser V_a or, in turn, a lesser V_a/V_w ratio and vice versa. Therefore, the bracketed term in the denominator of equation (3.139) needs to be revised or updated (increased or reduced) in proportion to 10/P to describe the actual V_a/V_w ratio that corresponds to P. The actual V_a/V_w ratio computed in the denominator of equation (3.139) and the inverse of the resulting sum of the denominator yields the actual degree of saturation that corresponds to P, which equals C to form the proportional equality equivalent to $C = S_r$. Such a description leads to defining the runoff factor C (equation (3.138)) as the degree of saturation, S_r, of the watershed for the actual rainfall P. It supports the validity of the above described $C = S_r$ concept. The variation of C with CN for varying P is shown in Figure 3.11. The implication of such an assertion is that for CN to represent a watershed characteristic, S/P should form as a basic parameter of the SCS-CN model while deriving CN from rainfall-runoff data, rather than S, as shown later. For practical applications, another interpretation of the S-CN mapping relation (equation (2.7)) available in the literature (Mishra and Singh, 1999b) is also provided.

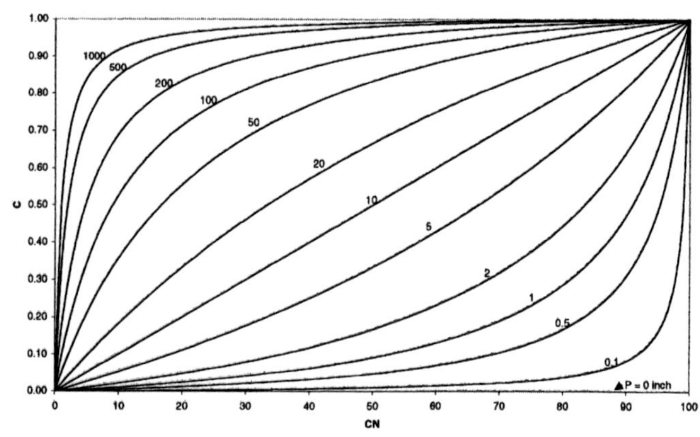

Figure 3.11. Variation of runoff factor C with curve number CN and precipitation P (inch).

3.7.4 ANOTHER INTERPRETATION OF S-CN MAPPING RELATION

The relation between S and CN is an empirical one, supposedly based on field experience and scaling. It is not, however, entirely clear as to the extent to which S could assume a value that is practically meaningful. Equation (2.7) can be recast as

$$\frac{S}{10} = \frac{100}{CN} - 1 \tag{3.140}$$

From equation (2.7) or equation (3.140) the value of 10 appearing in the denominator of its left-hand side can be interpreted as follows.

Using daily experimental data, the absolute maximum retention of 10 inches, within the confines of available experimental data, was examined. It is of common knowledge that the minimum possible S could be 0. Therefore, S was varied to a larger scale through CN taking non-linear scaling (equation (3.140)). An examination of equation (3.140) reveals that the stated range of S (0, 10) inches is completely exhausted in the CN range of (50, 100). Therefore, the CN values that are less than 50 should fall in the extrapolated range of experimental data. If such an assertion holds, the application of CN < 50 is unwarranted. While assessing CN for its practical application, Ponce (1989) recommended CN values that also fall in the aforementioned range. Incorporating the absolute maximum retention (S_{abs}), a new parameter, in equation (3.140), the general form of the equation can be written as

$$\frac{S}{S_{abs}} = \frac{100}{CN} - 1 \tag{3.141}$$

Here, S varies between 0 and S_{abs} inches (or any other unit). The importance of such a manipulation can be found in Williams and LaSuer (1976), Rallison and Miller (1982) and Ponce (1989). As an example, taking S_{abs} = 10 inch (= 254 mm) and using equation (3.141), a relationship between CN and b (equation (3.1)) can be derived as:

$$b = \frac{1.7098 \text{x} 10^{-3} CN}{100 - CN} \tag{3.142}$$

As CN → 0 or 100, b → 0 or ∞; for CN = 55, b = $1.7098 \text{x} 10^{-3}$, and for CN = 95, b = 0.0325.

3.8 Antecedent Moisture Conditions

NEH-4 identified three antecedent moisture conditions (AMC): AMC I, AMC II, and AMC III for dry, normal, and wet conditions of the watershed, respectively. As shown in Figure 3.12, AMC I corresponds to the lower enveloping CN, AMC II the median CN, and AMC III the upper enveloping CN. NEH-4 provides a conversion table (Table 2.7) from CN for AMC II to corresponding CNs for AMC I and AMC III. The original values (SCS, 1956) of this table were smoothened in the later versions of NEH-4 (for example, SCS, 1985). Based on the refined values, Sobhani (1975) and Hawkins et al. (1985) linked CNs of different AMCs in terms of the potential maximum retention as below:

$$S_I = 2.281\ S_{II}\ ;\ r^2 = 0.999 \text{ and } SE = 0.206 \text{ inch} \tag{3.143}$$

and

$$S_{III} = 0.427\ S_{II} \quad ; r^2 = 0.994 \text{ and SE} = 0.088 \text{ inch} \tag{3.144}$$

where SE is the standard error and subscripts I through III correspond to AMC I through III, respectively. Equations (3.143) and (3.144) are applicable in the CN range (55, 95). Substitution of equations (3.143) and (3.144) into equation (2.7) leads to

$$CN_I = \frac{CN_{II}}{2.281 - 0.01281CN_{II}} \; ; r^2 = 0.996 \text{ and SE} = 1.0 \text{ CN} \tag{3.145}$$

and

$$CN_{III} = \frac{CN_{II}}{0.427 + 0.00573CN_{II}} \; ; r^2 = 0.994 \text{ and SE} = 0.7 \text{ CN} \tag{3.146}$$

which are the same as described by equations (2.34) and (2.35), respectively, in Chapter 2.

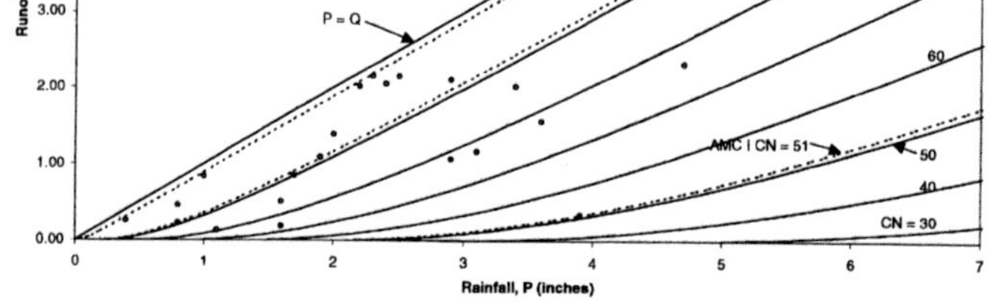

Figure 3.12. Determination of CN for AMC I through AMC III using existing SCS-CN method.

For practical applications, NEH-4 provides the derivation of CN based on the amount of the antecedent 5-d rainfall, which forms an index of the initial soil moisture (Chapter 2). The term initial stands for the state before the start of the storm. Equations (3.145) and (3.143) can be derived from equation (2.7), written in general form, as:

$$S = \frac{100X}{CN} - X \tag{3.147a}$$

or

$$CN = \frac{100X}{S + X} \tag{3.147b}$$

In equation (2.7), X = 10 inches of rainfall, which corresponds to the normal condition. Equations (3.147a) and (3.147b) are valid for all AMCs according to NEH-4. For example, if $S = S_{II}$ for normal AMC, then $CN = CN_{II}$ in these equations. Alternatively, CN for AMC I and AMC III can be derived from equation (3.147b), respectively, as

$$CN_I = \frac{100(X - P_I)}{S_{II} + (X - P_I)} \tag{3.148}$$

and

$$CN_{III} = \frac{100(X + P_{III})}{S_{II} + (X + P_{III})} \tag{3.149}$$

These equations imply that the normal rainfall amount X is reduced by P_I if CN_{II} converts to CN_I. Similarly, an additional rainfall amount of P_{III} over and above the normal X-value is required to raise CN_{II} to CN_{III}. A substitution for S_{II} in equations (3.148) and (3.149) leads, respectively, to

$$CN_I = \frac{CN_{II}}{\dfrac{X}{X - P_I} - \dfrac{P_I}{100(X - P_I)}} \tag{3.150}$$

and

$$CN_{III} = \frac{CN_{II}}{\dfrac{X}{X + P_{III}} + \dfrac{P_{III}}{100(X + P_{III})}} \tag{3.151}$$

which are the general expressions for CN according to any antecedent moisture condition for a given amount of the normal and the two extreme antecedent rainfalls X, P_I and P_{III}, respectively. A comparison of equations (3.150) and (3.151) with respective equations (3.145) and (3.146) leads, respectively, to

$$X = (2.281/1.281)\, P_I = 1.7806\, P_I \tag{3.152}$$

and

$$X = (0.427/0.573)\ P_{III} = 0.7452\ P_{III} \tag{3.153}$$

For given P_I and P_{III} values in NEH-4 (Table 2.6) equal, respectively, to 0.5 (1.3 cm) and 1.1 inch (2.8 cm) for a dormant season and 1.4 (3.6 cm) and 2.1 inches (5.3 cm) for a growing season, the normal antecedent rainfall, X, can be computed as:

Dormant season. X = 0.89 inch for AMC I and X = 0.82 inch for AMC III.

Growing season. X = 2.49 inches for AMC I and X = 1.56 inches for AMC III.

The inference drawn from the above calculations is that the normal amount of rainfall, X, varies with both AMC and season. The variation in X can be interpreted in terms of the initial abstraction amount, I_a, which can be neglected in AMC III. Thus, I_a = 0.89 – 0.82 = 0.07 inch for the dormant season and I_a = 2.49 – 1.56 = 0.93 inch for the growing season. Similarly, variation of X with the season can be attributed to the evapotranspiration that can be neglected in the dormant season. Thus, the evapotranspiration amount can be computed as equal to 0.74 (= 1.56 – 0.82) inch for the growing season under AMC III, and equal to 1.60 (= 2.49 – 0.89) inches under AMC I.

3.8.1 VARIATION OF CN WITH AMC

Following the above C = S_r concept, equation (2.2) (for I_a = 0) can be modified for the antecedent moisture (M) as

$$\frac{Q}{P - I_a} = \frac{F + M}{S + M} = \frac{F + M}{S_o} \tag{3.154}$$

where S_o represents the volume of air equal to the volume of voids (for complete antecedent dry condition) and M is computed as

$$M = \frac{P_5 S_o}{P_5 + S_o} \tag{3.155}$$

where P_5 is the antecedent 5-d rainfall amount and M is derived assuming the completely dry antecedent condition. Here, it is noted that the introduction of I_a leads to replacing P by P_e = P - I_a in Figure 3.11 for describing the C-CN relationship so that it holds for P_e. Thus, CN can be described from equations (3.154) and (3.155) for P_e = 10 inch as

$$CN = \frac{100}{S_o}\left[\frac{10S}{10 + S} + \frac{P_5\,S_o}{P_5 + S_o}\right] = \frac{1000}{10 + S} \tag{3.156}$$

Equation (3.156) leads to CN for AMC I, AMC II, and AMC III, respectively, as

$$CN_I = \frac{100}{S_o}\left[\frac{10S_I}{10+S_I} + \frac{P_{5(I)}S_o}{P_{5(I)}+S_o}\right] = \frac{1000}{10+S_I} \tag{3.157a}$$

$$CN_{II} = \frac{100}{S_o}\left[\frac{10S_{II}}{10+S_{II}} + \frac{P_{5(II)}S_o}{P_{5(II)}+S_o}\right] = \frac{1000}{10+S_{II}} \tag{3.157b}$$

$$CN_{III} = \frac{100}{S_o}\left[\frac{10S_{III}}{10+S_{III}} + \frac{P_{5(III)}S_o}{P_{5(III)}+S_o}\right] = \frac{1000}{10+S_{III}} \tag{3.157c}$$

where subscripts I, II, and III again refer to the same AMC I, AMC II, AMC III, respectively. From equations (3.157a-c) P_5 for AMC I – AMC III can be derived, respectively, as

$$P_{5(I)} = S_o\left[\frac{M_I}{1-M_I}\right] \tag{3.158a}$$

$$P_{5(II)} = S_o\left[\frac{M_{II}}{1-M_{II}}\right] \tag{3.158b}$$

$$P_{5(III)} = S_o\left[\frac{M_{III}}{1-M_{III}}\right] \tag{3.158c}$$

where M_I through M_{III} refer to AMC I through AMC III, respectively, and these can be derived as:

$$M_I = \left(1-\frac{S_I}{S_o}\right)\frac{10S_o}{10+S_I} \tag{3.159a}$$

$$M_{II} = \left(1-\frac{S_{II}}{S_o}\right)\frac{10S_o}{10+S_{II}} \tag{3.159b}$$

$$M_{III} = \left(1-\frac{S_{III}}{S_o}\right)\frac{10S_o}{10+S_{III}} \tag{3.159c}$$

which stand for AMC I through AMC III, respectively. It is apparent from equations (3.156)-(3.159) that S_o is required a priori for computing $P_{5(I)}$ through $P_{5(III)}$. Its derivation using the NEH-4 data is shown below:

As described above by equations (3.143) and (3.144), S_I, S_{II}, and S_{III} are related with each other. However, there does not exist a relationship between S_o and S_I. For this reason, trials were made for S_o and its relation with S_I such that the derived P_5-

values approximated the average of AMC II defined by NEH-4 (Table 3.8). The derived relationship, shown in Figure 3.13, which includes I_a (= 0.2 S), yields the limiting P_5-values for various AMCs, as shown in Table 3.8. The derived S_o values are equal to 0.11 inch for the dormant season and 0.56 inch for the growing season and S_o/S_I = 1.6. For the normal AMC II, the derived limiting P_5-values were equal to 0.8 and 1.75 for dormant and growing seasons, respectively. An assumption of soil porosity equal to 0.3 leads to the soil depth above the impeding layer (Huggins and Monke, 1966) equal to 0.36 and 1.87 inches for dormant and growing seasons, respectively. The determination of the soil depth above the impeding layer using infiltration data is discussed in Chapter 4 and its estimation using the universal soil loss equation is described in Chapter 5. Since the existing SCS-CN method accounts for only dynamic (or capillary) portion of infiltration and excludes its static (or gravitational) portion, these values of S_o represent the soil depths responsible for the dynamic infiltration process. It is consistent with the notion that only the top-soil-zone significantly affects the dynamic infiltration process (Fok and Chiang, 1984). In other words, the antecedent moisture model considers only the upper-most soil layer as a reservoir (Schaake et al., 1996), rather than the complete root zone depth. Figure 3.13 can be used for deriving AMC criteria for any other S_o-value that depends on the regional soil, vegetation, and land use characteristics. The following case study derives CN-values for the three AMCs using rainfall-runoff data.

Table 3.8 Antecedent soil moisture conditions (AMC)

AMC	Total 5-day antecedent rainfall, P_5 (inch) (NEH-4)[a]		Total 5-day antecedent rainfall, P_5 (inch) (derived from Figure 3.13)	
	Dormant season	Growing season	Dormant season[b]	Growing season[c]
I	Less than 0.5	Less than 1.4	Less than 0.22	Less than 0.49
II	0.5 to 1.1	1.4 to 2.1	0.22 to 2.28	0.49 to 5.02
III	More than 1.1	More than 2.1	More than 2.28	More than 5.02

[a]Source: McCuen (1982); [b]S_o = 0.11 inch; [c]S_o = 0.51 inch

3.8.2 CN DERIVATION FROM RAINFALL-RUNOFF DATA

a) *Application 1: Existing SCS-CN Method*

The daily data of the above-described Hemavati watershed were taken to represent the daily rainfall-runoff events considering the geography and the areal extent of the watershed in terms of the time of concentration, which is of the order of a few hours. It implies that the rainfall occurring on a day contributes fully to the surface runoff at the outlet of the catchment. The adoption of the average daily rainfall and runoff values refrains from incorporating a routing mechanism for computing routed flows.

The above assertion, however, leads to the runoff factor (C = Q/P) greater than 1 on days experiencing runoff greater than the rainfall amount because of baseflow contribution. Therefore, only those daily rainfall-runoff events yielding $C \leq 1$ were considered for deriving CN. Thus, 679 daily rainfall-runoff events shown in Figure 3.14 were selected for the CN-derivation. In Figure 3.14, the computed C-values correspond to the observed rainfall that is organized in descending order for assigning the probability of exceedance, which is derived using the following Weibull plotting position formula:

$$\text{Probability of exceedance} = \frac{m}{N+1} x100 \tag{3.160}$$

where m is the order or rank and N is the total number of rainfall-runoff events. It is apparent from Figure 3.14 that the observed runoff values assume a damping cyclic trend with generally increasing probability of exceedance or decreasing rainfall. The whole runoff spectrum appears to be reasonably bounded within two upper and lower envelopes, which can be reasonably simulated by the existing SCS-CN equation for fixed values of curve numbers, as shown in Figure 3.15. The upper and lower fitted bounds correspond to CN = 98 and CN = 60, respectively. Thus, the enveloping curves are distinguishable by the curve numbers. Here, it also provides a best fit, which can be taken to represent a normal state of the antecedent moisture of the rainfall-runoff phenomenon, for it falls in the middle of the two extremes denoted by upper and lower bounds. The curve number for such a normal state was found to be 84. The derived mathematical expressions for these fits are given below:

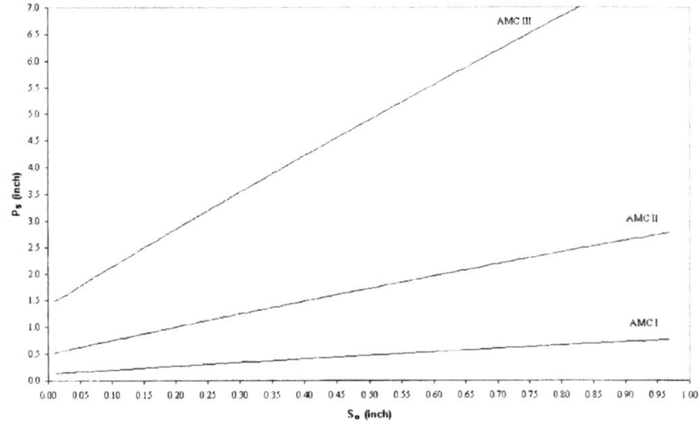

Figure 3.13. Variation of P_5 (inch) with S_o (inch). P_5 includes $I_a = 0.2S$.

Upper bound curve

$$y = -22.638\ln(x) + 98.23 \tag{3.161}$$

Lower bound curve

$$y = -5.3766\ln(x) + 23.239 \tag{3.162}$$

Best fit curve

$$y = -12.84\ln(x) + 55.611; \; r^2 = 0.6811 \tag{3.163}$$

where y is the runoff (mm/day) and x is the probability of exceedance (%) of rainfall (mm). From Figure 3.15 the limiting value of the probability of exceedance for CN = 60 is equal to 18.28%, which corresponds to the rainfall amount of 33.95 mm; for CN = 98, it is equal to 98.43%, corresponding to the rainfall magnitude of 1.04 mm; and for the normal state, CN = 84, the limiting value for which is 75.95%, equivalent to 5.06 mm of rainfall. The limiting rainfall amount of 33.95 mm for the lower bound represents the maximum amount of the initial abstraction taking place in the Hemavati watershed. Thus, from equation (3.161) the SCS-CN parameter S-value can be computed as 169.8 mm for $\lambda = 0.2$.

It is possible to describe the above CN-derivation for various AMCs using equation (3.59), which can be re-cast for convenience as:

$$CN = \frac{100P}{S_o + P} \tag{3.59}$$

For a given storm, P is constant and, therefore, CN is also constant since S_o is also a constant quantity for a given SVL of the watershed. On the other hand, if CN is assumed to be a constant quantity for characterization of a watershed for antecedent moisture, as in the above case study, the variation of S_o with P can be described from equation (3.59) as

$$S_o = \left(\frac{100 - CN}{CN}\right)P \tag{3.164}$$

Equation (3.164) describes S_o to be directly proportional to P for a given CN value. It suggests that S_o decreases with decreasing rainfall as shown in Figure 3.15 and vice versa.

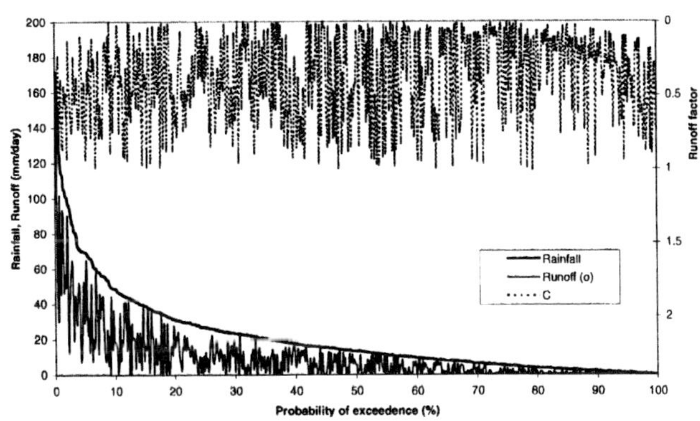

Figure 3.14. Variation of daily rainfall, runoff, and runoff factor for River Hemavati.

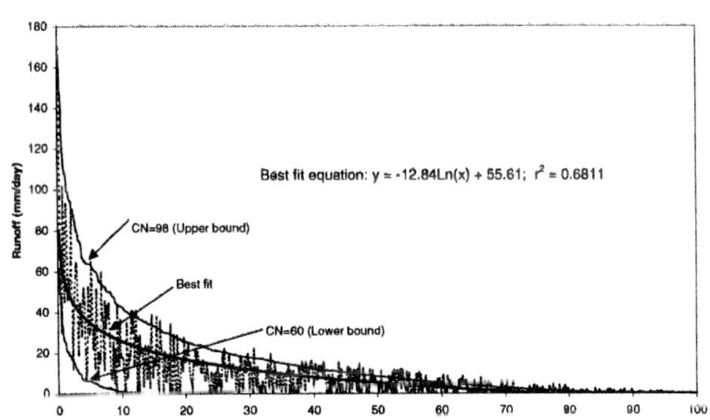

Figure 3.15. Variation of the observed runoff with the probability of exceedance.

b) *Application 2: Modified SCS-CN Method*

In this application, the CN-values for the modified method using the rainfall-runoff data of the earlier described sub-watersheds of Luni basin are derived from the above CN definition given by equation (3.137). According to this definition, the S-CN relationship for the modified method can be expressed as

$$CN = \frac{100F}{S} = \frac{100(S-5)}{S+5} \tag{3.165}$$

for P_e = 10 inches. In equation (3.165), S is in inches. This equation is valid for $S \geq 5$ inches, CN = 0 otherwise. In metric units, equation (3.165) can be modified as

$$CN = \frac{100(S-127)}{S+127} \tag{3.166}$$

where S is in mm. The CN values derived from equation (3.166) for the modified SCS-CN method using the rainfall-runoff data of the sub-watersheds of the Luni basin are shown in Table 3.7 as CN with superscript 'c'. The derived values range from 77.1-92.1. These values are fairly close to those derived from the NEH-4 tables as in the range of (61.3, 89.1) using the physical characteristics of the catchment.

3.9 SCS-CN Concept as an Alternative to Power Law

With the above analytical, conceptual and physical background of the SCS-CN method, it is now appropriate to explore if the SCS-CN concept could be a replacement of the widely used power law. In hydrology, the widely used steady-state stage-discharge (or velocity) relation generally expressed as a power function represents another form of the popular Manning equation (equation (1.79)). Here, it is necessary to describe first the basis for making such an attempt and to adapt an SCS-CN-based function that exhibits the growth or decay of a dependent variable with its independent variable.

Representing the spatial variation of interacting storage elements over a watershed by a power distribution function, Moore (1985) gave a relationship for computing surface runoff volume (or rainfall-excess) from given effective storm rainfall volume, as shown in Figure 3.16. Since the initial storage is indirectly related to the curve number in terms of the degree of saturation (see Section 3.7.3), Figure 3.16 appears to be close to Figure 2.8 which shows the variation of SCS-CN-generated runoff with rainfall. It implies that the power distribution is an indirect representation of the SCS-CN concept, as also shown in Chapter 1 while fitting equations (1.6) and (1.7).

Equation (3.61) can be re-written for describing the variation of cumulative infiltration F with the rainfall for a given S as

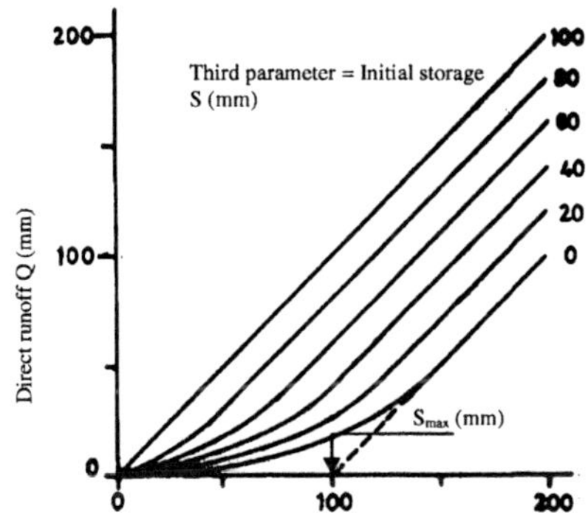

Figure 3.16. Rainfall-runoff relationship for the probability-distributed (power distribution) interacting storage capacity model. Maximum storage S_{max} = 100 mm.

$$F = \frac{SP}{S+P} \tag{3.167}$$

Replacing P by $i_e t$ (equation (3.13)) leads to

$$F = \frac{a\,t}{b+t} \tag{3.168}$$

where a = S and b = S/i_e, which is equal to the Horton decay parameter, α. Here, it is noted that these parameters 'a' and 'b' are different from those used earlier. Thus, parameter 'a' describes the system's potential maximum capacity, and b the decay of the phenomenon. Equation (3.168), however, describes the growth of the cumulative infiltration with time t. Thus, parameters 'a' and 'b' have a physical significance. A division of F by time t describes the decay of the average infiltration rate, f_{av}, with t as:

$$f_{av} = \frac{a}{b+t} \tag{3.169}$$

Equations (3.168) and (3.169) describing, respectively, the growth and decay functions can, in general, be expressed, respectively, as:

$$y = \frac{a\,x}{b+x} \tag{3.170}$$

and

$$y = \frac{a}{b + x} \tag{3.171}$$

where y and x are dependent and independent variables, respectively, and a and b are the parameters. Equation (3.170) is analogous to equation (1.7) (Chapter 1):

$$P = \frac{a\,t}{b + t} \tag{1.7}$$

which describes the growth of rainfall with rainstorm duration. In equation (1.7), parameter 'a' represents the maximum possible amount of P that can occur on a watershed in time duration t, similar to S representing the upper bound of F in equation (3.167). Thus, equations (3.170) and (3.171) are based on the capacity of the considered system. Parameter 'b' bears the dimension of time and describes the growth pattern of P with t.

Similarly, equation (3.171) is analogous to equation (1.10) (Chapter 1) relating the uniform rainfall intensity with the rainstorm duration as:

$$i_o = \frac{a}{b + t} \tag{1.10}$$

Parameters 'a' and 'b' of equation (1.10) can also be described, respectively, as the potential maximum rainfall intensity that can occur in time t on a watershed and decay factor of i_o. From equation (3.170) an expression for the steady-state rating curve for a unit-width rectangular channel can be given as:

$$v = \frac{a\,h}{b + h} \tag{3.172}$$

where v is the flow velocity, h is the corresponding stage of the flow, 'a' represents the potential maximum velocity, and 'b' is the decay factor of velocity. For a unit-width rectangular channel, discharge q can be given as:

$$q = \frac{a\,h^2}{b + h} \tag{3.173}$$

On the other hand, the steady-state rating curve in terms of the power law can be expressed as

$$q = a_1\, h^{1+m} \tag{3.174}$$

where q is the discharge, h is the depth of flow, and a_1 and m are the coefficient and the exponent, respectively. Parameters of equation (3.174) can be described by equating the wave celerity derived from the Seddon formula (Appendix C: equation (C.16)) as below.

From equations (3.173) and (3.174) the wave celerity, c, can be derived, respectively, as:

$$c = \left[\frac{2b + h}{b + h}\right] v \tag{3.175}$$

and

$$c = (1 + m)\, v \tag{3.176}$$

Here, the notation 'v' corresponds to 'u' in Appendix C. Coupling of equations (3.175) and (3.176) leads to the description of parameter m as:

$$m = \left(\frac{b}{a}\right)\frac{v}{h} \tag{3.177}$$

It implies that parameter m is a function of v and h, which is consistent with the wave derivations for m, shown in Appendix C, describing its dependence on the Froude Number and the kinematic wave number which, in turn, depend on v and h.

Furthermore, an assumption of v and h varying exponentially, as in Appendix C, leads to describing the v/h ratio for the kinematic wave as:

$$\frac{v}{h} = \frac{V_o}{H_o} \tag{3.178}$$

where V_o is the peak flow velocity and H_o is the corresponding depth of flow. Since m = 2/3 for kinematic wave (Appendix C), parameter 'a' relates to parameter 'b' as:

$$a = \frac{3V_o}{2H_o} b \tag{3.179}$$

V_o can be related to H_o using equation (3.172) as

$$V_o = \frac{a\,H_o}{b + H_o} \tag{3.180}$$

Coupling of equation (3.179) with (3.180) describes $a = 3V_o$ and $b = 2H_o$.

A replacement of 'h' by the hydraulic radius 'R' in equation (3.172) yields a general expression for velocity and, in turn, discharge as:

$$q = \frac{a R A}{b + R} \tag{3.181}$$

where A is the area of cross-section. Here, R is different from that in equation (3.36). The advantage of such a relationship is that its parameters are fully describable in terms of the SCS-CN concept. Including the Manning roughness and bed slope, a more general SCS-CN-based expression for the computation of discharge can be given as:

$$q = \frac{a R s_o A}{n(b + R)(c + s_o)} \tag{3.182}$$

where n is the Manning's roughness, s_o is the bed slope, and a, b, and c are the parameters.

CHAPTER 4

DETERMINATION OF 'S' USING VOLUMETRIC CONCEPT

In practice, the SCS-CN parameter, the potential maximum retention or postinitial abstraction retention (McCuen, 2002), S, is determined from scaled values of CN derived from the tables of National Engineering Handbook (NEH) for $\lambda = 0.2$ (Chapter 2). For computing S from rainfall-runoff data, equation (2.14) is used. Equation (2.14) is a specific form of equation (3.109). The existing SCS-CN method with $\lambda = 0.2$ is, therefore, a one-parameter model. In Chapter 3, the curve number CN is defined as the percent degree of saturation of the watershed soil by the 10-inch rainfall amount. As shown in Chapter 3, the generalization of the Mockus method leading to the SCS-CN method yields a relation between the Mockus parameter 'b' and S as: $b = 1/\{S \ln(10)\}$. Parameter 'b' depends on the antecedent moisture condition (AMC), vegetative cover, land use, time of the year, storm duration, and soil type; which describe CN. While attempting to theoretically justify the basis of the SCS-CN hypothesis, Yu (1998) described S as the product of the spatially averaged infiltration rate and the storm duration. In the previous chapter, the SCS-CN parameter S is distinguished from parameter s_{av} in the expression given by Yu (1998), for the former represents the volumetric capacity of the soil to retain water [L or L^3] and the latter represents the ratio of S to the time duration of the storm [LT^{-1}]. The description of S by Mockus (1964), given in Chapter 2, appeals most for S to be equivalent to the maximum difference of (P-Q), which corresponds to the maximum possible infiltration capacity of the soil, if other losses including initial abstractions are ignored. His further explanation is, however, ambiguous in that it compares the infiltration rate with the volumetric space available in the soil profile; the former represents the rate and the latter the volume, which are not comparable. In this chapter, S is distinguished from infiltration rate to represent the volumetric retention, consistent with the definition of McCuen (2002), and it is supported by the observed infiltration data. In addition to these, the runoff factor (C) = degree of saturation (S_r) concept (Chapter 3) is used to signify S and derive a hierarchy of the SCS-CN-based methods.

4.1 Analytical Derivation

The existing SCS-CN method consists of two external components, the antecedent moisture before the start of rain and the initial abstraction; excludes the static infiltration; and is valid for the time past ponding, as shown in Figure 4.1. In this figure, P_e = effective rainfall (cm) (= area covered by the rectangle DEGJ); F = cumulative infiltration (cm) (shaded by solid lines); Q = direct runoff (cm) = P_e-F (= area of DHJ); q = direct runoff rate (cm/hr); f_o = initial infiltration rate (cm/hr); f_c = minimum (or equilibrium) infiltration rate (cm/hr); t_p = time to ponding (min); $(t-t_p)$ = time past ponding; $i_e = f_o - f_c$ = effective rainfall intensity (cm/hr); and I_a = initial abstraction (cm). By definition, S = postinitial abstraction retention = maximum difference of (P_e-Q) (= area encompassed by DHMGE) that is equal to the total dynamic infiltration (F_d), and the proportionality concept (equation (2.2)) holds only for dynamic infiltration. The area of rectangle OAML represents the static portion (F_c) of the total infiltration.

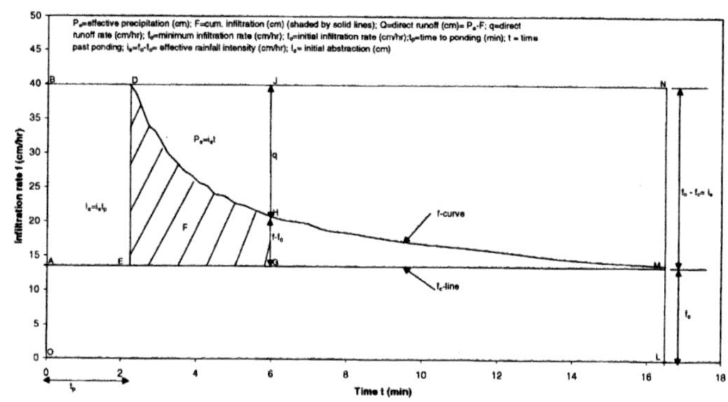

Figure 4.1. A descriptive infiltration curve of Columbia sandy loam (Project No. 34).

Utilizing the elementary definitions of soil physics given in Chapter 3 for describing the proportional equality by the $C = S_r$ equation (equation (3.133)), an analytical treatment is presented for investigating the existing SCS-CN method for antecedent moisture, initial abstraction I_a, and static infiltration F_c one by one using the same $C = S_r$ concept.

4.1.1 EQUIVALENCE BETWEEN SCS-CN PROPORTIONALITY AND C= S_r CONCEPTS

The basic SCS-CN hypothesis (equation (2.2)) does not explicitly account for the antecedent moisture (M) and the F_c-component of infiltration. Assuming an immediate ponding or $I_a = 0$, it can be expressed as:

$$\frac{Q}{P} = \frac{F}{S} \tag{4.1}$$

In equation (4.1), F represents the infiltrated amount of water ($= V_w$) and S is equal to the maximum amount of infiltration equal to the maximum (P-Q) difference, which, in turn, is equal to the volume of voids V_v. As described in Chapter 3, F/S represents the degree of saturation S_r (equation (3.130)) and Q/P represents the runoff factor C, which has a range of (0,1). Thus, the proportional equality is expressed as $C = S_r$ (equation (3.133)), which integrates the surface and subsurface flows linearly. It is, however, noted that under a fully dry condition, when $F \cong S \cong P$, then $Q \cong 0$. Here, $S_r \cong 1$ and $C \cong 0$. Similarly, a fully wet condition exhibiting $Q \cong P$, then $F \cong 0$. In this case, $C \cong 1$ and $S_r \cong 0$. Thus, these situations appear to violate the $C = S_r$ concept. However, the former dry condition can be circumvented as follows. The SCS-CN method partitions P into Q and F such that the proportional equality of equation (2.2) is maintained, as described in Chapter 2. It is for this reason that runoff in natural watersheds occurs even before the soil is fully saturated. Thus, the occurrence of the condition $F \cong S \cong P$ and $Q \cong 0$ is out of the realm of the SCS-CN method. Mathematically, since $F = PS/(P+S)$ according to the SCS-CN method, a substitution of $F = S = P$ leads to $P = P/2$, which is unrealistic. The other wet condition can be tackled by incorporating the antecedent moisture in equation (2.2) using $C = S_r$ concept and it is discussed in the subsequent section. Further analytical development proceeds with the analysis of the effects of antecedent moisture condition, initial abstraction, static infiltration, storm duration, rainfall intensity and turbidity, and land use treatment on S.

4.1.2 EFFECT OF ANTECEDENT MOISTURE CONDITION

The effect of antecedent moisture condition (AMC) on S can be described by taking $I_a = 0$ and assuming $M = V_w$, the moisture available in the soil column before the start of a storm, as shown in Figure 1.18. In such a situation, V_a in equation (3.126) represents the SCS-CN parameter, the potential maximum retention, S, and V_v represents the absolute potential maximum retention, S_{abs} (Mishra and Singh, 1999). Combining equation (3.129) with the assumption $S_{abs} = V_v$, the result is

$$M = (\frac{S_{abs}}{n})\theta \tag{4.2}$$

where n is the soil porosity and θ is the moisture content, described in Chapter 3. S_{abs} represents the maximum possible space available for water retention in a virgin, completely dry soil. It is noted that according to another interpretation of S in Chapter 3, S_{abs} was taken equivalent to 10 inches in equation (2.7) as a reference for deriving curve numbers. In equation (4.2), θ can be distinguished from θ_o to represent the initial moisture content, and thus, equation (4.2) is rewritten as

$$M = (\frac{S_{abs}}{n})\,\theta_o \tag{4.3}$$

Since n and S_{abs} are constant for a virgin soil, the amount of antecedent moisture, M, is directly proportional to θ_o. It is noted that S_{abs} is independent of 'M' as it represents the total volume of voids. However, S depends on 'M' as

$$S = S_{abs} - M \tag{4.4}$$

which can be derived from equation (3.126). From equation (4.4) $S = S_{abs}$ for $M = 0$, a completely dry condition, and $S = 0$ for $M = S_{abs}$, a fully saturated condition. Since the derivation of S from equation (2.14) using the P-Q data does not account for 'M', its derivation assumes $M = 0$, and thus the assumption of $S = V_v$ for $I_a = 0$ supports the validity of the $C = S_r$ concept. However, the assumption of $M = 0$ in equation (2.14) contrasts reality and exhibits contrasting features in respect of the SCS-CN application to gauged and ungauged watersheds. For a gauged watershed, the derived S (or CN) value corresponds to $M = 0$ and hence $S = V_v$, whereas S accounts for 'M' in terms of AMC when it is derived from the NEH-4 tables for ungauged watersheds and, therefore, $S = V_a$. The dependence of V_a on 'M' leads to an erroneous derivation of CN-values for gauged watersheds from the existing SCS-CN equation (equation (2.14)). It has led to the development of the concepts of (a) statistical error bands (Hawkins, 1996), frequently referred to as AMC I and AMC III (Chapters 2 and 3); (b) ordering P-Q data (Hjelmfelt, 1980; Hawkins, 1993; McCuen, 2002); and (c) derived distribution (Bonta, 1997). Therefore, to account for 'M' the basic hypothesis (equation (4.1)) can be modified using the above $C = S_r$ concept for $I_a = 0$ as follows (Chapter 3):

$$\frac{Q}{P}=\frac{F + M}{S + M} \tag{4.5}$$

A combination of equation (4.5) and equation (2.1) leads to

$$Q=\frac{P(P + M)}{P + S+ M} \tag{4.6}$$

or

$$\frac{Q}{P}=C=\frac{1+M^*}{1+S^* +M^*} \tag{4.7}$$

where $S^* = S/P$ and $M^* = M/P$. Thus, the runoff factor C depends on S, M, and P. Since (S+M) in equation (4.7) is constant (= S_{abs} or V_v) for a virgin soil and depends on its porosity, the runoff factor will increase with increasing 'M' and vice versa. Under initially fully wet situation, $M \cong S_{abs}$ and $S \cong 0$ and consequently, $F \cong 0$. Therefore, from equation (4.5) $C = S_r = 1$, supporting the validity of the $C = S_r$ concept and, in turn, eliminating the above perceived $C = S_r$ violation.

Assuming the total volume of soil profile per unit surface area equals the soil depth above the impeding layer (D_s), S_{abs} can be derived as

$$S_{abs} = n\, D_s \tag{4.8}$$

For an average porosity of 0.3, a value of S_{abs} equal to 10 inch void column yields 0.85 m soil column, the maximum effective depth. It is close to the usually described root zone depth of 1.0-1.5 m. Combination of equation (4.6) with equations (4.3) and (4.8) leads to

$$C = \frac{1+\theta_0(D_s/P)}{1+nD_s/P} = \frac{1/D_s+\theta_0/P}{1/D_s+n/P} \tag{4.9}$$

It implies that C depends on θ_0, n, D_s, and P. It is evident from equation (4.9) that for $D_s = 0$, C = 1; for $D_s = P$, $C = (1+\theta_0)/(1+n)$; and for $D_s = \infty$, $C = \theta_0/n$, from which C = 0 if $\theta_0 = 0$ and if $\theta_0 = n$, C = 1. Thus, for given n, D_s, and P, C increases with θ_0. It is emphasized that even if the soil profile is infinitely deep, C is not equal to zero if θ_0 is not equal to zero and, therefore, the SCS-CN concept further undermines the partial area concept as discussed in Chapter 2.

The effect of antecedent moisture on S can also be investigated using the mass-based definition of water content as follows:

$$\theta = \frac{\text{Mass of water}}{\text{Mass of solids}} \tag{4.10}$$

which leads to a relationship for S_r as:

$$S_r = \frac{(1-n)G\,\theta}{n} \tag{4.11}$$

where G is the specific gravity of the medium. Expressing $V_w = M$ and $V_v = S_{abs}$ in equation (3.129) and its coupling with equation (4.11) yield an expression for M as:

$$M = \frac{(1-n)S_{abs}G}{n}\theta \tag{4.12}$$

where n, S_{abs}, and G are constant for a virgin soil. Replacing equation (4.2) by equation (4.12) for $\theta = \theta_0$, an analysis can be performed to derive similar inferences. The present analysis, however, is based on the volumetric definition of θ (equation (3.127)).

4.1.3 EFFECT OF INITIAL ABSTRACTION

The initial abstraction, I_a, includes the interception loss, surface storage, evaporation, and infiltration (before the time to ponding after which runoff begins). The water that contributes to interception and surface storage is evaporated back to the atmosphere and contributes neither to runoff nor to infiltration. The infiltrated water before the time to ponding may be interpreted as to have satisfied the atmospheric demand of water absorption (molecular adsorption in particular) of the soil air column V_a, similar to evaporation. In other words, I_a primarily includes only short-term evaporation losses and neither contributes to runoff nor to infiltration, and thus, it is liable to be deducted from the amount of total rainfall P directly.

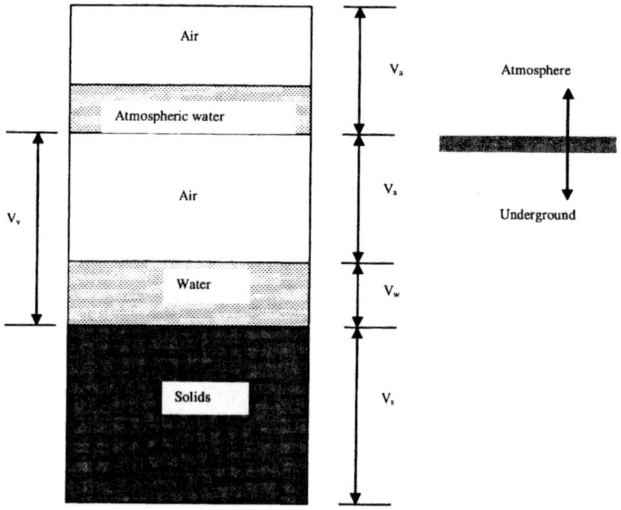

Figure 4.2. Explanatory diagram of the SCS-CN method.

The hypothesis (equation (2.3)) that I_a is some fraction (or multiple) λ of S (or S_{abs} with M = 0) actually represents the amount of water that goes to the atmosphere and, therefore, depends on the degree of saturation of the atmosphere. Assuming an air column of height V_a designated as $V_{a(atm)}$, λ can be expressed as the degree of atmospheric saturation, $S_{r(atm)}$, as

$$\lambda = \frac{I_a}{S} = \frac{I_a}{V_a} = \frac{V_{w(evap)}}{V_{a(atm)}} = S_{r(atm)} \tag{4.13}$$

In equation (4.13), $V_{w(evap)}$ represents the volume of water (in unit surface area) that contributes to $V_{a(atm)}$ as shown in Figure 4.2. $S_{r(atm)}$ can, however, vary from 0 to ∞, depending on the value of $V_{a(atm)}$ that can assume even a zero value. Thus, λ can also vary from 0 to ∞. However, the assumption $V_a = V_{a(atm)}$ forces λ to vary between 0 and 1. For practical applications, SCD (1972) reported λ to vary in the range of (0.1, 0.3), as discussed in Chapter 3. If the atmospheric temperature is high and humidity is low, more evaporation is likely to occur (Chapter 1), and higher will, thus, be the value of λ than otherwise. Similar to the pan evaporation, $S_{r\,(atm)}$ can be described as

$$\lambda = S_{r\,(atm)} = \frac{E_a}{E_p} \tag{4.14}$$

where E_a is the actual evaporation or evapotranspiration and E_p is the potential maximum evaporation. The right-hand side of equation (4.14) is taken to be directly proportional to temperature, as shown in Chapter 1. Details of evaporation in different climatic, geologic, and cover complexes are given by Shuttleworth (1992). Thus, λ depends on climatic conditions. Following Schaake et al. (1996), the initial abstraction coefficient, λ, or the ratio of E_a to E_p can also be determined as

$$\lambda = 1 - \frac{S}{S_{abs}} \tag{4.15}$$

The right-hand side of equation (4.15) is equal to degree of saturation, S_r. Thus, a relation can be derived among the runoff coefficient, C, the degree of saturation, S_r, and $S_{r(atm)}$ or λ as follows:

$$C = S_r = S_{r(atm)} = \lambda \tag{4.16}$$

It is again noted that equation (4.16) assumes the λ-variation in the range of (0, 1). Thus, the basic assumption of $V_a = V_{a(atm)}$ in equation (4.13) integrates in lumped form the atmospheric processes over and underneath the land surface. It can be inferred that the SCS-CN method, in general, integrates surface water, subsurface water, and atmospheric processes. The SCS-CN parameter λ can be interpreted in the context of infiltration with the aid of Figure 4.1, as follows.

From the derivations in Chapter 3 the following relations can be derived:

$$i_e = f_o - f_c = \alpha S \tag{3.10}$$

Replacement of P by P_e in equations (3.15) and (3.12) and their comparison lead to

$$\alpha t = P_e/S \tag{4.17}$$

The initial abstraction, I_a, is derived as

$$I_a = i_c t_p \tag{3.83}$$

which is consistent with infiltration observations (Figure 4.1). Equation (3.83) represents the area of the rectangle or the amount of infiltration before the time to ponding, t_p. Such a description of I_a signifies the SCS-CN parameter, λ, as follows.

Coupling of equation (2.3) with equation (3.83) leads to

$$\lambda = (i_e/S)\, t_p \tag{4.18}$$

A combination of equation (4.18) with equation (3.10) yields

$$\lambda = \alpha\, t_p \tag{4.19}$$

which is a specific form of equation (3.94). Thus, λ is a non-dimensional quantity and, within the framework of infiltration, it can be defined as the product of the Horton parameter, α $[T^{-1}]$, and time to ponding, t_p [T]. It is worth noting that as t_p increases, α decreases and vice versa. The workability of equation (4.19) is shown using a large set of infiltration data (Table 4.1) derived from laboratory tests published by Mein and Larson (1971). The λ-values computed from this equation and from equation (2.3) are compared in Figure 4.3. In this figure, the line of perfect fit (LPF) indicates that the values of λ derived from two distinct approaches are almost the same. All data points over the LPF (Figure 4.3) exhibit that equation (4.19) underestimates the λ-values and equation (2.3) overestimates them, which, however, is attributed to the computation of S-values that is partly affected by the approximate f_c-values derived from the data of infiltration tests.

Equation (4.19) helps describe λ physically as follows. If $\lambda = 0$, then either $\alpha = 0$ or $t_p = 0$. $\alpha = 0$ implies that there is no decay in infiltration (Figure 4.1), implying that the rate of infiltration is equal to the rainfall intensity or the whole rainfall is absorbed by the soil leading to no runoff condition. Such a situation describes $S = \infty$. On the other hand, $t_p = 0$ implies an immediate ponding or $I_a = 0$, for which S = maximum (P-Q) (equation (2.2)). If $\lambda = \infty$, then either $\alpha = \infty$ or $t_p = \infty$. $\alpha = \infty$ infers that the space for dynamic infiltration or the quantity of water retained in the soil profile is zero or $S = 0$ and $t_p = \infty$ implies that whole rainfall is absorbed as initial abstraction and, therefore, $Q = 0$. The latter condition also lends the postinitial abstraction retention $S = 0$, for the dynamic portion of infiltration is non-existent.

Table 4.1 Characteristics of laboratory-tested soils

Sl. No.	Soil	Number of tests	K_h (cm/sec)	Porosity	Reference
1	Plainfield Sand (PFS) (disturbed sample)	12	3.44×10^{-3}	0.477	Black et al. (1969)
2	Columbia Sandy Loam (CSL) (disturbed sample)	36	1.39×10^{-3}	0.518	Laliberte et al. (1966)
3	Guelph Loam (GL) (air dried, sieved)	12	3.67×10^{-4}	0.523	Elrick and Bowman (1964)
4	Ida Silt Loam (ISL) (undisturbed sample)	42	2.92×10^{-5}	0.530	Green (1962)
5	Yolo Light Clay (YLC) (disturbed sample)	12	1.23×10^{-5}	0.499	Moore (1939)

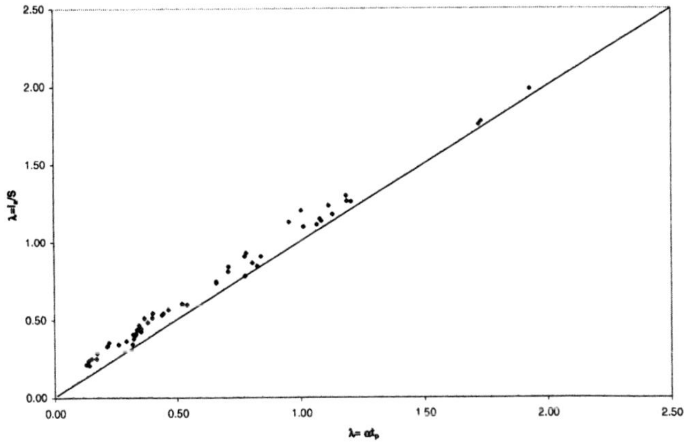

Figure 4.3. Comparison of λ from two approaches: $\lambda = I_a/S$ and $\lambda = \alpha t_p$; S=maximum (P_e-Q).

From the above discussion it is evident that an air column of λ times $V_{a(atm)}$ or I_a over the land surface is available for absorbing water from the total rainfall P. Therefore, the total space available for the water storage (S_T) is

$$S_T = (1+\lambda)\,S \qquad (4.20)$$

Thus, equation (2.5) (the existing SCS-CN method) describes the maximum possible water loss from an observed P-Q data set equal to S_T, as follows.

For S << (P-I_a), equation (2.5) reduces to

$$Q = P - I_a - S = P - (1+\lambda)S \qquad (4.21)$$

which represents equation (2.1) for F = S. Thus, the maximum possible amount of water lost through infiltration and initial abstraction is equal to (1+λ)S or S_T (equation (4.20)). The final value of S_T can be obtained by adding 'M' as:

$$S_T = (1+\lambda)\,S + M \qquad (4.22)$$

and for λ = 0.2,

$$S_T = 1.2\,S + M \tag{4.23}$$

which describes the maximum possible space available for water storage in the soil, according to the existing SCS-CN concept. Since I_a is assumed to be a part of molecular adsorption in a soil column and is treated separately, it does not form a part of the void space ($V_v = S_{abs}$) of the soil and, therefore, does not affect S. However, I_a affects λ since S is a constant for a given soil under a given antecedent moisture. Furthermore, since α depends on i_o and S, t_p is sized according to equation (4.19). α and t_p, in turn, help describe S implicitly, as above. Here, S_T is distinguished from S_{abs} in that the latter considers only the void space of the soil (= S+M), responsible for dynamic infiltration whereas the former defines the total volumetric loss of water. It might be coincidental but interesting to note that in unit surface area, the depth of antecedent 5-day rain water considered for the AMC computation (Table 2.6) was taken to be 21% of 10" of the absolute potential maximum retention and the value of λ was taken to be 20% of V_a, which is close to 21%. According to the present volumetric concept, S_T can be derived as follows.

Combination of equations (4.3) and (4.8) leads to

$$M = \theta_o D_s \tag{4.24}$$

Substitution of equation (4.8) into equation (4.4) yields

$$S = nD_s - M \tag{4.25}$$

and coupling of equation (4.25) with equation (4.24) further leads to

$$S = (n - \theta_o) D_s \tag{4.26}$$

Thus, S is directly proportional to D_s for a given n and θ_o of a virgin soil. Substitution of equation (4.26) into equation (4.23) describes S_T physically as

$$S_T = D_s\,[1.2\,n - 0.2\theta_o] \tag{4.27}$$

For $\theta_o << n$, equation (4.27) reduces to

$$S_T = 1.2\,nD_s = 1.2\,S_{abs} \tag{4.28}$$

In equation (4.27), S_T depends on 'M' because of the presence of θ_o, attributed to the inclusion of I_a (equation (2.3)).

Since the inclusion of I_a does not alter the physical significance of S as described above, the existing SCS-CN hypothesis (equation (2.2)) holds for $M = F_c = 0$. It can be modified for 'M' following the derivation of equation (4.6) as:

$$Q = \frac{(P - I_a)(P - I_a + M)}{P - I_a + S + M} \tag{4.29}$$

Equation (4.29) reduces to equation (2.5) for M = 0. Substitution of conditions $0 \leq C \leq 1$ into equation (4.29) leads to

$$0 \leq \frac{M}{S_{abs}} \leq 1 \tag{4.30a}$$

and

$$P \geq I_a \tag{4.30b}$$

Since the ratio M/S_{abs} represents the degree of saturation that can range from 0 to 1, $P > I_a$ forms the necessary condition for applicability of equation (4.29), which can also be expressed in non-dimensional form as:

$$\frac{Q}{P} = C = \frac{1 - I_a^*}{1 + S^*/(1 - I_a^* + M^*)} \tag{4.31}$$

where $I_a^* = I_a/P$. It can be shown that for C = 1, both I_a and S should be equal to zero (or $\lambda = 0/0$ from equation (2.3)) and for C = 0, I_a should be equal to P. Combining equation (4.31) with equations (4.3) and (4.4) for $\lambda = 0.2$ leads to

$$C = \frac{1 - 0.2D_s(n - \theta_o)/P}{1 + \{(n - \theta_o)D_s/P\}/\{1 - 0.2D_s n/P + 1.2D_s\theta_o/P\}} \tag{4.32}$$

Thus, the runoff factor C depends on the initial moisture content, soil porosity, soil depth above the impeding layer, and the total rainfall amount (or rainfall intensity and duration). For describing the behaviour of equation (4.29), S/P can be derived as

$$\frac{S}{P} = \frac{1}{2\lambda^2}[(2 + M^*)\lambda + (1 - \lambda)C - \sqrt{(\lambda M^*)^2 + (1 - \lambda)^2 C^2 + 2\lambda C\{2 + (1 + \lambda)M^*\}}] \tag{4.33}$$

Analogous to the description of the existing SCS-CN method (Chapter 3), it can be shown that for the condition $1 \geq I_a^* \geq 0$, $0 \leq C \leq 1$ and $\lambda \geq 0$,

4.1.4 EFFECT OF F_c

Using the water balance equation (equation (2.1)) and equation (2.5), an expression for the cumulative infiltration F can be derived as:

$$F = \frac{P_e S}{P_e + S} \tag{4.34}$$

Substitution of equation (4.17) into equation (4.34) leads to

$$F = \frac{\alpha S t}{(1 + \alpha t)} \tag{4.35}$$

Differentiation of equation (4.35) with respect to t yields

$$f = \frac{\alpha S}{(1 + \alpha t)^2} \tag{4.36}$$

Introduction of the non-zero f_c term in equation (4.36) results in

$$f = f_c + \frac{i_o - f_c}{(1 + \alpha t)^2} \tag{4.37}$$

where

$$f_o - f_c = i_o - f_c = \alpha S \tag{4.38}$$

which is the same as equation (3.10).

Equation (4.37) is the infiltration loss equation that is based on the SCS-CN method. It appears to be a variation of the models of Horton, modified Kostiakov, Philip, and Overton. Since the quantity αS (= i_o-f_c) is constant, f declines with time and approaches f_c asymptotically as $t \to \infty$. It is noted that the derivation assumes $I_a = 0$. The inclusion of I_a yields a general expression for computing infiltration rate as

$$f = f_c + \frac{i_o - f_c}{(1 - \lambda + \alpha t)^2} \tag{4.39}$$

or

$$f = f_c + \frac{i_o - f_c}{[1 + \alpha(t - t_p)]^2} \tag{4.40}$$

where the Horton decay parameter α = (i_o-f_c)S. Equation (4.39) or equation (4.40) appears to be an advanced form of the Horton method (equation (1.58)), modified Kostiakov model (equation (1.61)), Philip model (equation (1.62)), and Overton model (equation (1.67)).

Since equation (4.36) does not contain the final infiltration rate (f_c) or static infiltration, the existing SCS-CN method excludes this component and, therefore, underestimates infiltration. For inclusion of the cumulative static portion of infiltration (F_c) in the existing SCS-CN expression following the above C = S_r concept (equation (3.133)), F_c can be deducted directly from the cumulative rainfall amount as shown in Figure 4.1, leading to

$$\frac{Q}{P - I_a - F_c} = \frac{F_d}{S} \tag{4.41}$$

The sum of F_d and F_c yields the total infiltration, F, expressed as:

$$F = F_c + F_d \tag{4.42}$$

Thus, the water balance equation (equation (2.1)) is recast as

$$P = I_a + F_c + F_d + Q \tag{4.43}$$

Equation (4.41) can also be derived from the Horton equation (equation (1.58)), which is written for f_c and the time to ponding, t_p, as

$$f = f_c + (f_o - f_c)e^{-\alpha(t-t_p)} \tag{4.44}$$

Integration of equation (4.44) for F at time t leads to

$$\frac{F - F_c}{S} = \frac{F_d}{S} = [1 - e^{-\alpha(t-tp)}] \tag{4.45}$$

where F_c is the static portion of infiltration, F_d is the dynamic portion of infiltration, and

$$S = \frac{f_o - f_c}{\alpha} \tag{4.46}$$

which can be derived from equation (4.38). Here, P_e is defined by equation (3.82) as

$$P_e = P - F_c - I_a = \alpha\,(t - t_p)S \tag{3.82}$$

Therefore, the second SCS-CN hypothesis (equation (2.3)) is also modified as

$$I_a + F_c = \lambda S = \alpha\, t_p S \tag{4.47}$$

Here, I_a is, however, distinguished from F_c that occurs during the whole period of the rainstorm whereas I_a represents the initial abstraction occurring in the beginning of the storm. Coupling equation (4.47) with equation (4.38) leads to

$$\lambda = [(i_o - f_c)/S]\ t_p \tag{4.48}$$

or

$$\lambda = \alpha\ t_p \tag{4.49}$$

which is the same as equation (4.19).

Coupling of equation (3.82) with equation (4.45) yields

$$\frac{F_d}{S} = 1 - e^{-P_e/S} \tag{4.50}$$

Equation (4.50) can be generalised, as in Chapter 3, to

$$\frac{F_d}{S} = \frac{P_e}{P_e + S} \tag{4.51}$$

Further coupling of equation (4.51) with the expanded water balance equation (equation (4.43)) yields

$$\frac{Q}{P_e} = \frac{P_e}{P_e + S} \tag{4.52}$$

which is the same as equation (2.5), except that equation (2.5) ignores F_c, whereas equation (4.52) accounts for it. A comparison of equation (4.52) with equation (4.51) yields equation (4.41), describing the $C = S_r$ concept.

A useful form of the model incorporating I_a and F_c can be obtained by combining equations (4.41) and (4.43) as

$$Q = \frac{(P - I_a - F_c)^2}{P - I_a - F_c + S} \tag{4.53}$$

Thus, the effect of F_c on Q is similar to that of I_a; the greater the value of F_c, the lesser will be the value of Q and vice versa. Inclusion of 'M' in equation (4.53) on similar lines, as above, leads to the general form of the SCS-CN equation as follows:

$$Q = \frac{(P - I_a - F_c)(P - I_a - F_c + M)}{P - I_a - F_c + M + S} \tag{4.54}$$

The behaviour of equation (4.54) can be described by the following expression:

$$\frac{S}{P} = \frac{1}{2\lambda^2}[\lambda M^* + 2\lambda(1 - F_c^*) + (1-\lambda)C - \sqrt{(\lambda M^*)^2 + (1-\lambda)^2 C^2 + 4\lambda C(1 - F_c^*) + 2\lambda(1+\lambda)CM^*}\,] \tag{4.55}$$

where $M^* = M/P$, $F_c^* = F_c/P$, and $C = Q/P$. The S/P-expressions for other possible forms of equation (4.54) are given in Table 4.2. In this table, the last two columns provide conditions for various model formulations for $0 \leq I_a^* \leq 1$, consistent with the description for the functional behaviour of the existing SCS-CN method in Chapter 3. For example, the employment of the condition $I_a^* \geq 0$ to the general model (Model 8) (equation (4.55)) yields $M^* \leq F_c^* \leq 1$ for $C \leq 1$, and the condition $I_a^* \leq 1$ yields $F_c^* \geq M^* \geq 0$ for $C \geq 0$.

To exhibit the effect of F_c on S, an expression for Q identical to equation (4.21) is derived from equation (4.53) as:

$$Q = P - I_a - F_c - S = P - F_c - (1+\lambda)S \tag{4.56}$$

valid for $S << (P-I_a-F_c)$. Although equation (4.56) yields $S_T = F_c+(1+\lambda)S$ that is equal to the maximum possible water loss in the soil (Figure 4.1), F_c does not affect S for the reason that both are separate identities in this equation. Furthermore, according to the volumetric concept, because f_c is not a part of the void space (V_v) of a soil, S is independent of F_c. Physically, for a deep soil with fairly deep water table, F_c represents the amount of water percolated or drained to the saturated medium and, therefore, does not affect the available soil space or S. However, for a shallow soil with shallow water table, F_c will vary with the level of water table and will affect the available soil space or S.

The applicability of the SCS-CN-based infiltration equation (equation (4.40)) is now discussed. For the condition that $f = i_o$ at $t = t_p$, implying that equation (4.40) is valid for the time past ponding, equation (4.49) should hold and for $\lambda = 0.2$ (the existing SCS-CN method), equation (4.48) yields an expression for f_c as

$$f_c = i_o - \frac{0.2S}{t_p} \tag{4.57}$$

Equation (4.57) shows that f_c depends on the uniform rainfall intensity that is a function of storm rainfall and its duration, SVL complex and antecedent moisture condition affecting S, and the time to ponding. Alternatively, for a constant known f_c-value, the time to ponding, t_p, can be described as

$$t_p = \frac{0.2S}{i_o - f_c} \tag{4.58}$$

Table 4.2 SCS-CN inspired models

No.	Case	Basic Hypothesis	Q=	S/P=	Condition for $I_a^* \ge 0$	Condition for $I_a^* \le 1$
1	$M=F_c=I_a=0$	$\frac{Q}{P}=\frac{F}{S}$	$\frac{P^2}{P+S}$	$\frac{1-C}{C}$	As $C \to 0$, $S \to \infty$ and As $C \to 1$, $S \to 0$	
2	$F_c=I_a=0$	$\frac{Q}{P}=\frac{F+M}{S+M}$	$\frac{P(P+M)}{P+M+S}$	$\frac{(1-C)(1+M^*)}{C}$		
3	$M=F_c=0$	$\frac{Q}{P-I_a}=\frac{F}{S}$	$\frac{(P-I_a)^2}{P-I_a+S}$	$\frac{1}{2\lambda^2}[2\lambda+(1-\lambda)C-\sqrt{(1-\lambda)^2C^2+4\lambda C}]$	$C \le 1$	$C \ge 0$
4	$F_c=0$	$\frac{Q}{P-I_a}=\frac{F+M}{S+M}$	$\frac{(P-I_a)(P-I_a+M)}{P-I_a+M+S}$	$\frac{1}{2\lambda^2}[(2+M^*)\lambda+(1-\lambda)C-\sqrt{(\lambda M^*)^2+(1-\lambda)^2C^2+2\lambda C\{2+(1+\lambda)M^*\}}]$	$C \le 1$	$\lambda \ge 0$, $C \ge 0$
5	$M=I_a=0$	$\frac{Q}{P-F_c}=\frac{F_d}{S}$	$\frac{(P-F_c)^2}{P-F_c+S}$	$\frac{(1-F_c^*)(1-C-F_c^*)}{C}$	$F_c^* \le 1$ for $C \ge 0$	$F_c^* \ge 0$ for $C \le 1$
6	$M=0$	$\frac{Q}{P-I_a-F_c}=\frac{F_d}{S}$	$\frac{(P-I_a-F_c)^2}{P-I_a-F_c+S}$	$\frac{1}{2\lambda^2}[2\lambda(1-F_c^*)+(1-\lambda)C-\sqrt{(1-\lambda)^2C^2+4\lambda C(1-F_c^*)}]$	$F_c^* \le 1$ for $C \le 1$	$F_c^* \ge 0$ for $C \ge 0$
7	$I_a=0$	$\frac{Q}{P-F_c}=\frac{F_d+M}{S+M}$	$\frac{(P-F_c)(P-F_c+M)}{P-F_c+M+S}$	$\frac{(1-F_c^*+M^*)(1-C-F_c^*)}{C}$	For $S \ge 0$, $0 \le F_c^* \le 1$	
8	$M=F_c=I_a \ne 0$	$\frac{Q}{P-I_a-F_c}=\frac{F_d+M}{S+M}$	$\frac{(P-I_a-F_c)(P-I_a-F_c+M)}{P-I_a-F_c+M+S}$	$\frac{1}{2\lambda^2}[\lambda M^*+2\lambda(1-F_c^*)+(1-\lambda)C-\sqrt{(\lambda M^*)^2+(1-\lambda)^2C^2+4\lambda C(1-F_c^*)+2\lambda(1+\lambda)CM^*}]$	$M^* \le F_c^* \le 1$ for $C \le 1$	$F_c^* \ge M^* \ge 0$ for $C \ge 0$

Notations: $F_c^*=F_c/P$; $M^*=M/P$; $I_a^*=I_a/P$

Thus, the implication of the assumption for a constant λ-value forces t_p to rely on the rainfall intensity and the type of soil, characterized for f_c and S. As rainfall intensity increases, the time to ponding decreases and vice versa. It is consistent with the characteristics of the infiltration phenomenon (Mein and Larson, 1971). Here, a brief discussion of the variation of the minimum infiltration during the rainstorm is in order.

The minimum infiltration rate actually represents the final equilibrium infiltration rate, when the soil is fully saturated. As discussed in Chapter 1, the hydraulic conductivity, K_h, increases with the moisture content increasing continuously with the progressing rainstorm. Thus, the draining capacity of the soil increases with the increase in the soil moisture. It follows that for a completely dry soil, the minimum infiltration rate is equal to zero in the beginning of the rainfall, for the initial part of rainfall goes to meet the initial abstraction (or molecular adsorption in an infiltration phenomenon) requirements. Therefore, the minimum infiltration rate is likely to increase with time in such a fashion that it finally attains the equilibrium condition, where it is equal to f_c. Similar to the SCS-CN-based growth functions described in Chapters 1 and 3, a conceptual model can be given for the variation of the minimum infiltration rate with time as:

$$f_c(t) = \frac{a\,t}{b+t} \tag{4.59}$$

where $f_c(t)$ is the minimum infiltration rate at time t during infiltration and a and b are parameters, dependent on the antecedent moisture condition. In the foregoing discussion, such a variation of minimum infiltration rate is, however, ignored for the reasons of simplicity.

4.1.5 EFFECT OF STORM DURATION, RAINFALL INTENSITY, AND TURBIDITY

As discussed in Chapter 2, for a given rainfall amount, the storm duration, T_s, affects the rainfall intensity which influences α that, in turn, affects t_p for a given λ (equation (4.19)). These α and t_p help the implicit description of S. The impact of T_s on S can be further investigated taking rainfall as the product of rainfall intensity and storm duration, leading to an inverse relation between rainfall intensity and its time duration for a given rainfall amount. The greater the rainfall intensity, the lesser will be the time duration and vice versa. The greater intensity rainfall of lesser storm duration renders lesser time for rain water to stay over the land surface leading to less infiltration and consequently, greater direct runoff and vice versa. The NEH 4 table shows higher curve numbers for straight row ploughings than those for contoured ones (see Table 2.1) because the former yield lesser time for rain water to infiltrate than do the latter. A similar inference can be conveniently arrived if the impact of hydrologic condition on the amount of direct runoff is investigated in terms of the magnitude of surface roughness, which retards flow and, therefore, renders more time for infiltration to occur. Thus, the higher the roughness, the greater the amount of infiltration because of overstay of rainwater and vice versa. The resistance to flow through either contour ploughing or surface roughness actually increases the time lag (t_L) (Chapter 1), a transmission property of flow over land surfaces, frequently described in SCS-CN perspective as a function of S as:

$$t_L = \frac{\ell^{0.8}(S+1)^{0.7}}{1900\,Y^{0.5}} \tag{4.60}$$

where 'ℓ' is the hydraulic length in feet, Y is the percent slope of the watershed, and S is the potential maximum retention. Equation (4.60) is another form of equation (1.78) described in Chapter 1. In equation (4.60), S is intuitively taken as an index of surface roughness, for it represents the soil-vegetation-land use complex that affects the flow velocity over land surfaces. On the other hand, overstay of water on the land surface when $t_L > T_s$ leads to additional infiltration that causes reduction of Q. Thus, the storm duration affects F and Q, rather than S.

Equation (4.26) describes S to be directly proportional to D_s for a virgin, dry soil. Thus, S depends on n, D_s, and θ_0 and represents the postinitial abstraction retention before the start of a storm. Thus, it is independent of the amount of infiltration (Figure 4.1), as described above. Such an assertion holds if 'n' and D_s are taken constant for the above straight row and contoured conditions. It is worth noting that the main concept underlying the development of NEH-4 CN table (Table 2.1) is based on the assumption that the two complexes of same CN (or S)-values will exhibit the same kind of hydrologic behaviour. Such an assumption requires CN to be derived for a constant rainfall amount, for example, P = 10 inches (Chapter 3). Equation (4.26) also lends an alternative way to determination of D_s from n, θ_0, and infiltration data. Since the existing SCS-CN method (equation (2.5)) does not account for T_s and t_L separately, their impact on the resulting direct runoff appears to have been integrated with S (or CN) for reasons of simplicity.

From the view point of the application of the SCS-CN method, one can derive the limits of applicability for T_s. To this end, the rainfall-excess rate q (= i_o - f) can be derived from equation (4.39) as:

$$q = \left[1 - \frac{1}{(1-\lambda+\alpha t)^2}\right](i_o - f_c) \tag{4.61}$$

or

$$q = \left[\frac{(\alpha t - \lambda)(2 - \lambda + \alpha t)}{(1-\lambda+\alpha t)^2}\right](i_o - f_c) \tag{4.62}$$

Thus, for a given rainfall intensity, q increases with increasing T_s (= t) and vice versa. The total amount of rainfall also increases with T_s, and so does the amount of direct runoff. From equation (4.62) $\alpha t \geq \lambda$ or $t \geq t_p$ (time to ponding), as described above, for q

≥ 0. Thus, the SCS-CN method computes the runoff time past ponding. Similarly, as $t \to \infty$, $q \to (i_o\text{-}f_c)$. Replacing $\alpha(t\text{-}t_p)$ by P_e/S (equation (3.82)), equation (4.62) can, alternatively, be expressed in terms of the cumulative effective rainfall, P_e, and the potential maximum retention, S, as

$$q = \left[1 - \frac{1}{(1+P_e/S)^2}\right](i_o - f_c) \tag{4.63}$$

or

$$q = \left[\frac{P_e(P_e+2S)}{(P_e+S)^2}\right](i_o - f_c) \tag{4.64}$$

Equation (4.64) can be expanded as

$$q = \left[\frac{(P-I_a-F_c)(P-I_a-F_c+2S)}{(P-I_a-F_c+S)^2}\right](i_o - f_c) \tag{4.65}$$

which is analogous to the rational formula valid for rainfall-runoff rates and is a more useful form of the rainfall-runoff model. Taking P as a finite quantity, as $S \to 0$ (or $CN \to 100$), then $q \to (i_o\text{-}f_c)$.

Similarly, if $S \to \infty$ (or $CN \to 0$), then $q \to 0$. Therefore, for the dynamic infiltration to be significant, the following condition can be derived from equation (4.61):

$$t = T_s \leq [1/\sqrt{\varepsilon} - 1 + \lambda]/\alpha \tag{4.66}$$

or

$$S \geq \frac{P_e}{1/\sqrt{\varepsilon} - 1 + \lambda} \tag{4.67}$$

where ε is the tolerance limit, such as 0.01, 0.05, or 0.1, describing the relative magnitude of the dynamic infiltration compared to the effective rainfall intensity ($= i_o\text{-}f_c$). For example, if $\varepsilon = 0.1$ and $\lambda = 0.2$, $T_s \leq 2.36/\alpha$ or $S > 0.42\ P$. Similarly, for $\varepsilon = 0.01$, $S \geq 0.11\ P$. The above described condition $\alpha t \geq \lambda$ or $t \geq t_p$ is equivalent to $P \geq I_a$, a necessary condition for the applicability of the existing SCS-CN method.

The impact of rainfall intensity on S can be assessed from equation (4.61), in which q increases proportionally with rainfall intensity for a given storm duration and λ. It is again noted that an increase in the rainfall intensity also increases the total rainfall amount and, in turn, the amount of direct runoff. Theoretically, the rainfall intensity does not affect S, rather α is influenced, as described above. However, in reality, a high intensity rainfall or raindrop breaks down the upper most soil structure to make soil fines move into the soil surface or near-surface pores, leading to the formation of crust that impedes infiltration. The crust formation actually decreases D_s and n, and thus, decreases S or increases CN (equation (4.26)). Materials covering the surface (such as residues) protect the soil surface and play an otherwise role. The water turbidity results in choking of the upper soil pores, thus, reducing D_s or decreasing S or increasing CN.

4.1.6 EFFECT OF AGRICULTURAL MANAGEMENT PRACTICES

As described in Chapter 2, agricultural management systems involve different types of tillage, vegetation, and surface cover. Tillage practices, such as moldboard plough, chisel plough, and no till significantly affect infiltration. For example, the moldboard plough may increase soil porosity from 10-20%, depending on the soil texture, and increases infiltration rates over non-tilled soils. An increase in organic matter in the soil lowers the bulk density, or increases porosity, and hence increases infiltration and, in turn, increases S or decreases CN (equation (4.26)). The hydraulic conductivity, K_h, of a bare soil is reported to decrease between planting and midseason and then to stay stable because of crusting. The residue cover maintains a high steady state rate until harvest while the canopy and canopy-residue combination increase K_h. Thus, the residue cover as such increases D_s and 'n' and hence reduces CN. The combined effect of these factors on bulk density (or soil porosity) can be interpreted quantitatively from equations (2.13a) and (2.13b) (Chapter 2). The porosity, n, is further related with the soil sorptivity (S_s) as follows (Young, 1964):

$$S_s = \sqrt{2(n - \theta_o) K_h h_e} \tag{4.68}$$

where h_e is the Green-Ampt effective wetting front suction [L] which can be computed from the soil data using the relation:

$$h_e = \frac{2 + 3d}{1 + 3d} \frac{h_b}{2} \tag{4.69}$$

where d is the Brooks-Corey pore-size distribution index and h_b is the Brooks-Corey bubbling pressure head [L]. Substitution of equation (4.26) into equation (4.68) leads to a description of S as

$$S = \frac{1}{2} \frac{S_s^2}{K_h} \frac{D_s}{h_e} \tag{4.70}$$

Equation (4.70) also lends a way to determination of D_s using physical characteristics of the soil and rainfall-runoff data.

4.2 Verification of Existing AMC Criteria

As described in Chapters 2 and 3, NEH-4 described three antecedent moisture conditions: AMC I, AMC II, and AMC III, based on the amount of 5-day antecedent rainfall, as shown in Table 2.6. Based on the assumption that the soil is either dry (AMC I) or wet (AMC III), the S-values corresponding to these three conditions are functionally related with each other (Chapter 3: equations (3.143) and (3.144)). The above description, however, does not show the variation of S (or CN) between these AMCs, and thus, allows S to jump from one condition to the other. Alternatively, NEH-4 does not provide a functional relationship between 5-day antecedent rainfall, designated as P_5, and the corresponding antecedent moisture (M) of the soil, which represents the amount of moisture available in the soil column before the start of a storm. Therefore, it is necessary to derive a regular function for describing the variation of the antecedent moisture M with the antecedent rainfall P_5. To this end, an examination of the above relations (equations (3.143) and (3.144)) is in order.

Assume a soil is in a completely dry condition on a day, describing S to be equivalent to S_I on that day, and a part of the following 5-day rainfall (P_5) infiltrates into the soil, which forms 'M' for the next 6th day's storm considered for the SCS-CN analysis. According to the volumetric concept, S for the sixth day's actual storm will be equal to S_I minus infiltration that occurred during the above 5 days. Assuming equation (2.5) holds for this 5-d storm, the computed infiltration using S_I will equal the amount of the antecedent moisture M. Utilising equations (3.143) and (3.144) along with this assumption, the magnitudes of the antecedent moisture for AMC II and AMC III were computed. For example, the magnitude of the antecedent moisture corresponding to AMC II, designated as M_{II}, will be equal to S_I minus S_{II}. Similarly, the amount of the antecedent moisture corresponding to AMC III, designated as M_{III}, can be computed. Taking these M values equal to the amount of infiltration during 5 days, the values of P_5 were computed for AMC II and AMC III, which yielded the following relations respectively for AMC II and AMC III:

$$P_5=2.6372\ M_{II};\ r^2=1 \tag{4.71a}$$

$$P_5=5.5881\ M_{III};\ r^2=1 \tag{4.71b}$$

In equations (4.71a) and (4.71b), $r^2=1$ indicates a perfect fit. A completely dry soil of a given curve number exhibits two different values of 'M' (from equations (4.71a) and (4.71b)) for the same P_5, which contrasts physical reality, for a soil does not recognize these conditions. Instead, as described in Chapter 3, it is appropriate to compute the antecedent moisture amount, M, using the existing SCS-CN equation (equation (2.5)), which also holds for P_5 as it does for P. Therefore, the general equation describing the variation of 'M' with P_5 can be given as:

$$M=\frac{(P_5-0.2S_I)S_I}{P_5+0.8S_I} \tag{4.72}$$

which is another form of equation (3.155) and can be derived from equation (2.5) for F = M and S = S_I.

4.3 Determination of S

4.3.1 HOMOGENEOUS GAUGED WATERSHEDS

A step-by-step procedure for determining S_I (or CN_I), M, and S for a homogeneous gauged watershed using the general Model 8 (Table 4.2) is as follows:

1. Determine f_c, α, and t_p from infiltration test data.
2. Select storm events to determine P, Q, P_5, and T_s for each event.
3. Compute $\lambda = \alpha t_p$ or assume $\lambda = 0.2$.
4. Compute $F_c = f_c T_s$.
5. Compute $S_I = P(P-Q)/Q$
6. Determine M from equation (4.72) using S_I.
7. Determine $S = S_I - M$ and, in turn, compute Q from equation (4.54).
8. Compare the computed Q with the observed one. If both are close to each other, S_I of step 5 is the desired value for the event.
9. If the observed and computed values of Q do not match each other, vary S_I in proportion to the computed and observed Q-values.
10. Repeat steps 6 through 10 for all events and determine S_I for each storm event to obtain an average S_I that represents the watershed.

The application of the above procedure is illustrated in the following example.

Example 4.1: Compute S, M, and S_I for a watershed that produced 1.5 inches of runoff on a day from 3 inches of rainfall. The antecedent 5-day rainfall was equal to 1.0 inch. Assume the watershed has a soil group D for which minimum infiltration rate is equal to 0.042 inch/hr and assume $\lambda = 0.2$.

Solution: Using a trial and error approach as shown in the spread-sheet form in Tables 4.3 and 4.4, computations are as follows:

Assume an initial value of S_I as

$S_I = P(P-Q)/Q = 3 \times (3-1.5)/1.5 = 3.0$ inches

Compute M from equation (4.72) as:

$$M = \frac{(P_5 - 0.2S_I)S_I}{P_5 + 0.8S_I} = \frac{(1-0.2x3.0)x3.0}{1+0.8x3.0} = 0.353 \text{ inch}$$

From S_I and M compute $S = S_I - M = 3.0 - 0.353 = 2.647$ inches.

Compute the F_c-component of the total infiltration as

$F_c = 0.042 \times 24 = 1.00$ inch

Using S = 3.0 inches and F_c = 1.00 inch, compute Q from equation (4.54) as

$$Q = \frac{(P - I_a - F_c)(P - I_a - F_c + M)}{P - I_a - F_c + M + S}$$

$$= \frac{(3.0-0.2\text{x}2.647-1)(3.0-0.2\text{x}2.647-1+0.353)}{3.0-0.2\text{x}2.647-1+0.353+2.647} = 0.60 \text{ inch}$$

The computed Q given in the last column of Tables 4.3 and 4.4 (= 0.60 inch) is less than the observed Q (= 1.5 inches). Therefore, S_I needs to be increased for increasing the computed Q. The following two approaches can be employed:

Approach 1

As shown in Table 4.3, S_I for trial 2 is increased in proportion to Q as follows:
$S_{I\,(revised)} = S_{I\,(previous)}$ x [Q(comp.)/Q(obs.)] = 3.0 x 0.60/1.5 = 1.2 inches
where subscripts "revised" and "previous" correspond to the revised and previous trial-values of S_I and subscripts "comp." and "obs." refer to the computed and observed Q-values, respectively. Other spread-sheet calculations can be easily performed until the computed Q is close to the observed Q. This approach requires 12 iterations as shown in Table 4.3.

Table 4.3 Trial computations for S_I, M, and S using approach 1

Trial	P	Q (obs.)	P_5	F_c	λ	S_I	M	S	Q (comp.)
	(inch)	(inch)	(inch)	(inch)		(inch)	(inch)	(inch)	(inch)
1	3	1.5	1	1	0.2	3.000	0.353	2.647	0.600
2	3	1.5	1	1	0.2	1.200	0.465	0.734	1.407
3	3	1.5	1	1	0.2	1.126	0.459	0.667	1.451
4	3	1.5	1	1	0.2	1.089	0.455	0.633	1.473
5	3	1.5	1	1	0.2	1.069	0.453	0.616	1.484
6	3	1.5	1	1	0.2	1.058	0.452	0.606	1.491
7	3	1.5	1	1	0.2	1.051	0.451	0.600	1.495
8	3	1.5	1	1	0.2	1.048	0.451	0.597	1.497
9	3	1.5	1	1	0.2	1.046	0.450	0.595	1.498
10	3	1.5	1	1	0.2	1.044	0.450	0.594	1.499
11	3	1.5	1	1	0.2	1.044	0.450	0.594	1.499
12	3	1.5	1	1	0.2	1.043	0.450	0.593	1.500

Approach 2

In this approach, S_I is varied (Table 4.3) non-linearly as:

$S_{I\,(revised)} = S_{I\,(previous)}$ x $[Q(comp.)/Q\,(obs.)]^2$ = 3.0 x $[0.6/1.5]^2$ = 0.48 inch

The calculations for other trials are shown in Table 4.4. Apparently, the convergence is achieved in 8 iterations.

The above approaches are used only for demonstration purpose. Any other more suitable technique can also be employed instead.

4.3.2 HETEROGENEOUS GAUGED WATERSHEDS

A heterogeneous watershed can be distinguished to have more than one type of soil represented by different curve numbers or S-values that, in a way, represent storage capacities of soils for water retention. Thus, these spatially varying storage capacities can be assumed to be analogous to non-interacting buckets of depths equal to their capacities (Yu, 2000). For spatial averaging, several methods, such as arithmetic, geometric, harmonic, weighted means, are advocated and their application led Haverkamp and Vauclin (1979) to recommend the geometric mean method for computing the mean hydraulic conductivity for reasons of flexibility, precision, and feasibility. The existing SCS-CN method, however, utilises the area-weighted CN (Chapter 2) for computing the average S, as illustrated below.

Table 4.4 Trial computations for S_I, M, and S using approach 2

Trial	P	Q (obs.)	P_5	F_c	λ	S_I	M	S	Q (comp.)
	(inch)	(inch)	(inch)	(inch)		(inch)	(inch)	(inch)	(inch)
1	3	1.5	1	1	0.2	3.000	0.353	2.647	0.600
2	3	1.5	1	1	0.2	0.480	0.313	0.166	1.833
3	3	1.5	1	1	0.2	0.716	0.390	0.326	1.697
4	3	1.5	1	1	0.2	0.917	0.432	0.485	1.576
5	3	1.5	1	1	0.2	1.012	0.446	0.566	1.519
6	3	1.5	1	1	0.2	1.037	0.449	0.588	1.503
7	3	1.5	1	1	0.2	1.042	0.450	0.592	1.501
8	3	1.5	1	1	0.2	1.043	0.450	0.593	1.500

Example 4.2: Compute the area-weighted CN for a watershed of two major soils covering 100 and 75 ha (say) and exhibiting respective CN-values of 74 and 65.

Solution: The area-weighted CN is computed as

$$CN = \frac{100x74 + 75x65}{100 + 75} = 70$$

4.3.3 UNGAUGED WATERSHEDS

As discussed earlier, it is possible to derive S from in-situ infiltration data, as shown in Figure 4.1. For a given soil porosity n and initial moisture content θ_0, D_s can be computed from equation (4.26). The product of θ_0 and D_s represents the antecedent moisture amount M which, in turn, allows determination of S_I (= S+M). The S-values computed for each ungauged sub-watershed can be averaged, as described above. The computation of S for an ungauged watershed is illustrated below.

Example 4.3: Compute the potential maximum retention S, curve number CN, S_I (= S for AMC I), antecedent moisture amount M, and the depth of soil profile above the impeding layer D_s using the infiltration data of Figure 4.1. Assume soil porosity n = 0.518 and antecedent moisture content θ_o = 0.125.

Solution: By definition, the maximum dynamic portion of the infiltration amount represents S and it is computed from Figure 4.1 as equal to 1.31 cm. CN is derived from equation (2.7) and converted in consistent metric units as

$$CN = \frac{2540}{25.4 + S} = \frac{2540}{25.4 + 1.31} = 94$$

Using n = 0.518 and θ_o = 0.125 in equation (4.26), D_s can be derived as:

$$D_s = \frac{S}{n - \theta_o} = \frac{1.31}{0.518 - 0.125} = 3.33\,\text{cm}$$

Therefore,

$M = \theta_o D_s = 0.125 \times 3.33 = 0.42$ cm

and

$S_I = S + M = 1.31 + 0.42 = 1.73$ cm

4.4 Use of NEH-4 Tables

4.4.1 WORKABILITY OF MODEL 4

Since equation (2.5) of the existing SCS-CN method does not account for the antecedent moisture 'M' directly, it is appropriate to show the workability of the proposed Model 4 (Table 4.2) accounting for 'M'. To this end, the values of 'M' are determined using two different approaches illustrated in Table 4.5 for curve numbers of a residential lot of 1/8 acre or less size with an average of 65% impervious area under different soil conditions for which CNs are given in Table 2.1 (Chapter 2). The M-computation from the model, for an example AMC II, is expressed as

$$M_{II} = [Q_{II}S_I\text{-}(P\text{-}I_a)(P\text{-}I_a\text{-}Q_{II})]/(P\text{-}I_a);\ I_a = 0.2S_{II} \tag{4.73}$$

where Q_{II} stands for Q under AMC II. Although S in equation (4.73) corresponds to S_I, I_a is computed for S_{II} since it represents the current state of AMC. The M-computations using two approaches are illustrated below.

Example 4.4: Taking curve numbers (CN) equal to 77, 85, 90, and 92 for AMC II for hydrologic groups A thorough D, respectively, of a residential lot of 1/8 acre or less with the average 65% imperviousness, compute for all soil groups the antecedent moisture amounts that are equivalent to AMC II and AMC III. Also compute the antecedent moisture amounts using Model 4 assuming the storm rainfall of 5 inches.

Solution: The computations using the two approaches follow:

Approach 1

For computing the antecedent moisture amounts, first compute S_{II} from equation (2.7) for the given values of CN_{II} that correspond to AMC II. For example, for the soil of hydrologic group A, S_{II} is computed as

$$S_{II(A)} = \frac{1000}{CN_{II(A)}} - 10 = \frac{1000}{77} - 10 = 2.99 \text{ inch}$$

In this example, subscripts I, II, and III correspond to AMC I through AMC III, respectively, and subscript 'A' corresponds to hydrologic group 'A'. Similarly, S_{II} for

B, C, and D groups can be computed as 1.76, 1.11, and 0.87, respectively, as shown in Table 4.5.

Using equations (3.143) and (3.144), compute S_{III} and S_I for all soils as shown for soil group A below, respectively, as:

Table 4.5 Example computations for antecedent moisture (M) and f_c from NEH-4 CN-values. P = 5 inches assumed.

Sl. No.	Variable/parameter	Hydrologic Soil Groups			
		A	B	C	D
1	$^*CN_{II}$	77	85	90	92
2	$S_{II}=1000/CN_{II}-10$	2.99	1.76	1.11	0.87
3	$S_{III}=0.427S_{II}$	1.28	0.75	0.47	0.37
4	$S_I=5.342S_{III}$	6.81	4.03	2.53	1.98
5	$M_{II}=S_I-S_{II}$	3.83	2.26	1.42	1.11
6	$M_{III}=S_I-S_{III}$	5.54	3.27	2.06	1.61
f_c-computation using volumetric concept					
7	Q_I (equation (2.5))	1.27	2.14	2.87	3.22
8	$F_I=P-0.2S_I-Q_I$	2.37	2.05	1.62	1.38
9	$F_{II}=P-0.2S_{II}-Q_{II}$	1.78	1.28	0.90	0.74
10	$F_{III}=P-0.2S_{III}-Q_{III}$	1.01	0.65	0.43	0.35
11	$^aF_{c\,I}$	0.99	0.67	0.23	0.00
12	$^aF_{c\,II}$	1.04	0.54	0.16	0.00
13	$^aF_{c\,III}$	0.66	0.31	0.09	0.00
14	$f_{c\,I}=F_{c\,I}/24$	0.04	0.03	0.01	0.00
15	$f_{c\,II}=F_{c\,II}/24$	0.04	0.02	0.01	0.00
16	$f_{c\,III}=F_{c\,III}/24$	0.03	0.01	0.00	0.00
M- and f_c-computation using Model 8 (Table 4.2)					
17	Q_{II} (equation (2.5))	2.62	3.37	3.88	4.09
18	Q_{III} (equation (2.5))	3.74	4.20	4.47	4.58
19	$^bM_{II}$	2.28	1.64	1.15	0.94
20	$^bM_{III}$	4.36	2.83	1.88	1.50
21	$^cF_{c\,I}$	1.30	0.80	0.27	0.00
22	$^dF_{c\,II}$	1.25	0.70	0.30	0.13
23	$^eF_{c\,III}$	0.76	0.40	0.17	0.08
24	$f_{c\,I}=F_{c\,I}/24$	0.05	0.03	0.01	0.00
25	$f_{c\,II}=F_{c\,II}/24$	0.05	0.03	0.01	0.01
26	$f_{c\,III}=F_{c\,III}/24$	0.03	0.02	0.01	0.00

Note:

Subscripts I, II, and III represent AMC I, AMC II, AMC III, respectively.

Dimensions of S, M, F, F_c, Q, are in inch; f_c is in inch/hr; and CN is dimensionless.

$^*CN_{II}$ corresponds to residential lot of 1/8 acre or less size with average 65% impervious.

[a] $F_{c(A)}=F_{(A)}-F_{(D)}$; $F_{c(B)}=F_{(B)}-F_{(D)}$; $F_{c(C)}=F_{(C)}-F_{(D)}$; $F_{c(D)}=0$.

[b] $M_{II}=[Q_{II}S_I-(P-I_a)(P-I_a-Q_{II})]/(P-I_a)$; $I_a=0.2S_{II}$; [b] $M_{III}=[Q_{III}S_I-(P-I_a)(P-I_a-Q_{III})]/(P-I_a)$; $I_a=0.2S_{III}$

[c] $F_{cI(A)}=(P-0.2S_{I(A)})-0.5Q_{I(A)}-0.5\sqrt{[(Q_{I(A)})^2+4Q_{I(A)}S_{I(D)}]}$

[c] $F_{cI(B)}=(P-0.2S_{I(B)})-0.5Q_{I(B)}-0.5\sqrt{[(Q_{I(B)})^2+4Q_{I(B)}S_{I(D)}]}$

[c] $F_{c\,I\,(C)}=(P-0.2S_{I(C)})-0.5(Q_{I(C)}-0.5\sqrt{[(Q_{I(C)})^2+4Q_{I(C)}S_{I(D)}]}$

[c] $F_{c\,I\,(D)}=(P-0.2S_{I(D)})-0.5Q_{I(D)}-0.5\sqrt{[(Q_{I(D)})^2+4Q_{I(D)}S_{I(D)}]}$

[d] $F_{c\,II(A)}=(P-0.2S_{II(A)})-0.5(Q_{II(A)}-M_{II(D)})-0.5\sqrt{[(Q_{II(A)}-M_{II(D)})^2+4Q_{II(A)}S_{I(D)}]}$

[d] $F_{c\,II\,(B)}=(P-0.2S_{II(B)})-0.5(Q_{II(B)}-M_{II(D)})-0.5\sqrt{[(Q_{II(B)}-M_{II(D)})^2+4Q_{II(B)}S_{I(D)}]}$

[d] $F_{c\,II(C)}=(P-0.2S_{II(C)})-0.5(Q_{II(C)}-M_{II(D)})-0.5\sqrt{[(Q_{II(C)}-M_{II(D)})^2+4Q_{II(C)}S_{I(D)}]}$

[d] $F_{c\,II\,(D)}=(P-0.2S_{II(D)})-0.5(Q_{II(D)}-M_{II(D)})-0.5\sqrt{[(Q_{II(D)}-M_{II(D)})^2+4Q_{II(D)}S_{I(D)}]}$

[e] Expressions are similar to those bearing superscript 'd'.

$S_{III(A)} = 0.427S_{II(A)} = 0.427x2.99 = 1.28$ inches

and

$S_{I(A)} = 5.342S_{III(A)} = 5.342x1.28 = 6.81$ inches

Similarly, S_{III} and S_I for other soil groups can be computed, as shown in Table 4.5. Thus, for soil group A, the antecedent moisture (M) for AMC II and AMC III can be computed, respectively, as

$M_{II(A)} = S_{I(A)} - S_{II(A)} = 6.82 - 2.99 = 3.83$ inches

and

$M_{III(A)} = S_{I(A)} - S_{III(A)} = 6.82 - 1.28 = 5.54$ inches

Similarly, M-values for other soils can be computed, as shown in Table 4.5 at Sl. Nos. 5 and 6.

Approach 2

For the curve numbers used in Approach 1, compute Q from equation (2.5) for AMC II and AMC III, respectively, for soil type A as:

$$Q_{II(A)} = \frac{(P - 0.2S_{II(A)})^2}{P + 0.8S_{II(A)}} = \frac{(5.0 - 0.2x2.99)^2}{5.0 + 0.8x2.99} = 2.62 \text{ inches}$$

and

$$Q_{III(A)} = \frac{(P - 0.2S_{III(A)})^2}{P + 0.8S_{III(A)}} = \frac{(5.0 - 0.2x1.28)^2}{5.0 + 0.8x1.28} = 3.74 \text{ inches}$$

Similarly, compute Q values for all soil types and all AMCs. M_{II} can be computed using equation (4.73), which is recast for soil 'A' as

$M_{II(A)} = [Q_{II(A)}S_{I(A)}-(P-I_a)(P-I_a-Q_{II})]/(P-I_a)$

which is valid for $I_a = 0.2S_{II(A)}$. Using the $S_{I(A)}$ values computed in Approach 1, $M_{II(A)}$ can be computed as:

$M_{II(A)} = [2.62 \times 6.81 - (5.0 - 0.2 \times 2.99) \times (5.0 - 0.2 \times 2.99 - 2.62)]/(5.0 - 0.2 \times 2.99)$
$= 2.28$ inches

Computations for other AMCs and soils are shown in Table 4.5 at Sl. Nos. 17-20.

The approaches illustrated in the above examples assume $M_I = 0$ for all soils. Thus, as shown in Sl. Nos. 5 & 6 and 19 & 20 in Table 4.5, the M values were computed for 16 land uses and soil groups using P = 100 inches, which is taken for Q greater than zero corresponding to the minimum reported CN_{II} equal to 6 in NEH-4 table (Table 2.1), requiring $P \geq 71.5$ inches for Q_{III} (= Q under AMC III) ≥ 0. The former approach is based on the above volumetric concept and the latter derives 'M' from equation (4.73). Figures 4.4 through 4.7 depict the correspondence of these M-values for soil groups A and D for AMC II and AMC III. Except for Figures 4.4 and 4.5 in which the most deviating data point from the line of perfect fit (LPF) corresponds to the above $CN_{II} = 6$, most data points fall near LPF, and thus underscore the validity of the above M-computation.

4.4.2 INVERSE COMPUTATION OF f_c FROM NEH-4 CN-VALUES

NEH-4 describes the hydrologic soil groups based on the ranges of final infiltration rates of soils, as shown in Table 4.6. Since the existing SCS-CN equation (equation (2.5)) does not account for these infiltration rates, rather these are incorporated in S. Therefore, an inverse problem of determining f_c-values from the CN data of NEH-4 tables can be attempted. To this end, the f_c-computation using two different approaches based on the volumetric concept and Model 8 (Table 4.2) is explained below.

Table 4.6 Description of hydrologic soil groups

Soil	${}^{\gamma}f_c$ (inch/hr)	Computed f_c for AMC I (inch/hr)			Computed f_c for AMC II (inch/hr)			Computed f_c for AMC III (inch/hr)		
		Min.	Max.	Ave.	Min.	Max.	Ave.	Min.	Max.	Ave.
A	0.30-0.45	0.0	1.404	0.459	0.0	1.881	0.273	0.0	1.526	0.142
B	0.15-0.30	0.0	0.969	0.189	0.0	0.541	0.092	0.0	0.259	0.041
C	0.05-0.15	0.0	0.132	0.052	0.0	0.063	0.024	0.0	0.028	0.011
D	0-0.05	0.0	0.0	0.0	0.0	0.0	0.0	0.0	0.0	0.0

[γ]Source: McCuen (1982)

The volumetric derivation of F_c is based on the following hypothesis. For a given cover complex and antecedent moisture condition, as the soil group changes from D to A, CN reduces, as also seen from the CN_{II} values given in Table 4.5 for various soil groups. Since equation (2.5) does not account for F_c, continually increasing infiltration for soil groups B, C, and D can be attributed to the F_c-contribution. Such a hypothesis is, however, based on the assumption of equal porosity and D_s of all soils. Therefore, if the f_c-values of soils of hydrologic group D are known, these values can be derived for other soil groups from the difference in cumulative infiltration values. Mathematically, F_c for all soil groups can be computed as:

$$F_{c(A)} = F_{(A)} - F_{(D)} \tag{4.74a}$$

$$F_{c(B)} = F_{(B)} - F_{(D)} \tag{4.74b}$$

$$F_{c(C)} = F_{(C)} - F_{(D)} \tag{4.74c}$$

$$F_{c(D)} = 0 \tag{4.74d}$$

where $F_{c(A)}$ and $F_{(A)}$ represent, respectively, F_c and F of soil group A. Similarly, other variables of equation (4.74) can be defined. Equation (4.74) assumes F_c (or f_c) equal to zero for all cover complexes of the soils in hydrologic group D, a reasonable assumption (Table 4.6).

The F_c–values can also be derived from Model 8 (Table 4.2), assuming that for a given land use and AMC, the hydrologic behaviour of various soils will be the same. In other words, the porosity and the depth above the impermeable layer are the same for all soils. Thus, for a known S-value for the soil of hydrologic group D of a particular land use, the resulting infiltration F for a given storm rainfall will be affected by the change of static infiltration rates of soils. Equation (4.54) can be written for F_c as

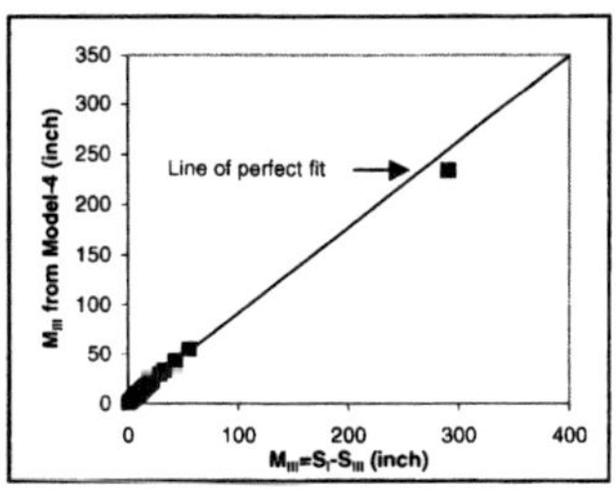

Figure 4.4. M_{III} computation for the soil of Hydrologic Group A.

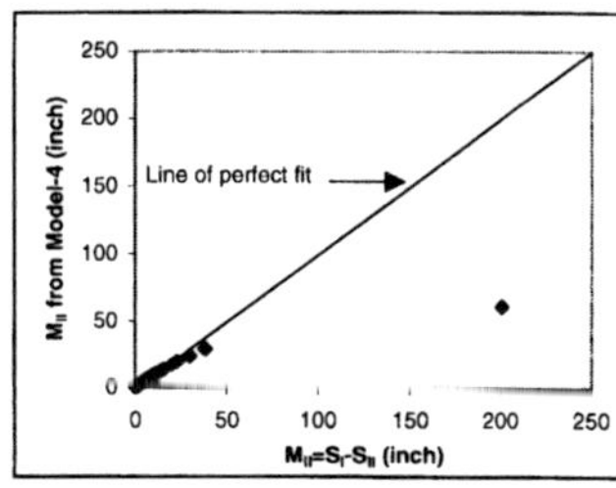

Figure 4.5. M_{II} computation for the soil of Hydrologic Group A.

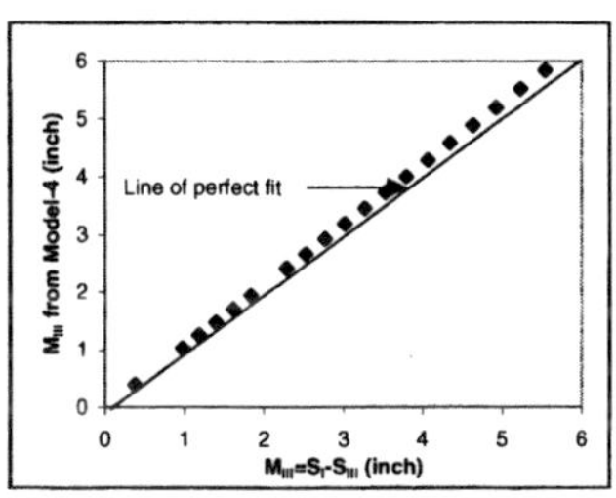

Figure 4.6. M_{III} computation for the soil of Hydrologic Group D.

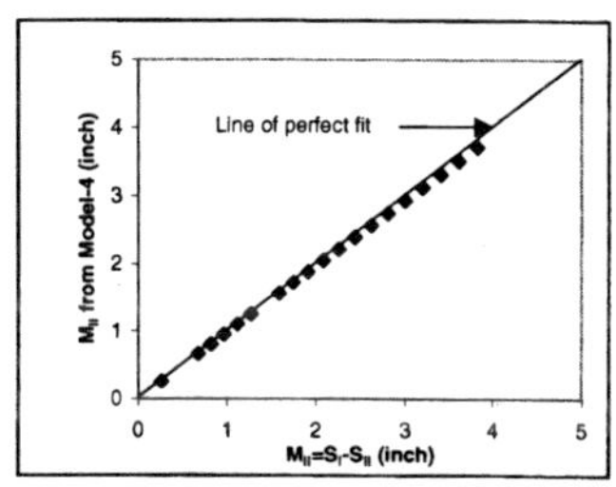

Figure 4.7. M_{II} computation for the soil of Hydrologic Group D.

$$F_c=(P-0.2S)-0.5(Q-M)-0.5\sqrt{[(Q-M)^2+4QS]} \tag{4.75}$$

Equation (4.75) implies that for given P and M, Q will vary because of changing S with the soil type. If Q is computed using the existing SCS-CN equation (equation (2.5)) that includes the impact of F_c through S, substitution of this Q in equation (4.75) along with S for $S_{I(D)}$ will yield F_c for the soil of corresponding Q. Thus, replacing S in equation (4.75) by $S_{I(D)}$ in the square root expression of equation (4.75) yields F_c, for example, soil group 'A' as:

$$F_{c\,I(A)}=(P-0.2S_{I(A)})-0.5Q_{I(A)}-0.5\sqrt{[(Q_{I(A)})^2+4Q_{I(A)}S_{I(D)}]} \tag{4.76a}$$

It is emphasized that for AMC I, M = 0. For AMC II and AMC III, the F_c-expressions for hydrologic soil group A are derived, respectively, as:

$F_{c\,II(A)}$=(P-0.2$S_{II(A)}$)-0.5($Q_{II(A)}$-$M_{II(D)}$)-0.5√[($Q_{II(A)}$-$M_{II(D)}$)^2+4$Q_{I(A)}S_{I(D)}$] (4.76b)

and

$F_{c\,III(A)}$=(P-0.2$S_{III(A)}$)-0.5($Q_{III(A)}$-$M_{III(D)}$)-0.5√[($Q_{III(A)}$-$M_{III(D)}$)^2+4$Q_{I(A)}S_{I(D)}$] (4.76c)

In this equation, $S_{I(A)}$ and $Q_{I(A)}$ represent, respectively, S_I and Q_I (= Q under AMC I) of soil group A. The initial abstraction, I_a, in equations (4.76a-c) is computed for the current level of S for the soil under consideration. Similar expressions for other soil groups can also be given. The final infiltration rates, f_c, can be computed as equal to F_c/24 inch/hr for daily rainfall-runoff values. The computation of f_c-values is illustrated below.

Example 4.5: Taking the curve numbers (CN) equal to 77, 85, 90, and 92 for AMC II for hydrologic groups A through D, respectively, of a residential lot of 1/8 acre or less with the average 65% imperviousness, compute for all soil groups and AMCs the minimum infiltration capacities of these soils. Assume storm rainfall = 5 inches.

Solution: The minimum infiltration capacity, f_c, using the NEH-4 CN-values can be derived using the following two approaches:

Approach 1

In this approach, compute Qs from equation (2.5) for all AMCs and soil groups, for example, AMC I and soil 'A' as

$$Q_{I(A)} = \frac{(P - 0.2S_{I(A)})^2}{P + 0.8S_{I(A)}} = \frac{(5.0 - 0.2x6.81)^2}{5.0 + 0.8x6.81} = 1.27 \text{ inches}$$

The cumulative infiltration F is computed using the water balance equation (equation (2.1)) as for AMC I and soil type 'A':

$F_{I(A)}$ = P - 0.2$S_{I(A)}$ - $Q_{I(A)}$ = 5.0 – 0.2 x 6.81 – 1.27 = 2.37 inches

Computations for all other soils and AMCs are shown in Table 4.5. F_c is computed from equation (4.74) as

$F_{c(A)} = F_{(A)} - F_{(D)}$

In this approach, $F_{c(D)}$ = 0. Therefore, for AMC I, for example, $F_{c(A)}$ is computed as:

$F_{c\,(A)} = F_{(A)} - F_{(D)}$ = 2.37 – 1.38 = 0.99 inch

The final infiltration rate, f_c, = 0.99/24 = 0.04 inch/hr. Other f_c-computations are shown in Table 4.5.

Approach 2

In this approach, F_c is computed using equation (4.76a) written for soil group 'A' and AMC I as

$F_{c(A)}$ = (P - 0.2$S_{I(A)}$) - 0.5$Q_{I(A)}$ - 0.5√[($Q_{I(A)}$)^2 + 4$Q_{I(A)}S_{I(D)}$]

= (5.0 - 0.2 x 6.81) – 0.5 x 1.27 – 0.5 √[1.27 x 1.27 + 4 x 1.27 x 1.98]

=1.30 inches

Similarly, F_c for all other AMCs and soils can be computed, as shown in Table 4.5.

Using the above two approaches, the f_c-values computed for 16 land uses of Table 2.1 for the same P = 100 inches, as above, are plotted for soil groups A, B, and C in Figures 4.8 through 4.10, respectively, and their statistic is shown in Table 4.6. It is seen from these figures that almost all the data points in all these figures fall on LPF and, therefore, verify the workability of both approaches. Furthermore, the f_c-statistic of Table 4.6 shows that the average values of the computed f_c's (for AMC I) are in close agreement with those given by NEH-4. Here, it is emphasized that for the inverse computation of f_c, P = 100 inches is assumed for the reason described earlier. However, it is possible to compute f_c-values for any other rainfall amount using the above procedure, as shown in the above example.

4.4.3 VERIFICATION OF AMC CRITERIA FOR f_c-VALUES

Although the hydraulic conductivity depends on the initial moisture content or the antecedent moisture (M) amount, the minimum infiltration rates are not reported to be dependent on 'M' in the literature. Assuming for a moment that they are independent of each other, it is possible to further verify the logic of the above AMC criterion. To this end, f_c-values for a soil group were computed for AMC II and AMC III, as given in Table 4.5 through the above example. These are plotted against each other in Figures 4.11-4.13 for soil groups A, B, C, respectively. Apparently, the computed f_c-values for Q_{II} are much higher than those derived for Q_{III} and the deviation from LPF increases with increasing Q_{II}. A large deviation of the data points from LPF actually undermines the AMC criterion, if the validity of F_c (or f_c)-computation for AMC I holds, as shown above. Table 4.6, however, shows that the average f_c decreases as M (or AMC) increases.

The above inference of f_c being dependent on AMC can also be asserted in view of equation (4.59) that describes the variation of f_c with time. Within the framework of the SCS-CN concept, as described in Chapter 3, parameters of equation (4.59) can be shown to be dependent on AMC analogous to the SCS-CN parameter S in the cumulative infiltration equation (equation (3.167)), which depends on AMC. However, the above inference of decreasing f_c with increasing AMC is in contrast with the increasing trend of the hydraulic conductivity, K_h, with the moisture content, as described in Chapter 1, for K_h describes the ease with which the soil water can be drained.

4.4.4 APPLICABILITY OF NEH-4 TABLES TO EXISTING AND GENERAL MODELS

Since the existing SCS-CN equation (equation (2.5)) does not account for AMC and F_c, the derivation of S from equation (2.14) corresponds to dry (or AMC I) and impermeable (f_c = 0) states of the soil. The impact of the latter is misleadingly, for reasons of simplicity, integrated with the S (or CN) values given in the NEH-4 tables. The proposed forms (Table 4.2) of the general model with M = f_c = 0 are valid for the same dry, impermeable conditions. Thus, the curve numbers that correspond to AMC I and hydrologic group D in NEH-4 table are applicable to such situations only. It leads to infer that the applicability criterion ($S \le 2P$ or $CN \ge 1000/(10+2P)$) of the existing SCS-CN method (Chapter 3: equation (3.115)) also corresponds to the fully dry condition of the soil or AMC I. Therefore, employing equations (3.143) and (3.144), criteria for AMC I, II, and III can be written in terms of CN_I, CN_{II}, and CN_{III}, respectively, as

$$CN_I \ge \frac{1000}{10+2P} \tag{4.77a}$$

$$CN_{II} \ge \frac{1000}{10+0.88P} \tag{4.77b}$$

$$CN_{III} \ge \frac{1000}{10+0.37P} \tag{4.77c}$$

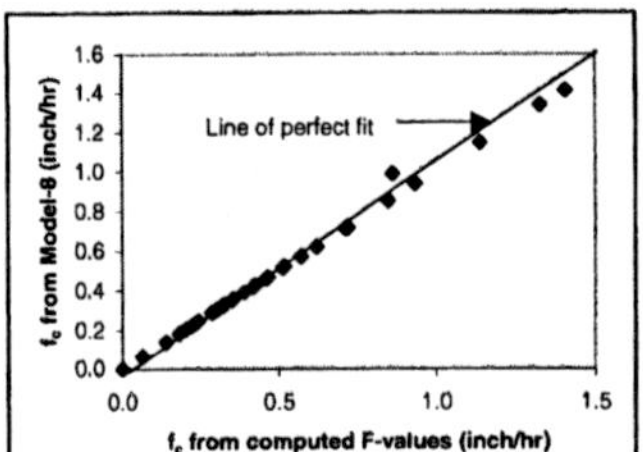

Figure 4.8. f_c computation from two approaches for Hydrologic Group A soil.

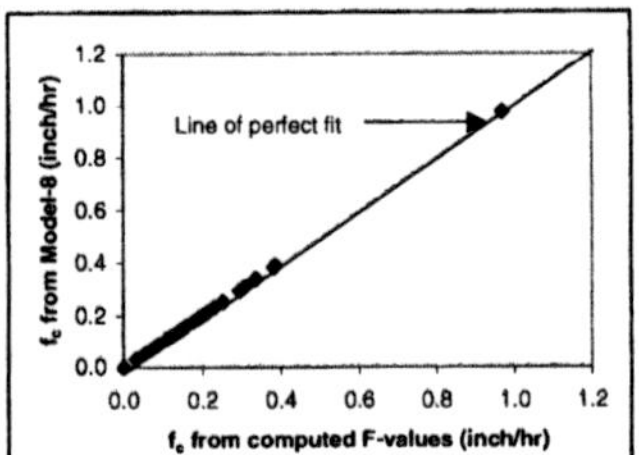

Figure 4.9. f_c computation from two approaches for hydrologic group B soil.

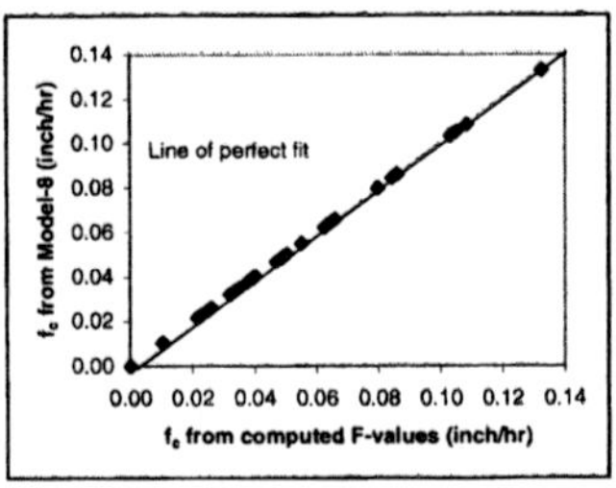

Figure 4.10. f_c computation from two approaches for hydrologic group C soil.

As an example, if P = 1 inch, $CN_{II} \ge 92$; if P = 5 inches, $CN_{II} \ge 69$; and if P = 10 inches, $CN_{II} \ge 50$ for the applicability of the existing SCS-CN method. Thus, it is possible to ascertain the applicability of the existing SCS-CN method only from the given amount of rainfall and the watershed characteristics yielding a CN-value from the NEH-4 table (Table 2.1). For computing Q from the general Model 8 (Table 4.2), F_c can be derived from field data (f_c and T_s); 'M' can be computed from the observed P_5 using equation (4.72), in which S_I for a cover complex corresponds to the NEH-4 table values for hydrologic group D; and $S = S_I - M$. Such an approach also eliminates sudden jumps from one AMC to another AMC or one hydrologic group to another group in the CN determination.

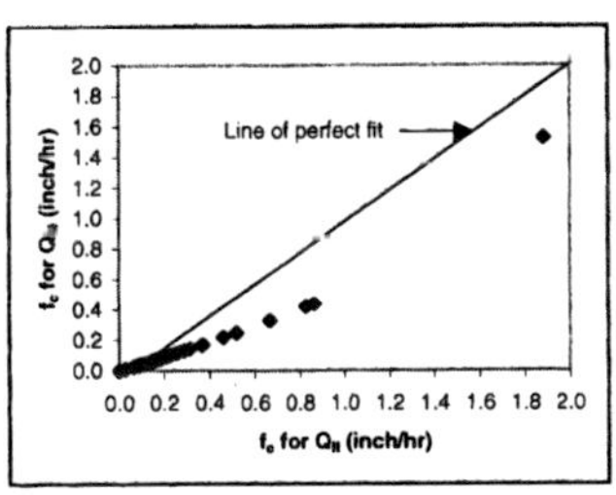

Figure 4.11. Computed f_c (inch/hr) for Q_{II} and Q_{III} for hydrologic group A.

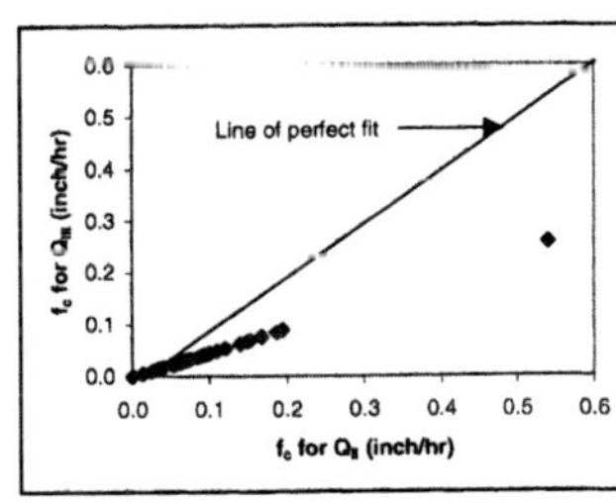

Figure 4.12. Computed f_c (inch/hr) for Q_{II} and Q_{III} for hydrologic group B.

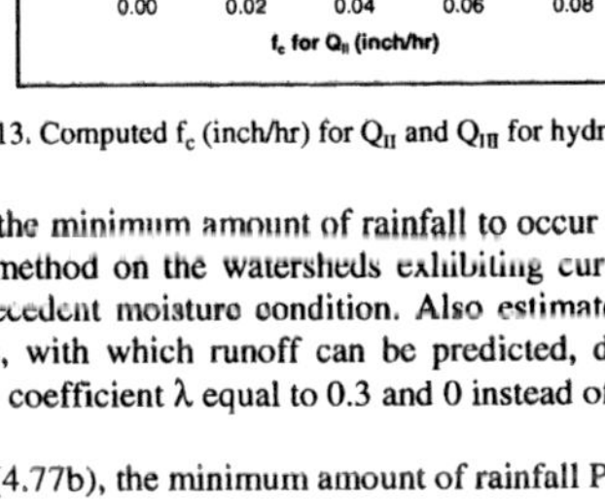

Figure 4.13. Computed f_c (inch/hr) for Q_{II} and Q_{III} for hydrologic group C.

Example 4.6: Determine the minimum amount of rainfall to occur for the applicability of the existing SCS-CN method on the watersheds exhibiting curve numbers 92, 67, and 32 under normal antecedent moisture condition. Also estimate the corresponding maximum possible errors, with which runoff can be predicted, due to an erroneous guess of initial abstraction coefficient λ equal to 0.3 and 0 instead of 0.2.

Solution: From equation (4.77b), the minimum amount of rainfall P can be given as:

$$P \geq 1.14\left[\frac{1000}{CN_{II}} - 10\right]$$

Thus, for $CN_{II} = 92$,

$P \geq 1.14 \times (1000/92 - 10) = 0.99$ inch.

Similarly, for $CN_{II} = 67$, $P \geq 5.61$ inches and for $CN_{II} = 32$, $P \geq 24.23$ inches.

For computing errors in runoff prediction due to error in λ, equation (2.28) can be used. Here, $\lambda = 0.2$; $P = 0.99$, 5.61, and 24.23 inches. Thus, I_a (= 0.2S)-values can be computed for the corresponding CN values as 0.174 inch, 0.985 inch, 4.25 inches, respectively. The corresponding I_a/P-value for all the cases is approximately equal to 0.175. Therefore, from equation (2.28)

$$\frac{\partial C}{\partial \lambda}\frac{\lambda}{C} = -\frac{\frac{I_a}{P}\left[1 + \frac{2(1-\lambda)}{\lambda}\frac{I_a}{P} - \frac{2-\lambda}{\lambda}\left(\frac{I_a}{P}\right)^2\right]}{\left(1 - \frac{I_a}{P}\right)^2\left(1 + \frac{1-\lambda}{\lambda}\frac{I_a}{P}\right)}$$

$$= -\frac{0.175\{1 + (2x0.8/0.2)x0.175 - (1.8/0.2)x(0.175)^2\}}{(1 - 0.175)^2(1 + (0.8/0.2)x0.175)}$$

$$= -\frac{0.175[1 + 1.40 - 0.276]}{0.681(1 + 0.70)} = -0.321$$

Thus, $\partial C/C$ = -0.321 x $(\partial\lambda/\lambda)$ = -0.321 x 0.1/0.2 = -0.161, which represents the maximum possible error in the runoff estimation due to the erroneous λ-value equal to 0.3, as shown in Chapter 2. Therefore, the computed runoff can be less than the actual by 16.1% for $\lambda = 0.3$. Similarly, if λ is taken equal to 0 instead of 0.2, the error will be equal to [–0.321 x (-0.2/0.2) x 100], which is simplified as equal to 32.1%. It implies that runoff will be overestimated by 32.1% if λ is taken equal to 0.

Example 4.7: Test the validity of the SCS-CN method to the data of Waco and Amicacola Creek watersheds shown in Tables 4.7a & b, respectively. The corresponding AMC I through AMC III CN values for the respective watersheds are 76, 89, 96 and 45, 65,82.

Solution: For testing the validity of the SCS-CN method, the applicability criteria (equations (4.77a through c)) are employed for the rainfall (P)-runoff (Q) data as shown in Tables 4.8a & b. In these tables, the runoff values have been computed for various events using the corresponding AMC dependent CN-values (Table 2.6). The corresponding SCS-CN applicability is shown in terms of 'yes' and 'no' for each event. The root mean square error (RMSE) value is 0.53 for the Waco data (Table 4.8a) and 1.22 for Amicalola Creek data (Table 4.8b). According to the criterion, the SCS-CN method is applicable to all of the events of the Waco watershed. On the other hand, the method is not applicable to any of the events of Amicalola Creek data. To test the criterion, the runoff values computed by the SCS-CN method are plotted against the observed values, as shown in Figures 4.14a & b for Waco and Amicalola watersheds, respectively. Apparently, the data points of the Waco watershed (Figure 4.14a) fall close to the line of perfect fit (LPF). For the Amicalola Creek watershed, the data points significantly deviate from LPF (Figure 4.14b), indicating a poor match between the observed and computed values. This shows that the SCS-CN method is applicable to the Waco watershed events, whereas it is not applicable to the events of the Amicalola watershed. It is noted that SCS (1971) provided a similar argument for negating the applicability of the SCS-CN method to the Amicalola watershed.

4.4.5 CONDENSATION OF NEH-4 TABLE

Within the premise of the volumetric concept, the CN-values for any soil type are derived with reference to soil group D, for which f_c-values are known or assumed to be zero. To this end, the above S_I-values of the hydrologic soil group D of NEH-4 table excluding the value corresponding to the above-said CN = 6 are related to those of A, B, and C soils, as shown in Figure 4.15, exhibiting the following relations:

$$S_{I(A)} = 1.9590[S_{I(D)}]^{1.6439}; r^2 = 0.9089 \quad (4.78a)$$

$$S_{I(B)} = 1.5577[S_{I(D)}]^{1.2519}; r^2 = 0.9382 \quad (4.78b)$$

$$S_{I(C)} = 1.2061[S_{I(D)}]^{1.0799}; r^2 = 0.9779 \quad (4.78c)$$

The workability of these empirical relations is illustrated below.

Table 4.7a Rainfall-runoff events of Waco watershed

Event	Date			Observed		AMC
	Year	Month	Day	Rainfall (inch)	Runoff (inch)	
1	1940	Nov.	22	9.35	5.99	I
2	1941	June	10	2.39	2.05	III
3	1942	Sept.	7	8.03	2.83	I
4	1943	June	5	1.58	0.51	I
5	1944	April	29	13.74	10.76	I
6	1945	March	2	3.27	2.38	I
7	1946	May	12	3.85	2.95	III
8	1947	March	18	1.74	0.85	I
9	1948	April	25	3.1	1.17	I
10	1949	July	4	2.86	1.07	I
11	1950	Feb.	12	1.94	1.09	III
12	1951	June	16	1.64	0.19	II

Table 4.7b Rainfall-runoff events of Amicacola Creek watershed

Event	Date			Observed		AMC
	Year	Month	Day	Rainfall (inch)	Runoff (inch)	
1	1940	Aug.	13	4.99	0.81	I
2	1941	July	5	5.72	1.4	III
3	1942	Feb.	17	5.24	1.74	I
4	1943	Dec.	29	4.31	1.65	III
5	1944	Mar.	19	3.8	1.16	I
6	1945	Feb.	13	1.95	0.36	I
7	1946	Feb.	10	5.39	2.33	III
8	1947	Jan.	20	4.05	1.59	III
9	1948	Aug.	4	5.69	1.36	I
10	1949	Nov.	28	5.59	1.85	III
11	1950	Mar.	13	3.77	1.15	III
12	1951	Mar.	29	4.71	1.33	I
13	1952	Mar.	11	3.83	2.01	I

Table 4.8a. Evaluation of the validity of the SCS-CN method to Waco watershed (λ=0.2 and CN = 76, 89, 96 for AMC I through AMC III, respectively).

Event	Observed		AMC	CN	Computed Runoff (inch)	SCS-CN applicability	
	Rainfall (inch)	Runoff (inch)				CN>	Applicability
1	9.35	5.99	I	76	6.4	35	Yes
2	2.39	2.05	III	96	1.95	92	Yes
3	8.03	2.83	I	76	2.02	38	Yes
4	1.58	0.51	I	76	0.22	76	Yes
5	13.74	10.76	I	76	10.56	27	Yes
6	3.27	2.38	I	76	1.20	60	Yes
7	3.85	2.95	III	96	3.39	88	Yes
8	1.74	0.85	I	76	0.29	74	Yes
9	3.1	1.17	I	76	1.08	62	Yes
10	2.86	1.07	I	76	0.92	64	Yes
11	1.94	1.09	III	96	1.52	93	Yes
12	1.64	0.19	II	89	0.74	87	Yes
Root Mean Square Error (inch)					0.53		

Table 4.8b Evaluation of the validity of the SCS-CN method to Creek watershed (λ=0.2 and CN = 45, 65,82 for AMC I through AMC III, respectively)

Event	Observed		AMC	CN	Computed Runoff (inch)	SCS-CN Validity	
	Rainfall (inch)	Runoff (inch)				CN>	Applicability
1	4.99	0.81	I	45	0.44	50	No
2	5.72	1.4	III	82	3.73	83	No
3	5.24	1.74	I	45	0.52	49	No
4	4.31	1.65	III	82	2.47	86	No
5	3.8	1.16	I	45	0.14	57	No
6	1.95	0.36	I	45	0.02	72	No
7	5.39	2.33	III	82	3.43	83	No
8	4.05	1.59	III	82	2.25	87	No
9	5.69	1.36	I	45	0.68	47	No
10	5.59	1.85	III	82	3.61	83	No
11	3.77	1.15	III	82	2.01	88	No
12	4.71	1.33	I	45	0.35	51	No
13	3.83	2.01	I	45	0.14	57	No
Root Mean Square Error					1.22		

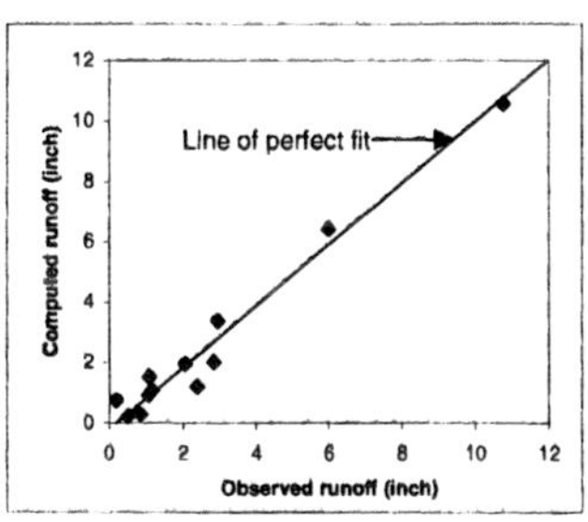

(a)

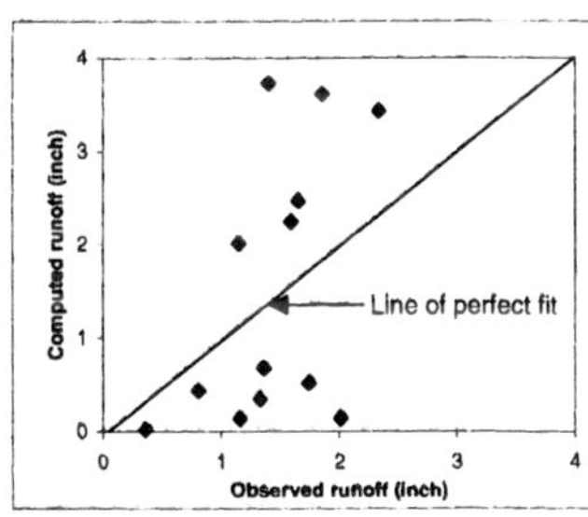

(b)

Figure 4.14. Correlation between obeserved and computed runoff values of (a) Waco watershed and (b) Amicacola Creek watershed.

Example 4.8: Taking the curve number CN for AMC II equal to 92 for the hydrologic group D of a residential lot of 1/8 acre or size with the average 65% imperviousness, compute CNs for hydrologic groups A, B, and C without the aid of the NEH-4 table, compare the derived CN values for AMC II with those from NEH-4 table for the same AMC, and discuss the implication.

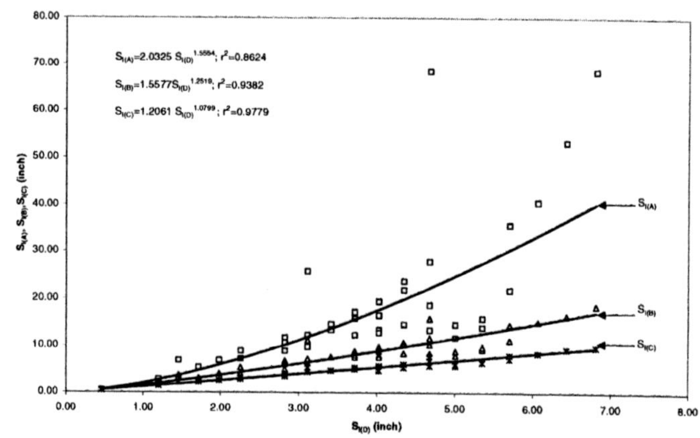

Figure 4.15. Relation between $S_{I(D)}$ and $S_{I(A)}$, $S_{I(B)}$, and $S_{I(C)}$.

Solution: For $CN_{II(D)} = 92$, $S_{I(D)}$ can be computed from equation (2.7) as equal to 1.98 inches, as shown in Table 4.5. Using equations (4.78a-c), S_I-values for the hydrologic groups A, B, and C can be derived, respectively, as

$$S_{I(A)} = 2.0325[S_{I(DD)}]^{1.5554} = 2.0325 \times 1.98^{1.5554} = 5.88 \text{ inches}$$

$$S_{I(B)} = 1.5577[S_{I(D)}]^{1.2519} = 1.5577 \times 1.98^{1.2519} = 3.66 \text{ inches}$$

$$S_{I(C)} = 1.2061[S_{I(D)}]^{1.0799} = 1.2061 \times 1.98^{1.0799} = 2.52 \text{ inches}$$

Using equation (3.143), S_{II} for soil groups A, B, and C can be computed, respectively, as

$$S_{II(A)} = S_{I(A)} / 2.281 = 5.88 / 2.281 = 2.58 \text{ inches}$$

$$S_{II(B)} = S_{I(B)} / 2.281 = 3.66 / 2.281 = 1.61 \text{ inches}$$

$$S_{II(C)} = S_{I(C)} / 2.281 = 2.52 / 2.281 = 1.11 \text{ inches}$$

Using equation (2.7), the corresponding CN_{II}-values for hydrologic groups A, B, and C can be determined as

$$CN_{II(A)} = \frac{1000}{10 + S_{II(A)}} = \frac{1000}{10 + 2.58} = 79$$

$$CN_{II(B)} = \frac{1000}{10 + S_{II(B)}} = \frac{1000}{10 + 1.61} = 86$$

$$CN_{II(C)} = \frac{1000}{10 + S_{II(C)}} = \frac{1000}{10 + 1.11} = 90$$

Corresponding to the land use characteristics of the watershed, the CN values for soil groups A, B, and C for AMC II can be obtained from the given NEH-4 values (Table 2.1). These are respectively as 77, 85, and 90. These CN values for respective soils are fairly close to those computed from the empirical formulae as 79, 86, and 90, respectively. It is, however, noted that as soil group changes from C to A, the deviation (although small) of these CN-values increases. The implication of such a CN-derivation is that within the premise of the workability of equations (3.143) and (3.155) and equations (4.78a-c), the CN-values of the NEH-4 table condense to the basic curve numbers CN_I (or S_I or S_{abs}) that correspond to dry soils of the hydrologic group D.

4.5 Advantages and Limitations of the Modified Model

The following are the advantages of using the modified method (equation (4.54)) over the existing method (equation (2.5)):

1. The existing version is a specific case of the modified version and thus, the latter is more general than the former.
2. The model parameters can be physically determined.
3. Since median S_I is more reliable than the median S that varies wildly with the antecedent moisture, the modified method is more reliable.
4. The modified version eliminates, to a great extent, the problem of accounting for the antecedent moisture.
5. For gauged watersheds, the method relies on the rainfall-runoff data, which are the observed input and output for deriving a single S_I-value for a watershed.
6. For ungauged watersheds, NEH-4's S (or CN)-values corresponding to AMC I and hydrologic soil group D are equivalent to S_I-values and, therefore, the NEH-4 tables are also useful for the modified method.

The following are the disadvantages of the modified version:

1. The existing SCS-CN method relies on only a single parameter S (or CN) that is derived from the NEH-4 tables for a wide range of geophysical settings. On the other hand, the modified version requires a prior knowledge of the minimum infiltration rates.
2. The modified version is a 2-parameter (for $\lambda = 0.2$) model whereas the existing version is a 1-parameter model, and thus, the former is of increased complexity.
3. The existing version is supported by myriad field applications and professional experience whereas the modified method is not.

CHAPTER 5

DETERMINATION OF 'S' USING PHYSICAL PRINCIPLES

In the previous chapter, the SCN-CN hypothesis was described using the C (runoff factor) = S_r (degree of saturation) concept derived from the volumetric concept described in Chapter 3. The SCS-CN parameter S was shown to represent the maximum possible retention of water in the soil profile or the maximum cumulative amount of dynamic infiltration. This description is closely associated with the discussion of Mockus (1964) (Chapter 2) on the physical significance of S. The Mockus description, however, does not distinguish the dynamic portion of infiltration from the static one and takes into account only the volumetric properties of the soil, excluding its transmission characteristics. His description also limits S to the value of the maximum infiltration rate occurring at the beginning of the infiltration phenomenon. Except for an unreasonable comparison between infiltration rate and volumetric space, Mockus attempted to account for both the volumetric and transmission properties of soils, which are fundamental to the description of a dynamic system. Defining S by the amount of water storage available in the soil profile, one excludes the transmission property of the soil.

To describe a dynamic system, it is necessary to account for both the storage and transmission properties of the medium. For water transmission through soil systems, soil sorptivity, S_s, (Philip, 1957, 1969) and hydraulic conductivity, K_h, describe, respectively, the volumetric and transmission characteristics of the soil. Both S_s and K_h depend on (1) soil characteristics (type, texture, and structure) and initial moisture condition; (2) land use; (3) rainfall intensity; and (4) turbidity and temperature. A detailed description of these properties is available in the textbooks by Hillel (1976) and Marshall et al. (1996).

Succinctly, S_s and K_h can be described by the ψ (negative pressure)-θ (moisture content) and K_h (hydraulic conductivity)-θ relations (Chapter 1), respectively, which characterize the moisture movement in unsaturated porous media (Collis-George, 1977). In such media, the adsorption and adhesion-cohesion forces resulting in concave air-water interfaces (also known as matric forces) lead to the capillary action responsible for water retention. The matric suction is a function of soil type and moisture content, θ; the greater the moisture content, the lower will be the suction force and vice versa (Moore, 1939). This phenomenon can also be explained in terms of the potential defined as the work required to transfer a unit quantity of water from a standard reference state, viz. zero, to the other. Whether water transfer actually takes place depends on the conductivity of the unsaturated soil or capillary conductivity that depends not only on the pore geometry but also on the water content.

In this chapter, the potential maximum retention, S, of the SCS-CN method is discussed in terms of ψ-θ and K_h-θ relations that describe the physical characteristics of the soil-vegetation-land use complex. The physical significance of S is explained using the diffusion term of the linear Fokker-Planck equation for infiltration, which relates S to the storage and transmission properties of the soil. Based on the physical description of S/P relations, the modified version of the SCS-CN model including the antecedent moisture and static infiltration, described in Chapter 4, is characterized. Finally, a procedure based on the universal soil loss equation is suggested for determination of the soil depth above the impeding layer, D_s, described in Chapters 2, 3, and 4.

5.1 Fokker-Planck Equation of Infiltration

For describing S in terms of the storage and transmission properties of soils, the linear Fokker-Planck equation for flow in unsaturated media (Philip, 1974), given in Chapter 1, is utilized. This is re-written here:

$$\frac{\partial \theta}{\partial t} + \frac{dK_h}{d\theta}\frac{\partial \theta}{\partial z} = D\frac{\partial^2 \theta}{\partial z^2} \tag{1.55}$$

The non-linear form of equation (1.55) (with its right-hand side equal to $\partial[D(\theta)(\partial\theta/\partial z)]/\partial z$) is derived from the combination of Darcy law, continuity equation, energy equation (or Buckingam law), and $\psi = \psi(\theta)$ relation (Philip, 1969; Feddes et al., 1988) (Chapter 1). In equation (1.55), z and t are the space and time coordinates, respectively. The value of z is measured from the soil surface positive down below, and D is the moisture diffusivity defined as (Philip, 1969) (Chapter 1):

$$D = K_h\, d\psi/d\theta \tag{1.56}$$

The phenomenon of soil water flow is highly nonlinear and is described by the non-linear Fokker–Planck equation. The nonlinearity is caused by the dependence of both the hydraulic conductivity and soil water pressure on the soil moisture content. This nonlinear equation has not yet been solved under general conditions. It has, however, been solved only for simplified flow conditions under a number of restrictive assumptions (Feddes et al., 1988). Therefore, numerical modelling is frequently resorted to, for example, CREAMS (Knisel, 1980), LEACHM (Wagenet and Hutson, 1987), MACRO (Jarvis et al., 1991), and WAVE (Vanclooster et al., 1994) are a few popular numerical models among many others. Owing to the fact that the SCS-CN method is a simple, one-parameter, event-based, lumped model that works with the mean values (Ponce and Hawkins, 1996), its parameter S is derived analytically from linear Fokker-Planck equation (Eq. 6) using the linear perturbation theory. Here, the diffusion term D is assumed constant in accordance with the SCS-CN assumption of the constant parameter S (or CN) for a rainstorm. It is acknowledged that for soils, D is seldom constant and as a result, linearization of the Fokker-Planck equation is only an assumption which has long been employed in hydrology. A brief discussion of other limitations of such an analysis follows.

Harmonic analysis of a wave phenomenon forms an initial step to gain insight to the mechanism of a system, whether linear or non-linear, in several branches of mechanics and physics (Crawford, 1965; Pedloskey, 1979) and the resulting linear solutions often prove adequate for practical purposes (Kundzewicz and Dooge, 1989). In open channel hydraulics, linear analyses have been carried out assuming either complex frequency and real wave number (Ponce and Simons, 1977) or real frequency and complex wave number (Vreugdenhill, 1972; De Souza and Price, 1980) and both yield an incomplete description of wave characteristics (Kundzewicz and Dooge, 1989). A feasible solution for complex frequency and wave number is not possible because of the unknown relation between these complex parameters. In the following analytical derivation, the approach of Ponce and Simons (1977) is utilised for the reason that the derived wave criteria (Ponce et al., 1978) perform satisfactorily in field applications (Mishra and Seth, 1996; Mishra et al., 1997; Mishra and Singh, 1999, 2001a,b; and others).

Assuming that K_h-θ and ψ-θ relations follow a power relation (Campbell, 1974):

$$K_h = a\,\theta^m \tag{5.1}$$

$$\psi = a_1\,\theta^{m_1} \tag{5.2}$$

where a and a_1 are coefficients, and m and m_1 are exponents. Equation (5.2) was proposed by Visser (1969). For other forms ψ-θ relations, the work of Mallants et al. (1998) can be referred. Campbell (1974) tested these relations on five soils, Botany, Guelph, Cecil Ap, Cecil B2, and Cecil B3, in θ ranges: 0.06-0.36, 0.25-0.53, 0.05-0.40, 0.28-0.47, and 0.23-0.40 (volume/volume), respectively. Gardener et al. (1970) also found equation (5.2) to be sufficiently accurate over a limited range. Following Ponce and Simons (1977), the characteristic length χ (Todd, 1980) can be defined as

$$\chi = \frac{\psi_o}{\partial\psi/\partial z} \tag{5.3}$$

where ψ_o is the negative pressure at the initial moisture content, θ_o. In equation (5.3), $\partial\psi/\partial z$ can be replaced by $\partial\theta/\partial z$ using the ψ-θ relation (equation (5.2)). From equations (5.1) and (5.2) the derivatives for K_h and ψ can, respectively, be derived as

$$\frac{dK_h}{d\theta} = m\,a\,\theta^{m-1} = m\frac{K_h}{\theta} \tag{5.4}$$

and

$$\frac{d\psi}{d\theta} = m_1\,a_1\,\theta^{m_1-1} = m_1\left(\frac{\psi}{\theta}\right) \tag{5.5}$$

Substitution of equation (5.5) into equation (1.56) leads to

$$D = m_1\left(\frac{K_h}{\theta}\right)\psi \tag{5.6}$$

Equation (5.6) implies that for a given θ, D is directly proportional to the energy expended in unit time (or power) per unit weight of the absorbate, and the proportionality constant m_1 is a ψ-θ characteristic as mentioned above.

Using equations (5.4) and (5.6), equation (1.55) can be linearized as

$$\frac{\partial\theta}{\partial t}+\frac{m K_{ho}}{\theta_o}\frac{\partial\theta}{\partial z}=\frac{m_1 K_{ho}\Psi_o}{\theta_o}\frac{\partial^2\theta}{\partial z^2} \tag{5.7}$$

where K_{ho} and ψ_o correspond to θ_o, a reference moisture content. For a linear perturbation analysis, the perturbed quantity of θ is defined as:

$$\theta' = \theta - \theta_o \tag{5.8}$$

where θ' is the perturbation about θ_o. Equation (5.7) can be re-written as:

$$\frac{\partial\theta'}{\partial t}+\frac{m K_{ho}}{\theta_o}\frac{\partial\theta'}{\partial z}=\frac{m_1 K_{ho}\Psi_o}{\theta_o}\frac{\partial^2\theta'}{\partial z^2} \tag{5.9}$$

For the foregoing wave analysis, (θ'/θ) is assumed to follow a sinusoidal form as

$$\frac{\theta'}{\theta_o}=\Omega\ \exp[i(\hat{\sigma}\hat{z}-\hat{\beta}\hat{t})] \tag{5.10}$$

where

$$\hat{\sigma}=\frac{2\pi}{L}\chi \tag{5.11a}$$

$$\hat{\beta}=\frac{2\pi}{T}\left(\frac{\chi}{K_{ho}}\right) \tag{5.11b}$$

$$\hat{z}=\frac{z}{\chi} \tag{5.11c}$$

$$\hat{t}=\frac{t K_{ho}}{\chi} \tag{5.11d}$$

Ω is the non-dimensional wave amplitude, L is the wave length, T is the time period, $\hat{\sigma}$ is the dimensionless wave number, $\hat{\beta}$ is the dimensionless frequency factor, and 'i' is the imaginary unit.

From equations (5.9), (5.10), and (5.11a-d) the following characteristic equation can be derived:

$$\hat{\beta}=\frac{m}{\theta_o}\hat{\sigma}-i\left(\frac{m_1\Psi_o}{\theta_o\chi}\right)\hat{\sigma}^2 \tag{5.12}$$

from which real and imaginary components can be extracted, respectively, as:

$$\hat{\beta}_R = \frac{m}{\theta_o}\hat{\sigma} \tag{5.13a}$$

and

$$\hat{\beta}_I = -\frac{m_I \Psi_o}{\theta_o \chi}\hat{\sigma}^2 \tag{5.13b}$$

By definition, the non-dimensional wave celerity ($\hat{c}$) and logarithmic decrement (δ) of the wave can, respectively, be derived as follows:

$$\hat{c} = \frac{\hat{\beta}_R}{\hat{\sigma}} = \frac{m}{\theta_o} \tag{5.14}$$

and

$$\delta = 2\pi\frac{\hat{\beta}_I}{\hat{\beta}_R} = -2\pi(\frac{m_I}{m})(\frac{\Psi_o}{\chi})\hat{\sigma} \tag{5.15}$$

where θ_o is the moisture content at $t = 0$; $\hat{c} = c/K_{ho}$, c is the dimensional wave celerity; the '-' sign indicates that the wave propagates in the z-direction and attenuates; and the '+' sign indicates that the wave travels opposite to the z-direction and amplifies. From the above, c and δ can also be expressed, respectively, as

$$c = \frac{m\,K_{ho}}{\theta_o} \tag{5.16}$$

and

$$\delta = -2\pi(\frac{D}{c\chi})\hat{\sigma} \tag{5.17}$$

The derivation of equation (5.16) is analogous to deriving the Seddon speed from the stage-discharge relation in open channel flow (Mishra and Singh, 2001b) (Appendix C). Here K_{ho}/θ_o represents the velocity of the moisture movement (Young, 1957). Equation (5.17) can also be written, alternatively, as

$$\delta = -4\pi^2\frac{m_I \Psi_o \theta_o}{m^2 K_{ho} T} \tag{5.18}$$

For a physical description, δ can also be defined as (Ponce et al, 1978):

$$\delta = \ln(\theta_T) - \ln(\theta_i) \tag{5.19}$$

or

$$\theta_T = \theta_i e^{\delta} \tag{5.20a}$$

where θ_i and θ_T are the initial moisture content and the moisture content at time, t = T, respectively.

Equation (5.20a) can, alternatively, be written in terms of the water column equivalent to F as

$$F_T = F_i e^{\delta} \tag{5.20b}$$

where F_T and F_i represent the cumulative infiltration at times T and i (initial), respectively. Similarly, for the maximum cumulative infiltration values or S, equation (5.20b) can be recast as

$$S_T = S_i e^{\delta} \tag{5.20c}$$

where S_T is the potential maximum retention after the time period, T, and S_i is the initial value of S.

Since equation (1.55) describes a diffusion process, δ will be a negative quantity and, therefore, $S_T < S_i$. Equations (5.20b) and (5.20c) help describe the cyclic effects of the absorption-desorption process of cumulative infiltration. In an absorption process, F starts from 0 and ends at S and the following desorption process takes F to 0. In the subsequent absorption cycle, F again reaches S and so on. Here, S_i corresponds to S of the first cycle and S_T stands for S of the second cycle. It is for this reason that the maximum cumulative infiltration or S is of prime interest here, analogous to the flood peak description in flood routing. Thus, the same soil will yield different S-values in the absorption-desorption cyclic process owing to the hysteretic effect (Chang Kun and Jung Ho, 1993). The difference in S, however, depends on the magnitude of δ, which, in turn, depends on K_h-θ, ψ-θ, initial moisture content, and the time period.

Parallel to the hysteresis-based studies dealing with open channel flows (Mishra and Seth, 1996; Mishra and Singh, 1999, 2001a; and others), the time period, T, can be described approximately equal to twice the time base of infiltration, T_b, or $T_b = T/2$. Thus, δ can be re-written in terms of T_b as

$$\delta = -2\pi^2 \frac{m_1 \psi_o \theta_o}{m^2 K_{ho} T_b} \tag{5.21}$$

which can be derived from soil characteristics: ψ-θ and K_h-θ relations, initial moisture content, and infiltration data, as illustrated below.

Example 5.1. Compute the logarithmic decrement using the infiltration data of Columbia sandy loam, Project No. 34 of Minnesota soil depicted in Figure 4.1. ψ-θ and K_h-θ relations are given in Table 5.1 and initial moisture content = 0.125 (vol./vol.).

Solution: Using the data of Table 5.1, K_h-θ and ψ-θ power-relations are developed, as shown in Figure 5.1, as:

$y = 0.9107\, x^{9.0273}$; $r^2 = 0.9339$

and

$y = -6.4616\, x^{-1.482}$; $r^2 = 0.9213$

where y is K_h (cm/sec) in the first relation and it is ψ (cm) in the second relation, and x is θ (vol/vol). A comparison of these relations with those given by equations yield m = 9.0273 and m_l = -1.482. From Figure 4.1, the time base of the infiltration curve, T_b = 914 sec and from Table 5.1 ψ_o = -209.9 cm and $K_{ho} = 2.14\text{x}10^{-10}$ cm/sec for θ_o = 0.125. Substituting these values into equation (5.21), the logarithmic decrement, δ, can be computed as

$$\delta = -2\pi^2 \frac{m_l \psi_o \theta_o}{m^2 K_{ho} T_b} = -2\pi^2 \frac{(-1.482)(-209.9)\text{x}0.125}{(9.0273)^2 \text{x} 2.14\text{x}10^{-10} \text{x} 914} = -4.815\text{x}10^7$$

The resulting negative sign indicates that wave subsides during its travel.

Table 5.1 ψ-θ and K_h-θ relations for Columbia sandy loam (Project no. 34) (Data source: Mein and Larson (1971))

Sl. No.	θ (vol./vol.)	ψ (cm)	K_h (cm/sec)	Sl. No.	θ (vol./vol.)	ψ (cm)	K_h (cm/sec)
1	0.121	-281.3	3.35E-10	22	0.367	-28.9	0.000178
2	0.124	-218.8	2.28E-10	23	0.383	-27.9	0.000228
3	0.132	-156.3	1.39E-10	24	0.401	-26.9	0.000293
4	0.139	-130.0	1.38E-08	25	0.423	-25.7	0.000395
5	0.145	-104.4	3.29E-08	26	0.445	-24.7	0.000513
6	0.152	-95.9	5.75E-08	27	0.481	-22.8	0.000816
7	0.160	-89.4	9.23E-08	28	0.496	-21.9	0.001006
8	0.165	-82.0	1.64E-07	29	0.504	-21.3	0.001154
9	0.170	-75.2	2.95E-07	30	0.509	-20.7	0.001226
10	0.178	-69.4	5.32E-07	31	0.513	-20.2	0.001250
11	0.186	-64.4	8.30E-07	32	0.5145	-19.4	0.001279
12	0.197	-59.4	1.58E-06	33	0.5155	-18.7	0.001302
13	0.212	-53.5	2.89E-06	34	0.516	-17.6	0.001322
14	0.230	-48.1	6.67E-06	35	0.5162	-15.6	0.001350
15	0.249	-43.8	1.10E-05	36	0.5165	-13.1	0.001364
16	0.263	-40.4	1.89E-05	37	0.517	-8.6	0.001375
17	0.277	-38.1	2.88E-05	38	0.5175	-4.1	0.001383
18	0.288	-36.2	3.99E-05	39	0.518	0.0	0.001390
19	0.301	-34.8	5.34E-05	40	0.5185	3.1	0.001390
20	0.315	-33.1	7.30E-05	41	0.519	6.3	0.001390
21	0.338	-31.0	0.000112				

Note: Data are linearly interpolated

5.2 Description of S

5.2.1 USE OF S_s AND K_h

Collis-George (1977) related the maximum dynamic infiltration amount with the soil sorptivity, S_s, and hydraulic conductivity, K_h. This relation can be described in terms of S as:

$$S = \frac{\pi}{4}\frac{S_s^2}{K_h} \tag{5.22}$$

which can be derived from equation (1.55). According to Collis-George (1977), equation (5.22) is difficult to use over the whole time range of infiltration unless the experimentally derived maximum cumulative infiltration (or S) is appropriate for the S_s and K_h values. At small values of time, equation (5.22) gives a good description of the maximum cumulative infiltration or S. Therefore, he generalized this equation in terms of a parameter N (Philip, 1974) that integrates the effects of capillary and gravitational phenomena as:

$$N = \frac{S}{K_h t_c} = \frac{S_s^2}{SK_h} = \frac{S_s}{K_h (t_c)^{1/2}} \tag{5.23}$$

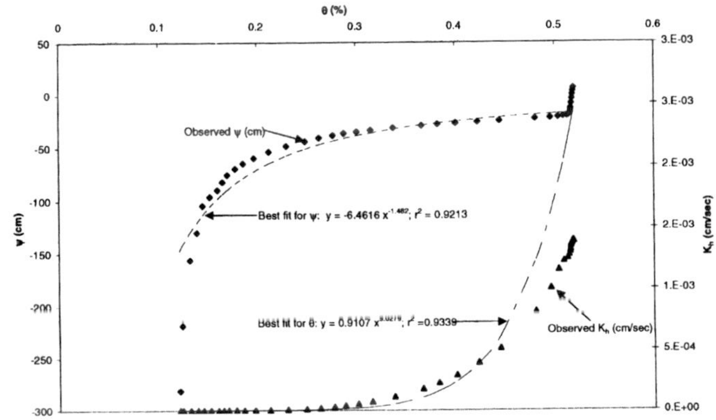

Figure 5.1. ψ-θ and K_h-θ relations for Columbia sandy loam.

where S represents the integrated effects of capillary phenomenon, and the product of K_h and t_c of the gravitational phenomenon. Thus, using equation (5.23), equation (5.22) can be generalized as

$$S = \frac{1}{N}\frac{S_s^2}{K_h} \tag{5.24}$$

For N = 4/π, equation (5.24) reduces to equation (5.22).

5.2.2 USE OF K_h-θ AND ψ-θ RELATIONS

For defining the SCS-CN parameter S in terms of K_h-θ and ψ-θ relations, S_s is investigated using the moisture diffusivity, D, defined by Philip (1957) as

$$\int_{\theta_0}^{\theta_1} D d\theta = \frac{S_s^2}{2} \tag{5.25}$$

Coupling of equation (5.25) with equation (1.56) yields

$$\int_{\psi_0}^{\psi_1} K_h d\psi = \frac{S_s^2}{2} \tag{5.26}$$

Equation (5.26) along with equations (5.1) and (5.2) leads to defining S_s as

$$S_s = \frac{2\alpha}{\beta+1}[\psi_1^{\beta+1} - \psi_o^{\beta+1}] \tag{5.27}$$

where

$$\alpha = \frac{a}{(a_1)^{m/m_1}} \tag{5.28a}$$

and

$$\beta = \frac{m}{m_1} \tag{5.28b}$$

The right-hand side of equation (5.27) represents twice the area below the K_h-ψ relation.

The hysteresis in the ψ-θ relation (Mein and Larson, 1971) leads to a reduction in the sorptivity of the soil or in the above-described S in the absorption-desorption cyclic process.

Other approximate expressions for S_s and, in turn, S can be derived taking D as a constant quantity. To this end, integration of equation (5.25) yields

$$D(\theta_1 - \theta_o) = \frac{S_s^2}{2} \tag{5.29}$$

where $\theta_1 > \theta_o$. Here, it is noted that in the context of a rainstorm, θ_o and θ_1 can be viewed as the moisture contents at the beginning and end of the storm, respectively. Substitution of equation (5.29) into equation (5.6) leads to

$$S_s^2 = 2\, m_1 [\underbrace{\overline{\psi}(K_{h1} - K_{ho})}_{\text{I}} + \underbrace{\overline{K}_h(\psi_1 - \psi_o)}_{\text{II}}] \tag{5.30}$$

where (K_{h0}, ψ_0) and (K_{h1}, ψ_1) correspond to θ_o and θ_1, respectively, and $\overline{\psi}$ and $\overline{K}_h$ correspond to the average ψ and K_h, respectively, for the θ-range of (θ_o, θ_1). In equation (5.30), the I and II parts implicitly represent energy expenditure in unit time for unit weight of the absorbate, respectively, at constant pressure and at constant volume (which can be derived from Darcy's law, equation (1.51), following Rankine (1870)), which, in turn, are analogous to specific heats at constant pressure (c_p) and constant volume (c_v), respectively (Mishra and Singh, 2001b). Thus, S_s is analogous to the hysteresis defined by Mishra and Seth (1996), Mishra and Singh (1999), and others as the energy expenditure in wave propagation during one time period.

In equation (5.30), the first part dominates when $\theta_1 \rightarrow \theta_{sat}$, as shown by the ψ-θ relation (Philip, 1969), whereas K_h varies rapidly with θ. However, the second part dominates when $\theta_1 \rightarrow \theta_o$, where the variation in K_h with θ is insignificant and ψ varies rapidly. Thus, equation (5.30) covers the whole range of the θ-variation.

The above feature of the first part of equation (5.30) leads to

$$S_s^2 = 2\, m_1 \overline{\psi}(K_{h1} - K_{ho}) \tag{5.31}$$

which, for $K_{h1} >> K_{ho}$, is analogous to the following relation (Young, 1964) (Chapter 4: equation (4.68)):

$$S_s^2 = 2\,(n - \theta)\, K_h\, \psi_e \tag{5.32}$$

Here, ψ_e (= h_e) is the Green-Ampt effective wetting front suction. A comparison of equation (5.31) with equation (5.32) yields $m_1 \overline{\psi} = (n - \theta_o)\, \psi_e$ or $\psi_e = m_1 \overline{\psi} / (n - \theta_o)$. It is noted here that ψ_e and $\overline{\psi}$ represent the Green-Ampt effective wetting front suction and average suction, respectively; and m_1 and $(n - \theta_o)$ represent the pore size distribution index

(Singh, 1997) and the effective soil porosity, respectively. Then, an expression for the SCS-CN parameter S can be derived from equation (5.24) for $K_{h0} << K_{h1}$ as

$$S = \frac{1}{N}\left(\frac{S_s^2}{K_{h1} - K_{ho}}\right) \tag{5.33}$$

Substitution of equation (5.31) into equation (5.33) yields

$$S = \frac{2}{N} m_l \overline{\psi} \tag{5.34}$$

where m_l represents the ψ-θ characteristic. From equation (5.34), it can be inferred that for a given soil, S is directly proportional to the average head and, therefore, this equation describes the variation of S with time conforming to the infiltration decay pattern experienced in field or laboratory observations. As time progresses, infiltration rate decreases or, in other words, θ increases or ψ decreases, and thus, $\overline{\psi}$ decreases. Reduction in $\overline{\psi}$ decreases S or increases CN, consistent with the storage hypothesis described in Chapter 3 (equation (3.17a)). Mishra et al. (1999) also described a similar temporal variation of S using the storage concept of the general infiltration model (Singh and Yu, 1990). As a result, it is possible to quantify the S (or CN)-variation within the storm duration.

Furthermore, using the volumetric concept, S is defined as (Chapter 4):

$$S = (n - \theta_o)\, D_s \tag{4.26}$$

A comparison of equation (4.26) with equation (5.34) yields

$$D_s = (2/N) m_l \overline{\psi} / (n - \theta_o) \tag{5.35}$$

It follows that D_s depends on pore size distribution, average suction, and effective porosity, and thus, it is a physically determinable, but θ-dependent, quantity. Such an analysis also helps describe physically the S/P (ordinate) versus the runoff factor (abscissa) relationship used for describing the behavior of the SCS-CN method (Chapter 3) as below.

For $\theta_o = 0$ (dry soil), $S = S_I = nD_s$, where subscript 'I' refers to antecedent moisture condition I implying dry condition of the soil. The runoff factor, as described above, equals the degree of saturation, S_r, and the product of S_r and the soil porosity, n, defines the moisture content, θ (equation (3.128)). Thus, the product of the runoff factor and n represents θ on the abscissa of the plot and S_I/P on the ordinate represents the product of n and D_s for a unit rainfall amount. Therefore, the (S_I/P)-runoff factor plot describes the variation of 'nD_s' with 'θ/n' for a unit rainfall amount. From equation (5.34), this plot can be interpreted as an indirect representation of the $\overline{\psi}$-θ relationship.

The general form of equation (5.34) can also be derived using equation (5.30) as

$$S = \frac{2}{N} m_l [\bar{\psi} + \frac{\overline{K_h}(\psi_l - \psi_o)}{(K_{hl} - K_{ho})}] \quad (5.36)$$

which can be derived by combining equations (5.30) and (5.33). From equation (5.36) it is evident that S is also a function of ψ_0 and K_{h0}, which themselves depend on the initial moisture content, θ_o, and, in turn, S also depends on θ_o. It is consistent with the concept of AMC that is identical to θ_o (Chapter 4: Section 4.1.2). Equation (5.36) actually quantifies all the four runoff producing watershed characteristics, viz., soil type, land use, hydrologic condition of land surface, and AMC, used for derivation of the curve number, CN, from the NEH-4 tables. These watershed properties eventually affect K_h-θ and ψ-θ relations, and thus, affect the S or CN computation.

For practical purposes, an average value of S can be derived for $\psi_l << \psi_o$ and $K_{ho} << K_{hl}$ (here, ψ_l and K_{hl} correspond to θ_{sat}) and, therefore, equation (5.36) converges to

$$S = \frac{2}{N} m_l [\bar{\psi} - \frac{\overline{K_h}}{K_{hl}} \psi_o] \quad (5.37)$$

Since ψ is influenced by the magnitude of the depth of storm precipitation (Philip, 1969), S also depends on the precipitation magnitude.

It is apparent from equation (5.24) that the determination of parameter S requires a determination of N a priori for field applications. From several infiltration data sets derived from laboratory and field tests, Collis-George (1977) determined the value of N ranging from 0.54 to 5.8 for various soils. The computation of N is illustrated below.

Example 5.2. Compute N and the characteristic time for the infiltration data of Columbia sandy loam (Project No. 34) given in Figure 4.1 assuming the K_h-θ and ψ-θ relations are given, respectively, as $K_h = 0.9107\ \theta^{9.0273}$ and $\psi = -6.4616\ \theta^{-1.482}$. Given are the initial moisture content, θ_o, = 0.125 (vol./vol.); saturated moisture content, θ_{sat}, = 0.518 (vol./vol.); and saturated hydraulic conductivity = $1.39\text{x}10^{-3}$ cm/sec.

Solution: From Figure 4.1 time base, T_b = 914 sec. The saturated hydraulic conductivity, $K_{h(sat)} = 1.39\text{x}10^{-3}$ cm/sec. The total area under the infiltration curve (Figure 4.1) (F_{max}) is estimated as 5.678 cm. Therefore,

$S = F_{max} - K_{h(sat)}T_b = 5.678 - 1.39\text{x}10^{-3} \text{ x } 914 = 4.408$ cm

From the given power relations for K_h-θ and ψ-θ relations, a = 0.9107, m = 9.0273, a_l = -6.4616, and m_l = -1.482. Therefore, α and β can be computed from equations (5.28a&b), respectively, as

$$\alpha = \frac{a}{(a_1)^{m/m_l}} = \frac{0.9107}{(6.4616)^{-9.0273/1.482}} = 7.87\text{x}10^4$$

and

$$\beta = \frac{m}{m_1} = -6.0917$$

ψ_{sat} (= ψ_1) can be computed from the ψ-θ relation as

$\psi_{sat} = -6.4616\ \theta^{-1.482} = -6.4616x(0.518)^{-1.482} = -17.128$ cm.

Similarly, ψ_o can be computed as

$\psi_o = -6.4616x(0.125)^{-1.482} = -140.834$ cm.

Thus, the soil sorptivity, S_s, can be computed from equation (5.27) as

$$S_s = \frac{2\alpha}{\beta+1}[\psi_1^{\beta+1} - \psi_o^{\beta+1}]$$

$$= \frac{2x7.87x10^4}{-6.0917+1}[(17.128)^{-6.0917+1} - (140.834)^{-6.0917+1}] = 1.616x10^{-2}\ \text{cm/sec}^{1/2}$$

This S_s-value is the absolute value of above computation. Therefore, N can be computed from equation (5.23) as

$$N = \frac{S_s^2}{SK_{h(sat)}} = \frac{(1.616x10^{-2})^2}{4.408x1.39x10^{-3}} = 0.043$$

The characteristic time, t_c, of the infiltration phenomenon can be computed from equation (5.23) as

$$t_c = \frac{S^2}{S_s^2} = \frac{4.408x4.408}{(1.616x10^{-2})^2} = 7.44x10^4\ \text{sec}.$$

In the above computations, S includes I_a. However, for S to exclude I_a, N can also be computed by subtracting the I_a amount that is computed for the given time to ponding, t_p = 136 sec. and uniform rainfall intensity, $i_o = 1.110x10^{-2}$ cm/sec. The initial abstraction, I_a, is computed as

$I_a = i_o t_p = 1.110x10^{-2} x\ 136 = 1.51$ cm

and thus, S = 4.408 – 1.51 = 2.90 cm. Following the above procedure, N = 0.065 and $t_c = 3.22x10^4$ sec. Following the computations as in the above example, Mishra and Singh (2000d) computed N-values in the range from 0.005-0.094 for the Minnesota soils (Table 4.1), of which a brief description follows.

Table 5.2 Fitted parameters of the ψ-θ, K_h-θ, and K_h-ψ relations

Sl. No.	Soil type	Fitted parameters for relations							
		K_h-θ			ψ-θ			K_h-ψ	
		a	m	r^2	a_1	m_1	r^2	α	β
1	PFS	0.8399	6.7160	0.9842	-2.0654	-1.7034	0.9530	4.81E-02	-3.94
2	CSL	0.9107	9.0279	0.9339	-6.4616	-1.4820	0.9213	1.05E-05	-6.09
3	GL	1.3824	12.359	0.9602	-1.5532	-4.0433	0.9896	3.60E-01	-3.06
4	ISL	0.3280	14.845	0.9982	-6.6796	-3.3730	0.9998	7.69E-05	-4.40
5	YLC	0.0041	8.6778	0.9934	-0.5435	-4.8768	0.9988	1.21E-02	-1.78

Note: PFS = Plainfield sand, CSL = Columbia sandy loam, GL = Guelph loam, ISL = Ida silt loam, YLC = Yololight clay

The infiltration tests were conducted for varying uniform rainfall intensity and initial moisture contents. For the given magnitudes of rainfall intensity and initial moisture content, the available information on the time to ponding, time base, and maximum cumulative infiltration amount from infiltration tests were averaged to characterize the infiltration process of a soil in a given situation. The parameters of the power relations fitted to the available K_h-θ, ψ-θ, and K_h-ψ relations for these soils are described in Table 5.2 and the computed information is listed in Table 5.3 along with the hydraulic conductivity and magnitudes of suction for initial (ψ_o) and saturated (ψ_{sat}) conditions. The time base of an infiltration curve was taken as the last value of the published data. Notably, the derived ψ-θ curves for the first three soils in Table 4.1 were desorption data. Since infiltration is an absorption process, Mein and Larson (1971) divided the drying curve by a factor of 1.6 for converting to the absorption or wetting curve of the ψ-θ curve.

In Table 5.3, K_{ho} and ψ_o values were computed using the developed power-relation K_{ho}-θ and ψ_o-θ; S-values were derived following Collis-George (1977) as

$$S = F_{max} - K_{h(sat)}\, T_b \tag{5.38}$$

where F_{max} is the maximum infiltration including the static portion of infiltration and S_s-values were derived from equation (5.27) that allows variation of D with θ. The computed S-values are shown in Table 5.3. Apparently, S varies from 2.87 to 4.21 cm for Plainfield sand, from 3.12 to 6.09 cm for Columbia sandy loam, from 2.45 to 3.46 cm for Guelph loam, from 0.52 to 1.55 cm for Ida silt loam, and from 0.97 to 2.18 cm for Yololight clay. On the whole, S-values tend to decrease with the soil type from Plainfield sand to Yololight clay in the given order. Table 5.3 also indicates the variation of the depth above the impeding layer, D_s, (cm) (derived using equation (4.26)) with the soil type. To distinguish the S-computations utilizing linear and non-linear assumptions the S-values determined from equation (5.34) using $N = 4/\pi$ were plotted against each other and it was found that the assumption of D not varying with θ allowed several fold over-estimation of S compared to that computed using D varying with θ. Furthermore, the adoption of N equal to $4/\pi$ exhibited a contrasting trend of S-variation with the soil type; in general, S should decrease with the soil changing from Plainfield sand to Yololight clay.

The above-derived N-values were examined for their correlation with the available physical parameters and these were found to show a good correlation with the initial moisture content, θ_o, rainfall intensity, i_o, and time base, T_b, as expressed below:

$$y = 1782.5\, x^{-1.8887};\ r^2 = 0.9580 \tag{5.39}$$

where $y = Ni_o/\theta_o$ (cm/sec) and x is the time base (sec). The derived S-values (equation (5.39)) for both the observed and predicted ones plotted in terms of respective CNs in Figure 5.2. Although there exists a large scatter in the data points, the figure indicates that CN-values can be derived from physical characteristics of the soil.

The above computations, however, require information on T_b a priori. Therefore, it is useful to derive an empirical relation for determination of T_b. To this end, the product of i_o and θ_o was found to exhibit a correlation with the time base, as shown in Figure 5.3 and expressed mathematically as

Table 5.3 Computation of parameters

Soil	i_o	θ_o	$\theta_{sat}=\theta_1$	t_p	T_b	K_{bo}	ψ_o	$\psi_{sat}=\psi_1$	F_{max}	I_a	δ	S_s^2	N	t_c	$S=F_{max}-K_bT_b$	CN	D_s	KW parameter
	(cm/sec)	(vol/vol)	(vol/vol)	(sec)	(sec)	(cm/sec)	(cm)	(cm)	(cm)	(cm)		(cm²/sec)		(sec)	(cm)		(cm)	
Plainfield sand	1.38E-02	0.13	0.472	104	1145	9.41E-07	-66.7	-7.4	8.154	1.44	-0.66	2.73E-02	0.051	2.4E+04	4.21	85.8	12.15	3.29E-03
	1.38E-02	0.23	0.472	77	789	4.34E-05	-25.2	-7.4	5.583	1.06	-0.96	2.66E-02	0.072	1.2E+04	2.87	89.9	11.61	1.56E-01
	2.75E-02	0.13	0.472	25	1070	9.41E-07	-66.7	-7.4	7.954	0.68	-0.71	2.73E-02	0.051	2.4E+04	4.27	85.6	12.31	3.07E-03
	2.75E-02	0.23	0.472	60	796	4.34E-05	-25.2	-7.4	5.738	1.64	-0.95	2.66E-02	0.069	1.3E+04	3.00	89.4	12.14	1.58E-01
Columbia sandy loam	2.78E-03	0.25	0.518	2636	3034	3.34E-06	-50.4	-17.1	8.312	7.33	-0.76	1.61E-02	0.045	6.5E+04	4.10	86.1	15.28	4.42E-02
	5.56E-03	0.125	0.518	599	1540	6.40E-09	-140.8	-17.1	7.183	3.33	-1.49	1.62E-02	0.037	9.7E+04	5.04	83.4	12.83	3.08E-05
	5.56E-03	0.15	0.518	553	2090	3.32E-08	-107.5	-17.1	8.377	3.08	-1.10	1.62E-02	0.034	1.1E+05	5.47	82.3	14.87	2.37E-04
	5.56E-03	0.2	0.518	490	1676	4.46E-07	-70.2	-17.1	7.110	2.72	-1.37	1.61E-02	0.039	8.8E+04	4.78	84.2	15.03	2.93E-03
	5.56E-03	0.25	0.518	420	1523	3.34E-06	-50.4	-17.1	6.270	2.34	-1.50	1.61E-02	0.045	6.7E+04	4.15	85.9	15.50	2.22E-02
	5.56E-03	0.318	0.518	324	1219	2.93E-05	-35.3	-17.1	4.817	1.80	-1.88	1.57E-02	0.057	3.9E+04	3.12	89.1	15.61	1.75E-01
	8.34E-03	0.125	0.518	244	970	6.40E-09	-140.8	-17.1	5.699	2.04	-2.36	1.62E-02	0.043	7.3E+04	4.35	85.4	11.07	1.94E-05
	8.34E-03	0.25	0.518	170	1495	3.34E-06	-50.4	-17.1	6.490	1.42	-1.53	1.61E-02	0.042	7.5E+04	4.41	85.2	16.46	2.18E-02
	1.11E-02	0.125	0.518	136	914	6.40E-09	-140.8	-17.1	5.678	1.51	-2.51	1.62E-02	0.043	7.4E+04	4.41	85.2	11.22	1.83E-05
	1.11E-02	0.15	0.518	127	1971	3.32E-08	-107.5	-17.1	8.827	1.40	-1.16	1.62E-02	0.031	1.4E+05	6.09	80.7	16.54	2.23E-04
	1.11E-02	0.2	0.518	117	1737	4.46E-07	-70.2	-17.1	7.731	1.30	-1.32	1.61E-02	0.035	1.1E+05	5.32	82.7	16.72	3.03E-03
	1.11E-02	0.25	0.518	90	1457	3.34E-06	-50.4	-17.1	6.479	1.00	-1.57	1.61E-02	0.042	7.7E+04	4.45	85.1	16.62	2.12E-02
	1.11E-02	0.318	0.518	75	1039	2.93E-05	-35.3	-17.1	4.606	0.83	-2.21	1.57E-02	0.056	4.0E+04	3.16	88.9	15.81	1.49E-01
Guelph loam	1.47E-03	0.3	0.523	1702	3668	4.77E-07	-202.0	-21.4	4.592	2.50	-4.33	9.43E-03	0.075	1.2E+05	3.25	88.7	14.55	1.09E-03
	1.47E-03	0.35	0.523	1246	2855	3.20E-06	-108.3	-21.4	3.496	1.83	-5.57	9.19E-03	0.094	7.1E+04	2.45	91.2	14.15	9.12E-03
	2.94E-03	0.3	0.523	376	3187	4.77E-07	-202.0	-21.4	4.628	1.11	-4.99	9.43E-03	0.070	1.3E+05	3.46	88.0	15.51	9.47E-04
	2.94E-03	0.35	0.523	305	2513	3.20E-06	-108.3	-21.4	3.540	0.90	-6.33	9.19E-03	0.088	8.1E+04	2.62	90.7	15.13	8.02E-03

Table 5.3 Continued

Soil	i_c	θ_o	$\theta_{sat}=\theta_1$	t_p	T_b	K_{bo}	ψ_o	$\psi_{sat}=\psi_1$	F_{max}	I_a	δ	S_s^2	N	t_c	$S=F_{max}-K_bT_b$	CN	D_s	KW parameter
	(cm/sec)	(vol/vol)	(vol/vol)	(sec)	(sec)	(cm/sec)	(cm)	(cm)	(cm)	(cm)		(cm²/sec)		(sec)	(cm)		(cm)	
Ida silt loam	5.84E-05	0.25	0.53	32495	39597	3.79E-10	-717.0	-56.9	2.255	1.90	-7.87	8.85E-04	0.024	1.5E+06	1.10	95.9	3.92	5.47E-06
	5.84E-05	0.4	0.53	14232	39715	4.06E-07	-146.9	-56.9	1.843	0.83	-7.85	8.50E-04	0.036	6.5E+05	0.68	97.4	5.26	1.79E-02
	1.17E-04	0.25	0.53	7449	37689	3.79E-10	-717.0	-56.9	2.651	0.87	-8.27	8.85E-04	0.017	3.1E+06	1.55	94.2	5.54	5.20E-06
	1.17E-04	0.3	0.53	5907	38962	5.67E-09	387.6	-56.9	2.500	0.69	-8.00	8.84E-04	0.020	2.4E+06	1.36	94.9	5.92	1.24E-04
	1.17E-04	0.35	0.53	5123	18976	5.59E-08	-230.5	-56.9	1.473	0.60	-16.43	8.77E-04	0.029	1.1E+06	0.92	96.5	5.11	8.59E-04
	1.17E-04	0.4	0.53	2963	39734	4.06E-07	-146.9	-56.9	1.967	0.35	-7.85	8.50E-04	0.031	9.0E+05	0.81	96.9	6.21	1.79E-02
	1.17E-04	0.43	0.53	2515	19304	1.19E-06	-115.1	-56.9	1.082	0.29	-16.15	8.05E-04	0.043	4.2E+05	0.52	98.0	5.18	3.03E-02
	1.75E-04	0.25	0.53	3745	18501	3.79E-10	-717.0	-56.9	1.841	0.66	-16.85	8.85E-04	0.021	2.2E+06	1.30	95.1	4.65	2.55E-06
	1.75E-04	0.4	0.53	1853	19281	4.06E-07	-146.9	-56.9	1.272	0.32	-16.17	8.50E-04	0.035	7.0E+05	0.71	97.3	5.45	8.70E-03
	2.33E-04	0.25	0.53	2384	17530	3.79E-10	-717.0	-56.9	1.909	0.56	-17.79	8.85E-04	0.019	2.5E+06	1.40	94.8	4.99	2.42E-06
	2.33E-04	0.3	0.53	1674	19035	5.67E-09	-387.6	-56.9	1.778	0.39	-16.38	8.84E-04	0.022	1.9E+06	1.22	95.4	5.31	6.07E-05
	2.33E-04	0.35	0.53	1248	19260	5.59E-08	-230.5	-56.9	1.553	0.29	-16.19	8.77E-04	0.027	1.3E+06	0.99	96.2	5.50	8.72E-04
	2.33E-04	0.4	0.53	763	19926	4.06E-07	-146.9	-56.9	1.315	0.18	-15.65	8.50E-04	0.034	7.4E+05	0.73	97.2	5.64	8.99E-03
	2.33E-04	0.43	0.53	671	19447	1.19E-06	-115.1	-56.9	1.097	0.16	-16.03	8.05E-04	0.042	4.3E+05	0.53	98.0	5.29	3.05E-02
Yolo light clay	4.92E-05	0.25	0.499	38783	48342	2.44E-08	-469.2	-16.1	2.313	1.91	-17.30	3.78E-04	0.007	2.1E+07	1.72	93.7	6.90	1.56E-04
	4.92E-05	0.35	0.499	20782	29127	4.53E-07	-90.9	-16.1	1.326	1.02	-28.71	3.01E-04	0.008	1.0E+07	0.97	96.3	6.49	6.41E-03
	9.84E-05	0.25	0.499	9874	48017	2.44E-08	-469.2	-16.1	2.766	0.97	-17.42	3.78E-04	0.005	3.3E+07	2.18	92.1	8.74	1.55E-04
	9.84E-05	0.35	0.499	5048	28549	4.53E-07	-90.9	-16.1	1.525	0.50	-29.29	3.01E-04	0.006	1.5E+07	1.17	95.6	7.88	6.28E-03

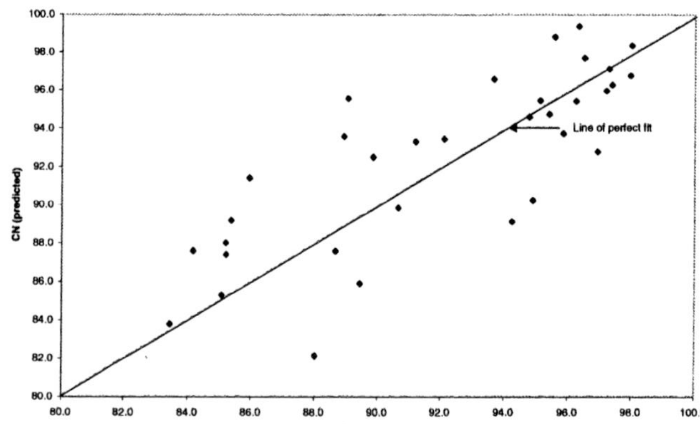

Figure 5.2. Observed versus predicted CN-values.

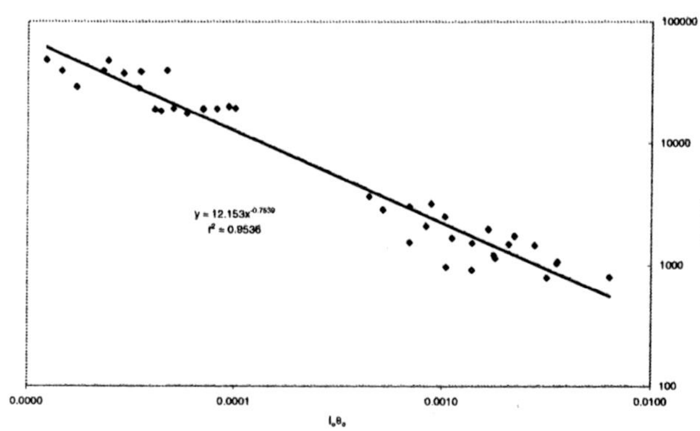

Figure 5.3. Relation between rainfall intensity, i_o (cm/sec), initial moisture content, θ_o (vol./vol.), and time base, T_b (sec).

$$y = 12.153\, x^{-0.7539} ; \; r^2 = 0.9536 \tag{5.40}$$

where $y = T_b$ (sec) and $x = i_o\theta_o$ (cm/sec).

The N-values derived using the approximate relation (equation (5.36)) were compared with the above derived from the non-linear analysis. The former were found to be much higher than the latter. It implies that the approximate average relations do not have the efficacy to capture the characteristics of the K_h-θ and ψ-θ relations. They are, however, useful for understanding the physical process. It is noted that in the above computations (Table 5.3), S has included I_a whereas in the water balance equation (equation (3.1)), F is separate from I_a. Alternatively, for F = S (equation (2.2)), S excludes I_a in equation (2.1), as also described in Chapter 3. Therefore, an investigation of S for I_a is also in order.

It is evident from the above that the initial abstraction amount, I_a, is equal to the product of the rainfall intensity and time to ponding, t_p. To investigate S for I_a these were plotted against each other for four cases (1) S included I_a as well as the static portion of infiltration, F_c, (2) S included I_a and excluded F_c, (3) S excluded I_a and included F_c, and (4) S excluded both I_a and F_c. Here, F_c is taken equal to the product of $K_{h(sat)}$ and T_b (equation (5.38)). The fitted relations for these cases are as follows:

Case 1

$$y = 0.2749\, x ; \; r^2 = 0.3561 \tag{5.41a}$$

Case 2

$$y = 0.4240\, x ; \; r^2 = 0.4103 \tag{5.41b}$$

where $y = I_a$ (cm) and x-axis represents the S values for the respective cases, as described above. It is worth emphasizing that Cases 1 and 2 show spurious relations for both the x- and y-axes include I_a. Furthermore, the above relations (equations (5.41a) and (5.41b)) exhibit a poor fit, for the resulting r^2-values are very low in all the cases. Cases 3 and 4 yielded negative value of r^2 indicating no relation. The ratio of y to x in these equations represent the magnitudes of initial abstraction coefficient, λ. Evidently, λ is equal to 0.275 and 0.42 (approximate) for Cases 1 and 2, respectively. For Cases 3 and 4, these were 0.33 and 0.57 (approximate), respectively. The value for Case 1 is close to 0.25 derived by Chen (1982).

In the NEH-4 tables, S includes both I_a and F_c whereas, as shown in Chapters 3 and 4, S excludes both. Therefore, two cases including and excluding both I_a and F_c were considered for evaluating the impact of the initial moisture content (AMC) on S. Since AMC represents the amount of antecedent moisture before the start of a storm, the initial moisture content, θ_o, is a variant of AMC, as described in Chapter 4. The plot between S and θ_o for S excluding both I_a and F_c is shown in Figure 5.4. Apparently, the figure exhibits a decreasing trend of S with increasing, θ_o, which is consistent with the general notion that S or CN decreases or increases with the increasing θ_o. Furthermore, it also depicts a loop in the S-θ_o relation. The apparent lower and upper enveloping curves are analogous to AMC III and AMC I, respectively. Similar inferences were derived from the plot (not shown) depicting the relation between θ_o and S including both I_a and F_c.

5.2.3 USE OF INTRINSIC SORPTIVITY

S can also be expressed in terms of the intrinsic sorptivity (ς), defined by Philip (1969), as

$$\varsigma = \left(\frac{\mu}{\sigma}\right)^{1/2} S_s \tag{5.42}$$

where μ is the dynamic viscosity and σ is the surface tension (different from $\hat{\sigma}$ which is the wave number) of the absorbate. Here ς is analogous to the intrinsic permeability, a characteristic of the internal geometry of the medium and is independent of the properties of the absorbate (Philip, 1969). Equation (5.42) can be expressed for S_s as

$$S_s^2 = \frac{\varsigma^2 \sigma}{\mu} \tag{5.43}$$

Therefore, S can also be derived from equations (5.43) and (5.33) as

$$S = \frac{1}{N} \frac{\varsigma^2 \sigma}{\mu (K_{hl} - K_{ho})} \tag{5.44}$$

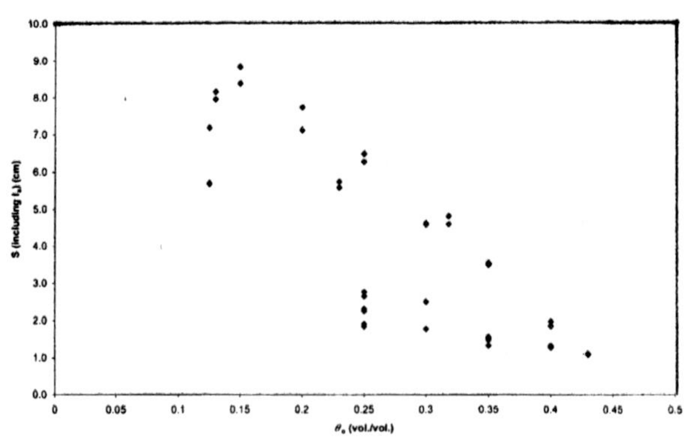

Figure 5.4. Relation between S (including I_a) and initial moisture content, θ_o.

Alternatively, the intrinsic sorptivity can be expressed using equations (5.30) and (5.42) as

$$\varsigma = \sqrt{\frac{2 m_l \mu}{\sigma}[\bar{\psi}(K_{hl} - K_{ho}) + \overline{K_h}(\psi_l - \psi_o)]} \tag{5.45}$$

and can be explained in terms of c_p and c_v, as above. It is important to note that S is inversely proportional to μ implying that CN increases with μ and vice versa. Lee (1983) found that the freezing of the soil of high moisture content decreased infiltration to zero whereas freezing at low moisture content increased infiltration to twice its normal rate. Such behaviour is, however, attributed to the variation of μ with temperature and moisture content.

5.2.4 VERTICAL INFILTRATION

The vertical infiltration phenomenon is described by the equation derived from equation (1.55) eliminating the product of two derivatives on the left-hand side of this equation (Philip, 1969). Here gravity forces are insignificant compared to matric forces. Similar to the above wave solutions, it can be shown that the wave celerity in vertical infiltration is equal to zero implying that 'm' is equal to zero in equation (5.16) since neither K_h nor will θ be zero. It infers that for such a soil exhibiting only vertical infiltration, K_h should be independent of θ. It is possible when $K_h = K_{h(sat)}$ or $\theta = \theta_{sat}$. In this case, the moisture profile moves with a velocity of K_h/θ with $\delta = \infty$, implying that the whole moisture is absorbed up to the distance the profile travels.

The physical significance of the SCS-CN parameter S is described for this case following Philip (1957; 1969) who described S_s using the relation:

$$S_s = x\, t^{-1/2} \tag{5.46}$$

where x is the vertical depth of flow from the ground surface. Substitution of equation (5.46) into equation (5.24) leads to

$$S = \frac{1}{N}\frac{x^2 t^{-1}}{K_h} \tag{5.47}$$

or, alternatively, K_h can be defined from equation (5.23) as

$$K_h = \frac{x}{\sqrt{t\, t_c}} \tag{5.48}$$

Equation (5.48) describes the variation of K_h with x and t in the vertical infiltration. At a given location x, K_h initially reduces more sharply than it does at longer times, for t_c

represents a characteristic time. An alternative form of equation (5.47) describing S is as below:

$$S = \frac{1}{N}\left(\frac{x\,t^{-1}}{K_h}\right) x \tag{5.49}$$

Under free fall situations:

$$x = \frac{1}{2} g\, t^2 \tag{5.50}$$

where g is the gravitational acceleration. Substituting 't' from equation (5.50) into equation (5.49) leads to

$$S = \frac{1}{2N}\left(\frac{\sqrt{2\,g\,x}}{K_h}\right) x \tag{5.51}$$

The ratio in the bracketed part of equation (5.51) is the ratio of two speeds: That in the numerator describes the velocity of flow under free fall and that in the denominator is the average velocity of flow under the unit hydraulic gradient (Darcy's law, equation (1.51)). Here, the inverse of the ratio is defined by M as follows:

$$M = \frac{K_h}{\sqrt{2\,g\,x}} \tag{5.52}$$

which appears to be a hydraulic number (Mishra and Singh, in press) characterizing flow transmission in unsaturated porous media, analogous to the Froude number defining the hydraulic properties of flow in open channels. Since K_h can assume any value ($\leq$ or > $\sqrt{2gx}$), M can also assume any value ($\leq$ or > 1). As the minimum value that K_h can assume is equal to zero, M can have a range of $(0, \infty)$. Therefore, equation (5.52) can be written as

$$S = \frac{1}{2N}\frac{x}{M} \tag{5.53}$$

Equation (5.53) implies that in a given flow regime (or constant M), S increases linearly or CN decreases with the depth of flow. Near the ground surface ($x \rightarrow 0$) $S \rightarrow 0$ and CN $\rightarrow 100$ implying that during a storm, surface runoff can commence even before the soil storage is fully exhausted, which is fundamental to the SCS-CN concept and contrasts the partial or source area concept, as discussed in Chapter 2.

5.2.5 KINEMATIC WAVE

Another hypothetical case is one of the kinematic wave which can be derived from equation (1.55) that represents the linear form of a diffusion wave. The kinematic wave form can be derived by ignoring the second-order term containing D at the right-hand side of equation (1.55). From a linear analysis it can be shown that the kinematic wave celerity, c_k, is the same as that of equation (5.16) as

$$c_k = \frac{m\,K_{ho}}{\theta_o} \tag{5.54}$$

It implies that such a wave propagates in the z-direction, and $\delta = 0$ implies that these waves do not attenuate. Since $D = 0$ or $S_s = 0$, $S = 0$. Following Ponce et al. (1978), a criterion for the kinematic wave application for 95% accuracy can be derived as follows;

$$\frac{T_b\,K_{ho}\,m^2}{m_l\,\psi_o\,\theta_o} \geq 385 \tag{5.55}$$

The applicability of inequality (equation (5.55)) is demonstrated using the infiltration data observed from laboratory tests on the above Minnesota soils (Table 4.1). The kinematic wave parameter that is equivalent to the left-hand side of equation (5.55) is computed for all infiltration data sets and given in Table 5.3. Its values range between 2.42×10^{-6} and 0.175, which are far less than 385. It implies that the infiltration process is not kinematic in nature, rather it is either a linear (equation (1.55)) or non-linear (Philip, 1969) diffusion or higher order dynamic wave phenomenon.

5.3 S/P Relations for the Modified Model

Using the physical description of S/P relations, the modified version of the SCS-CN model (Chapter 4) can be the characterized in terms of K_h-θ and ψ-θ relations. To this end, twenty cases listed in Table 5.4 for various combinations of λ, M^*, and F_c^* are considered. λ is varied from 0.01 to 0.5, M^* from 0.0 to 5.0, and F_c^* from 0.0 to 1.0, covering a reasonable range of variability of these parameters/variables for practical applications. Here $M^* = 0$ is taken for the reason of non-availability of a valid θ-range for equations (5.1) and (5.2). A specific case (Case 12) of Table 5.4 taking runoff factor C = Q/P is described. The subsequent discussion explains the features of the model using the S_t/P versus C [= Q/(P-I_a-F_c)] relation for the reason that (a) the plots derived using both the definitions of C exhibit a similar pattern, and (b) the rainfall-runoff process described in terms of (nD_s) and θ/n is physically realizable.

The variation of S/P with C (= Q/P) for $\lambda = 0.2$; $M^* = 0.2$; and $F_c^* = 0.0, 0.2, 0.4, 0.6, 0.8$, and 0.99 is depicted in Figure 5.5. It is apparent from this figure that the ratio S/P decreases almost exponentially with C for a given F_c^*. For a given C, S/P increases with the reduction in F_c^*. Alternatively, for a given rainfall amount, S decreases with Q for a given F_c and it increases with a decrease in F_c. The former implies that I_a decreases with increasing Q for the reason that S decreases for a given λ and $(F + I_a)$ also decreases with Q to ensure the water balance (equation (2.1)). Taking $\lambda = 0.0$ for a moment, it can be shown that F decreases as Q increases. Thus, it can be inferred that both F and I_a decrease with increasing Q for given P, F_c, and λ. Its impact on the infiltration capacity curve (Figure 4.1) derived from the infiltration data (Project No. 34) of Columbia sandy loam (Mein and Larson, 1971) is demonstrated below.

Table 5.4 Cases for describing the model features

Case	λ	M^*	F_c^*					
1	0.01	0.0	0.0	0.2	0.4	0.6	0.8	0.99
2	0.01	0.2	0.0	0.2	0.4	0.6	0.8	0.99
3	0.01	0.5	0.0	0.2	0.4	0.6	0.8	0.99
4	0.01	1.0	0.0	0.2	0.4	0.6	0.8	0.99
5	0.01	5.0	0.0	0.2	0.4	0.6	0.8	0.99
6	0.1	0.0	0.0	0.2	0.4	0.6	0.8	0.99
7	0.1	0.2	0.0	0.2	0.4	0.6	0.8	0.99
8	0.1	0.5	0.0	0.2	0.4	0.6	0.8	0.99
9	0.1	1.0	0.0	0.2	0.4	0.6	0.8	0.99
10	0.1	5.0	0.0	0.2	0.4	0.6	0.8	0.99
11	0.2	0.0	0.0	0.2	0.4	0.6	0.8	0.99
12	0.2	0.2	0.0	0.2	0.4	0.6	0.8	0.99
13	0.2	0.5	0.0	0.2	0.4	0.6	0.8	0.99
14	0.2	1.0	0.0	0.2	0.4	0.6	0.8	0.99
15	0.2	5.0	0.0	0.2	0.4	0.6	0.8	0.99
16	0.2	0.0	0.0	0.2	0.4	0.6	0.8	0.99
17	0.5	0.2	0.0	0.2	0.4	0.6	0.8	0.99
18	0.5	0.5	0.0	0.2	0.4	0.6	0.8	0.99
19	0.5	1.0	0.0	0.2	0.4	0.6	0.8	0.99
20	0.5	5.0	0.0	0.2	0.4	0.6	0.8	0.99

As apparent from Figure 4.1, the reduction in I_a due to the reduction in S will lead to a shift of line DE towards $t \rightarrow 0$, for providing an additional space to augment S. For S to reduce with an increase in Q for a given rainfall amount, the decay of the curve DHM should increase with increasing Q. Similarly, the effect of reducing Q on infiltration for given values of P, λ, and F_c can be explained. Such an example corresponds to a special case of F = S. Mathematically, from the water balance equation (2.1), Q can be computed for $\lambda = 0.2$ as

$$Q = P - 1.2\,S \tag{5.56}$$

For a given P, Q increases when S reduces, leading to the reduction in both I_a and S. The impact of F_c on Q can be explained as follows. As f_c (or F_c) increases, both I_a and F decrease and vice versa. However, the rate of decay of infiltration capacity curve (curve DHM) should increase or decrease with the reduction or increase in F_c.

The features of the modified SCS-CN model can be described in terms of the impacts of F_c^*, λ, M^*, and P on S_I utilizing Figures 5.6-5.9. In these figures, the ordinate represents S_I/P instead of S/P and the abscissa represents the runoff factor equal to $Q/(P-I_a-F_c)$ [$= C/(1-I_a^*-F_c^*)$, where $C = Q/P$ and $I_a^* = I_a/P$]. To derive S_I/P, S/P-values were computed from equation (4.55) for various values of F_c^*, λ, M^*, C (= Q/P) and added to respective M^*-values. It is again emphasized that the (S_I/P)-$(C/(1-I_a^*-F_c^*))$ relation depicts the (nD_s)-(θ/n) (or, as above, $\bar{\psi}$ -θ) relation for a unit rainfall amount.

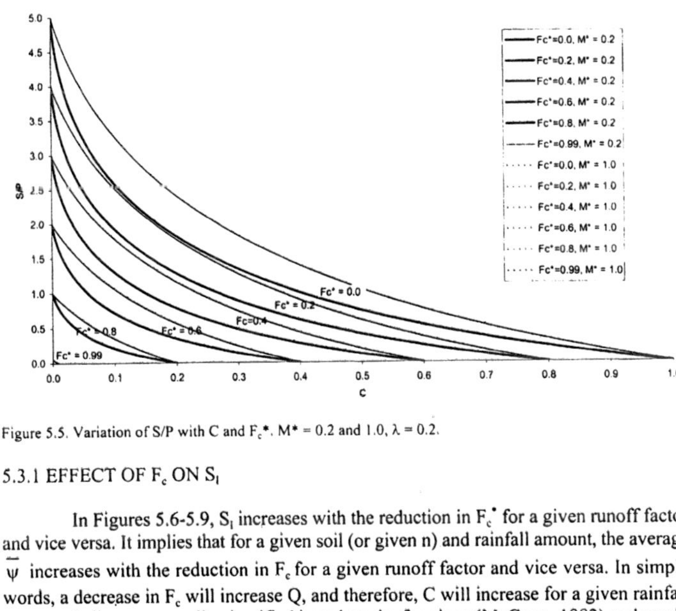

Figure 5.5. Variation of S/P with C and F_c^*. M^* = 0.2 and 1.0, λ = 0.2.

5.3.1 EFFECT OF F_c ON S_I

In Figures 5.6-5.9, S_I increases with the reduction in F_c^* for a given runoff factor and vice versa. It implies that for a given soil (or given n) and rainfall amount, the average $\bar{\psi}$ increases with the reduction in F_c for a given runoff factor and vice versa. In simple words, a decrease in F_c will increase Q, and therefore, C will increase for a given rainfall amount. Soils are generally classified based on the f_c-values (McCuen, 1982) and sandy soils generally exhibit higher f_c-values than do clay soils. The ψ-θ data of the Minnesota soils (Mein and Larson, 1971) shown in Table 5.3 exhibits a higher value of the average ψ for Yololight clay (average f_c = 0.0023 cm/hr and $\bar{\psi}$ = 242.7 cm) and Ida silt clay (average f_c = 0.0027 cm/hr and $\bar{\psi}$ = 387.0 cm) than those of the Plainfield sand (average f_c = 0.303 cm/hr and $\bar{\psi}$ = 16.3 cm), Columbia sandy loam (average f_c = 0.176 cm/hr and $\bar{\psi}$ = 38.7 cm), and Guelph loam (average f_c = 0.055 cm/hr and $\bar{\psi}$ = 111.7 cm). Such a description generally asserts the validity of the model formulation for the inclusion of F_c in the model. On the other hand, the NEH-4 tables exhibit higher S-values for hydrologic group A soils (for which f_c ranges from 0.30-0.45 inch/hr) than the group C soils (f_c range

0.15-0.30 inch/hr) or group C soils (f_c range 0.05-0.15 inch/hr) or group D soils (f_c range 0-0.05 inch/hr). This is because the effect of f_c on infiltration is also accounted for by the NEH-4 S-values.

5.3.2 EFFECT OF M ON S_I

To investigate the model behaviour for varying antecedent moisture amounts, a test example of an agricultural watershed characterized by small grain, contoured crop, poor condition, clay soil (hydrologic group D) is considered. The CN values corresponding to the three antecedent moisture conditions (AMC) for this watershed are given by NEH-4 as 70, 85, and 97 for AMC I through AMC III, respectively, as shown in Table 5.5. Based on the computed Q values using equation (2.5) for P = 10 inches, the resulting C (= Q/P) values are 0.62, 0.82, and 0.96 for AMC I through AMC III, respectively. Assuming that the soil of AMC I is completely dry (or M = 0), M-values for AMC II and III can be computed, respectively, following equation (4.4) as:

$$M_{II} = S_I - S_{II} \tag{5.57a}$$

and

$$M_{III} = S_I - S_{III} \tag{5.57b}$$

where subscripts I through III refer to AMC I through AMC III, respectively. Here, S_{abs} is taken equivalent to S_I. Further, assuming that $f_c = 0$ for the soils of soil group D as described above and taking the M-values derived using equations (5.57a,b), the computed Q-values using equation (4.29) for the same P = 10 inches yield C (= Q/P) as 0.62, 0.84, and 0.97 for AMC I through AMC III, respectively, as shown in Table 5.5. These C-values are fairly close to those derived above using the existing SCS-CN method (equation (2.5)). The small deviation is attributed to the different approaches of the antecedent moisture computation.

Table 5.5 Computations of C (=Q/P) for three AMC levels using the existing and the modified SCS-CN methods

AMC	Existing SCS-CN method				Modified SCS-CN method		
	CN	S	Q	C (=Q/P)	M	Q	C(=Q/P)
I	70	4.29	6.22	0.62	0.00	6.22	0.62
II	85	1.76	8.16	0.82	2.52	8.43	0.84
III	97	0.31	9.64	0.96	3.98	9.72	0.97

The above example shows that the functioning of the modified version of the SCS-CN method for the antecedent moisture is the same as that of the existing version of the method. Another outcome of the example is that for given P, F_c, and S_I, the runoff factor C (= Q/P) increases with M. With these conclusions, the effect of M on S_I is described with the help of Figures 5.6-5.9.

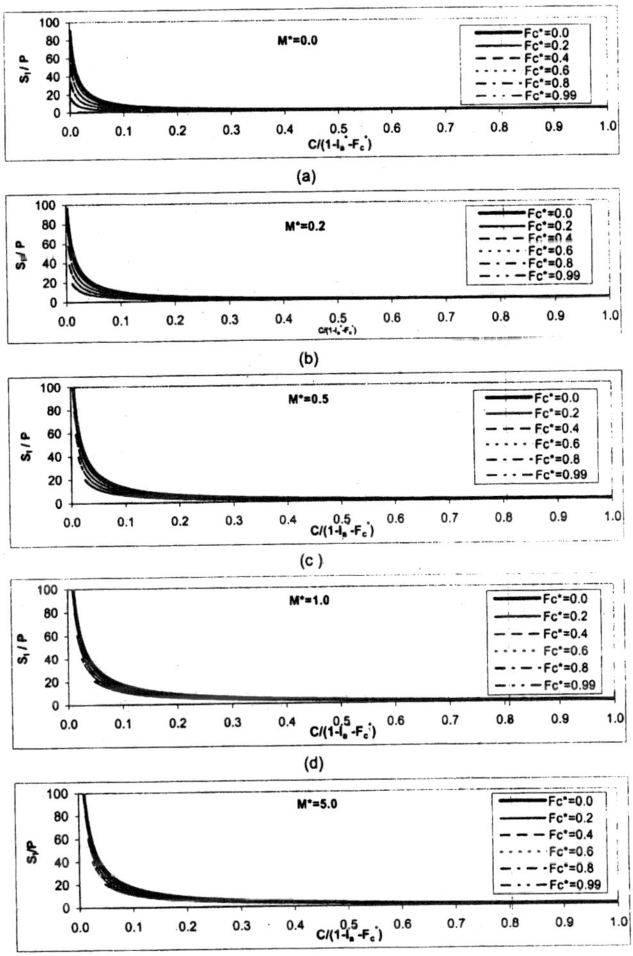

Figure 5.6. Variation of S_l/P with $C/(1-I_a^*-F_c^*)$, F_c^*, and M*. $\lambda = 0.01$.

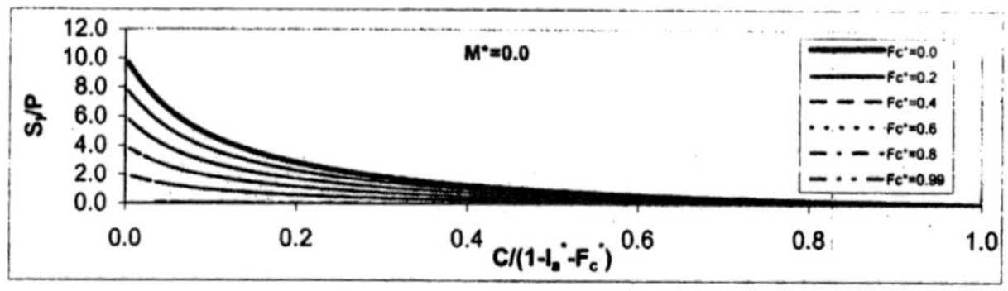

(a)

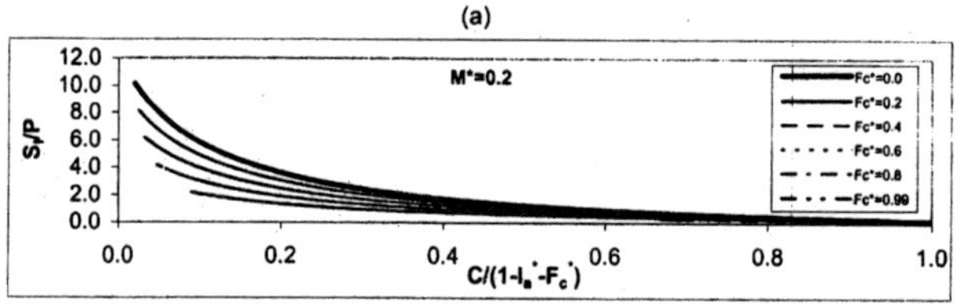

(b)

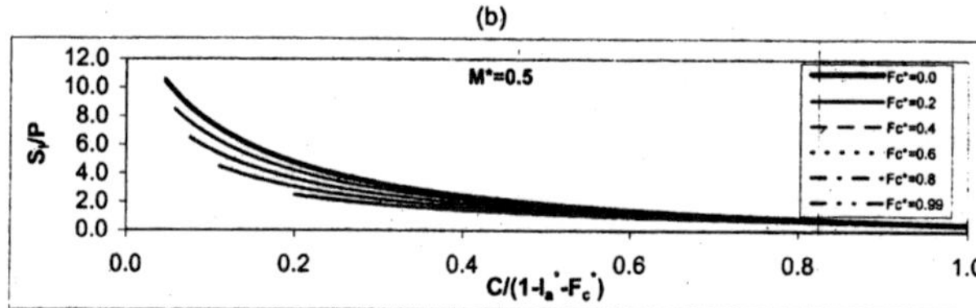

(c)

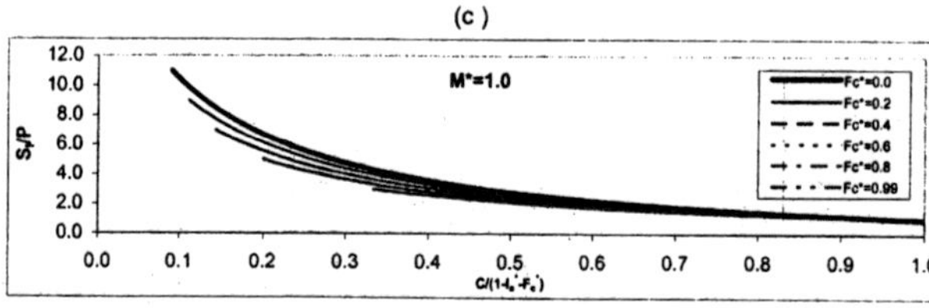

(d)

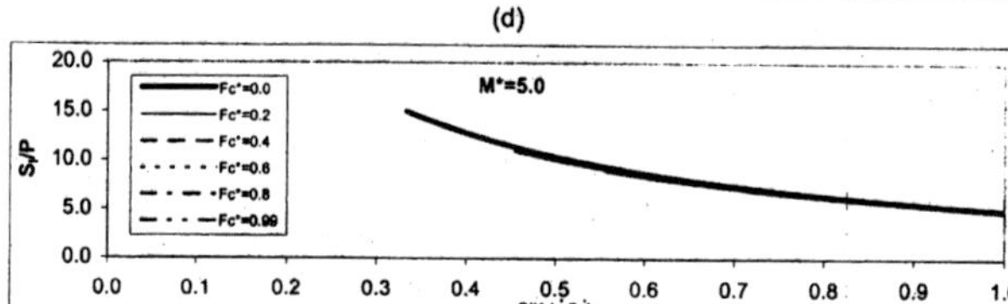

(e)

Figure 5.7. Variation of S_I/P with $C/(1-I_a^*-F_c^*)$, F_c^*, and M*. $\lambda = 0.1$.

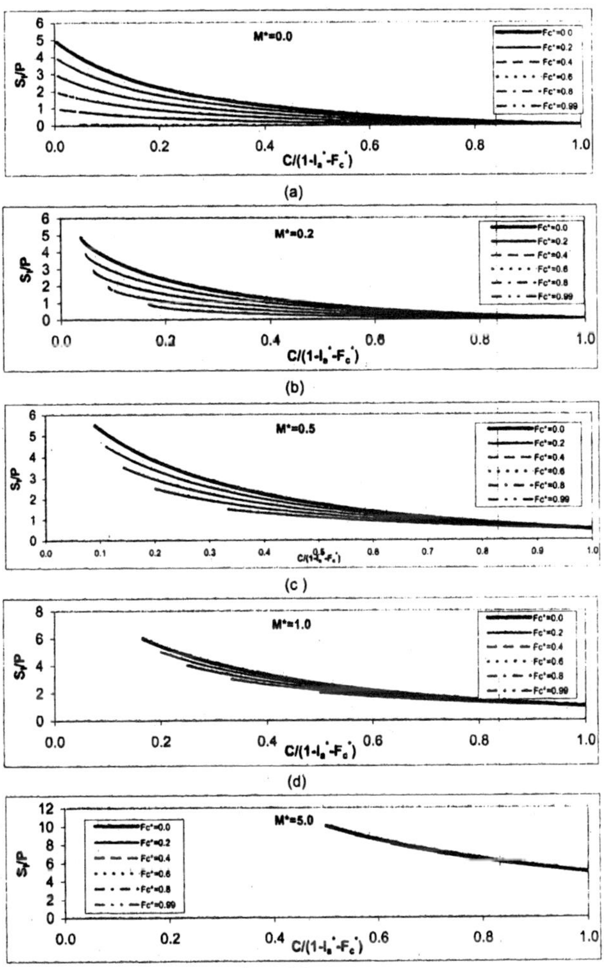

Figure 5.8. Variation of S_l/P with $C/(1-I_a^*-F_c^*)$, F_c^*, and M^*. $\lambda = 0.2$.

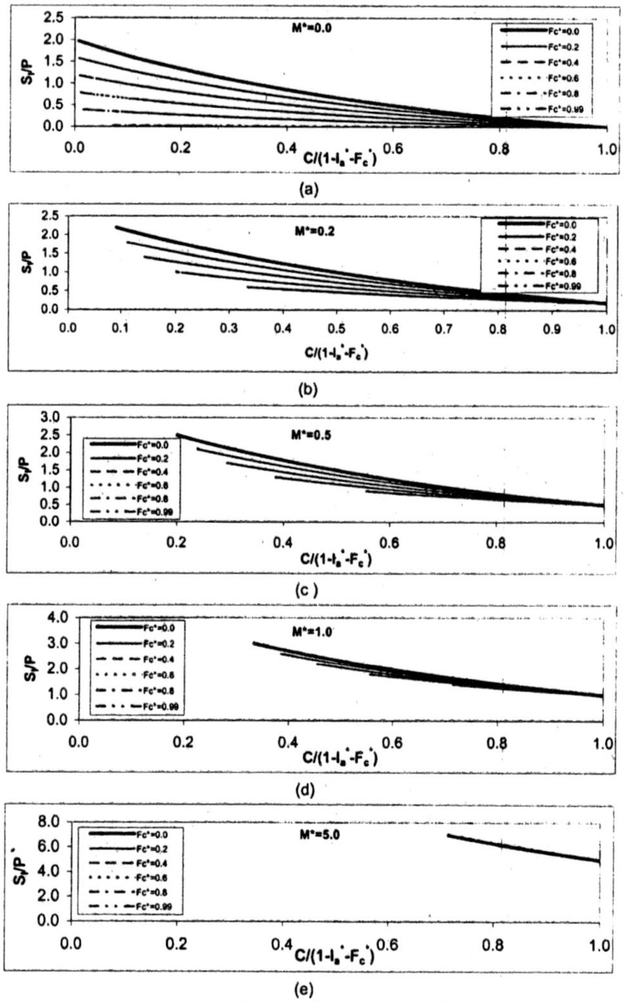

Figure 5.9. Variation of S_l/P with $C/(1-I_a^*-F_c^*)$, F_c^*, and M*. $\lambda = 0.5$.

It is apparent from Figure 5.5 that for given C, F_c^*, and λ values, S/P increases with M. Similarly, from Figures 5.6-5.9 the effect of increasing M^* on S_l/P is to shift diagonally right all the curves bearing different F_c^*'s. It is due to the fact that as M increases, C (=Q/P) also increases, as shown above. The increase in C results in the reduction in S and vice versa. In other words, an increase in the AMC level or M of a watershed increases the runoff-producing characteristic explained by CN (equation (2.7)) or C (=Q/P). Thus, for given P, λ, and F_c, a watershed exhibits its inefficacy to maintain C (=Q/P) at a value under varying AMCs. Therefore, an increase in M leads to an increase in S or S_l to maintain C at a lower value than that corresponding to the increased M.

In a $\overline{\psi}$ -θ relationship, the moisture content increases the average ψ decreases and vice versa. Thus, an increase in the amount of antecedent moisture of a soil of given porosity will increase the soil's moisture content and decrease both S and S_l (equation (5.34) or equation (5.35)). However, if C is kept constant, the result will be reverse.

Another important feature of the Modified SCS-CN model as apparent from Figures 5.6-5.9 is that for a given porosity and θ, the magnitude of increase in S_l/P due to the increase in F_c^* reduces with increasing M, leading to the narrowing of the range of variation of S_l/P due to variation in F_c^*. Thus, an increase in AMC or M acts to subside the impact of f_c–variation on S_l or S.

5.3.3 EFFECT OF λ ON S_l

Prior to the investigation of the impact of λ on S_l, it is appropriate to emphasize the role of λ in the description of the behaviour of the existing SCS-CN method (equation (2.5)). For given values of P and Q, F depends on S and since I_a is taken as a function of S (equation (2.3)), I_a also depends on S. Adoption of a constant λ-value forces I_a to compensate for F for varying magnitudes of S. Such an assumption also requires I_a to be a part of F or, in turn, S. This, however, is in contrast with the fundamental SCS-CN proportional equality (equation (2.2)), describing F to vary from 0 to S that represents a watershed characteristic. Furthermore, the water balance (equation (2.1)) also separates I_a from F (or S for F = S). For S including I_a, λ = 0.2 (SCS, 1971) and λ = 0.25 (Chen, 1982) for S excluding I_a. It is worth mentioning that λ can vary from 0 to ∞ (Chapter 3).

For given P, Q, and antecedent moisture amounts, I_a will increase with increasing λ-values and consequently, F will decrease. Such a reduction in F requires reduction in S according to the proportionality concept of equation (2.2). In other words, an increase in I_a due to the increase in λ-value increases the left-hand side of equation (2.2) and consequent reduction in F forces S to decrease, for balancing the proportionality. Such an impact is seen in Figures 5.6-5.9. For example, at the degree of saturation, S_r (= $C/(1-I_a^*-F_c^*)$) approaching to 0, $S_l/P \approx 90$ for $M^* = 0.0$, $\lambda = 0.01$, and $F_c^* = 0.0$ (Figure 5.6a) and $S_l/P \approx 10$ for $\lambda = 0.1$, and the same values of other variables (Figure 5.7a). Similarly, for the same values of other variables, $S_l/P \approx 5$ for $\lambda = 0.2$ (Figure 5.8a) and $S_l/P \approx 2.0$ for λ = 0.5 (Figure 5.9a). Thus, S_l/P -values reduce drastically as λ changes from 0.01 to 0.1. However, this reduction becomes mild with further increase in the λ-value. It becomes further milder with the increasing F_c^*-values. For example, for the same S_r approaching 0 and the same values of the other variables but $F_c^* = 0.8$, $S_l/P \approx 15$ for $\lambda = 0.01$ (Figure

5.6a), $S_l/P \approx 2$ for $\lambda = 0.1$ (Figure 5.7a), $S_l/P \approx 1.0$ for $\lambda = 0.2$ (Figure 5.8a), and $S_l/P \approx 0.5$ for $\lambda = 0.5$ (Figure 5.9a). The impact of λ diminishes with increasing S_r.

5.3.3 EFFECT OF P ON S_l

Figures 5.6-5.9, showing the S_l/P variation, exhibit the variation of 'nD_s' with 'θ/n' or, as above, $\bar{\psi}$ with θ for a unit rainfall amount. Since D_s relies on θ-dependent $\bar{\psi}$ (equation (5.35)) that, in turn, depends on the amount of P, for the larger P will yield a larger F or θ_l-value, and therefore, a lower $\bar{\psi}$ and, from equation (5.35), a lower S_l-value. This infers that S_l depends on θ and is inversely related to P. A reduction in S_l with P will increase Q and, in turn, C. To compensate for this effect, S_l will increase with P to reduce C to its given value. Thus, S_l increases with P, which, according to the SCS-CN concept, varies linearly for a given C. It, however, requires a detailed investigation in future. The diminishing effect of P on S_l at high C (= S_r)-values, as shown in Figures 5.6-5.9, is indicative of the diminishing effect of $\bar{\psi}$ -variation at high θ-values, which is close to real observations (Mein and Larson, 1971).

5.4 Determination of D_s from Universal Soil Loss Equation

For the computation of the soil depth above the impeding layer D_s, procedures are suggested in Chapter 4 and in this chapter, D_s is signified by equation (5.35) as a function of ψ-θ characteristic of the soil. However, there exists a difficulty in its determination when employed to a watershed/catchment scale. Therefore, it is appropriate to explore an approximate method for its determination from easily available field data. To this end, its determination from the universal soil loss equation (USLE) described in Appendix D is suggested.

USLE is expressed as:

$$A = R\,K\,L_s V_c M_p \tag{D.1}$$

where A is the potential soil loss for a given storm or period (tonnes/ha), R is the rainfall factor, K is the soil erodibility factor, L_s is the slope-length factor, V_c is the vegetative cover (or cropping management) factor, M_p is the erosion control practice factor. Here, a brief description about the possible linkage between USLE and the SCS-CN methodology is in order.

The soil texture determines both permeability and erodibility of soils. Permeability describes infiltration, which, in turn, determines hydrological activeness of the soil surface in terms of both runoff generation and soil erosion. Erosion occurs only when surface runoff is generated, if wind effects are ignored, and according to the SCS-CN method, the runoff generation is closely linked with infiltration. Thus, the processes of runoff generation and soil erosion are closely interrelated. However, these two equations have not yet been investigated for their interrelationship, leading to a lack of understanding of their interdependency. These equations are coupled for better interpretation of the parameters of both methods.

USLE (equation (D.1)) computes the potential loss in terms of tonnes per hactare of the watershed. Thus, a division of both the sides of equation (D.1) by S (which, by definition, is equal to V_a, as shown in Figure 1.18) leads to

$$\frac{A}{S} = \frac{m}{A_w \times V_a} \tag{5.58}$$

where m is the mass of the maximum potential erosion (tonnes) and A_w is the watershed area (hectares). As above, for a completely dry soil, $V_a = V_v$. Thus, substitution of V_w for V_a and a multiplication of both the numerator and denominator of the right-hand side of equation (5.58) by the total volume, V, yields

$$\frac{A}{S} = \frac{m \times V}{A_w \times V \times V_v} \tag{5.59}$$

Considering V and V_v in terms of depths (for a unit surface area) leads to defining the product (A_w x V) as the total volume of the potentially eroded soil from the watershed. In equation (5.59), the ratio, m/(A_w x V), represents the bulk density, ρ_s, of eroded sediments (tonnes/ha-mm) and V_v/V stands for the soil porosity, n (dimensionless). Here, S is in mm. It follows that

$$\frac{A}{S} = \frac{\rho_s}{n} \tag{5.60}$$

Thus,

$$S = \frac{n}{\rho_s} A \tag{5.61}$$

Coupling of equation (5.61) with equation (D.1) leads to

$$S = (n/\rho_s)\, R\, K\, L_s\, V_c\, M_p \tag{5.62}$$

Equations (5.61) and (5.62) define S to be directly proportional to A and n and inversely proportional to ρ_s. It implies that S depends on all the factors affecting the potential erosion of the watershed; the higher the potential erosion from watershed, the higher will be S and vice versa. It is of common knowledge that highly porous soils (of high n), such as sands, yield high S values whereas nonporous soils (of low n), such as loams and clays, yield low S values. Such a description is also consistent with the discussion of the factors affecting S in Chapters 2 and 4. According to equation (5.61), heavy sediments (of high mass density) will yield low S values and vice versa.

Equation (5.62) can also be written in terms of the potentially erodible soil depth, V, in mm as:

$$V = (1/\rho_s)\, R\, K\, L_s\, V_c\, M_p \tag{5.63}$$

Thus, USLE indirectly measures the potential erodible depth of the watershed. Since equation (5.61) also describes 'A' in terms of the watershed characteristic, described by S and the soil type, described by n and ρ_s, 'A' can be better described as the potential maximum erodible depth of the watershed. Such an interpretation of equation (5.62) or equation (5.63) can also be supported by the following observation of Novotny and Olem (1994):

"For unconsolidated geological materials (soils, river deposits, sand dunes, etc.), the erodibility depends on the particle size and the texture of the material, water content, composition of the material, and the presence or absence of protective surface cover such as vegetation. Furthermore, loose soils with low chemical and clay content have the highest erodibility."

Coupling of equation (5.61) with equation (4.26) yields

$$D_s = \frac{n}{n - \theta_o} \frac{A}{\rho_s} \tag{5.64}$$

An assumption of $\theta_o = 0$ under AMC I directly yields D_s from equation (5.64) as:

$$D_s = \frac{A}{\rho_s} \tag{5.65}$$

A comparison of equation (5.64) with equation (5.35) lends the physical description of potential erosion 'A' as:

$$A = \frac{2\rho_s}{N} \frac{m_1}{n} \overline{\psi} \tag{5.66}$$

Equation (5.66) describes 'A' to depend on density of soil particles, soil porosity, ψ-θ characteristics, and capillary and gravitational characteristic of infiltration. In simple words, 'A' depends on the density of soil particles and ψ-θ characteristics of the soil. Such a relationship, however, assumes that the whole watershed contributes to erosion. Secondly, USLE is generally used for estimating the long-term (annual) potential soil loss. Therefore, its applicability to storm rainfall-runoff events needs a detailed investigation in future.

Example 5.3: For given rainstorm, determine the effective soil depth above the impeding layer for both the land use alternatives given in Example D.1 (Appendix D). Assume the bulk density of the eroded soil = 2.65 g/cm^3.

Solution: First, compute the bulk density, ρ_s, in tonnes/ha-mm as below:
$\rho_s = 2.65 \times 10^{-6}$ tonnes/($10^{-6} \times 10^{-4} \times 10^{3}$ ha-mm) = 26.5 tonnes/ha-mm

For alternative (a), A is computed as equal to 4.60 tonnes/ha. Therefore, from equation (5.65) the effective soil depth above the impeding layer, D_s, is computed as

$D_s = 4.60/26.5 = 0.17$ mm

For alternative (b), A = 0.046 tonnes/ha. Thus, from equation (5.65)

$D_s = 0.046/26.5 = 0.002$ mm

Example 5.4: Using the data and results of Example 5.3 and assuming a soil porosity of 0.3, evaluate the universal soil loss equation in terms of the SCS-CN methodology assuming that the soil erodibility factor corresponds to the following composition of the soil: Clay 20%, silt 35%, fine sand 20%, (silt + fine sand) 55%, and coarse sand and gravel 25%.

Solution: For given alternatives, the curve numbers for alternative (a) for AMC I can be computed from equation (4.26) as: $S_I = nD_s = 0.3 \times 0.17 = 0.051$ mm
Therefore, from equation (3.143) $S_{II} = S_I/2.281 = 0.051/2.281 = 0.022$ mm. The corresponding CN_{II} can be computed using equation (2.7) as: $CN_{II} = 25400/(0.022 + 254) = 100$

Similarly, CN is computed as 100 for alternative (b). To evaluate the universal soil loss equation (USLE) with reference to the SCS-CN methodology, it is necessary to determine the corresponding CN values from the NEH-4 table.

For the given soil composition, it is possible to categorize it as soil type D. The land use of alternative (a) is defined as "All vegetation is stripped from the soil surface (100% pervious)". Use Table 2.1 to determine the CN value for the given land use, which fairly matches with the land use described as "Developing areas: Newly graded areas (pervious areas only, no vegetation)" in Sl. No. 6 of Table 2.1. Thus, the NEH-4 CN is equal to 94. The average curve number for alternative (b) can be derived using the area-weighted method described in Chapter 2. Here, 25% area of the plot is impervious for which CN = 98 (Table 2.1: Sl. No. 2) and the rest of the area has a CN-value equal to 94. Thus, the average CN is computed as

$CN_{II} = 0.25 \times 98 + 0.75 \times 94 = 95$

The CN-values derived from NEH-4 are quite high, suggesting a high runoff factor of the 50-ha plot and, in turn, the low depth of the soil above the impeding layer, D_s. The CN-values derived from USLE for both alternatives are however equal to 100, which implies that USLE yields higher CN-values or lower D_s–values than those derived from the corresponding NEH-4 CN-values. It infers that USLE needs to be further improved for yielding a higher A-value than that due to the present USLE. Such an improvement will, however, lead to a reduction in the delivery ratio of the watershed. This assertion is based on the assumption that the SCS-CN methodology is superior to USLE, which is not unreasonable because the science of rainfall-runoff modelling is more advanced than the science of modelling erosion or sedimentation processes.

CHAPTER 6

INFILTRATION AND RUNOFF HYDROGRAPH SIMULATION

As discussed in Chapter 2, the SCS-CN method was originally developed for computing the direct surface runoff (or rainfall-excess) amount from a given amount of storm rainfall occurring on agricultural sites, where it performs best (Ponce and Hawkins, 1996). It performs fairly on range sites and poorly in applications to forest sites (Hawkins, 1984; 1993). The method is also recommended for use in urban environments (SCS, 1975). One of the several problems associated with the SCS-CN method (Mockus, 1964; Ponce and Hawkins, 1996; and others) is that the method does not contain any expression for time and ignores the impact of rainfall intensity and its temporal distribution. Since the SCS-CN method is also construed as an infiltration model (Aron et al., 1977; Chen, 1982; Ponce and Hawkins, 1996; Mishra, 1998; Mishra and Garg, 2000), there exists a possibility for incorporating time in the SCS-CN method for predicting infiltration rates (Chapter 4) and consequently, the runoff hydrograph.

As indicated by W.L. Cowan in a personal communication to Rallison and Miller (1982) (Chapter 2), time was not incorporated in the method because (a) sufficiently reliable data were not available to describe infiltration rates for a wide range of soil-vegetation-land use (SVL) complex and (b) there was no reliable method available for distributing rainfall in time. NEH-4 also does not provide an explicit provision for the spatial scale effects. According to Simanton et al. (1973), CN for areas less than 227 ha in southeastern Arizona tended to decrease with increasing watershed size, exhibiting a significant role of channel transmission losses. Although Mockus (in: SCS, 1971) indicated that the method was applicable to even large catchments, a clear guidance still does not exist if it can reliably be applied to small, mid, or large size catchments. According to Ponce and Hawkins (1996), it should be applied cautiously to the watersheds larger than 250 sq. km.

The objective of this chapter is to present an SCS-CN-based time distributed runoff model for event-based hydrologic simulation, including computation of infiltration and runoff rates using field/laboratory data, and evaluate its performance in various applications.

6.1 SCS-CN-Based Infiltration and Runoff Models

Using the Horton equation (equation (1.58)) and equation (2.2), the derivation of the Mockus method (Chapter 3: equation (3.1)) yielded the following parametric relation (equation (3.10)):

$$i_o - f_c = \alpha S \tag{3.10}$$

where i_o is the uniform rainfall intensity, α is the Horton decay parameter, and f_c is the minimum infiltration rate. An assumption of the cumulative rainfall varying linearly with time enables differentiation (with respect to time) of the resulting expression for F from equations (2.5) and (2.1). Since the resulting expression does not contain the f_c-term, it assumes $f_c = 0$ and therefore underestimates infiltration. An incorporation of 'f_c' yields the infiltration rate, f, as (equation (4.39)):

$$f = f_c + \frac{i_o - f_c}{(1 - \lambda + \alpha t)^2} \tag{4.39}$$

where t is the time.

The direct surface runoff rate q is computed by subtracting the infiltration rate (equation (4.39)) from the uniform rainfall intensity, i_o. The resulting rainfall-excess rate, q, is expressed as (equation (4.61)):

$$q = [1 - \frac{1}{(1 - \lambda + \alpha t)^2}](i_o - f_c) \tag{4.61}$$

Equation (4.61) is valid for $\alpha t \geq \lambda$ or $t \geq t_p$ (time to ponding), $q = 0$ otherwise. Thus, the SCS-CN method computes the rainfall-excess time past ponding. Similarly, as $t \rightarrow \infty$, $q \rightarrow (i_o - f_c)$. It is further emphasized that the derivation of equation (4.61) assumes a linear growth of cumulative rainfall with time. The routing of the rainfall-excess by a single linear reservoir or any other technique (Chapter 1) yields the direct surface runoff hydrograph at the outlet of the watershed. In finite difference form, the continuity equation (equation (1.83)) is expressed as:

$$q - O = \Delta V / \Delta t \tag{6.1}$$

where V is the reservoir storage, Δt is the time interval, and O is the outflow or runoff rate at the outlet of the watershed. Coupling of the finite difference form of equation (1.84), representing a single linear reservoir, with equation (6.1) yields

$$O_{(j+1)} = d_0 q_j + d_1 q_j + d_2 O_j \tag{6.2}$$

where j and j+1 are the time steps at Δt interval and

$$d_0 = \frac{\Delta t / K}{2 + \Delta t / K} \tag{6.3a}$$

$$d_1 = d_0 \tag{6.3b}$$

$$d_2 = \frac{2 - \Delta t / K}{2 + \Delta t / K} \tag{6.3c}$$

In equations (6.3a-c), K is the storage coefficient. Using finite differences, O at different time steps can be computed as follows. The routed flows, O, represent the flow ordinates of the direct surface runoff hydrograph at different times at the outlet of the watershed. The addition of baseflow O_b, if any, yields the total runoff hydrograph.

Table 6.1 Soil characteristics and other infiltration parameters

Sl. No.	Soil type	Proj. No.	n_e (v/v)	θ_o (v/v)	K_b (cm/hr)	$^*\psi$ (cm)	Computations from infiltration data: NOR	i_e (cm/hr)	f_c (cm/hr)	t_p (min)	α-fit statistic: α (min^{-1})	α-fit statistic: r^2	$^{**}S$ (cm)	CN	D_s (cm)	$\lambda = \alpha t_p$	$^{***}S_s$ (cm $s^{-1/2}$)	M (cm)
1	PFS	1	0.35	0.13	12.38	11.73	19	31.51	18.12	1.73	0.2572	0.95	2.09	92.4	5.98	0.44	0.0958	0.78
2	PFS	2	0.35	0.13	12.38	11.73	19	31.64	18.00	1.73	0.2535	0.95	2.14	92.2	6.11	0.44	0.0968	0.79
3	PFS	3	0.35	0.13	12.38	11.73	18	31.29	18.34	1.74	0.2675	0.95	2.01	92.7	5.74	0.47	0.0938	0.75
4	PFS	8	0.35	0.13	12.38	11.73	24	80.55	18.34	0.41	0.3635	0.73	2.73	90.3	7.79	0.15	0.1093	1.01
5	PFS	9	0.35	0.13	12.38	11.73	17	80.65	18.10	0.42	0.7544	0.86	2.05	92.5	5.85	0.32	0.0947	0.76
6	PFS	10	0.25	0.23	12.38	11.73	12	81.33	17.88	0.28	0.4569	0.78	2.21	92.0	8.84	0.13	0.0984	2.03
7	PFS	11	0.25	0.23	12.38	23.83	12	80.80	17.54	0.32	0.4287	0.76	2.25	91.9	9.00	0.14	0.0992	2.07
8	PFS	12	0.25	0.23	12.38	23.83	11	77.61	18.17	0.36	0.4770	0.74	2.02	92.6	8.10	0.17	0.0942	1.86
9	CSL	13	0.39	0.13	5.00	23.83	14	7.65	12.37	9.84	0.1750	0.98	0.89	96.6	2.29	1.72	0.0398	0.30
10	CSL	14	0.39	0.13	5.00	23.83	14	7.78	12.23	10.01	0.1731	0.98	0.91	96.5	2.34	1.73	0.0402	0.30
11	CSL	15	0.39	0.13	5.00	23.83	13	7.47	12.54	10.13	0.1904	0.98	0.80	97.0	2.04	1.93	0.0375	0.27
12	CSL	16	0.37	0.15	5.00	23.83	12	9.16	10.86	9.17	0.1234	0.99	1.49	94.5	4.01	1.13	0.0513	0.60
13	CSL	17	0.37	0.15	5.00	23.83	12	9.37	10.65	9.22	0.1179	0.99	1.59	94.1	4.29	1.09	0.0530	0.64
14	CSL	18	0.37	0.15	5.00	23.83	12	9.52	10.50	9.29	0.1149	0.99	1.66	93.9	4.48	1.07	0.0541	0.67
15	CSL	28	0.39	0.13	5.00	23.83	24	16.13	13.88	4.03	0.2680	0.99	1.18	95.6	3.02	1.08	0.0457	0.39
16	CSL	29	0.39	0.13	5.00	23.83	23	15.66	14.34	4.06	0.2974	0.98	1.05	96.0	2.70	1.21	0.0431	0.35
17	CSL	30	0.39	0.13	5.00	23.83	23	15.86	14.15	4.12	0.2888	0.98	1.08	95.9	2.77	1.19	0.0437	0.36
18	CSL	31	0.27	0.25	5.00	23.83	9	18.99	11.01	2.68	0.2019	0.97	1.78	93.5	6.59	0.54	0.0561	1.65
19	CSL	33	0.27	0.25	5.00	23.83	9	19.62	10.20	3.00	0.1736	0.95	2.04	92.6	7.54	0.52	0.0600	1.88
20	CSL	34	0.39	0.13	5.00	23.83	29	26.09	13.78	2.21	0.2981	0.96	1.63	94.0	4.19	0.66	0.0537	0.54
21	CSL	35	0.39	0.13	5.00	23.83	29	26.28	13.60	2.26	0.2913	0.96	1.66	93.9	4.26	0.66	0.0542	0.55
22	CSL	36	0.39	0.13	5.00	23.83	28	25.97	13.90	2.31	0.3061	0.96	1.55	94.3	3.97	0.71	0.0523	0.52
23	CSL	37	0.37	0.15	5.00	23.83	24	29.15	10.77	2.07	0.1702	0.88	2.82	90.0	7.63	0.35	0.0707	1.14
24	CSL	38	0.37	0.15	5.00	23.83	24	29.36	10.55	2.10	0.1626	0.88	2.93	89.7	7.92	0.34	0.0720	1.19
25	CSL	39	0.37	0.15	5.00	23.83	24	29.54	10.35	2.16	0.1553	0.87	3.03	89.3	8.19	0.34	0.0732	1.23
26	CSL	40	0.32	0.20	5.00	23.83	17	29.45	10.53	1.76	0.1853	0.88	2.63	90.6	8.22	0.33	0.0682	1.64
27	CSL	41	0.32	0.20	5.00	23.83	17	29.69	10.24	1.83	0.1743	0.87	2.76	90.2	8.63	0.32	0.0699	1.73
28	CSL	42	0.32	0.20	5.00	23.83	16	28.99	10.81	1.92	0.1979	0.87	2.39	91.4	7.48	0.38	0.0651	1.50
29	CSL	43	0.27	0.25	5.00	23.83	12	29.74	10.23	1.47	0.1984	0.91	2.49	91.1	9.23	0.29	0.0664	2.31
30	CSL	44	0.27	0.25	5.00	23.83	11	28.92	11.02	1.49	0.2361	0.93	2.10	92.4	7.79	0.35	0.0610	1.95

Table 6.1 Continued

Sl. No.	Soil type	Proj No.	n_e (v/v)	θ_o (v/v)	K_h (cm/hr)	*ψ (cm)	Computations from infiltration data: NOR	i_e (cm/hr)	f_c (cm/hr)	t_p (min)	α-fit statistic: α (min^{-1})	r^2	**S (cm)	CN	D_s (cm)	$\lambda = \alpha t_p$	***S_s (cm s$^{-1/2}$)	M (cm)
31	CSL	45	0.27	0.25	5.00	23.83	11	29.31	10.60	1.53	0.2168	0.91	2.27	91.8	8.39	0.33	0.0633	2.10
32	CSL	46	0.20	0.32	5.00	23.83	8	29.24	10.75	1.07	0.3009	0.93	1.71	93.7	8.56	0.32	0.0550	2.74
33	CSL	47	0.20	0.32	5.00	23.83	8	29.44	10.16	1.22	0.2699	0.90	1.85	93.2	9.25	0.33	0.0572	2.96
34	ISL	67	0.28	0.25	0.11	7.43	23	0.25	0.17	121.29	0.0092	0.98	0.51	98.0	1.83	1.12	0.0045	0.46
35	ISL	68	0.28	0.25	0.11	7.43	23	0.26	0.16	124.04	0.0077	0.94	0.60	97.7	2.13	0.96	0.0048	0.53
36	ISL	69	0.28	0.25	0.11	7.43	23	0.26	0.16	127.11	0.0079	0.94	0.57	97.8	2.04	1.00	0.0047	0.51
37	ISL	70	0.23	0.30	0.11	7.43	15	0.27	0.15	95.80	0.0074	0.96	0.64	97.5	2.79	0.71	0.0050	0.84
38	ISL	71	0.23	0.30	0.11	7.43	14	0.27	0.15	98.11	0.0079	0.97	0.61	97.7	2.65	0.78	0.0049	0.79
39	ISL	72	0.23	0.30	0.11	7.43	14	0.27	0.15	101.44	0.0077	0.96	0.61	97.6	2.67	0.78	0.0049	0.80
40	ISL	76	0.13	0.40	0.11	7.43	15	0.29	0.13	45.28	0.0088	0.94	0.53	97.9	4.10	0.40	0.0046	1.64
41	ISL	77	0.13	0.40	0.11	7.43	15	0.29	0.12	48.42	0.0071	0.93	0.63	97.6	4.82	0.34	0.0049	1.93
42	ISL	78	0.13	0.40	0.11	7.43	15	0.29	0.12	52.94	0.0069	0.88	0.62	97.6	4.80	0.37	0.0049	1.92
43	ISL	96	0.18	0.35	0.11	7.43	8	0.65	0.18	23.83	0.0168	0.89	0.59	97.7	3.29	0.40	0.0048	1.15
44	ISL	97	0.13	0.40	0.11	7.43	11	0.67	0.16	11.74	0.0181	0.89	0.50	98.1	3.83	0.21	0.0044	1.53
45	ISL	98	0.13	0.40	0.11	7.43	11	0.68	0.15	12.72	0.0169	0.89	0.54	97.9	4.16	0.21	0.0046	1.66
46	ISL	99	0.13	0.40	0.11	7.43	11	0.66	0.15	13.71	0.0161	0.86	0.53	98.0	4.07	0.22	0.0045	1.63
47	ISL	100	0.13	0.40	0.11	7.43	8	0.70	0.14	7.50	0.0188	0.92	0.52	98.0	4.03	0.14	0.0045	1.61
48	ISL	101	0.10	0.43	0.11	7.43	8	0.66	0.14	9.60	0.0177	0.90	0.52	98.0	5.23	0.17	0.0045	2.25
49	ISL	102	0.10	0.43	0.11	7.43	8	0.59	0.14	13.25	0.0196	0.94	0.47	98.2	4.75	0.26	0.0043	2.04
50	YLC	109	0.25	0.25	0.04	22.36	10	0.23	0.12	155.69	0.0054	0.99	0.82	96.9	3.29	0.84	0.0034	0.82
51	YLC	110	0.25	0.25	0.04	22.36	9	0.22	0.13	163.44	0.0062	0.99	0.68	97.4	2.73	1.01	0.0031	0.68
52	YLC	111	0.25	0.25	0.04	22.36	9	0.22	0.13	174.58	0.0068	0.97	0.62	97.6	2.47	1.19	0.0030	0.62
53	YLC	112	0.15	0.35	0.04	22.36	7	0.23	0.12	78.40	0.0099	0.99	0.48	98.1	3.21	0.78	0.0026	1.12
54	YLC	113	0.15	0.35	0.04	22.36	7	0.23	0.12	83.40	0.0099	0.99	0.47	98.2	3.16	0.83	0.0026	1.11
55	YLC	114	0.15	0.35	0.04	22.36	7	0.24	0.11	90.59	0.0089	0.98	0.52	98.0	3.50	0.81	0.0027	1.22

Notations: n_e = effective porosity; NOR = No. of ordinates; *ψ = capillary suction; **$S=1.25F_{max}$; *** $S_s=\sqrt{(4/\pi)(1.25F_{max}K_h)}$. The Sl. Nos. in Col. 1 are different from Project numbers.

To avoid the problem of negative outflows, K is kept greater than or equal to $\Delta t/2$. The single linear reservoir method of routing assumes a linear storage relationship with the outflow, as shown by equation (1.84). The experience, however, supports the use of an improved technique such as the Muskingum-Cunge method (Ponce, 1989) or any other physically based method. In this chapter, the single linear reservoir technique is used for simplicity reasons.

The parameters of the SCS-CN-based time-distributed runoff model include the Horton decay parameter, α; minimum infiltration rate, f_c; and routing parameter, K. The initial abstraction coefficient, λ, can be determined from equation (4.19) using the infiltration data or can be assumed as 0.2 (a standard value). Equation (4.19), however, is a specific form of equation (3.87) and, therefore, ignores other factors affecting λ, which are discussed in Chapters 3 and 4. Parameter f_c can be determined from infiltration data or obtained from text books (for example, Singh, 1992; Ponce, 1989) for various soil types. Parameter K represents the lag time between the rainfall and outflow volumes and can be derived from rainfall-runoff records. In the absence of this information, the model parameters can, in general, be optimized using a suitable optimization algorithm, for example, the Marquardt least square approach (Appendix B).

6.2 Application of Infiltration and Runoff Models

6.2.1 INFILTRATION DATA

The infiltration data due to Mein and Larson (1971) given in Table 4.1 were used for application of the SCS-CN-based infiltration model and the existing model. These data were derived from laboratory tests for the Plainfield sand (PFS), Columbia sandy loam (CSL), Ida silt loam (ISL), and Yololight clay (YLC) of U.S.A. The characteristics of these soils are further given in Table 6.1.

6.2.2 ARS WATERSHEDS

Four Agricultural Research Service (ARS) watersheds along with one urban paved watershed were selected for model application: Y-8, W-1, W-3, Goodwin Creek, and an experimental watershed described below.

The Y-8 (no. 42106) experimental watershed is located in Waco, Riesel, Texas, U.S.A. The watershed is 20.8 acres in area and its outlet is located at latitude 31° 28' 22" and longitude 96° 52' 54". The watershed has 3% pasture, 95% fall planted oats, and 2% gravel roads. The cropland is terraced and contour-tilled. The runoff data are available from 1968 to 81 and rainfall data recorded at the raingauge no. RG00075A are available from 1939-81.

The gauging site of W-1 watershed (no. 42006) is located in Riesel, Texas, U.S.A. at latitude 31° 27' 27" and longitude 96° 52' 48". The watershed is characterized by 85% pasture, 11% cotton, 3% gravel and paved roads, and 1% other land use. Approximately 90% of the other land use is Johnson grass and weeds in conservation reserve, but is neither tilled nor grazed. Runoff and rainfall (raingauge no. RG000W2A) data, respectively, are available from 1940 to 1981 and from 1939 to 1981.

The outlet of the Stillwater (Oklahoma, U.S.A.) W-3 watershed (no. 37002) is located at a latitude of 36° 21' 00" and a longitude of 97° 04' 00". The watershed is characterized by native grass cover: 32% in hay meadow and 78% in pasture. Runoff and rainfall (raingauge no. RG000003) data are, respectively, available for the period 1959-79 and 1969-1979.

The Goodwin Creek Experimental Watershed is a 15-acre watershed near Batesville, Mississippi. The watershed is located in the bluff hills of the Yazoo River basin of northern Mississippi, with its outlet at a latitude of 89° 54' 50" and a longitude of 34° 13' 55". The land use and management practices that influence the rate and amount of stream flows from the uplands range from timbered areas to row crops. About 13 percent of the total watershed area is under cultivation and the rest lies as idle pasture and forest land. Terrain elevation ranges from 71 to 128 m above the mean sea level, with an average channel slope of 0.004 in Goodwin Creek. The runoff and rainfall (raingauge station no. RG000057) data are available for the period 1980 to 1990 and 1981 to 1990, respectively.

An experimental urban watershed in Cincinnati (U.S.A.) of 15x20 m asphalt pavement of I-75 (at milestone 2.6) that is a major north-south interstate was selected for which rainfall-runoff data were collected during 1995-97 by Sansalone and Buchberger (1997). The runoff on highway site located at the upstream end of an urban watershed finally drains into Mill Creek. The runoff from the selected stretch of 300 sq. m is contributed by the four southbound lanes, an exit lane, and a paved shoulder, all draining to a grassy v-section median at a transverse pavement cross-slope of 0.020 m/m, as shown in the plan view of Figures 6.1a & b. The flow on the longitudinal pavement with a slope of 0.004 is primarily characterized by sheet flow and the land use as urban (industrial, commercial, residential). The storm water runoff diverted through the epoxy-coated converging slab, a 2.54 cm diameter Parshall flume, and a 2 m long 25.4 cm diameter PVC pipe to a 2000 liter storage tank was measured at a regular 1 min interval using an automated 24-bottle sampler with polypropylene bottles. Rainfall was recorded using a tipping bucket gauge in increments of 0.254 mm.

6.2.3 ERROR CRITERIA FOR SIMULATION

The following error criteria were considered for application of the Marquardt algorithm (Appendix B):

$$SE = \sqrt{\sum_{j=1}^{N} (Y_o - Y_c)_j^2 / (N - m + 1)} \tag{6.4}$$

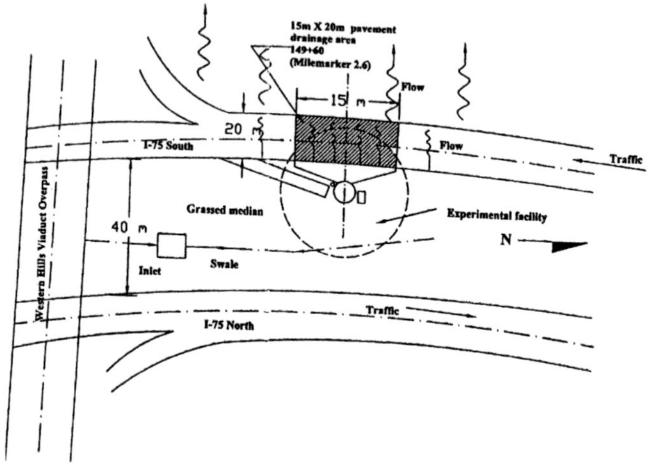

Figure 6.1a. Plan view of I-75 highway site at Cincinnati.

$$r^2 = 1 - \frac{\sum_{j=1}^{N} (Y_o - Y_c)_j^2}{\sum_{j=1}^{N} (Y_o - Y_{mean})_j^2} \tag{6.5}$$

where SE is the standard error, r^2 is the coefficient of determination, N is the number of ordinates in an event, j is an integer varying from 1 to N, m is the number of model parameters, Y_o is the observed quantity, Y_c is the computed quantity, and Y_{mean} is the mean quantity. The higher values of SE and lower values of r^2 show a poor fit and the reverse holds in the otherwise situation; SE = 0 exhibits a perfect fit. The value of r^2 varies on a scale of 0-1; where 0 indicates a mean model fitting better than the proposed model whereas 1 exhibits a perfect fit.

6.2.4 MODEL APPLICATION TO INFILTRATION DATA

Before the SCS-CN method is applied to infiltration data and evaluated for performance, it is appropriate now to discuss the computation of its parameters from infiltration data and other related variables affecting these parameters.

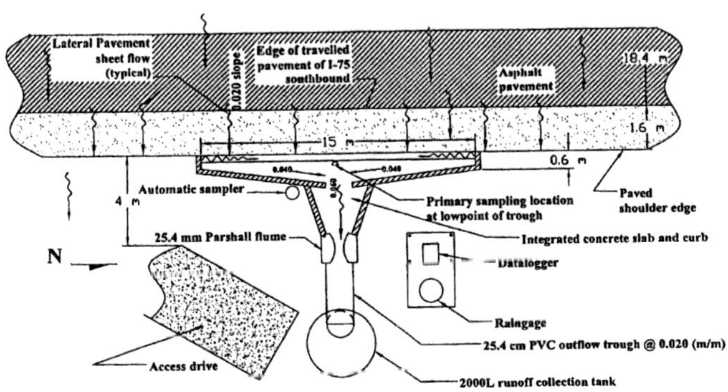

Figure 6.1b. Plan view of experimental setup at I-75 site.

(a) *Computation of Parameters*

Computation of S. For computation of S, the data of Figure 4.1 are considered in the following illustrative example.

Example 6.1: Investigate the behavior of the SCS-CN model parameter potential maximum retention, S, during an infiltration decay process using the following infiltration data of Project No. 34, also shown in Figure 4.1.

Time (min)	2.21	2.38	2.54	2.72	2.87	3.05	3.22	3.42	3.66	3.87
f (cm/hr)	39.87	38.54	36.37	33.94	33.2	31.63	30	28.86	27.43	26.71

Time	4.08	4.28	4.5	4.75	5.03	5.31	5.62	5.93	6.32	6.91
f	25.63	25.17	24	23.68	22.71	22.25	21.52	20.9	20.24	19.76

Time	7.43	8.24	8.97	9.85	10.84	12.14	13.43	14.54	16.49
f	18.81	18.22	17.55	16.91	16.37	15.65	14.85	14.36	13.78

Solution: It is evident from the given infiltration data set that the time to ponding is approximately equal to 2.21 min. To exclude the initial abstraction, I_a, or the amount of infiltration to the time to ponding, P, Q, and F are computed at a given time past ponding, as shown in Table 6.2.

Table 6.2 Computation of S (Example 6.1)

$t-t_p$ (min)	P_e (cm)	Q (cm)	F (cm)	$S=P_eF/Q$ (cm)
0.00	0.00	0.00	0.00	0.00
0.17	0.07	0.00	0.07	2.83
0.33	0.14	0.01	0.14	2.33
0.51	0.22	0.02	0.20	1.97
0.66	0.29	0.04	0.25	1.87
0.84	0.37	0.06	0.30	1.84
1.01	0.44	0.09	0.35	1.80
1.21	0.53	0.12	0.41	1.76
1.45	0.63	0.17	0.46	1.74
1.66	0.72	0.21	0.51	1.73
1.87	0.81	0.26	0.55	1.72
2.07	0.90	0.31	0.59	1.72
2.29	1.00	0.36	0.63	1.72
2.54	1.10	0.43	0.67	1.72
2.82	1.23	0.51	0.72	1.72
3.10	1.35	0.59	0.76	1.73
3.41	1.48	0.68	0.80	1.73
3.72	1.62	0.78	0.84	1.74
4.11	1.79	0.91	0.88	1.74
4.70	2.04	1.10	0.94	1.75
5.22	2.27	1.28	0.99	1.76
6.03	2.62	1.57	1.05	1.76
6.76	2.94	1.84	1.10	1.77
7.64	3.32	2.17	1.15	1.77
8.63	3.75	2.55	1.20	1.77
9.93	4.32	3.07	1.25	1.76
11.22	4.88	3.60	1.28	1.74
12.33	5.36	4.06	1.30	1.71
14.28	6.21	4.90	1.31	1.66

It is noted that the uniform rainfall intensity is taken as 39.87 cm/hr and it is used for computation of P_e (= $P - I_a - F_c$). The cumulative infiltration, F, is computed for the same time past ponding and Q is computed from the water balance equation (equation (2.1)) using P_e and F. Using the proportionality hypothesis (equation (2.2)), S is computed at each time, as shown in the last column of the table. These values when plotted in Figure 6.2 depict the behavior of S during the infiltration process. It is apparent from this figure that at the outset (t = 0), S increases with t, reaches a maximum value, and then decreases gradually. It is noted that at t = 0, $P_e = Q = F = 0$. Therefore, S = 0/0, interpreted as 0 in Figure 6.2. The trend, however, shows that $S \rightarrow \infty$ as $t \rightarrow 0$ and $F \rightarrow F_{max} = S$ as $t \rightarrow \infty$. On the other hand, F increases with t and reaches a maximum value F_{max} at $f = f_c$. It is again noted that the trend of reducing S with time t corresponds to a real data set, whereas according to the Hortonian concept, $F \rightarrow F_{max}$ as $t \rightarrow \infty$. Thus, F and S will increase and decrease, respectively, with time such that the minimum S-value for a data set corresponds to F_{max}. Here, the minimum S stands for the value computed at $f = f_c$. However, S and F_{max} will never converge for the real data because of the SCS-CN hypothesis (equation (2.2)), as shown below.

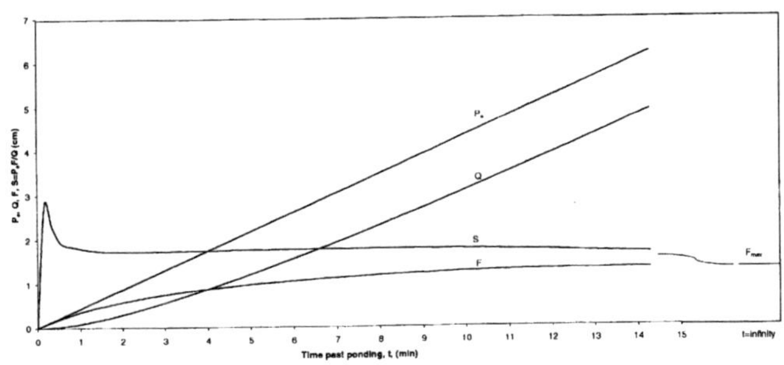

Figure 6.2. Variation of S, P_e, F, and Q with time past ponding t for Columbia sandy loam (Project No. 34).

For $I_a = 0$, equation (2.2) can be rewritten as:

$$S = F(1+F/Q) \tag{6.6}$$

The Hortonian or SCS-CN infiltration concept allows much faster growth of Q than F (or Q>>F) at larger t (Figure 6.2). It is consistent with the description of Chen (1982). Thus, as t increases, S approaches F_{max}. Absolutely, the minimum S will always be higher than F_{max} for real data, consistent with equation (6.6).

Following the computations of Example 6.1, the final S-values corresponding to the maximum F-values were computed for each infiltration data set. For example, these are equal to 1.66 cm and 1.31 cm, respectively, in the above example. These computed values are plotted against each other in Figure 6.3. Apparently, there exists a relationship between S and F_{max}, expressed mathematically as $S = 1.25F_{max}$. For this reason, almost all the data points hover around the line of perfect fit (LPF) drawn for S and $1.25F_{max}$. The deviation of the data points from LPF can be attributed to the computation of F_{max}-values, which were derived using the trapezoidal formula assuming the minimum infiltration rate as available from the data as f_c-values. A higher f_c-value will yield a lower F_{max} and vice versa and the reverse will hold for S. Therefore, refinement of f_c-values was possible for improving the F_{max}-minimum S fit. The close proximity of the data points to the line of minimum $S = 1.25F_{max}$ however asserts that S is the maximum difference of (P_e-Q), consistent with the McCuen's (2002) definition of S as potential postinitial abstraction retention. The significance of such an inference lies in the fact that it lends an approach for deriving CN and λ from easily available infiltration data for any soil-vegetation-land cover (SVL) setting.

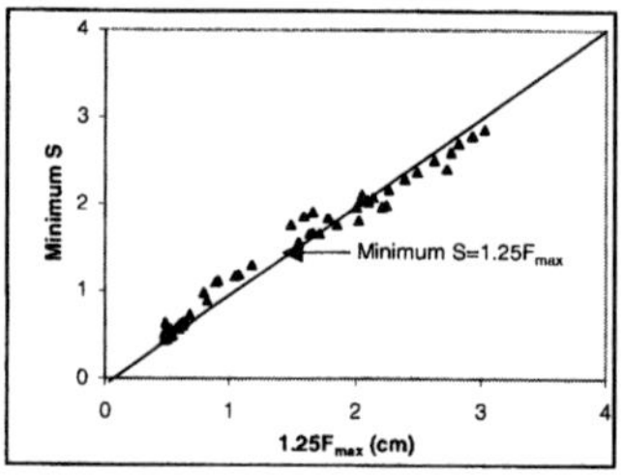

Figure 6.3. Maximum cumulative infiltration, F_{max}, versus minimum S.

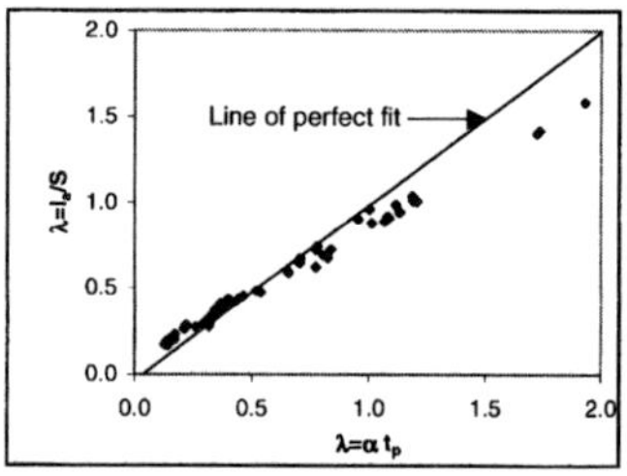

Figure 6.4. Comparison of λ from two appraches: $\lambda = I_a/S$ and $\lambda = \alpha t_p$; $S = 1.25F_{max}$.

Computation of λ-values. To show the applicability of the relation, $\lambda = \alpha t_p$ (equation (4.19)), the λ-values were computed for the U.S.A. data using this relation and using equation (2.3) taking $S \approx 1.25F_{max}$. α-values were derived from the best-fit statistic of each infiltration data set, as discussed in Chapter 1 and later in this chapter. Thus, these values represent the best-ft values of the whole data set. The time-to-ponding values were determined for each data set, as shown in Example 6.1. The computed parameter values of all infiltration data sets of U.S.A. soils are summarized in Table 6.1. These values are compared with those computed using equation (2.3) in Figure 6.4. In this figure, the data points lying on LPF indicate that the values of λ derived from the two approaches are exactly the same. Apparently, all the data points for λ ($= \alpha t_p$) < 0.5 fall on LPF supporting the validity of the relation for $\alpha t_p < 0.5$. However, the relation $\lambda = \alpha t_p$ over-estimates λ-values for $\alpha t_p > 0.5$.

Variation of effective soil depth, soil sorptivity, CN, λ, and 'M' values. The effective soil depth (D_s) computed using equation (4.26), soil sorptivity (S_s) using equation (5.22), and curve numbers (CN) using $S = 1.25F_{max}$ and equation (2.7) for various soils are summarized in Table 6.1, and their range of variation is given in Table 6.3. It is evident from Table 6.3 that D_s–values vary from 1.46 to 7.40 cm. These small magnitudes support the notion that only the top-soil-zone affects significantly the dynamic infiltration (Fok and Chiang, 1984). Collis-George (1977) reported the values of soil sorptivity equal to 0.216, 0.0497, and 0.041 cm $s^{-1/2}$ for Pialligo sand, Barton clay loam, and Red clay, respectively. Except for the sandy soil, the other values are of the same order as those derived for CSL, ISL, and YLC and shown in Table 6.3. However, the high values of CN (> 90) of these soils were not comparable with those from NEH-4 tables, a plausible subject of future research. The computed λ values range from 0.13 to 1.93 for all the soils, in contrast with the earlier described range (0.1, 0.3) (Chapters 3 and 4). In general, the PFS and YLC exhibit two extremes, the former exhibits λ-values in the lowest range (0.13, 0.47) and the latter in the highest range (0.78, 1.19), except for CSL that shows the highest λ-value equal to 1.93. CSL shows the widest range of variation (0.29, 1.93), followed by ISL for which λ varies between

0.21 and 1.12. Thus, the λ-values of loamy soils (CSL and ISL) show similar ranges of variation, whereas the ranges for PFS and YLC set the two opposite extremes. The significance of computed M-values is to show the approximate height of the water column corresponding to the initial moisture content, useful in future for assessing rainfall requirements of a soil for defining AMC. These M-values are approximate since the S-values of Table 6.1 or Table 6.3 do not correspond to AMC-I, leading to the lower M-values than the actual ones.

Table 6.3 Range of effective soil depth, D_s, soil sorptivity, S_s, curve number, CN, and antecedent moisture, M, derived from Table 6.1

Sl. No.	Soil type	D_s (cm)		S_s ($cm\ s^{-1/2}$)		CN		M (cm)	
		Min.	Max.	Min.	Max.	Min.	Max.	Min.	Max.
1	PFS	5.74	9.00	0.0938	0.1093	90.3	92.7	0.75	2.07
2	CSL	2.04	9.25	0.0375	0.0732	89.3	97.0	0.27	2.96
3	ISL	1.83	5.23	0.0043	0.0050	97.5	98.2	0.46	2.25
4	YLC	2.47	3.50	0.0026	0.0034	96.9	98.2	0.62	1.22

(b) Performance of the Existing SCS-CN Method

Using the above infiltration data, the performance of the existing SCS-CN method (equation (2.5)) is evaluated. To this end, infiltration rates were computed excluding f_c, and the total rainfall P including I_a (= P_e+I_a) and the cumulative infiltration F (= F_d) were derived for all discrete times for each infiltration data set. Similarly, the Q-values were computed taking the potential maximum retention S = 1.25F_{max} and λ = 0.2. The computation of Q is demonstrated for Project No. 34 (Figure 4.1) in the following example.

Example 6.2: Using the infiltration data of Project No. 34 employed in Example 6.1, examine the applicability of the existing SCS-CN method for computing infiltration.

Solution: For examining the applicability of the existing SCS-CN method to infiltration data of Project No. 34, infiltration values are computed using the method as follows. The effective rainfall amount is added to the initial abstraction (before the time to ponding) to compute the total rainfall and initial abstraction is computed as equal to 0.2S. Thus, the direct surface runoff Q from the existing SCS-CN method is computed as

$$Q = (P_e+I_a-0.2S)^2/(P_e+I_a+0.8S)$$

where $S = F_{max}$ = 1.31 cm. The computations for Q values are shown in Table 6.4. Then, using water balance equation (2.1), infiltration F is computed as

$$F = P_e + I_a - 0.2S - Q$$

The computed values of F are shown in Table 6.4. It is seen that the computed F values are generally lower than the observed but quite close to the observed values, as shown in Figure 6.5a. It is attributed to the minimum infiltration rate, as discussed in Chapters 3 and 4. The under-prediction of F-values (Figure 6.5a) leads to the over-prediction of Q values (Figure 6.5b). However, the closeness of the data points to the line of perfect fit supports the applicability of the SCS-CN method to the given set of infiltration data.

Table 6.4 Computation for infiltration using SCS-CN method using the data of Project No. 34 (Example 6.2)

Time past ponding (min)	P_e (cm)	I_a (cm)	F (observed) (cm)	Q (observed) (cm)	Q (computed)* (cm)	F (computed) (cm)
0.00	0	0.26	0	0	0	0
0.17	0.07	0.26	0.07	0	0	0.07
0.33	0.14	0.26	0.14	0.01	0.01	0.13
0.51	0.22	0.26	0.20	0.02	0.03	0.19
0.66	0.29	0.26	0.25	0.04	0.05	0.24
0.84	0.37	0.26	0.30	0.06	0.08	0.29
1.01	0.44	0.26	0.35	0.09	0.11	0.33
1.21	0.53	0.26	0.41	0.12	0.15	0.38
1.45	0.63	0.26	0.46	0.17	0.20	0.43
1.66	0.72	0.26	0.51	0.21	0.26	0.46
1.87	0.81	0.26	0.55	0.26	0.31	0.50
2.07	0.90	0.26	0.59	0.31	0.37	0.53
2.29	1.00	0.26	0.63	0.36	0.43	0.57
2.54	1.10	0.26	0.67	0.43	0.50	0.60
2.82	1.23	0.26	0.72	0.51	0.60	0.63
3.10	1.35	0.26	0.76	0.59	0.69	0.66
3.41	1.48	0.26	0.80	0.68	0.79	0.69
3.72	1.62	0.26	0.84	0.78	0.90	0.72
4.11	1.79	0.26	0.88	0.91	1.03	0.76
4.70	2.04	0.26	0.94	1.10	1.24	0.80
5.22	2.27	0.26	0.99	1.28	1.44	0.83
6.03	2.62	0.26	1.05	1.57	1.75	0.87
6.76	2.94	0.26	1.10	1.84	2.03	0.91
7.64	3.32	0.26	1.15	2.17	2.38	0.94
8.63	3.75	0.26	1.20	2.55	2.78	0.97
9.93	4.32	0.26	1.25	3.07	3.31	1.01
11.22	4.88	0.26	1.28	3.60	3.85	1.03
12.33	5.36	0.26	1.30	4.06	4.31	1.05
14.28	6.21	0.26	1.31	4.90	5.13	1.08

* $Q = (P_e+I_a-0.2S)^2/(P_e+I_a+0.8S)$; $S = F_{max} = 1.31$ cm.

The Q-values computed using the above approach for all the infiltration data sets are compared with the observed Q (= P_e-F) values in Figure 6.6. The comparison reveals that most data points lie over and above, but near the LPF, and thus, generally support the applicability of the method although the method, in general, over-estimates runoff. It is because the existing SCS-CN methodology assumes the minimum infiltration rate, f_c, equal to zero.

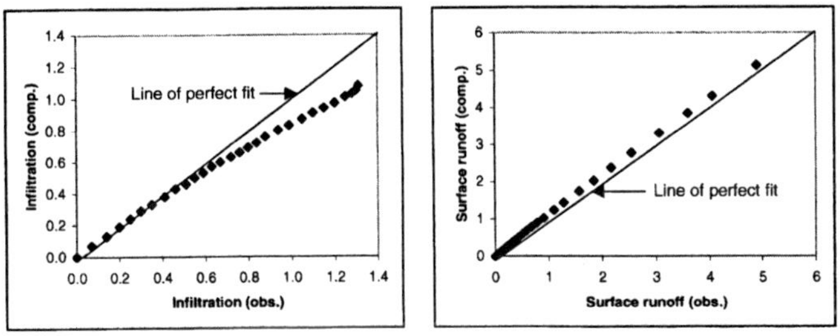

(a) (b)

Figure 6.5. Computed and observed (a) infiltration, F, (cm) and (b) direct surface runoff, Q, (cm) values for Project No. 34.

(c) Performance of the SCS-CN-Based Infiltration Model

Employing the above 55 sets of infiltration data, the validity of the SCS-CN-based infiltration model (equation (4.39)) is also evaluated. To this end, the available infiltration data were first processed for application of the SCS-CN method, which holds for $f_c = 0$ and $t > t_p$ (Figure 4.1). The $(f - f_c)$ values were derived from infiltration data and taken as observed and α-values for each data set were optimized using the least squares approach, as discussed in Chapter 1 and shown in Figure 6.7 for the data set of Project No. 34. The α-values computed for all data sets are summarized in Table 6.1. In Figure 6.7, the slope of the fitted line (= 0.2981) represents the α-value (min^{-1}). For this data set, $i_e = f_0 - f_c = 26.09$ cm/hr is taken as the effective uniform rainfall intensity. Using the SCS-CN-based infiltration model (equation (4.39)) and the Horton model, the $f - f_c$ values were computed and compared with the observed values in Figure 6.8. As seen from this figure, the computed values using the SCS-CN-based infiltration model match each other fairly closely with the observed ones and more closely than those computed using the Horton model. A little scatter of the data points may, however, be attributed to the assumption of a single best-fit α-value for each data set. The computed $(f - f_c)$ values for all 55 data sets are compared with the observed ones in Figure 6.9. The data points hovering around LPF exhibit a reasonably good performance of the above described SCS-CN-based infiltration model.

6.2.5 MODEL APPLICATION TO RAINFALL-RUNOFF DATA

The above time-distributed runoff model was applied to the rainfall-runoff events of the micro urban watershed located in Cincinnati and the other four agricultural watersheds, described above. Details of the rainstorms of ARS watersheds are given in Table 6.5. Before discussing the simulation results of the actual rainfall-runoff events, the computation of infiltration rates from temporally varying rainfall intensities, instead of the uniform rainfall that was considered in model application to infiltration data, is first elaborated. The following describes the computation from a typical rainstorm.

Figure 6.10 shows the variation of the rainfall intensity, i, with time (t) during a typical rainstorm. Here, t corresponds to the time past ponding. In the present SCS-CN-based infiltration model, the infiltration rate f_1 (Figure 6.10) at the end of time interval Δt is computed as

$$f_1 = f_c + \frac{i_1 - f_c}{(1+\alpha\Delta t)^2} \tag{6.7}$$

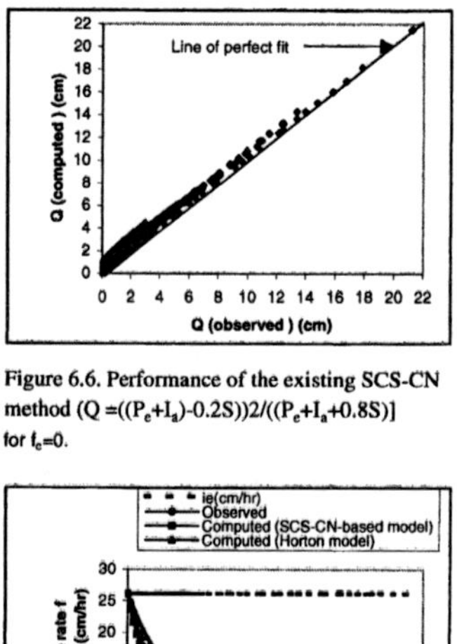

Figure 6.6. Performance of the existing SCS-CN method (Q =((P$_e$+I$_a$)-0.2S))2/((P$_e$+I$_a$+0.8S)] for f_c=0.

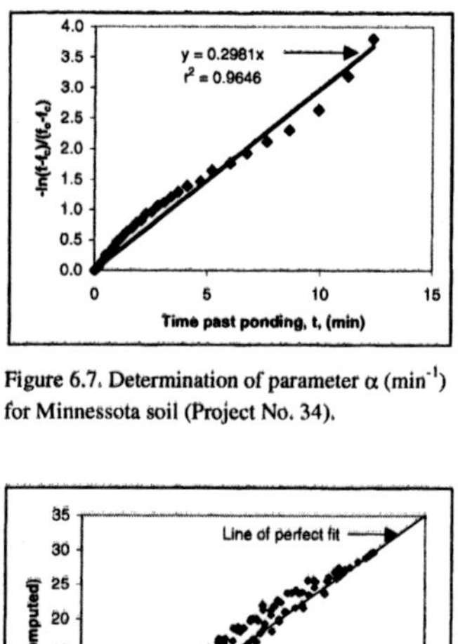

Figure 6.7. Determination of parameter α (min^{-1}) for Minnessota soil (Project No. 34).

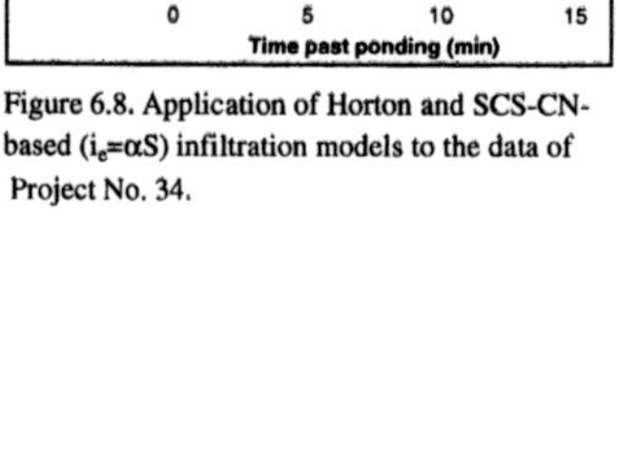

Figure 6.8. Application of Horton and SCS-CN-based (i_e=αS) infiltration models to the data of Project No. 34.

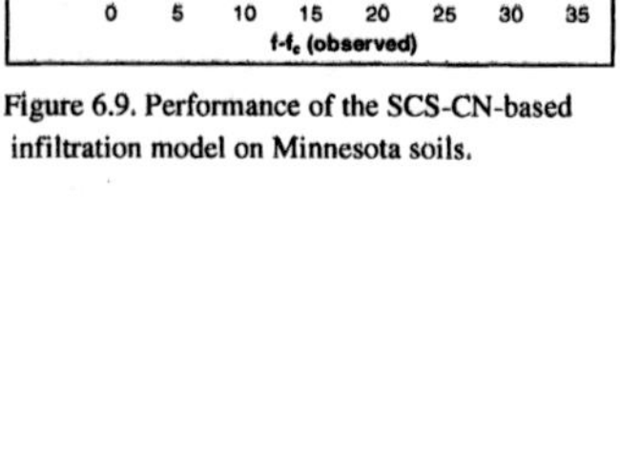

Figure 6.9. Performance of the SCS-CN-based infiltration model on Minnesota soils.

Table 6.5 Simulation of runoff from varous watersheds

Event No.	Date	Δt	Runoff Characteristics			Computed Parameters			Computed Runoff				S.E.	r^2
			'q_p'	't_p'	Time base	α	K	f_c	'q_p'	Rel. Error	't_p'	Rel. Error		
		(min)	(l/min)	(min)	(min)	(min^{-1})	(min)	(mm/hr)	(l/min)	%	(min)	%	(l/min)	
CINCINNATI WATERSHED														
1	08.08.1996	1	391.5	19	60	2470.0	1.94	2.14	432.2	-10.40	20	-5.26	15.74	0.9761
2	16.12.1996	5	2.1	45	80	6010.0	72.30	1.17	2.1	0.00	60	-33.33	0.29	0.8818
3	16.12.1996	5	2.0	135	170	6.13	43.90	0.73	1.9	5.00	170	-25.93	0.56	0.3646
4	07.07.1996	1	315.7	37	58	506.0	1.71	10.60	327.7	-3.80	37	0.00	22.66	0.9649
5	12.06.1997	1	77.2	15	30	0.43	1.46	1.48	63.2	18.13	16	-6.67	6.19	0.9251
6	18.06.1996	1	237.5	43	77	0.11	1.43	2.09	248.6	-4.67	45	-4.65	11.25	0.9615
7	21.05.1996	2	1.7	12	28	530.0	2.93	0.35	1.4	17.65	20	-66.67	0.24	0.8263
8	25.11.1996	5	4.9	45	70	1520.0	13.10	2.00	4.2	14.29	45	0.00	0.48	0.9233
9	25.11.1996	5	3.0	25	65	132.0	20.40	0.80	3.2	-6.67	25	0.00	0.31	0.9342
Y-8 WATERSHED OF TEXAS														
		(min)	(in/hr)	(min)	(min)	(min^{-1})	(min)	(in/hr)	(in/hr)	%	(min)	%	(in/hr)	
1	10.13.1981	30	1.0698	540	630	0.0033	18.80	0.60	0.9422	11.93	540	0.00	0.0569	0.9571
2	05.25.1973	30	0.1608	300	570	0.0070	27.60	0.64	0.1445	10.14	300	0.00	0.0108	0.9172
3	01.18.1968	30	0.0422	210	1020	0.0033	12.30	0.14	0.0377	10.66	180	14.29	0.0029	0.8791
4	10.30.1979	30	0.0190	150	570	0.0029	51.10	0.09	0.0172	9.47	150	0.00	0.0010	0.9701
5	03.06.1978	30	2.0114	570	1050	0.0022	7.40	0.18	1.3980	30.50	570	0.00	0.1452	0.8349
6	05.31.1976	30	0.4019	1080	1290	0.0010	27.70	0.81	0.3739	6.97	1080	0.00	0.0155	0.9665
7	02.01.1975	30	0.0965	600	1590	0.0006	33.80	0.00	0.0822	14.82	600	0.00	0.0054	0.9308
8	04.09.1970	30	0.0351	330	420	0.0007	9.65	0.14	0.0265	24.50	330	0.00	0.0047	0.7751

Table 6.5 Continued

Event	Date	Δt	Runoff Characteristics			Computed Parameters			Computed Runoff				S.E.	r^2
			'q_p'	't_p'	Time base	α	K	f_c	'q_p'	Rel. Error	't_p'	Rel. Error		
		(min)	(l/min)	(min)	(min)	(min^{-1})	(min)	(mm/hr)	(l/min)	%	(min)	%	(l/min)	
9	07.21.1971	30	0.1170	1290	2010	1.4300	49.70	0.51	0.1162	0.68	1290	0.00	0.0069	0.9154
W-1 WATERSHED OF TEXAS														
1	06.04.1981	60	0.1373	720	2340	0.0090	72.30	0.08	0.1188	13.47	720	0.00	0.0067	0.9482
2	04.19.1979	60	0.0074	180	780	2.2510	84.60	0.13	0.0065	12.16	180	0.00	0.0004	0.9443
3	06.26.1979	60	0.1299	360	2040	0.0017	59.95	0.04	0.1110	14.55	420	-16.67	0.0078	0.9228
4	02.12.1978	60	0.0229	300	1200	0.0035	107.9	0.37	0.0218	4.80	240	20.00	0.0014	0.9545
5	03.03.1977	60	0.0561	420	2500	0.0039	86.64	0.03	0.0542	3.39	420	0.00	0.0020	0.9803
6	10.29.1976	60	0.1238	600	2340	2.5704	94.56	0.13	0.1032	16.64	600	0.00	0.0066	0.9584
7	02.01.1975	60	0.4726	1680	5340	0.5899	37.34	0.06	0.4089	13.48	1680	0.00	0.0187	0.9259
OKLAHOMA WATERSHED														
1	05.08.1959	30	0.2603	300	4590	0.01	37.96	0.12	0.1801	30.81	300	0.00	0.0119	0.8538
2	05.05.1960	30	0.0393	240	4890	1299.5	90.87	0.27	0.0260	33.84	240	0.00	0.0016	0.8762
3	05.08.1961	30	0.3663	270	2160	17896	26.47	0.03	0.3288	10.24	270	0.00	0.0168	0.9070
4	09.27.1961	30	0.2411	90	660	4566.9	34.58	0.37	0.1966	18.46	90	0.00	0.0161	0.9346
5	02.15.1962	30	0.0124	210	1020	1566.5	95.93	0.28	0.0081	34.68	210	0.00	0.0015	0.7972
6	09.16.1963	30	0.0925	240	1440	54395	76.59	0.18	0.0568	38.59	240	0.00	0.0097	0.7886
7	07.23.1966	30	0.1392	390	840	2.35	40.17	1.58	0.1110	20.26	390	0.00	0.0071	0.9425
8	04.02.1968	30	1.2869	180	1530	1930.2	28.89	0.01	0.8382	34.87	180	0.00	0.0904	0.8406
9	09.15.1969	30	0.0106	90	1260	1971.5	21.48	1.02	0.0088	16.98	90	0.00	0.0008	0.7922
MISSISSIPPI WATERSHED														
1	02.01.1983	15	0.1844	105	615	536.11	33.36	0.06	0.1878	-1.84	135	-28.57	0.0135	0.9327
2	02.12.1984	15	0.1118	150	450	355.16	56.12	0.04	0.1092	2.33	150	0.00	0.0109	0.8998
3	02.26.1984	15	0.0322	135	675	94.36	92.28	0.05	0.0314	2.48	165	-22.22	0.0031	0.9155
4	04.23.1985	15	0.6458	60	660	111.16	29.74	0.16	0.4724	26.85	60	0.00	0.0495	0.8393
5	07.26.1985	15	1.0886	225	570	0.00	9.60	0.03	0.9176	15.71	240	-6.67	0.0658	0.9476
6	01.18.1988	15	0.1591	1455	2880	0.28	118.8	0.19	0.1019	35.95	1500	-3.09	0.0100	0.8563
7	03.03.1988	15	0.1244	135	795	13.35	39.35	0.53	0.1011	18.73	135	0.00	0.0072	0.9142
8	05.21.1990	15	0.2485	30	345	165.36	14.96	0.42	0.2246	9.62	30	0.00	0.0148	0.8880
9	05.21.1990	15	0.1215	390	450	1004.0	31.13	0.57	0.0879	27.65	405	-3.85	0.0108	0.8913

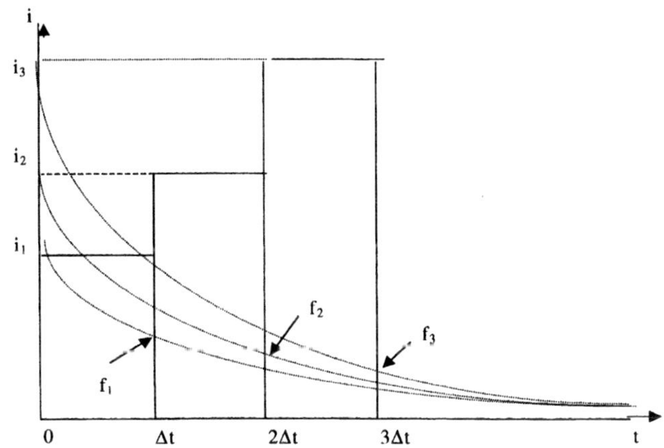

Figure 6.10. Variation of infiltration capacity during a typical rainstorm.

Similarly, the infiltration rates f_2 and f_3 (Figure 6.10) at the end of $2\Delta t$ and $3\Delta t$, respectively, are computed by replacing i_1 by respective i_2 and i_3 and corresponding t-values by $2\Delta t$ and $3\Delta t$, respectively. It is apparent from the figure that the difference between f-values reduces with increasing time. Such an approach supports the dependence of infiltration rate on rainfall intensity (Hawkins, 1982). It is worth emphasizing that for a possible case of $i_2 < i_1$, f_2 will also decrease and will take the value derived using the same above approach. Thus, the rainfall-excess rates can be computed by subtracting f-values from respective i-values. These rainfall-excess values can be routed using a suitable routing scheme to compute outflows at the outlet of the watershed. An application of this approach is illustrated below using the real data.

Example 6.3: Compute infiltration and the corresponding direct runoff rates for a rainfall-runoff event of Y-8 watershed given in the following table. Compare the computed runoff with the observed runoff. Assume $\alpha = 0.0034$, $f_c = 0.60$ inch/hr, and storage coefficient K = 18.80 min, $\lambda = 0.2$, and baseflow = 0. Consider 2 reservoirs in linear reservoir routing.

Time (min)	0	30	60	90	120	150	180	210	240	270	300	330
Rainfall (inch/hr)	0	0.97	0.03	0	0	0	0	0	0	0	0	0
Runoff (inch/hr)	0	0	0	0	0	0	0	0	0	0	0	0

Time	360	390	420	450	480	510	540	570	600	630
Rainfall	0	0	0	0	0.60	3.02	0.89	0	0	0
Runoff	0	0	0	0	0	0.20	1.07	0.65	0.19	0.06

Solution: For given $\alpha = 0.0034$ and $\lambda = 0.2$, the time to ponding, t_p, is computed as $t_p = 0.2/0.0034 = 58.82$ min. Thus, up to $t \approx 60$ min, the infiltration rate is equal to the rainfall intensity and consequent runoff rates are equal to 0. At t = 480 min, f is computed as

$$f = f_c + \frac{i_o - f_c}{(1-\lambda+\alpha t)^2}$$

$$= 0.60 + \frac{0.60 - 0.60}{(1-0.2+0.0034\text{x}480)^2} = 0.60 \text{ inch/hr}$$

Therefore, the direct rainfall-excess rate, q = 0.60 − 0.60 = 0. Thus, during the time period 60-480 min, both 'f' and 'q' are zero, for the rainfall intensities are equal to zero. At t = 510 min,

$f = 0.60 + (3.02\text{-}0.60)/(0.8+0.0034 \text{ x } 510)^2 = 0.98$ inch/hr

Therefore, q = 3.02 − 0.98 = 2.04 inch/hr

For routing of the rainfall-excess, routing coefficients d_0, d_1, and d_2 can be computed as

$$d_0 = d_1 = \frac{\Delta t / K}{2 + \Delta t / K} = \frac{30/18.80}{2+30/18.8} = 0.444$$

and

$$d_2 = \frac{2 - \Delta t / K}{2 + \Delta t / K} = \frac{2-30/18.80}{2+30/18.8} = 0.112$$

Table 6.6 Computation of infiltration and runoff rates for an event of Y-8 watershed (Example 6.3)

Time	i_o	'f'	Q	First reservoir			X	Second reservoir			O
(min)	(in/ hr)	(in/ hr)	(in/ hr)	d_0q_{j-1}	d_1q_j	d_2X_{j-1}	(in/hr)	d_0X_{j-1}	d_1X_j	d_2O_{j-1}	(in/ hr)
0	0	0	0				0				0
30	0.97	0.97	0	0	0	0	0	0	0	0	0
60	0.03	0.03	0	0	0	0	0	0	0	0	0
90	0	0	0	0	0	0	0	0	0	0	0
120	0	0	0	0	0	0	0	0	0	0	0
150	0	0	0	0	0	0	0	0	0	0	0
180	0	0	0	0	0	0	0	0	0	0	0
210	0	0	0	0	0	0	0	0	0	0	0
240	0	0	0	0	0	0	0	0	0	0	0
270	0	0	0	0	0	0	0	0	0	0	0
300	0	0	0	0	0	0	0	0	0	0	0
330	0	0	0	0	0	0	0	0	0	0	0
360	0	0	0	0	0	0	0	0	0	0	0
390	0	0	0	0	0	0	0	0	0	0	0
420	0	0	0	0	0	0	0	0	0	0	0
450	0	0	0	0	0	0	0	0	0	0	0
480	0.6	0.60	0	0	0	0	0	0	0	0	0
510	3.02	0.98	2.04	0.00	0.90	0.00	0.90	0.00	0.40	0.00	0.40
540	0.89	0.65	0.24	0.90	0.11	0.10	1.11	0.40	0.49	0.04	0.94
570	0	0	0	0.11	0.00	0.12	0.23	0.49	0.10	0.10	0.70
600	0	0	0	0	0	0.03	0.03	0.10	0.01	0.08	0.19
630	0	0	0	0	0	0	0	0.01	0.00	0.02	0.03

The rainfall-excess is routed through two reservoirs in series to compute outflow at the outlet of the watershed. Thus, routing is carried out in two steps as:

$$X_j = d_0\, q_{j-1} + d_1\, q_j + d_2\, X_{j-1}$$

and

$$O_j = d_0\, X_{j-1} + d_1\, X_j + d_2\, O_{j-1}$$

where X is the outflow from the first reservoir and inflow to the second reservoir. Here, since baseflow is nil, initial values of X and O are equal to zero. Computations for outflows at different time steps are shown in Table 6.6. It is apparent from this table that during this storm, the Y-8 watershed produced a direct runoff peak of 0.94 inch/hr whereas the rainfall peak was of 3.02 inch/hr. As shown in Figure 6.11 and Table 6.6, both computed and observed runoff peaks occur at 540 min and the former is lower than the observed by 12.15%, which is quite satisfactory.

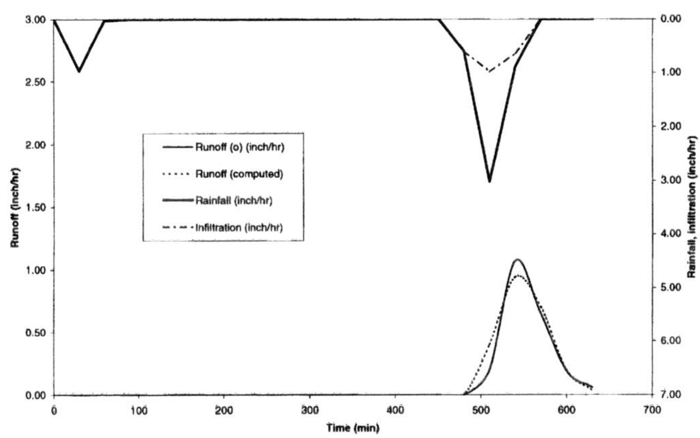

Fig 6.11. Observed and computed runoff hydrographs of Example 6.3.

a) *Model Performance in Runoff Peak Simulation*

The model performance in simulation of runoff from the above watersheds is evaluated using the standard error, SE, (equation (6.4)) and coefficient of determination, r^2, (equation (6.5)), which are listed in Table 6.5. It is apparent from this table that for the Cincinnati watershed, the resulting SE-values range from 0.24 to 22.66 l/min and r^2-values range from 0.83 to 0.98, except for event 3 for which r^2 is 0.36 which is very low because of low runoff values observed at the observation site (Figure 6.12). The regularly fluctuating runoff values for uniform rainfall intensities during a time interval lead to the low r^2-value. On the other hand, event 1 simulated with the highest r^2-value (= 0.98) and shown in Figure 6.13 exhibits a very good match between the observed and computed runoff-values. Since α occurs in the denominator of the SCS-CN-based infiltration equation (equation (4.39)), the variation of infiltration is almost constant throughout the rainstorm. It shows that this storm can reasonably be simulated using $\alpha = 0$. In such a situation, f_c may play an important role in the determination of infiltration rates and, in turn, the runoff rates. The above computed r^2-value is actually an indicator of the model performance in simulating the complete hydrograph.

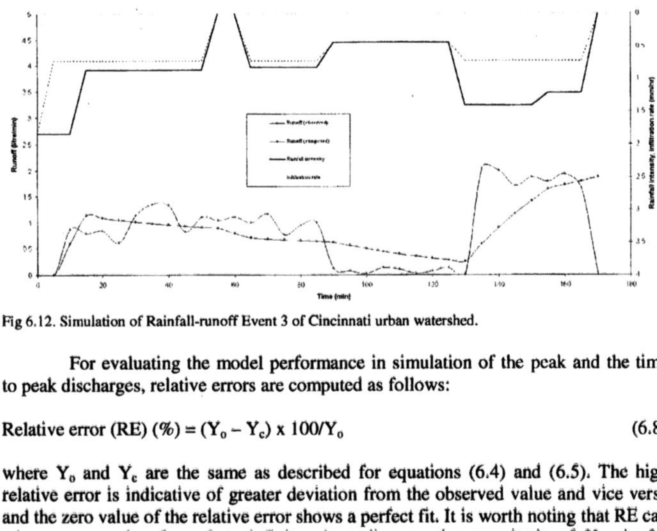

Fig 6.12. Simulation of Rainfall-runoff Event 3 of Cincinnati urban watershed.

For evaluating the model performance in simulation of the peak and the time to peak discharges, relative errors are computed as follows:

$$\text{Relative error (RE) (\%)} = (Y_o - Y_c) \times 100/Y_o \qquad (6.8)$$

where Y_o and Y_c are the same as described for equations (6.4) and (6.5). The high relative error is indicative of greater deviation from the observed value and vice versa and the zero value of the relative error shows a perfect fit. It is worth noting that RE can take on any value from 0 to infinity, depending on the magnitude of Y_o. As Y_o approaches zero, RE may approach infinity. As shown in Table 6.5, runoff peaks of the events of Cincinnati watershed are simulated with relative errors ranging from −10.40 to 18.13% and their occurrence with relative errors varying from 0 to 66.67%. Here, '+' sign indicates that the observed value is higher than the computed and the '-' sign corresponds to an otherwise situation.

Similarly, the model performance in runoff simulation on other watersheds can be explained. The values of r^2 range from 0.78 to 0.97, 0.92 to 0.98, 0.79 to 0.94, and 0.85 to 0.95 for Y-8, W-1, Oklahoma, and Mississippi watershed events, respectively. These values are reasonably high suggesting that the model performs well. Similarly, the peak discharges are simulated with relative errors ranging from 0.68 to 30.50%, from 3.39 to 16.64%, from 10.24 to 34.87%, from −1.84 to 35.95%, respectively, for Y-8, W-1, Oklahoma, and Mississippi watershed events. The generally occurring positive values of REs indicate that the SCS-CN-based runoff model has underestimated the peak runoff values. It may be attributed to the optimization based on the sum of errors between the observed and computed runoff ordinates in a complete rainfall-runoff event. The occurrence of peaks for all the events of all watersheds is simulated with REs ranging from –28.57 to +20.00%, which is an acceptable range of variation. For illustrative purposes, the simulation results exhibiting the highest and lowest r^2-values are depicted in Figures 6.14 through 6.21 for all the ARS watersheds.

b) Model Performance in Mass Conservation

It is important to note that the model parameters for all rainfall-runoff events of all the above described watersheds were computed using the error criteria described by equations (6.4) and (6.5) that do not directly connect to mass conservation. However, if the deviation between the computed and observed values is less, the volume of the computed runoff is more likely to match with the observed one. It can be observed from the volumetric statistic given in Table 6.7. For the Cincinnati watershed event 1, RE = -7.78% and its corresponding r^2 is equal to 0.98 whereas event 3 simulated with r^2 = 0.36 yields RE = 46.71%. Similar inferences can be drawn from the results of the other watersheds and it can be further substantiated from Figure 6.22 depicting the plot between these two types of errors.

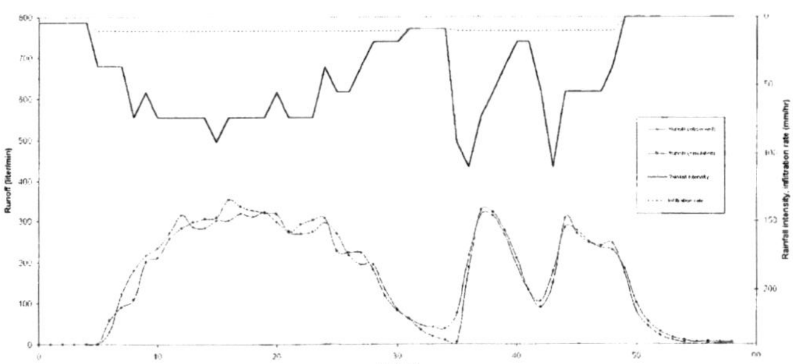

Figure 6.13. Simulation of Event 1 of Cincinnati urban watershed.

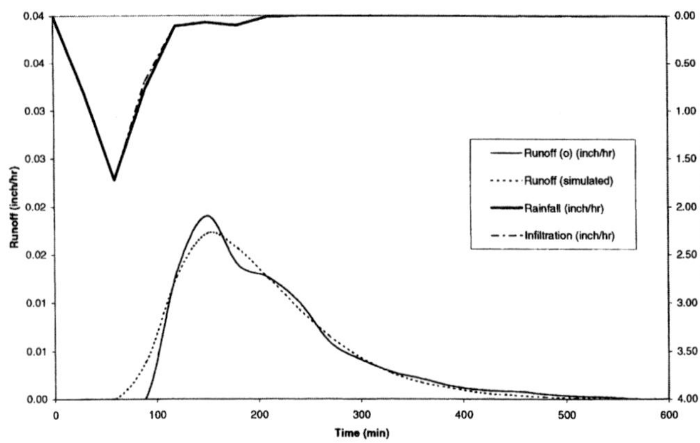

Figure 6.14. Simulation of event 4 of Y-8 (Texas) watershed.

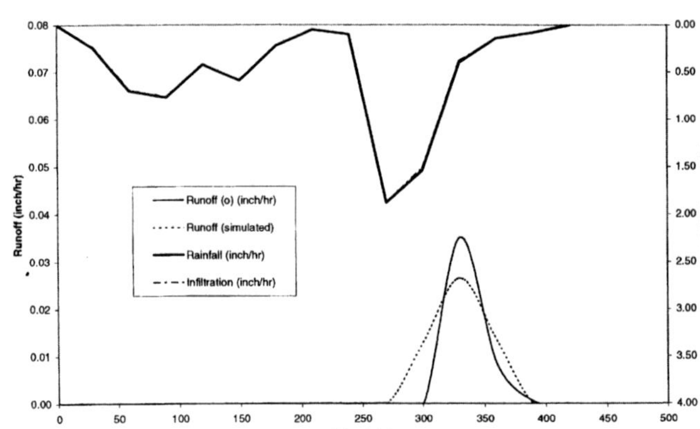

Figure 6.15. Simulation of event 8 of Y-8 (Texas) watershed.

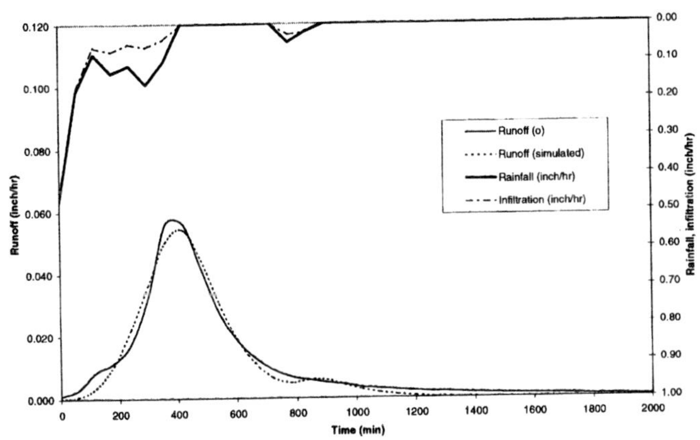

Figure 6.16. Simulation of event 5 of W-1 watershed.

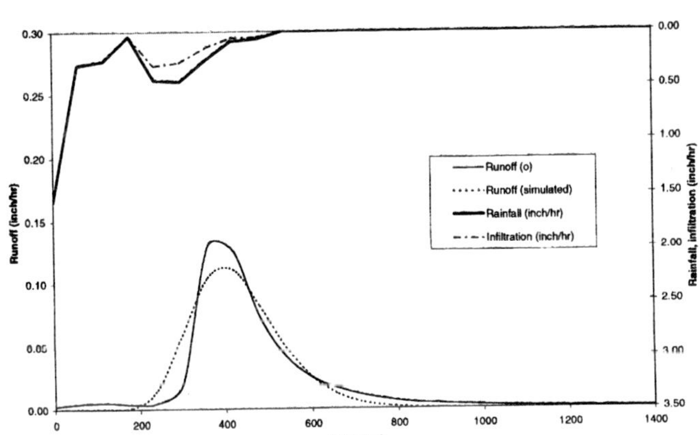

Figure 6.17. Simulation of event 3 of W-1 watershed.

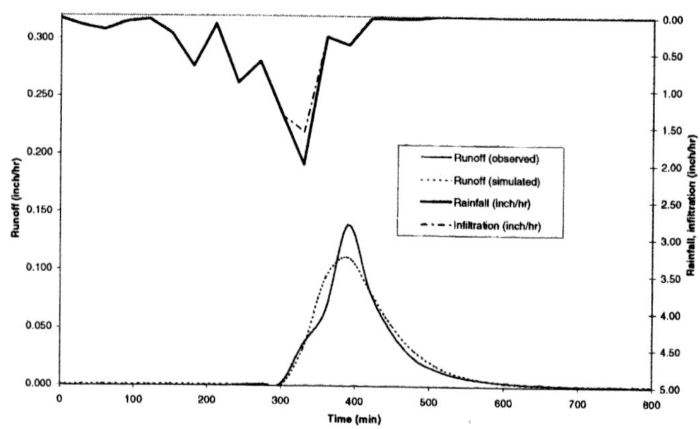

Figure 6.18. Simulation and prediction of event 7 of Oklahoma watershed.

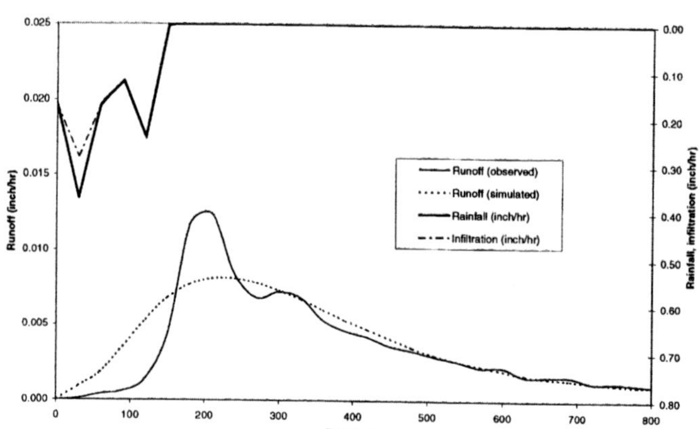

Figure 6.19. Simulation of event 5 of Oklahoma watershed.

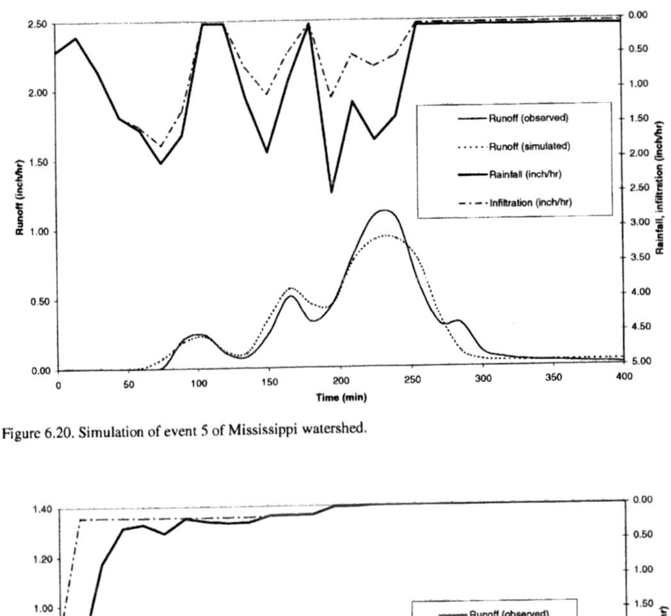

Figure 6.20. Simulation of event 5 of Mississippi watershed.

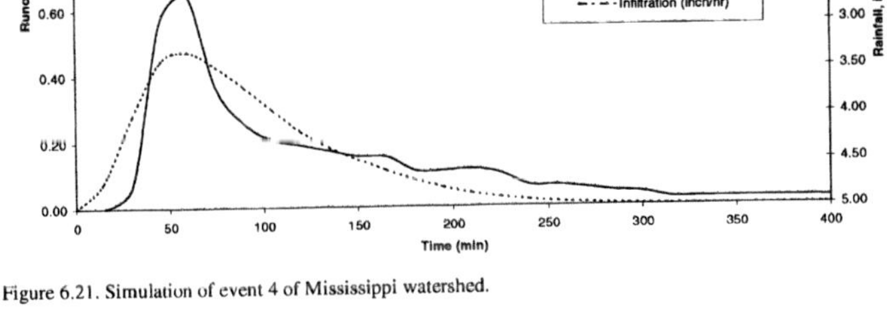

Figure 6.21. Simulation of event 4 of Mississippi watershed.

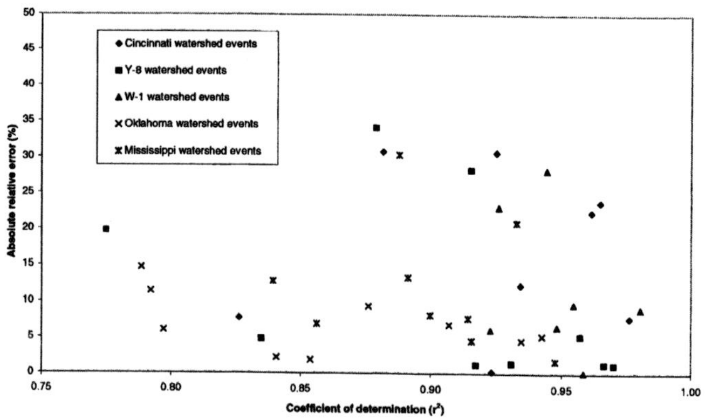

Figure 6.22. Variation of the volumetric relative error with the coefficient of determination.

The simulation results of Table 6.7 are discussed in terms of the percent (%) relative error between computed and observed runoff masses. The runoff volumes of 8 events of the Cincinnati watershed, except the one with r^2 = 0.36, are simulated with relative errors ranging from −30.68% to 30.76%. The events of Y-8 watershed are simulated with relative errors ranging from −19.73 to 34.07% and these range from −9.64 to 28.22, −14.68 to 11.43%, and −13.30 to 30.35% for the W-1, Oklahoma, and Mississippi watersheds, respectively. It is apparent from the table that 23 events out of 34 events of ARS watersheds are simulated with relative volumetric errors less than ±10%, which is quite satisfactory.

c) *Sensitivity of Model Parameters*

The sensitivity of model parameters, viz., α, K, and f_c, is evaluated in terms of the impact of their variation on the computed runoff of an event of a watershed. To this end, event 1 of Y-8 watershed, event 2 of W-1 watershed, event 7 of Oklahoma watershed, event 2 of Mississippi watershed, and event 1 of the Cincinnati watershed are selected for analysis. The parameters computed in simulation and shown in Table 6.5 were changed by ±25% of their values shown in Table 6.5 and the resulting runoff hydrographs are shown in Figures 6.23-6.27 for the events of the respective watersheds. In these figures, (a) stands for the sensitivity of parameter α, (b) stands for K, and (c) shows the sensitivity of f_c. The impact of the variation of these parameters on the runoff hydrograph can be physically interpreted as follows.

Table 6.7 Volumetric statistic of runoff simulation

Event	Rainfall	Infiltration	Runoff (o)	Runoff (c)	Relative Error
CINCINNATI WATERSHED					
	(mm)	(mm)	(mm)	(mm)	(%)
1	14.08	0.14	12.92	13.93	-7.78
2	1.22	0.91	0.35	0.24	30.76
3	2.14	1.82	0.54	0.29	46.71
4	40.47	0.07	32.15	39.78	-23.74
5	2.03	0.01	1.55	2.02	-30.68
6	11.32	0.00	9.25	11.32	-22.38
7	0.21	0.09	0.12	0.11	7.72
8	2.08	1.61	0.40	0.40	0.27
9	1.04	0.73	0.32	0.28	12.28
Y-8 WATERSHED					
	(inch)	(inch)	(inch)	(inch)	(%)
1	2.76	1.62	1.08	1.14	-5.21
2	3.03	2.76	0.27	0.27	1.23
3	0.97	0.84	0.20	0.13	34.07
4	1.79	1.75	0.04	0.04	-1.25
5	5.50	2.86	2.54	2.66	-4.79
6	2.47	1.85	0.60	0.60	-1.31
7	2.12	1.84	0.27	0.28	-1.43
8	3.57	3.54	0.02	0.03	-19.73
9	3.71	3.40	0.44	0.31	28.24
W-8 WATERSHED					
1	1.69	0.96	0.76	0.71	6.54
2	0.72	0.69	0.04	0.03	28.22
3	3.52	3.04	0.50	0.47	6.09
4	1.31	1.21	0.10	0.11	-9.64
5	1.30	0.98	0.35	0.32	9.01
6	1.81	1.17	0.64	0.64	-0.16
7	2.98	1.00	2.52	1.94	23.07
OKLAHOMA WATERSHED					
1	1.66	1.07	0.65	0.64	1.86
2	1.08	0.97	0.16	0.18	-9.38
3	1.58	0.62	1.06	0.98	6.8
4	1.19	0.81	0.36	0.37	-4.56
5	0.54	0.50	0.05	0.06	-6.05
6	0.90	0.66	0.22	0.25	-14.68
7	3.59	3.37	0.22	0.23	-5.24
8	1.77	0.09	1.74	1.70	2.19
9	1.20	1.19	0.03	0.02	11.43
MISSISSIPPI WATERSHED					
1	0.79	0.32	0.60	0.48	20.89
2	0.44	0.12	0.35	0.32	8.13
3	0.45	0.29	0.16	0.16	4.55
4	1.74	0.92	0.95	0.83	12.79
5	5.04	3.31	1.69	1.71	-1.75
6	0.80	0.26	0.63	0.59	6.92
7	0.88	0.69	0.22	0.20	7.67
8	0.51	0.32	0.29	0.20	30.35
9	0.77	0.62	0.11	0.13	-13.3

From equation (4.39) the role of parameter α can be determined in terms of the magnitude of dynamic infiltration for a given rainfall intensity and minimum infiltration rate. The value of $\alpha = 0$ implies that the time to ponding is equal to infinity for $\lambda = 0.2$, and thus, the whole rainfall intensity amount goes to infiltration. On the other hand, if α approaches infinity, the second term of equation (4.39) defining the dynamic portion of infiltration approaches zero and infiltration approaches the minimum infiltration rate, f_c. Thus, as α increases, the infiltration capacity of the watershed decreases, leading to lesser infiltration and greater runoff for a given rainfall amount, and vice versa. It implies that runoff will increase or decrease with the increase or decrease of the α-value. The same effect is seen in Figure 6.23a. An increase or decrease of α by 25% of the original parameter value (Table 6.5), denoted by alpha (o), increases or decreases the peak magnitude of the runoff hydrograph by 5.07% and 8.52%, respectively. Thus, all the ordinates of the hydrograph are accordingly increased or decreased. It is, however, noted that no change in the hydrographs is reflected despite the change in the α-value. It implies that in these events, the dynamic portion of infiltration does not play a significánt role, rather static infiltration governs the determination of the runoff magnitude, as shown later. In channel or reservoir routing, the parameter storage routing coefficient, K, is taken to represent the lag between input and output waves or hydrographs. This lag depends on the channel length and the wave celerity that, in turn, depends on the magnitude of the input wave. The larger the magnitude of the input wave, the greater will be the wave celerity and the lesser the time of travel (time lag) of the wave in a given reach. The greater the length of the channel, the larger will be the time of travel (or time lag). Thus, K is directly proportional to the channel length and inversely proportional to the wave celerity. In watershed routing, the rainfall-excess hyetograph can be taken equivalent to the input wave and the outflow hydrograph as the output wave. The wave celerity is inversely related to attenuation (also see Appendix C). Thus, the larger the K value, the larger will be the attenuation and consequently the lesser will be the spread of the hydrograph. Succinctly, with the increase of K, the hydrograph peak will subside and shift to the right, and the base of the runoff hydrograph will increase. The reverse will hold in an otherwise situation. Such an impact is clearly seen in Figures 6.23a-6.26a. A decrease or increase by 25% from the original K-value that is denoted by K(o) shifts the runoff peak to the left or right, increases or decreases the magnitude of peak runoff by 18.38% and 15.79%, respectively, and decreases or increases the base of the runoff hydrograph.

The impact of f_c on the resulting runoff hydrograph is the same as of α. A reduction or increase of the minimum infiltration rate will reduce or increase infiltration and, in turn, will increase or reduce runoff. The same impact of f_c is seen on the resulting hydrographs of the events shown in Figures 6.23c through 6.27c. A decrease or increase of the f_c-value by 25% from the original f_c-value, designated as f_c(o), increases or decreases the runoff peak magnitude by 8.60% and 11.74%. It is noted from these figures that event 1 of Y-8, the Texas watershed, is almost equally affected by both α and f_c (Figures 6.23a-c). However, the other watershed events shown in Figures 6.23-6.26 are least affected by α but greatly by f_c.

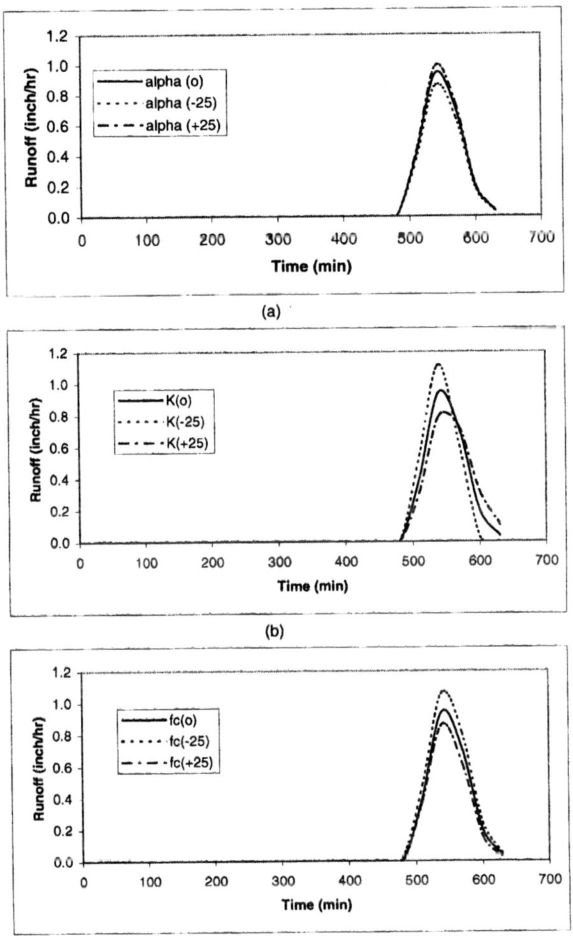

Figure 6.23. Sensitivity analaysis of (a) α(alpha), (b) K, and (c) f_c using event 1 of Y-8 (Texas) watershed.

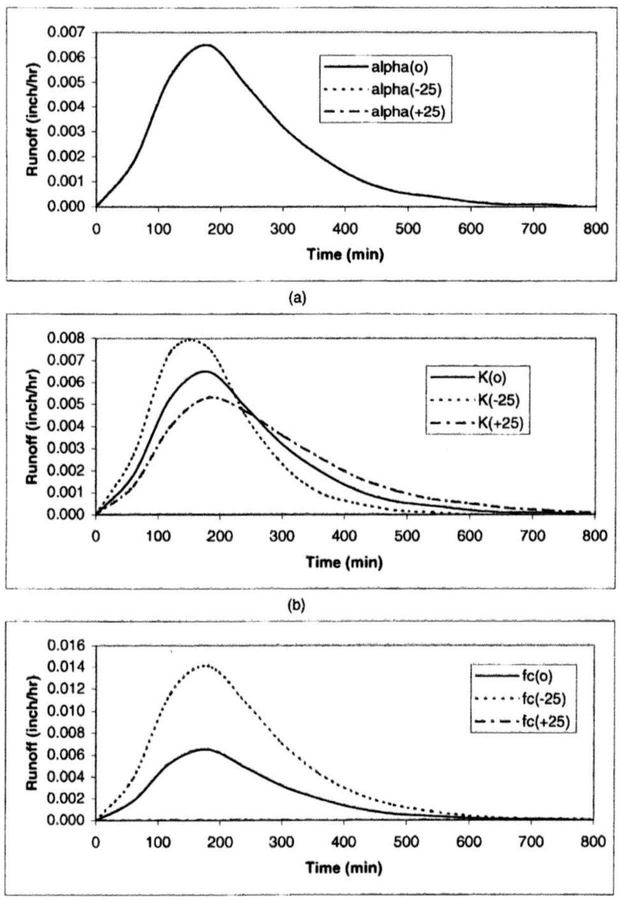

Figure 6.24. Sensitivity analaysis of (a) α (alpha), (b) K, and (c) f_c using event 2 of W-1 (Texas) watershed.

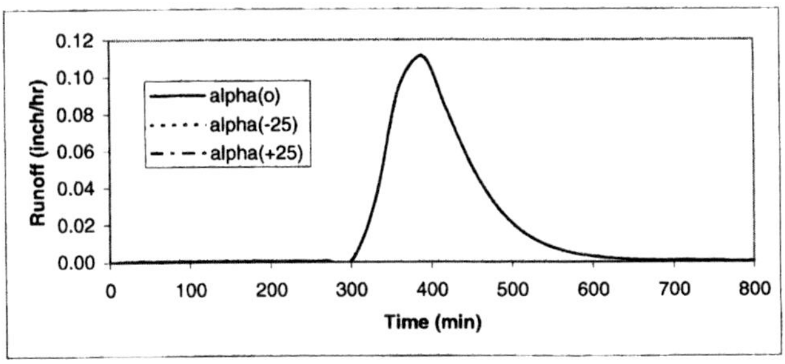

(a)

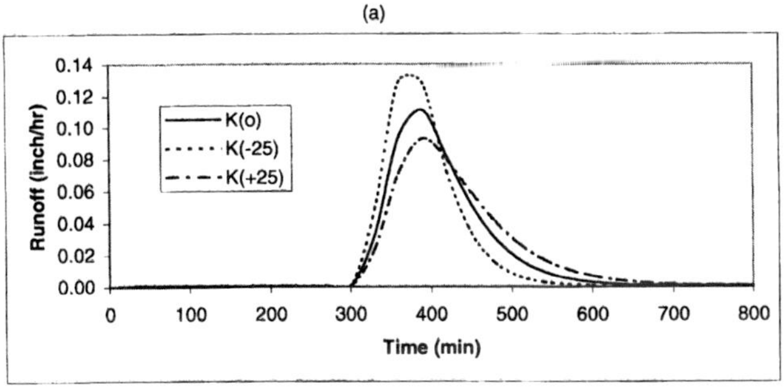

(b)

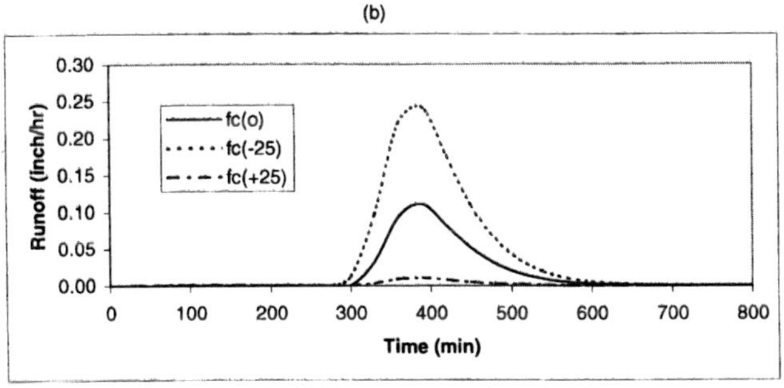

(c)

Figure 6.25. Sensitivity analaysis of (a) α (alpha), (b) K, and (c) f_c using event 7 of Oklahoma watershed.

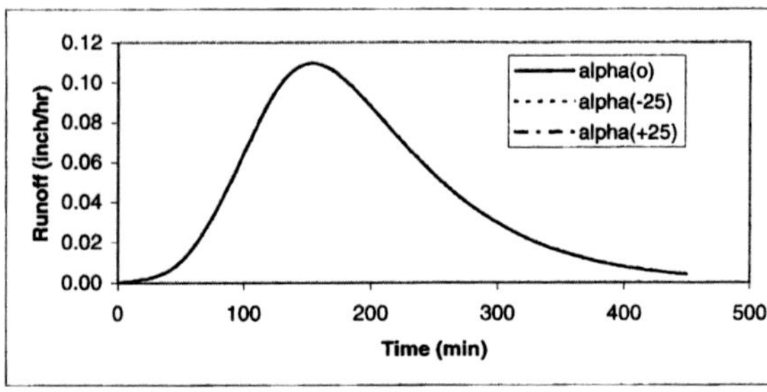

(a)

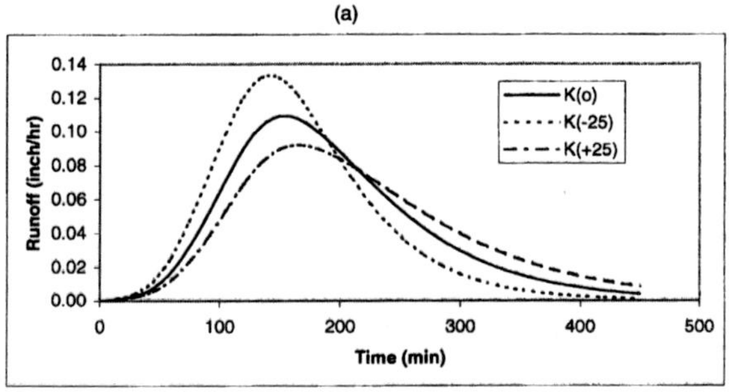

(b)

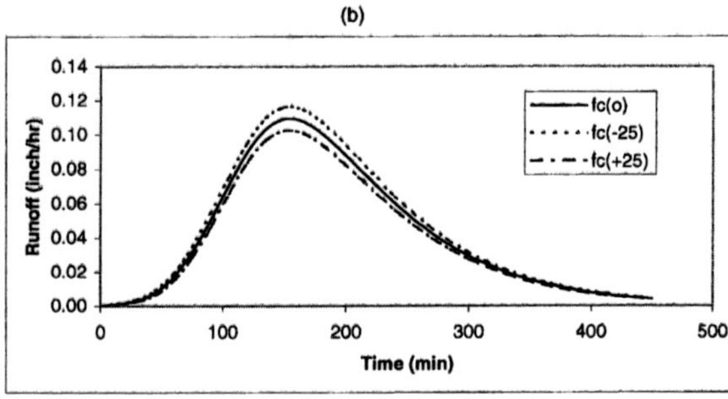

(c)

Figure 6.26. Sensitivity analaysis of (a) α (alpha), (b) K, and (c) f_c using event 2 of Mississippi watershed.

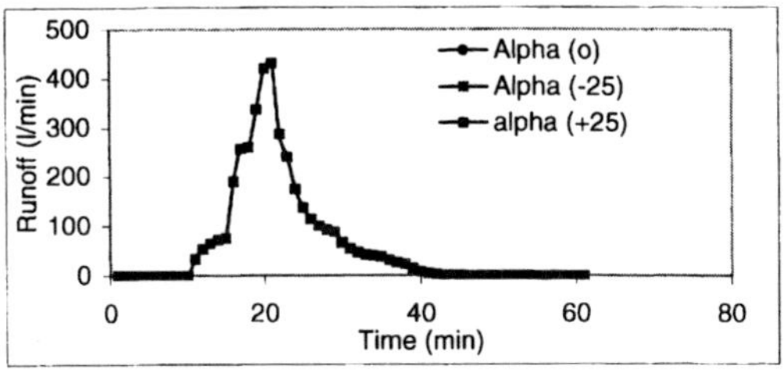

(a)

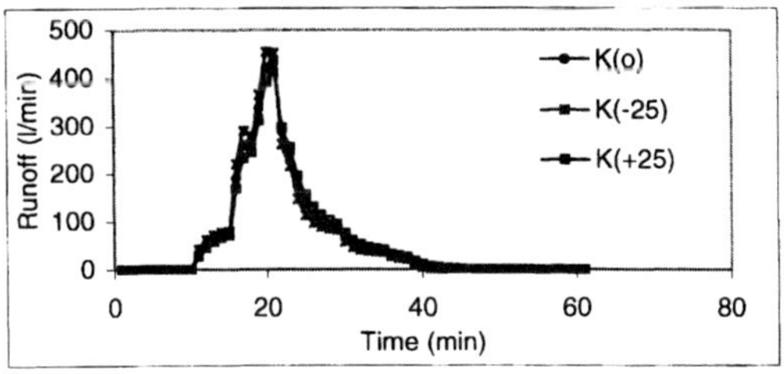

(b)

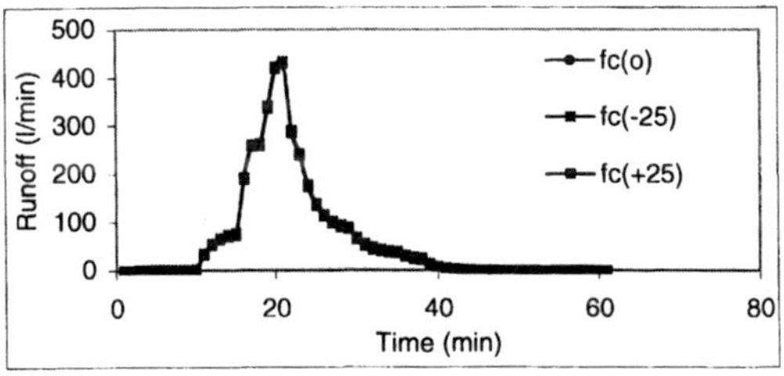

(c)

Figure 6.27. Effect of (a) α (alpha), (b) K, and (c) f_c on the runoff hydrograph of Cincinnati event 1.

The sensitivity analysis (Figure 6.27) of the Cincinnati event indicates that both α and f_c least affect the runoff hydrograph, whereas K most affects it. It is because the paved nature of the watershed, for which the potential maximum retention, S, is approximately equal to zero and, therefore, the Horton decay parameter, α, assumes a very high value, as seen in Table 6.5. Furthermore, owing to the impermeable nature of the watershed, the optimized values of the parameter f_c are significantly low, compared to the peak rainfall intensity.

d) *Empirical Determination of Parameters*

Several factors affect the process of runoff generation from rainfall (Hewlett and Hibbert, 1967; Mein and Larson, 1971; Singh, 1988a,b; HEC, 1981, 1990). Apart from physical characteristics of the watershed, rainfall forms the most easily discernible governing factor. For example, parameter α is significantly influenced by rainfall intensity (Mein and Larson, 1971). The variation of the above parameters with the rainfall intensity is now discussed.

It is known from the above that λ is described as the product of α and the time to ponding, t_p (equation (4.19)). Since λ in the above simulation is taken as 0.2, the variation of t_p from one event to the other can be described as:

$$t_p = 0.2/\alpha \tag{6.9}$$

The infiltration studies (Smith, 1972) indicate the existence of a power relationship between the time to ponding, t_p, and uniform rainfall intensity, i_o, expressed as:

$$t_p = a\, i_o^{-b} \tag{6.10}$$

where 'a' and 'b' are non-negative coefficient and exponent, respectively. These parameters depend on the initial soil moisture conditions. For U.S.A. soils (Table 6.1), a = 42.515 and b = 1.1182, as shown in Figure 6.28. In this figure, i_o is in cm/hr and t_p is in minutes. To derive a relationship for rainfall-runoff events under consideration, it is necessary to investigate the variation of the potential maximum retention, S, with time during rainstorms. To this end, utilizing the values of model parameters derived for the Y-8 watershed events (Table 6.5), the S-values were computed using the following relation:

$$S = \frac{P - F_c}{\alpha t} \tag{6.11}$$

where P is the cumulative rainfall and F_c is the cumulative minimum infiltration. The computed S-values are plotted against time for all 9 events of Y-8 watershed in Figure 6.29. Since at t = 0, S is actually equal to 0/0 that is indeterminate and, therefore, has been taken as zero. Thus, S first increases with time, generally attains a peak in a short time, and then apparently decreases exponentially. Notably, S increases at the time of occurrence of rainfall; it is valid for multiple peaked rainfall hyetographs. This is the characteristic of the SCS-CN method, as discussed in Chapter 5. Important to note here is that the first peak S-value of the graph primarily governs the runoff generation from rainfall. Furthermore, these peaks appear to correspond to t_p of respective rainfall-runoff events.

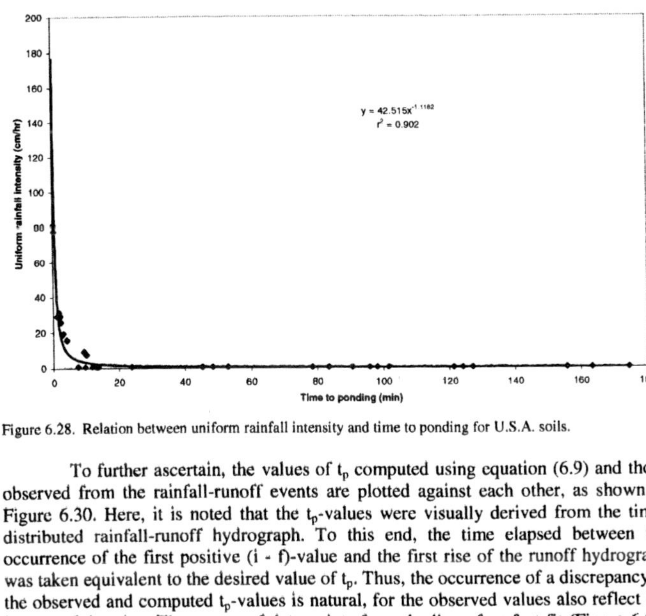

Figure 6.28. Relation between uniform rainfall intensity and time to ponding for U.S.A. soils.

To further ascertain, the values of t_p computed using equation (6.9) and those observed from the rainfall-runoff events are plotted against each other, as shown in Figure 6.30. Here, it is noted that the t_p-values were visually derived from the time-distributed rainfall-runoff hydrograph. To this end, the time elapsed between the occurrence of the first positive (i - f)-value and the first rise of the runoff hydrograph was taken equivalent to the desired value of t_p. Thus, the occurrence of a discrepancy in the observed and computed t_p-values is natural, for the observed values also reflect the impact of time lag. The scatter of data points from the line of perfect fit (Figure 6.30) also reflects the same. The compatibility of the above S-values (corresponding to the t_p-values) with those derived from the total rainfall-runoff volumes is also illustrated by plotting them against each other, as shown in Figure 6.31. Despite the likely scatter, the relationship between the S-values derived from two sources is encouraging for the reason that it supports the use of NEH-4 S-values for computing the runoff hydrograph with a marked difference that these S-values exclude the static portion of infiltration whereas NEH-4 S-values include it (Chapter 4). Thus, it can be safely assumed that for $\lambda = 0.2$, I_a is equivalent to the rainfall (excluding f_c) up to time to ponding, for deriving S as

$$S = 5\,\Sigma(i-f_c)t \tag{6.12}$$

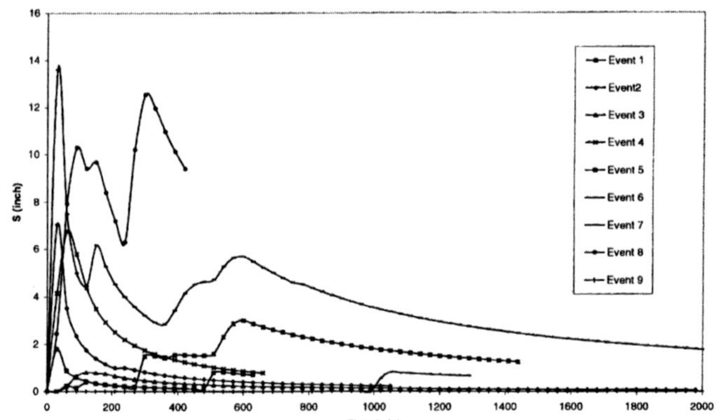

Figure 6.29. Variation of S during storm events of Y-8 watershed.

which is valid for $t \leq t_p$. Here, for determining S and α, it is necessary to determine a priori the time to ponding. To this end, a relation analogous to equation (6.10) is sought. The uniform rainfall intensity occurring at the time to ponding were plotted against the corresponding time to ponding, as shown in Figure 6.32. It is apparent from this figure that there exists a relation between these two quantities and it may be possible to derive relations for the time to ponding for each watershed. The advantage of such a relationship is that for a given f_c-value or the land use of a watershed, the time to ponding can be determined with a fairly reasonable accuracy.

It is evident from the above that the minimum infiltration rate, f_c, forms an important parameter of the infiltration model. It is noted that f_c is used to classify soils in hydrologic groups A, B, C, or D, as discussed in Chapter 2. To investigate the variation of f_c with the uniform rainfall intensity, i_o, these values (Table 6.1) were plotted against each other as shown in Figure 6.33. As seen from this figure, the data points are clustered according to the soil type. The Plainfield sand (PFS) exhibits the highest f_c-values and Ida silt loam (ISL) and Yololight clay (YLC) do the lowest f_c-values. The f_c-values of the Columbia sandy loam (CSL) fall between these two extremes. However, there appears to be a continuously increasing trend of f_c with the increasing uniform rainfall intensities and vice versa. Here, such a variation of f_c with i_o is distinguished with the description in Chapter 4 (Section 4.1.4) in that equation (4.59) describes the variation of minimum infiltration rates during the rainstorm, whereas Figure 6.33 shows the variation of equilibrium values of f_c with the corresponding i_o-values. In an attempt to derive an f_c-i_o relation using the rainfall-runoff data of the above watersheds, the most discernible peak rainfall intensity of the rainstorm was considered in place of the uniform rainfall intensity. The values representing the percent ratio of the minimum infiltration rate to peak rainfall intensity were plotted against the peak rainfall intensity, as shown in Figure 6.34. The derived empirical relation for the above ARS watersheds are given as

$$f_c^* = a_1\, i_p^{-b_1} \tag{6.13}$$

where f_c^* is the percent ratio of the minimum infiltration rate, f_c, (in/hr), to the peak rainfall intensity, i_p (in/hr); and a_1 and b_1 are the coefficient and exponent of the power relation (equation (6.13)), respectively. The values of a_1 and b_1 derived for the above watersheds are listed in Table 6.8.

It is noted that the plot (not shown) between the f_c and i_p exhibited approximately a similar rising trend of f_c-values with the peak rainfall intensity, as exhibited by the U.S.A. soils (Figure 6.33). It is an indicator of the dependence of f_c on the uniform rainfall intensity. Equation (6.13) also exhibits the same trend if f_c^* is converted to f_c.

Another model parameter is the storage routing coefficient, K, which primarily signifies the lag time of the watershed and depends on the watershed characteristics, such as the hydraulic length, slope, and land use of the watershed and, as discussed above, on the rainfall intensity. To physically relate K with the watershed characteristics, the Muskingum-Cunge approach is resorted to in what follows.

Equations (6.2) and (6.3a-c) also represent the Muskingum-Cunge method of flow routing if K in equations (6.3a-c) is replaced by the product of the Muskingum-Cunge parameters K and (1-ε), where K and ε are defined, respectively, as (Ponce, 1989):

$$K = \Delta x / c \tag{6.14}$$

and

$$\varepsilon = \frac{1}{2}\left[1 - \frac{q_o}{BYc\Delta x}\right] \tag{6.15}$$

Here, K represents the time of the wave travel, ε is the weighting parameter varying from 0 to 0.5, Δx is the length of the channel [L], c is the wave celerity [LT^{-1}], Y is the channel bed slope, B is the channel width [L], and q_o is the reference flow [L^3T^{-1}]. Equations (6.14) and (6.15) are also applicable to overland flows or to flows in wide rectangular open channels. Assuming $q_o = (i_o - f)B\Delta x$, where B and Δx are, respectively, the width and length of the watershed, and $c = (5/3)(i_o - f)$, the product K(1-ε) or parameter K of equations (6.3a-c) is expressed as:

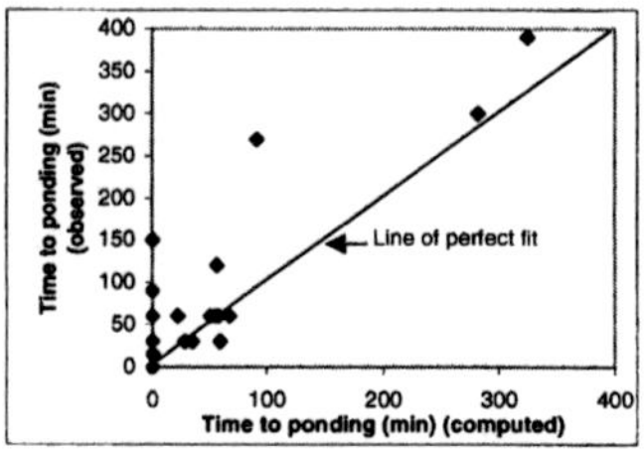

Figure 6.30. Observed and computed times to ponding (min).

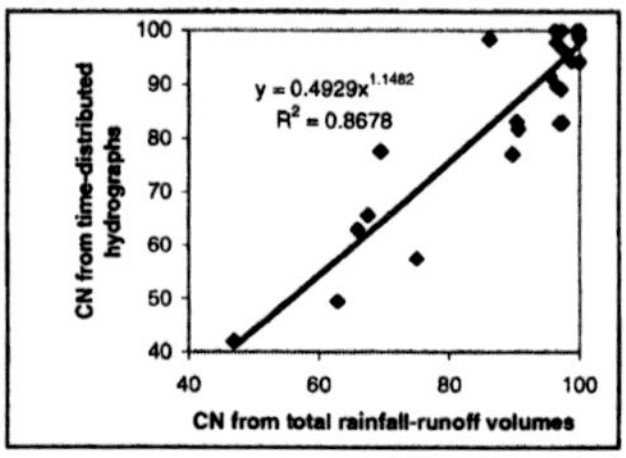

Figure 6.31. Relation between CN computed from observed total rainfall-runoff volumes (including f_c) and from time distributed hydrographs.

Table 6.8 Values of parameters a_1 and b_1

Watershed	a_1	b_1
Y-8 (Texas) watershed	3.60	1.91
W-1 (Texas) watershed	19.75	0.51
Oklahoma watershed	7.00	0.72
Mississippi watershed	0.11	0.09

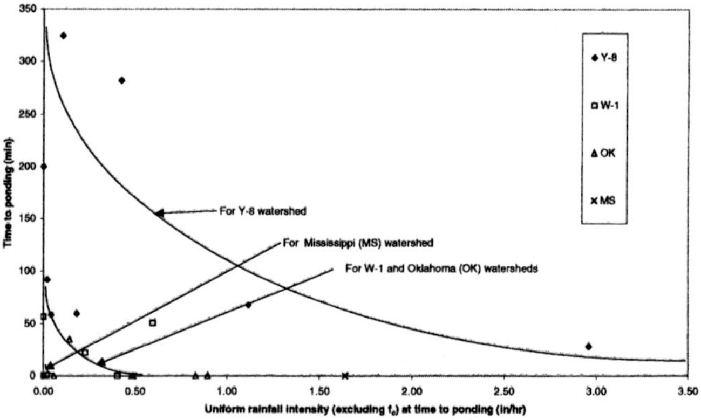

Figure 6.32. Average rainfall intensity (excluding f_c) at time to ponding versus time to ponding.

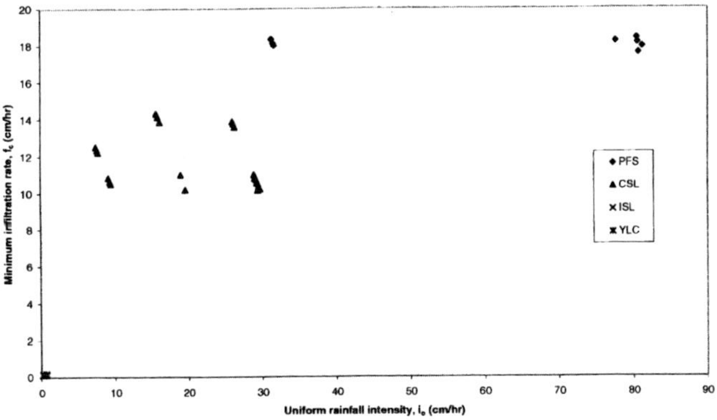

Figure 6.33. Relation between minimum infiltration rate, f_c, and uniform rainfall intensity, i_o (cm/hr) for U.S.A. soils.

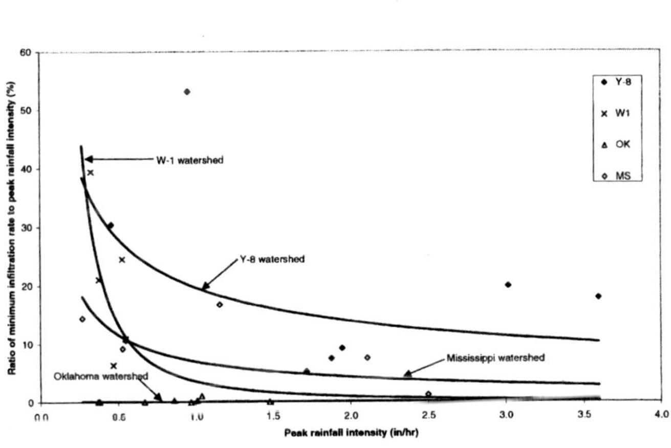

Figure 6.34. Peak rainfall intesity (in/hr) versus percent ratio of minimum infiltration rate, f_c, to peak rainfall intensity for the ARS watersheds.

$$K = \frac{\Delta x[1 + 0.6/Y]}{2(i_o - f_c)\{1 - 1/[1 - \lambda + (i_o - f_c)T_p/S]^2\}} \tag{6.16}$$

where T_p is the time of occurrence of the peak rainfall intensity.

Equation (6.16) describes physically parameter K (storage coefficient of the single linear reservoir) of equations (6.3a-c) as a function of rainfall intensity, potential maximum retention, S, λ, f_c, and length and slope of the watershed. If others are constant, parameter K increases or decreases as S increases or decreases. Equation (6.16), however, assumes the overland flow velocities equivalent to the rate of rainfall-excess and, therefore, limits the applicability of the Muskingum-Cunge method to overland flows. Since K also represents the time lag between the occurrence of rainfall peak and peak discharge, equation (6.16) is a more rational form of the empirical relation given by equation (4.60) for the time lag (t_L) (Chapter 4) as:

$$t_L = \frac{\ell^{0.8}(S+1)^{0.7}}{1900Y^{0.5}} \tag{4.60}$$

where ℓ is the hydraulic length in feet and Y is the watershed slope in percent, and S is taken to represent the surface roughness (SCS, 1971; McCuen, 1982), whereas in equation (6.16), it represents the potential maximum retention of water in a soil profile under a given antecedent moisture.

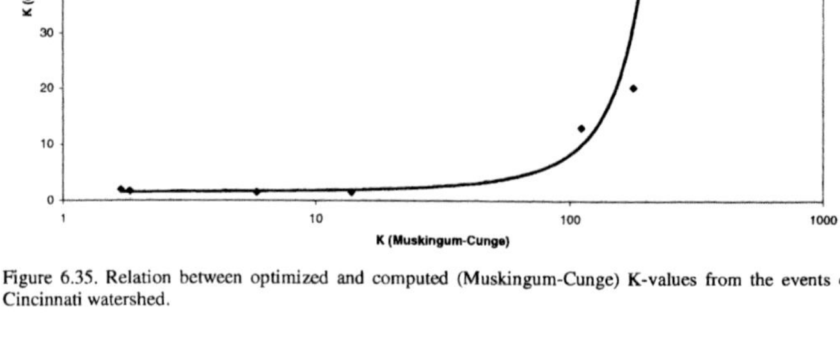

Figure 6.35. Relation between optimized and computed (Muskingum-Cunge) K-values from the events of Cincinnati watershed.

To support the above analytical derivation, the optimized K-values (Table 6.5) and those computed using equation (6.16) for the events of the Cincinnati watershed were plotted against each other, as shown in Figure 6.35. Apparently, the optimized K-values show a good correlation ($r^2 = 0.97$, which is a reasonably high value) with those computed using equation (6.16). It is noted that the deviation of K-values from 1:1 relation is primarily attributed to the above overland flow velocity assumption. However, the high correlation is the result of the watershed's known regular geometry and its physical characteristics. On the other hand, agricultural watersheds seldom exhibit such a logically consistent behavior, for the parameters depend on several other unaccounted characteristics, for example, the antecedent moisture amount, evaporation, surface detention storage, etc. Therefore, for the sake of simplicity, the variation of optimized K-values is investigated with the peak rainfall intensity, as shown in Figures 6.36 and 6.37 for Y-8 & Mississippi and W-1 & Oklahoma watersheds. The relations for the respective watersheds are expressed as:

$$K = 19.35\, i_p^{-0.9898}\ ;\ r^2 = 0.93 \qquad (6.17a)$$

and

$$K = 25.382\, i_p^{-1.2485}\ ;\ r^2 = 0.85 \qquad (6.17b)$$

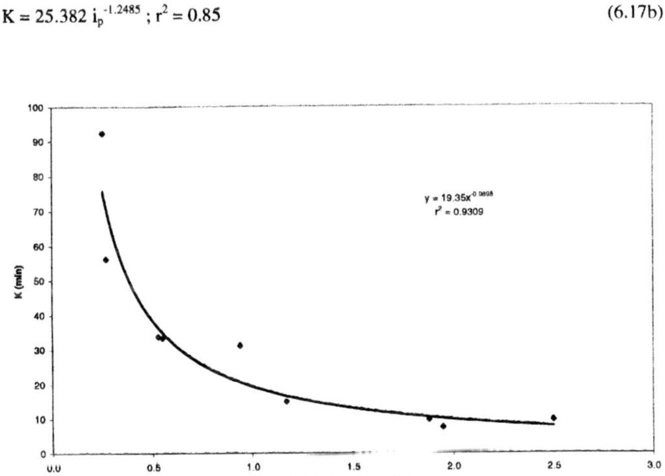

Figure 6.36. Relation between K and peak rainfall intensity for the Y-8 and Mississippi watersheds.

where K is in minutes and peak rainfall intensity, i_p, is in in/hr. These simple relations exhibit only an approximate variation of K with i_p. It is evident from equations

(6.17a,b) that K decreases with the increase of peak rainfall intensity and it, in turn, supports the above Muskingum-Cunge or wave celerity concept. Therefore, it may be possible to derive more complicated relations similar to equation (6.16) that utilize the watershed characteristics.

e) *Procedure for Application to Ungauged Watersheds*

Utilizing the relation derived above, a procedure for computing the runoff hydrograph using the SCS-CN-based infiltration model for ungauged watersheds is suggested in what follows:

1. Using the NEH-4 tables, determine a suitable curve number for the given watershed and compute S from the above CN from equation (2.7).
2. Determine the peak rainfall intensity from the given storm hyetograph.
3. Derive f_c from equation (6.13) using the peak rainfall intensity. If the derived value is much greater than that derived from watershed characteristics, the latter may be accepted.
4. Subtract the f_c-value from the given rainfall hyetograph. If the rainfall intensity in a given time interval is less than or equal to f_c, the difference is taken equal to zero.
5. Compute the cumulative effective rainfall values until it is approximately equal to the initial abstraction, I_a, which is equal to 0.2S. The corresponding time is taken equal to the time to ponding, t_p.
6. Compute SCS-CN-based infiltration model parameter, $\alpha = 0.2/t_p$. It is, however, valid for t_p greater than 0. If $t_p = 0$, then $\lambda \approx 0$ and α assumes a very high value such that the dynamic infiltration is insignificant compared to the static one.
7. Determine K-value for a watershed from the relations (equations (6.17a & b)), as shown in Figures 6.36 and 6.37, or any other relation derived from the physical characteristics of the watershed (equation (4.60)). Note that for the applicability of the above storage routing, $K \geq \Delta t$.

The model application for computing the runoff hydrograph is illustrated below.

Example 6.4: Compute the direct surface runoff hydrograph for the rainfall-runoff event of Y-8 (Texas) watershed given in Example 6.3. Watershed conditions are described by 3% pasture, 95% fall planted oats, and 2% gravel roads. Cropland is terraced and contour-tilled and the antecedent moisture condition of the watershed corresponds to AMC III.

Solution: From the given rainfall data, the peak rainfall intensity i_p = 3.02 in/hr. Compute f_c using equation (6.13) as

$$f_c^* = a_1\, i_p^{-b_1} = 19.75 \times 3.02^{-0.51} = 11.24\%$$

Therefore, f_c = 11.24 x 3.02/100 = 0.34 in/hr. The derived value suggests the watershed soils to generally fall in the hydrologic group A (Table 2.2). Using this information along with the watershed characteristics, CN can be taken equal to 61 (at Serial No. 9, the last row of Table 2.1). Thus, the CN corresponding to AMC III can be derived from Table 2.6 as equal to 78. From equation (2.7) compute S = 2.82 inches and the initial abstraction I_a = 0.2 x 2.82 = 0.564 inch. Since the computed S also accounts for f_c (Chapter 4) and, therefore, I_a also includes f_c. Therefore, according to equation (6.12), a cumulative sum of the rainfall values from the beginning (t = 0) can be taken to represent I_a for the purpose of determining the time to ponding. From the given rainfall data the sum of the rainfall is computed as equal to 0.5 inch occurring at t = 60 min. Therefore, the time to ponding can be taken as 60 min and the model parameter α can be computed from equation (6.9) as

$\alpha = 0.2/60 = 0.0033\ \text{min}^{-1}$

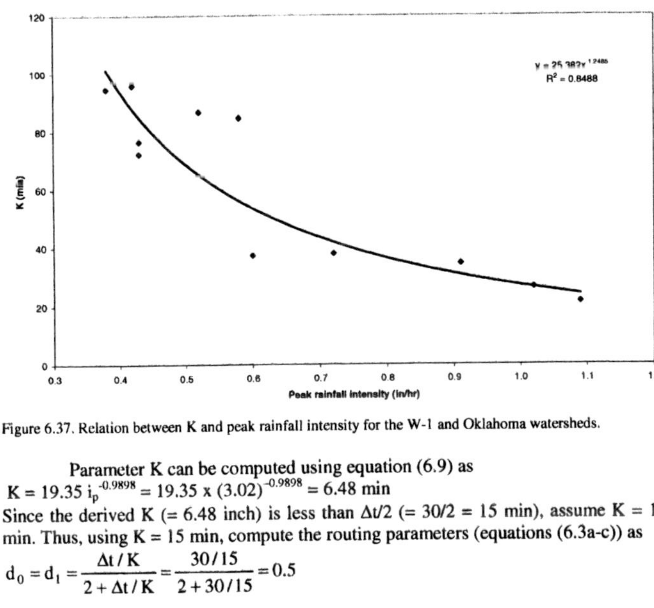

Figure 6.37. Relation between K and peak rainfall intensity for the W-1 and Oklahoma watersheds.

Parameter K can be computed using equation (6.9) as

$K = 19.35\ i_p^{-0.9898} = 19.35 \times (3.02)^{-0.9898} = 6.48$ min

Since the derived K (= 6.48 inch) is less than $\Delta t/2$ (= 30/2 = 15 min), assume K = 15 min. Thus, using K = 15 min, compute the routing parameters (equations (6.3a-c)) as

$$d_0 = d_1 = \frac{\Delta t / K}{2 + \Delta t / K} = \frac{30/15}{2 + 30/15} = 0.5$$

$$d_2 = \frac{2 - \Delta t / K}{2 + \Delta t / K} = \frac{2 - 30/15}{2 + 30/15} = 0$$

Using the above parameter values, the runoff hydrograph can be determined as shown in Table 6.9, similar to that in Example 6.3. Apparently, the peak of the resulting runoff hydrograph is equal to 1.29 in/hr, which is higher than that of the observed (=1.07 inch/hr) by 20.56%.

Table 6.9 Computation of infiltration and runoff rates for event 1 of Y-8 watershed

Time	i_o	'f'	Q	First reservoir			X	Second reservoir			O
(min)	(in/hr)	(in/hr)	(in/hr)	d_0q_{j-1}	d_1q_j	d_2X_{j-1}	(in/hr)	d_0X_{j-1}	d_1X_j	d_2O_{j-1}	(in/hr)
0	0	0	0				0				0
30	0.97	0.97	0	0	0	0	0	0	0	0	0
60	0.03	0.03	0	0	0	0	0	0	0	0	0
90	0	0	0	0	0	0	0	0	0	0	0
120	0	0	0	0	0	0	0	0	0	0	0
150	0	0	0	0	0	0	0	0	0	0	0
180	0	0	0	0	0	0	0	0	0	0	0
210	0	0	0	0	0	0	0	0	0	0	0
240	0	0	0	0	0	0	0	0	0	0	0
270	0	0	0	0	0	0	0	0	0	0	0
300	0	0	0	0	0	0	0	0	0	0	0
330	0	0	0	0	0	0	0	0	0	0	0
360	0	0	0	0	0	0	0	0	0	0	0
390	0	0	0	0	0	0	0	0	0	0	0
420	0	0	0	0	0	0	0	0	0	0	0
450	0	0	0	0	0	0	0	0	0	0	0
480	0.6	0.39	0.21	0	0.11	0	0.11	0	0.05	0	0.05
510	3.02	0.78	2.24	0.11	1.12	0	1.23	0.05	0.61	0	0.67
540	0.89	0.42	0.47	1.12	0.24	0	1.36	0.61	0.68	0	1.29
570	0	0	0	0.24	0	0	0.24	0.68	0.12	0	0.80
600	0	0	0	0	0	0	0	0.12	0	0	0.12
630	0	0	0	0	0	0	0	0	0	0	0

CHAPTER 7

LONG-TERM HYDROLOGIC SIMULATION

Long-term hydrologic simulation is required for augmentation of hydrologic data. It is useful for water resources planning and watershed management. Long-term hydrologic data are specifically required for analyses of water availability; computation of daily, fortnightly, and monthly flows for reservoir operation; and drought analyses. Since the rainfall data are generally available for a much longer period than are the stream flow data, long-term hydrologic simulation helps extend the gauged data required for the above applications. Since this book deals with the SCS-CN method, the available methods utilizing the SCS-CN method for hydrologic simulation are reviewed and their application is demonstrated using simple examples. Finally, the modified version of the SCS-CN method (Chapter 4) along with its variants is applied to the data set of Hemavati watershed (area = 600 sq. km) in India.

There exists a multitude of models for hydrologic simulation. In 1991, The U. S. Bureau of Reclamation prepared an inventory of 64 watershed models into four categories and the inventory is being currently updated. Burton (1993) compiled Proceedings of the Federal Interagency Workshop on Hydrologic Modeling Demands for the 1990's, which contain several important watershed hydrology models. Singh (1995) edited a book that summarized 26 popular models from around the globe. The Subcommittee on Hydrology of the Interagency Advisory Committee on Water Data (1998) published proceedings of the First Federal Interagency Hydrologic Modeling Conference which contains many popular watershed hydrology models developed by federal agencies in the United States. Wurbs (1998) listed a number of generalized water resources simulation models in seven categories and discussed their dissemination.

The available models vary in description of the components of the hydrologic cycle, degree of complexity of inputs, number of parameters to be determined, time interval used, and output generated. Some models like Hydrologic Simulation Package Fortran (HSPF), USDAHL (Holtan and Lopez, 1971) and its variants, Systeme Hydrologique Europeen (SHE), Hydrologic Engineering Centre (HEC)- Hydrologic Modeling System (HMS) (HEC, 2000), etc. have a number of parameters, usually use a short time interval, and produce hydrographs as well as water yield. The HSPF and SHE models are not applicable to ungauged watersheds for the reason that their application requires a priori calibration with measured runoff data for each watershed. The USDAHL model can, however, be used for ungauged watersheds, but the prediction accuracy is not commensurate with the input detail. These models are better suited for detailed scientific, hydrologic studies. Holtan and Lopez (1971) found the USDAHL model to explain about 90% of the variation in monthly runoff for four watersheds up to 40 sq. km. The Haan (1975) model has four parameters, uses a 1-d time interval (except for a 1-hr interval is used during rains), has simple inputs, and only outputs runoff volume. In testing, this model was reported to explain about 80% of the variation in monthly runoff from 46 watersheds of generally less than 100 sq. km. However, no provision exists for estimating the parameters of this model for its employment to ungauged watersheds. Woodward and Gburek (1992) compared some of the available models and found them widely varying in their degree of success.

Despite their comprehensive structure, many of these models have not yet become standard tools in hydrologic practice in developing countries, such as India, Pakistan, Nepal, and other countries of Asia as well as African countries. The reason is twofold. First, most basins in these countries are ungauged and there is little hydrologic data available. Second, these models contain too many parameters which are difficult to estimate in practice and which vary from basin to basin. Although some of these models have been applied to ungauged basins, the fact is that they are not easy for practical applications. Furthermore, when these models are compared on the same basin, they are found widely varying in their performance (Woodward and Gburek, 1992). Thus, what is needed in developing countries is simple models which can provide reasonable simulations and need little data. The SCS-CN-based simulation models seem to satisfy these criteria.

The SCS-CN method is an infiltration loss model and, therefore, its applicability is supposedly restricted to modelling storm losses (Ponce and Hawkins, 1996). Since the SCS-CN method is theoretically applicable to any watershed of any size as long as the measured runoff corresponds to the observed rainfall amount. Some restrictions regarding its application to watersheds of less than 250 sq. km. for practical reasons have however been reported in literature (for example, Ponce and Hawkins (1996)). Using theoretical arguments, it is possible to apply the SCS-CN method for long-term hydrologic simulation to any basin. It is for this reason that the SCS-CN method computes the rainfall-excess that equals the direct surface runoff. In large watersheds, routing plays an important role in converting the rainfall-excess to the surface runoff hydrograph produced at the outlet of the basin. On the other hand, small watersheds require minimal routing in long-term hydrologic simulation utilizing a time interval of 1-d or larger. Consequently, the SCS-CN method has been used in long-term hydrologic simulation and several models have been developed in the past two decades. The models of Williams and LaSeur (1976), Huber et al. (1976), and Knisel (1980) that have been applied with varying degrees of success (Woodward and Gburek, 1992) are notable among others. The model of Soni and Mishra (1985) is a variant of the Hawkins (1978) model. The generally available and frequently cited models of Williams and LaSeur (1976) and Hawkins (1978) along with the recent models of Pandit and Gopalakrishnan (1996) and Mishra et al. (1998) are described to help better understand the mathematical treatment of hydrological processes by the SCS-CN method.

7.1 SCS-CN-Based Hydrologic Models

7.1.1 WILLIAMS-LASEUR MODEL

Williams and LaSeur (1976) proposed a model based on the existing SCS-CN method (equations (2.5) and (2.7)), which for $\lambda = 0.2$ can be written as

$$Q = \frac{(P - 0.2S)^2}{P + 0.8S} \tag{7.1}$$

CN is varied as CN_I, CN_{II}, or CN_{III} according to the antecedent 5-d rainfall index (Table 2.6). Each storm is assigned a CN-value based on the antecedent 5-d rainfall amount and the corresponding S-value from equation (2.7) is used in equation (7.1) for computing the rainfall-excess or direct surface runoff.

The SCS-CN parameter S is linked with the soil moisture (M) according to equation (4.4) expressed as:

$$M = S_{abs} - S \tag{4.4}$$

where S_{abs} is the maximum potential maximum retention, which is taken as equal to 20 inches. M is depleted continuously between storms by evapotranspiration and deep storage. Depletion is high when soil moisture and lake evaporation are high, and most rapid immediately after a storm (high M). M is assumed to vary with the lake evaporation as

$$\frac{d(M)}{dt} = -b_c\, M^2\, E \tag{7.2}$$

where t is the time, b_c is the depletion coefficient, and E is the lake evaporation. Equation (7.2) represents a second-order process and is analogous to that described in Chapter 3 for deriving the SCS-CN method. The lake evaporation is used as a climatic index. According to Williams and LaSeur, equation (7.2) works well for the average monthly values for runoff predictions. They found their model to perform poorly when used daily pan evaporation and temperature as climatic indices. From equation (7.2) M is solved as

$$M_t = \frac{M}{1.0 + b_c\, M \sum_{t=1}^{T} E_t} \tag{7.3}$$

where M is the soil moisture index at the beginning of the first storm, M_t is the soil moisture index at any time t, E_t is the average monthly lake evaporation for day t, and T is the number of days between storms.

For model operation, the amount of water infiltrated during a rainstorm (= P – Q) is added to the soil moisture. The rainfall of the first day of the T-day period is added to M before equation (7.3) is solved. However, runoff is not abstracted from rainfall until the end of the T-day period, for the reason that runoff lags rainfall and may be subjected to depletion for several days on large watersheds. Thus, equation (7.2) is modified for P as

$$M_t = \frac{M + P}{1.0 + b_c\, M \sum_{t=1}^{T} E_t} - Q \tag{7.4}$$

where P and Q are, respectively, the rainfall and runoff for the first storm. The retention parameter S is computed from equation (4.4) for S_{abs} = 20 inches for computing runoff for the second storm using equation (7.1). The procedure is repeated for each storm in the rainfall series. Thus, the Williams-LaSeur model can also be applied to the pre-identified rainstorms other than 1 day.

The model is calibrated with data from a gauged watershed by adjusting the depletion coefficient, b_c, until the predicted average annual runoff matches closely with the measured average annual runoff. The initial estimate of b_c is derived from the average annual rainfall and runoff values as

$$DP = \frac{AVP - AVQ}{365} \tag{7.5}$$

where DP is the average daily depletion, AVP is the average annual rainfall, and AVQ is the average annual runoff. The value of b_c can be computed from equation (7.4) assuming that (a) T = 1 day; (b) M is the average soil moisture index, M_A; (c) E is the average lake evaporation; and (d) P = Q = 0 for the day. For this situation, equation (7.4) can be recast as

$$M_t = \frac{M_A}{1.0 + b_c M_A E_t} \tag{7.6}$$

in which M_A is computed from equations (2.7) and (4.4) for CN corresponding to AMC II. The average daily depletion computed from equation (7.5) is set equal to the change in soil moisture for 1 day as

$$DP = M_A - M_t \tag{7.7}$$

Combining equations (7.6) and (7.7), one obtains

$$DP = M_A - \frac{M_A}{1.0 + b_c M_A E_t} \tag{7.8}$$

from which b_c can be derived as

$$b_c = \frac{-DP}{E_t M_A (DP - M_A)} \tag{7.9}$$

The simulation begins 1 year before the actual calibration period because of a priori determination of the initial soil moisture index. At the end of 1 year, the soil moisture is taken to represent the actual soil moisture conditions. Here, the initial estimate of M is M_A.

For employing the model to ungauged watersheds, CN_{II} is determined using watershed characteristics: soil type, land use, and hydrologic condition. Other inputs are daily rainfall and monthly lake evaporation. Williams and LaSeur suggested adjusting CN for an ungauged watershed as

$$CN_{\mu} = \frac{CN_{II\mu}\, CN_{G}}{CN_{IIG}} \tag{7.10}$$

where all CNs represent the average curve numbers, CN_{II} is the CN corresponding to AMC II, and subscripts 'μ' and 'G' correspond to ungauged and gauged watersheds, respectively. CN_{II} is determined using the usual NEH-4 procedure (Chapter 2) and the average CN_G is determined by the model for the calibrated watershed for computing CN_{μ} from equation (7.10). The value of b_c is estimated initially as the value of b_c for the calibrated watershed. At the end of an iteration, the average curve number is computed by summing all curve numbers for individual storms. The computed average is then compared with the average curve number computed from equation (7.10). If the comparison is unsatisfactory, b_c is adjusted using Newton's method until the match is satisfactory. The runoff computed at the end of iterations is taken to represent the actual runoff for the ungauged watershed.

To summarize, the Williams-LaSeur model has one parameter, uses a 1-day time interval, has simple inputs, and only outputs runoff volume. It eliminates sudden jumps in the CN-values while changing from one AMC to the other. The input requirements are: (a) an estimate of the curve number (for AMC-II condition) for the watershed; (b) measured monthly runoff; (c) daily rainfall; and (d) average monthly lake evaporation. The model computes a soil moisture index depletion parameter that forces an agreement between the measured and the predicted average annual runoff. According to Williams and LaSeur, other optimization schemes, such as optimization using monthly or annual runoff, were not used because these did not predict consistently the appropriate average annual runoff, and thus, did not provide a good estimate of the average curve number. This model has the advantage that it can be applied to nearby ungauged watersheds by adjusting the curve number for the ungauged watershed in proportion to the ratio of the AMC-II curve number to the average predicted curve number for the calibrated watershed.

The model, however, has its own limitations that are of concern. It utilises an arbitrarily assigned value of 20 inches for S_{abs}; assumes a physically unrealisable decay of soil moisture with lake evaporation; and simulates runoff on monthly and annual bases, although runoff is computed daily, treating rainfall of a day as a storm. Since the lake evaporation was taken on a monthly basis, the daily averaged evaporation was used in the model calibration and validation. It is worth noting that the model efficiencies resulting from the use of a greater time interval may not be representative of the model performance for daily computations. Furthermore, iterations required for adjusting the important soil moisture decay coefficient loses the physical soundness of the model apart from the loss of 1-year rainfall-runoff information in computing the actual soil moisture amount in calibration. The loss of data may be viewed as a serious drawback of the model, for small watersheds generally lack the hydro-meteorological information (Singh et al., 2001). Due to the faulty assumption of decay of soil moisture, the Williams-LaSeur model describes the variation of direct runoff Q with P analogous to F, and thus, contrasts the existing SCS-CN approach, as shown below.

Taking $S_{abs} = S_o = S$ for AMC I (consistent with the notations of Chapter 4), equation (4.4) can be written for time t as

$$M_t = S_o - S_t \tag{7.11}$$

where subscript t refers to time t. If $S_t = 0$ at time $t = 0$, $M_t = S_o$. Substituting equation (7.11) into equation (7.3), one gets

$$\frac{S_o - S_t}{S_o} = \frac{1}{1.0 + b_c\, S_o\, \overline{E}\, t} \tag{7.12}$$

where $\overline{E}$ is the average rate of evapotranspiration. From the volumetric concept (Chapters 3 and 4), the left-hand side of equation (7.12) represents the F/S_o-ratio as:

$$\frac{F}{S_o} = \frac{S_o - S_t}{S_o} \tag{7.13}$$

This is based on the assumption that the soil is completely dry prior to the rainstorm so that the moisture M equals F. Relating the term containing 't' in the denominator of equation (7.12) with the storm rainfall P as (Chapter 3):

$$\frac{P}{S_o} = b_c\, S_o\, \overline{E}\, t \tag{7.14}$$

leads to describing the uniform rainfall intensity, i_o, as

$$i_o = b_c\, S_o^2 \overline{E} \tag{7.15}$$

Coupling of equations (7.13) and (7.14) with equation (7.12) yields

$$\frac{F}{S_o} = \frac{S_o}{S_o + P} \tag{7.16}$$

For $I_a = 0$, substitution of equation (7.16) into the water balance equation (equation (2.1)) yields

$$Q = \frac{PS_o}{S_o + P} \tag{7.17}$$

or

$$Q = \frac{PS}{S + P} \tag{7.18}$$

which describes the Q-variation with P. According to the existing SCS-CN approach, the right-hand side of equation (7.18) describes the variation of F, rather than Q, and thus, the

Williams-LaSeur model is in contrast with reality.

7.1.2 HAWKINS MODEL

Hawkins (1978) derived a daily simulation model by expressing equation (7.1) as:

$$Q = P - S\left(1.2 - \frac{S}{P + 0.8S}\right) \tag{7.19}$$

which is valid for $P \geq 0.2$ S. It is evident from this equation that as $P \rightarrow \infty$, the maximum possible water loss is equal to S_T and it is computed as (see Chapter 4):

$$S_T = 1.2\,S \tag{7.20}$$

which can be derived from equation (4.20) assuming $\lambda = 0.2$. Substitution of equation (2.7) into equation (7.20) yields a storage relation for any time t as

$$S_{T(t)} = 1.2 S_t = 1.2\left(\frac{1000}{CN_t} - 10\right) \tag{7.21}$$

where subscript 't' represents the time level. Taking into account the evapotranspiration (ET), the maximum water loss at a higher time level (t+Δt), where Δt is the storm duration, can be derived from the moisture balance as

$$S_{T(t+\Delta t)} = S_{T(t)} + [ET - (P - Q)]_{(t,t+\Delta t)} \tag{7.22}$$

where the last term in the bracket corresponds to the Δt duration between time t and (t+Δt), denoted by subscript (t, t+Δt). Following the above argument, equation (7.22) can alternatively be written as

$$S_{T(t+\Delta t)} = 1.2\,S_{(t+\Delta t)} \tag{7.23}$$

Here, it is noted that ET also intuitively accounts for the interim drainage, if any. Coupling of equation (7.22) with equation (7.23) and substitution of equation (2.7) into the resulting expression leads to

$$1.2\left(\frac{1000}{CN_t} - 10\right) + [ET - (P - Q)]_{(t,t+\Delta t)} = 1.2\left(\frac{1000}{CN_{t+\Delta t}} - 10\right) \tag{7.24}$$

which can be solved for $CN_{t+\Delta t}$ as:

$$CN_{t+\Delta t} = \frac{1200}{\frac{1200}{CN_t} + [ET - (P - Q)]_{(t,t+\Delta t)}} \tag{7.25}$$

Since ET, P, and Q in equation (7.25) correspond to the time duration Δt and these are known quantities, Q can be computed from equation (7.1) for a given CN_t. Input of these values along with the known value of ET yields CN at time level ($t+\Delta t$).

It is apparent from the above that the Hawkins model accounts for the site moisture on a continuous basis using the volumetric concept (Chapter 4). It is worth emphasizing here that the Hawkins model is analogous to a bottomless reservoir, implying that the reservoir never depletes fully or the reservoir is of infinite storage capacity. Such a description is, however, physically realizable in terms of ψ-θ relationship (Chapter 5), according to which S is directly proportional to the average ψ which approaches infinity as $\theta \to 0$. Under the situation that the soil is fully saturated or $\theta \to n$ (soil porosity), $\psi \to 0$. Thus, similar to S, S_T (equation (7.20)) will also vary from 0 to ∞. Following this argument, S_{abs} = 20 inches in the Williams-LaSeur model appears to be a forced assumption. While applying the Hawkins model, Soni and Mishra (1985) also employed a similar assumption by fixing the depth of soil profile to the root zone depth of 1.2 m for computing S.

The advantage of the Hawkins method is that it also eliminates sudden quantum jumps in the CN-values while changing from one AMC level to the other, similar to the Williams-LaSeur model. However, the Hawkins model also has the following limitations:

(i) It does not distinguish the dynamic infiltration from the static one. The water drained down to meet the water table may not be available for evapotranspiration.

(ii) The interim drainage is coupled with the evapotranspiration intuitively.

(iii) According to the model formulation (equation (7.19)), the term (I_a+S) takes part in the dynamic infiltration process, rather than the S alone. Since the initially adsorbed water (= I_a) due to very high capillary suction is not available for transpiration, I_a does not play a part in the dynamic infiltration process.

(iv) The follow up of the above (iii) leads to the assumption that the SCS-CN method is based on (I_a+S) scheme, whereas I_a is separate from S. The Hawkins model considers the maximum F amount equal to (I_a + S).

(v) A substitution of P = 0 in equation (7.19) yields Q = 0.05S, which is impossible. Although equation (2.1) is stated to be valid for $P \geq 0.2$ S, equation (7.25) carries its impact by allowing an additional storage space of 20% of S available for water retention at every time level. Mathematically, it can be shown as follows:

From equation (7.19), F is computed as

$$F = S\left(1 - \frac{S}{P + 0.8S}\right) \tag{7.26}$$

which is derived using the water balance equation (equation (2.1)). When $P \to 0$, equation (7.26) describes the maximum F to be equal to S, rather than 1.2S. Thus, S at time t (= S_t) corresponds to CN at time t (= CN_t). Accordingly, equation

(7.24) is modified as

$$\left(\frac{1000}{CN_t} - 10\right) + [ET - (P - Q)]_{(t,t+\Delta t)} = \left(\frac{1000}{CN_{t+\Delta t}} - 10\right) \qquad (7.27)$$

which solves for CN at time level (t+Δt) (= $CN_{t+\Delta t}$) as

$$CN_{t+\Delta t} = \frac{1000}{\frac{1000}{CN_t} + [ET - (P - Q)]_{(t,t+\Delta t)}} \qquad (7.28)$$

which is described as the modified Hawkins method.

The application of the modified Hawkins method is illustrated for typical test examples relevant to long-term hydrologic simulation and the results are compared with the existing SCS-CN method in what follows.

Example 7.1: A forested watershed in midsummer is at CN = 68. Over 7 days, a 2-inch storm occurs. Compare the CN values resulting from the Hawkins, modified Hawkins, and NEH-4 models taking time interval equal to 7 days. Assume the evapotranspiration (including drainage) equal to (a) 0.10 inch/day, (b) 0.05 inch/day, and (c) 0.15 inch/day. Also show the daily variation of CN using the modified Hawkins method for ET = 0.10 inch/day.

Solution: First, Δt = 7 days and the total evapotranspiration (including drainage) values during 7 days are: (a) ET = 0.70 inch, (b) ET = 0.35 inch, and (c) ET = 1.05 inch. Given the total rainfall P = 2 inches and initial CN (= CN_t) = 68 which, in turn, describes the corresponding S from equation (2.7) as

S_t = 1000/68 – 10 = 4.71 inches

Since P > 0.2S (= 0.94 inch), the total direct runoff Q can be computed using the modified Hawkins approach (equation (7.1)) as

$$Q = \frac{(2.0 - 0.2x4.71)^2}{2.0 + 0.8x4.71} = 0.19 \text{ inch}$$

Thus, CN_{t+7} can be computed from equation (7.28) as

$$CN_{t+7} = \frac{1000}{\frac{1000}{68} + [0.70 - (2.0 - 0.19)]} = 73.6$$

Similarly, for cases (b) and (c), CN_{t+7} values can be calculated as 75.5 and 71.7, respectively. On the other hand, following the Hawkins (1978) model (equation (7.25)), CN_{t+7} for the cases (a) through (c) are computed as 72.5, 74.1, and 71.0, respectively. Evidently, the Hawkins model yields lesser CN values than the modified model, implying that the former model provides an additional storage space of 0.2S than does the latter, which is the impact of including I_a with S, as described above. The midsummer can be taken as a growing season, and the antecedent 5-d rainfall amount equal to 2.0 x 5/7 = 1.43 inches. For these conditions, Table 2.6 of NEH-4 describes AMC II and, therefore, suggests no change in CN. Thus, according to NEH-4, CN will not change. Such a procedure, however, assumes that the CN value prior to the beginning of rainstorm was at AMC II.

For Δt = 1 day, daily rainfall P = 2/7 = 0.29 inch and ET = 0.10 inch/day. The computations for daily variation of CN using the modified Hawkins method (equation (7.28)) are shown in Table 7.1. In this table, runoff Q = 0 on each day because $P < I_a$. It is because of the computation of the initial abstraction in the SCS-CN method, ignoring the impact of time duration of the storm. When Δt = 7 days, the computation of initial abstraction holds for 7 days rainfall. On the other hand, in the daily computation, the initial abstraction is considered each day. As seen from Table 7.1, CN increases from 68 to 74.6 compared to 73.6 for Δt = 7 days.

Table 7.1 Daily variation of CN for ET = 0.10 inch/day (Example 7.1)

Day	P (inch)	ET (inch)	I_a (inch)	Q (inch)	CN
0	-	-	-	-	68
1	0.29	0.10	0.94	0	68.9
2	0.29	0.10	0.90	0	69.8
3	0.29	0.10	0.87	0	70.7
4	0.29	0.10	0.83	0	71.6
5	0.29	0.10	0.79	0	72.6
6	0.29	0.10	0.76	0	73.6
7	0.29	0.10	0.72	0	74.6

Example 7.2: In a dormant condition, a pasture has a CN = 82. Over 5 days, no rainfall occurs. Compute the resulting CN assuming a nominal drainage of 0.01 inch/day. Compare the computed CN values derived from the Hawkins, modified Hawkins, and NEH-4 models. Assuming a uniform daily variation of rainfall, also show the daily variation of CN using the modified Hawkins method.

Solution: Here, Δt = 5 days and CN_t = 82. Under dormant conditions, ET may be neglected. The 5-d drainage amount is equal to 0.01 x 5 = 0.05 inch. Thus, ET (including drainage) = 0.05 inch. Since P = 0, Q = 0. Using this information in equation (7.28), CN_{t+5} can be computed as equal to 81.66. The Hawkins model yields CN_{t+5} equal to 81.72. Thus, the deviation in the resulting CN values from the two models apparently reduces with the reduction in evapotranspiration and drainage. According to the NEH-4 table (Table 2.6) for no rainfall condition in the dormant season, the AMC is described as AMC I, for which CN_I = 66 (Table 2.7) corresponding to CN = 81.7. Thus, the NEH-4 model computes a significantly low CN-value than that due to either of the other two models.

For daily variation of CN, the computations using the modified Hawkins method (equation (7.28)) are shown in Table 7.2. Since daily rainfall P is zero each day, the reduction in CN (or increase in S) is attributed to the drainage; CN decreases with the increase in drainage and vice versa.

Table 7.2 Daily variation of CN (Example 7.2)

Day	P (inch)	Drainage (inch)	I_a (inch)	Q (inch)	CN
0	-	-	-	-	82
1	0.00	0.01	0.44	0	81.9
2	0.00	0.01	0.44	0	81.9
3	0.00	0.01	0.44	0	81.8
4	0.00	0.01	0.45	0	81.7
5	0.00	0.01	0.45	0	81.7

Example 7.3: Compute the amount of the storm rainfall that must fall on a watershed of CN = 80 to raise CN to 85 using the Hawkins, modified Hawkins, and NEH-4 models and compare them. Assume evapotranspiration (including drainage) during the rain period has been nil.

Solution: For the given curve numbers CN_t and $CN_{t+\Delta t}$, where Δt is the time duration of the rain, and ET = 0, equation (7.28) (the modified Hawkins model) solves for (P-Q) as

$$P - Q = \frac{1000}{CN_t} - \frac{1000}{CN_{t+\Delta t}} = \frac{1000}{80} - \frac{1000}{85} = 0.74 \text{ inch}$$

For $CN_t = 80$, $S_t = 2.5$ inches. Substitution of P-Q = 0.74 inch in equation (7.19) solves for P as

$$P = \frac{2.5}{1.2 - 0.74/2.5} - 0.8x2.5 = 0.76 \text{ inch}$$

which is greater than $0.2S_t = 0.2 \times 0.76 = 0.15$ inch and, therefore, the computed rainfall is the desired storm rainfall for raising CN from 80 to 85.

Similarly, using the Hawkins model, P is computed as 0.95 inch. This is higher than that derived using the above modified Hawkins model because the former approach also needs to satisfy higher water retention demands than is required for the mass balance. The NEH-4 table (Table 2.7) raises CN from 80 to 91 if the rainfall amount is greater than 1.10 inches (Table 2.6), which is comparable with those derived from the existing and modified Hawkins models for CN raise to 85, which is less than 91.

7.1.3 PANDIT AND GOPALAKRISHNAN MODEL

Pandit and Gopalakrishnan (1996) suggested a continuous simulation model for computing the annual amount of runoff for computing annual pollutant loads (see Chapter 8). This model is specifically useful for urban areas characterized primarily by the percent imperviousness (Chapter 2). It involves the following steps:

(i) Determine the pervious curve number for AMC II.

(ii) Determine the directly connected impervious area of the urban watershed.

(iii) Estimate daily runoff depth for both pervious and impervious areas separately using equation (7.1).

(iv) Determine the actual AMC based on the previous 5-day rainfall and modify CN using equations (3.145) and (3.146) such that CN does not exceed 98.

(v) Calculate the yearly storm runoff depth by summing the runoff for each day

The method is very simple, allows sudden jumps in CN values, and ignores evapotranspiration, drainage contribution, and watershed routing. Since routing is ignored, it is useful for small watersheds, where routing is minimal in daily runoff computation. This model is a specific form of the Mishra et al. (1998) model described subsequently. Its use in the computation of annual pollutant loads is discussed in Chapter 8.

7.1.4 MISHRA ET AL. MODEL

Mishra et al. (1998) suggested a model based on the existing SCS-CN method. They described the antecedent moisture conditions (AMC) based on the NEH-4 table (Table 2.6) and varied CN with time t according to equations (2.34) and (2.35), which are expressed, respectively, as:

$$CN_{t+\Delta t} = \frac{CN_t}{2.281 - 0.01281\,CN_t} \tag{7.29}$$

or

$$CN_{t+\Delta t} = \frac{CN_t}{0.427 + 0.00573\,CN_t} \tag{7.30}$$

Equations (7.29) and (7.30) are valid for AMC I and AMC III, respectively. In these equations, if $\Delta t = 1$ day, the term $(t+\Delta t)$ denotes the next day of t for $\Delta t = 1$ day. The initial $(t = 0)$ value of CN ($= CN_o$) is taken to correspond to AMC II. Thus, the variation of CN is primarily governed by AMC.

Since the computed RO_t represents the direct runoff amount or rainfall-excess, Q (equation 7.1)), at any time t corresponding to P_e of the same day, its transformation to direct runoff amount is carried out using the linear regression approach as

$$DO_t = d_1\,RO_t + d_2\,RO_{t-1} + d_3\,RO_{t-2} + \ldots\ldots \tag{7.31}$$

where DO_t is the direct runoff at the outlet and d_1, d_2, d_3,... are the nondimensional regression coefficients. This approach is analogous to the unit hydrograph scheme.

The baseflow (O_b) is assumed to be a fraction, b_f, of F lagged by NLAG days as below:

$$O_{b(t+NLAG)} = b_f\ F_t \tag{7.32}$$

The total daily flow, O_t, is taken as the sum of DO_t and O_b, as follows:

$$O_t = DO_t + O_{b(t)} \tag{7.33}$$

The parameters of the model are determined using the non-linear Marquardt algorithm (Appendix B) utilizing the objective function of minimizing the errors between the computed and observed data or maximizing the model efficiency.

The advantage of the Mishra et al. model is that it allows the transformation of rainfall-excess to direct runoff and takes into account the baseflow. Thus, the above model is applicable to large basins. However, the following are the limitations of this model.

(i) This model does not distinguish between dynamic and static infiltration, similar to the Williams-LaSeur and Hawkins models.

(ii) The model allows sudden jumps in CN values while changing from one AMC to another AMC level.

(iii) The model assumes the preceding day CN-value at normal AMC II level and converts it to other AMC level based on the previous 5-d antecedent rainfall amount, P_5, which is not rational. As shown in Example 7.2, the CN value would have been of the day preceding P_5-storm.

(iv) The employment of equation (7.31) invokes the problem of mass balance, for the sum of the regression coefficients is seldom equal to 1.0 in long-term hydrologic simulation.

(v) The baseflow is taken as a fraction of F, which is not rational. The water retained in the soil pores may not be available for baseflow, rather the water that percolates down to meet the water table may appear at the outlet as baseflow.

Example 7.4: In a dormant season, an urban watershed of the weighted CN = 87 (for AMC II) experienced a rainfall of 3 inches in 7 days. Assuming a no rainfall condition for about seven days prior to the rainstorm, compute the total runoff using time interval equal to (a) 1 day and (b) 7 days.

Solution: In the dormant season, the evapotranspiration can be ignored. The daily computation of runoff is shown in Table 7.3. In this table, daily rainfall = 3/7 = 0.43 inch. The cumulative P is used for determination of AMC using Table 2.6. At the beginning, it is AMC I, for there has been no rainfall for seven days prior to the rainstorm. Then CN is computed using equation (7.29) or equation (7.30) depending on the AMC. Notably, CN jumps from 74.6 to 87 at the end of 1 day and it further jumps to 94 at the end of the second day. S is computed from equation (2.7), $I_a = 0.2S$, and Q for each day is computed using equation (7.1). Apparently, the total runoff Q = 0.51 inch. On the other hand, if the storm duration is taken equal to 7 days, Q = 0.94 inch from equation (7.1) for CN = 74.6 which corresponds to AMC I. As stated earlier, the reduced runoff value in daily computation is largely attributed to the accounting of I_a each day.

Table 7.3 Computation of runoff (Example 7.4)

Day	P (inch)	Cumulative P (inch)	AMC	CN	S (inch)	I_a (inch)	Q (inch)
0	0	0	I	74.6	3.41	0.68	0
1	0.43	0.43	I	74.6	3.41	0.68	0
2	0.43	0.86	II	87.0	1.49	0.30	0.01
3	0.43	1.29	III	94.0	0.64	0.13	0.10
4	0.43	1.71	III	94.0	0.64	0.13	0.10
5	0.43	2.14	III	94.0	0.64	0.13	0.10
6	0.43	2.57	III	94.0	0.64	0.13	0.10
7	0.43	3.00	III	94.0	0.64	0.13	0.10
Total							0.51

It is evident from the above description that there exists a need for an improved model that eliminates for the most part the above limitations. An attempt is made to formulate a model based on the modified SCS-CN method (Chapter 4) in what follows.

7.2 Simulation Using the Modified SCS-CN Model

7.2.1 RAINFALL-EXCESS COMPUTATION

The modified SCS-CN model (equation (4.54)) can be written for time t as

$$RO_{(t,t+\Delta t)} = \frac{(P_{(t,t+\Delta t)} - I_{a(t)} - F_{c(t,t+\Delta t)})(P_{(t,t+\Delta t)} - I_{a(t)} - F_{c(t,t+\Delta t)} + M_t)}{P_{(t,t+\Delta t)} - I_{a(t)} - F_{c(t,t+\Delta t)} + M_t + S_t} \tag{7.34}$$

where M_t is the antecedent moisture amount, prior to the beginning of the storm. Equation (7.34) is valid for $P_{(t,\, t+\Delta t)} \geq (I_{a(t)} + F_{c(t,\, t+\Delta t)})$, $RO_{(t,\, t+\Delta t)} = 0$ otherwise. If $P_{(t,\, t+\Delta t)} \leq I_{a(t)}$, then $F_{c(t,\, t+\Delta t)} = 0$ and $RO_{(t,\, t+\Delta t)} = 0$. If $P_{(t,\, t+\Delta t)} \leq (I_{a(t)} + F_{c(t,\, t+\Delta t)})$, then $F_{c(t,\, t+\Delta t)} = (P_{(t,\, t+\Delta t)} - I_{a(t)})$. Here, $I_{a(t)} = \lambda S_t$.

7.2.2 SOIL MOISTURE BUDGETING

The dynamic infiltration component of infiltration (F_d) that occurred during the Δt period can be computed from the water balance equation as

$$F_{d(t,\, t+\Delta t)} = P_{(t,\, t+\Delta t)} - I_{a(t)} - F_{c(t,\, t+\Delta t)} - RO_{(t,\, t+\Delta t)} \tag{7.35}$$

which is valid for $RO_{(t,\, t+\Delta t)} \geq 0$, $F_{d(t,\, t+\Delta t)} = 0$ otherwise. The term $F_{d(t,\, t+\Delta t)}$ also represents an increase in the amount of soil moisture in the soil profile during Δt period, which when added to its antecedent moisture M_t leads to the antecedent moisture amount for the next storm as:

$$M_{(t,\, t+\Delta t)} = F_{d(t,\, t+\Delta t)} + M_t \quad - ET_{(t,\, t+\Delta t)} \tag{7.36}$$

where $M_{(t,\, t+\Delta t)}$ varies from 0 to S_{abs}. S_t can be modified for the next storm by balancing the soil moisture as

$$S_{(t+\Delta t)} = S_t - F_{d(t,\, t+\Delta t)} + ET_{(t,\, t+\Delta t)} \tag{7.37}$$

It is noted that the sum (S_t+M_t) for a watershed represents the absolute potential maximum retention, S_{abs}, which corresponds to the completely dry condition of the soil. It, in turn, represents the maximum possible available void space in the soil. Therefore, it represents the upper bound of S-variation. The minimum value of S is, however, equal to zero. Here it is noted that equation (7.36) is an alternative form of equation (4.4).

7.2.3 COMPUTATION OF EVAPOTRANSPIRATION

The potential evapotranspiration (PET) can be computed using the pan evaporation as

$$PET_{(t,\, t+\Delta t)} = PANC \times E_{(t,\, t+\Delta t)} \tag{7.38}$$

where PANC is the pan coefficient and $E_{(t,\, t+\Delta t)}$ is the pan evaporation during Δt period. Pan evaporation depends on several meteorological factors such as temperature, humidity, wind speed, solar radiation, etc. PANC depends on the vegetative cover and season, and thus, is a function of the time of the year. It can be computed as follows:

Assuming that the potential evaporation from the upper soil layer is equivalent to the pan evaporation, $E_{(t,\, t+\Delta t)}$, under the condition of the ample amount of water availability, the actual evaporation, $E_{a(t,\, t+\Delta t)}$, can be computed following Schaake et al. (1996) as:

$$E_{a(t,\, t+\Delta t)} = E_{(t,\, t+\Delta t)}\left(1 - \frac{S_t}{S_{abs}}\right) \tag{7.39}$$

The Schaake et al. model assumes a 2-layer (upper and lower) storage model, similar to the concept of the Tank model. The potential evapotranspiration from the lower level is computed as

$$PET_{(t,\, t+\Delta t)} = E_{(t,\, t+\Delta t)} - E_{a(t,\, t+\Delta t)} \tag{7.40}$$

A combination of equations (7.39) and (7.40) yields

$$PET_{(t,\, t+\Delta t)} = \frac{S_t}{S_{abs}} E_{(t,\, t+\Delta t)} \tag{7.41}$$

A comparison of equation (7.41) with equation (7.38) leads to

$$PANC = \frac{S_t}{S_{abs}} \tag{7.42}$$

Thus, PANC is defined as the ratio of the potential maximum retention at a time t to the absolute potential maximum retention. Following Schaake et al. (1996) further with the assumption that the ratio of moisture deficits for both the upper and lower layers is equal to each other leads to a description of the total amount of evapotranspiration (ET) as

$$ET_{(t,\, t+\Delta t)} = E_{(t,\, t+\Delta t)}[1 - (S_t / S_{abs})^2] \tag{7.43}$$

or

$$ET_{(t,\,t+\Delta t)} = E_{(t,\,t+\Delta t)}[1 - (PANC)^2] \tag{7.44}$$

The initial abstraction coefficient, λ, is taken equal to PANC, which varies from 0 to 1.

7.2.4 CATCHMENT ROUTING

The daily rainfall-excess rates computed using equation (7.34) are routed through the watershed using the single linear reservoir technique described in Chapter 6. Accordingly, the continuity and storage equations are written, respectively, as:

$$RO - DO = \Delta V/\Delta t \tag{7.45}$$

and

$$V = K\,(DO) \tag{7.46}$$

where V is the reservoir storage or the rainfall-excess detention storage, K is the storage coefficient, Δt is the time interval, RO is the rainfall-excess rate, and DO is the outflow or runoff rate at the outlet of the watershed. Using a finite difference scheme, DO at different time steps can be computed as:

$$DO_{(t+\Delta t)} = d_0\, RO_t + d_1\, RO_t + d_2\, DO_t \tag{7.47}$$

where t and t+Δt are the time steps at Δt interval (which equals 1-d in daily simulation), and d_o, d_1, and d_2 are the same as described by equations (6.3a-c), respectively. Here, any other suitable method can also be applied. The method of single linear reservoir is adhered to for simplicity reasons.

7.2.5 BASEFLOW COMPUTATION

It is known that infiltration depends on rainfall. Therefore, if P-I_a is less than F_c on a given day, then F_c = P-I_a. It is emphasized that under such a situation, $RO_t = 0$ or $F_{d(t)} = 0$. It implies that F_c exists even prior to the satisfaction of the capillary demands, which is in contrast with reality. This is because of the assumed equivalence between f_c and the minimum infiltration rate at a time approaching infinity, as discussed in Chapter 4. Considering that the water infiltrating after saturation through F_c percolates down to meet the water table and finally appears at the outlet of the basin. It further assumes that the basin boundary coincides with the aquifer boundary and no lateral flow contributes to the water table from across the defined watershed boundary. Thus, applying equations (7.45) through (7.47), baseflow (O_b) can be computed as

$$O_{b(t+\Delta t)} = g_0\, F_{c(t)} + g_1\, F_{c(t)} + g_2\, O_{b(t)} \tag{7.48}$$

where

$$g_0 = \frac{\Delta t / K_b}{2 + \Delta t / K_b} \tag{7.49a}$$

$$g_1 = g_0 \tag{7.49b}$$

$$g_2 = \frac{2 - \Delta t / K_b}{2 + \Delta t / K_b} \tag{7.49c}$$

In equations (7.49a-c), K_b is the baseflow storage coefficient [T].

Thus, the total runoff hydrograph, O, appearing at the outlet of the catchment is computed as the sum of the routed rainfall-excess, DO, and the baseflow, O_b. Expressed mathematically,

$$O_t = DO_t + O_{b(t)} \tag{7.50}$$

Similar to the above, the following examples deal with certain specific situations useful in the illustration of the application and evaluation of the above models in long-term hydrologic simulation. The last example deals with the hydrologic simulation of the Hemavati watershed.

Example 7.5: A forested watershed in midsummer is at CN = 68. Over 7 days, a 2-inch rainstorm occurs. Compare the CN values resulting from the Hawkins, modified Hawkins, NEH-4, and modified SCS-CN models. Assume f_c = 0.01 inch/day, λ = 0.25, and the evapotranspiration is equal to (a) 0.09 inch/day, (b) 0.04 inch/day, and (c) 0.14 inch/day. For evapotranspiration equal to 0.09 inch/day, also show the daily variation of CN using the modified SCS-CN model.

Solution: First consider Δt = 7 days. Assuming that CN = 68 corresponds to AMC II. Thus, CN corresponding to AMC I can be derived from NEH-4 table (Table 2.7) as equal to 48. Thus, the corresponding S for AMC II and AMC I can be derived from equation (2.7) as S_{II} = 4.706 inch and S_I = 10.833 inch, respectively. Thus, the antecedent moisture amount, M, associated with S_{II} is computed as

$M_{II} = M_t = S_I - S_{II} = 10.833 - 4.706 = 6.127$ inch

F_c can be computed from given f_c as equal to 7 x 0.01 = 0.07 inch. For given P = 2.0 inches, RO is computed from equation (7.34) as

$$RO_{(t,t+\Delta t)} = \frac{(P_{(t,t+\Delta t)} - I_{a(t)} - F_{c(t,t+\Delta t)})(P_{(t,t+\Delta t)} - I_{a(t)} - F_{c(t,t+\Delta t)} + M_t)}{P_{(t,t+\Delta t)} - I_{a(t)} - F_{c(t,t+\Delta t)} + M_t + S_t}$$

$$= \frac{(2.0 - 0.25 \text{x} 4.706 - 0.07)\text{x}(2.0 - 0.25\text{x}4.706 - 0.07 + 6.127)}{(2.0 - 0.25\text{x}4.706 - 0.07 + 6.127 + 4.706)}$$

$= 0.447$ inch

Therefore, F_d can be computed from the water balance equation (equation (2.1)) as

$F_{d(t,t+7)} = P_{(t,t+7)} - I_{a(t)} - F_{c(t,t+7)} - RO_{(t,t+7)}$

$= 2.0 - 0.25 \text{ x } 4.706 - 0.07 - 0.447$

= 0.307 inch

This F_d amount occupies the void space in the soil profile during the given period of 7 days. Including evaporanspiration, the S-values after 7 days can be computed from equation (7.37) for the three cases as:

Case (a)

$S_{t+7} = S_t - F_{d(t,\,t+7)} + ET_{(t,\,t+7)}$

$= 4.706 - 0.307 + 0.09 \times 7 = 5.029$ inches

Case (b)

$S_{t+7} = 4.706 - 0.307 + 0.04 \times 7 = 4.679$ inches

Case (c)

$S_{t+7} = 4.706 - 0.307 + 0.14 \times 7 = 5.379$ inches

The corresponding CN-values for cases (a)-(c) can be computed from equation (2.7) as 66.5, 68.1, and 65.0, respectively. The CN-values derived using other models for these cases that are approximately equivalent to those of Example 7.1 are compared in Table 7.4. Here, it is noted that in this example, the drainage rate is taken equal to the minimum infiltration rate, f_c. The derived CN-values are lower than those derived using the existing and modified Hawkins models and closer to those given by the NEH-4 model.

Table 7.4 Derivation of CN-values using different models (Example 7.5) (Δt = 7 days)

Case	NEH-4	Hawkins (1978)	Modified Hawkins	Modified SCS-CN
'a'	68	72.5	73.6	66.5
'b'	68	74.1	75.5	68.1
'c'	68	71.0	71.7	65.0

Table 7.5 Computation of CN-values using the modified SCS-CN method for Δt = 1 day (Example 7.5)

Day	P (inch)	F_c (inch)	I_a (inch)	RO (inch)	F_d (inch)	M (inch)	ET (inch)	S (inch)	CN
0						6.127		4.706	68.0
1	0.29	0.01	0.94	0	0	6.037	0.09	4.796	67.6
2	0.29	0.01	0.96	0	0	5.947	0.09	4.886	67.2
3	0.29	0.01	0.98	0	0	5.857	0.09	4.976	66.8
4	0.29	0.01	1.00	0	0	5.767	0.09	5.066	66.4
5	0.29	0.01	1.01	0	0	5.677	0.09	5.156	66.0
6	0.29	0.01	1.03	0	0	5.587	0.09	5.246	65.6
7	0.29	0.01	1.05	0	0	5.497	0.09	5.336	65.2

For Δt = 1 day, the computations using the modified SCS-CN method are shown in Table 7.5. In this table, the first col. stands for day, day 0 indicates the beginning of the storm when CN = 68 or S = 4.706 and the initial M (= 6.127 inches) is derived using S of AMC I, as described above. Notations in the table are consistent with the above notations of different variables. Since ($I_a + F_c$) is greater than P (= 2/7 = 0.29 inch), runoff RO = 0 and, therefore, F_d = 0 and M = 0. From equation (7.37), S at the end of 1 day is: S = 4.706 – 0 + 0.09 = 4.796 inches and the corresponding CN from equation (2.7) is equal to 67.6. It is apparent from Table 7.5 that the CN values decrease during the storm. It is because the moisture content (or the dynamic portion of infiltration) is zero on each day and ET supports the enhancement of the S-value.

Example 7.6: In a dormant season, a pasture has a CN = 82. Over 5 days, no rainfall occurs. 1) Compute the CN value resulting after 5 days using the modified SCS-CN method assuming a nominal drainage equal to 0.01 inch/day and $\Delta t = 5$ days. 2) Compare the CN values resulting from the Hawkins, modified Hawkins, and NEH-4 models using $\Delta t = 5$ days. 3) Show the daily variation of CN values derived from the modified SCS-CN method.

Solution: Consider first that $\Delta t = 5$ days. Under dormant conditions, ET may be neglected. The 5-d drainage is equal to 0.01 x 5 = 0.05 inch. Thus, the total evapotranspiration including drainage is equal to 0.05 inch. Given $CN_t = 82$. Since $P_t = 0$, $Q_t = 0$ and $F_d =$ 0.Therefore, $M_{(t, t+5)} = 0$. Thus, S_{t+5} can be computed from equation (7.37) as

$$S_{t+5} = S_t - F_{d\,(t,\,t+5)} + ET_{(t,\,t+5)}$$
$$= (1000/82 - 10) - 0 + 0.05 = 2.245 \text{ inch}$$

which is equivalent to $CN_{t+5} = 81.66$ and it is equal to the CN-values computed using both the existing and modified Hawkins models. According to NEH-4 table (Table 2.6) for no rainfall condition in dormant season, the AMC is described as AMC I, for which $CN_I = 66$ (Table 2.7), which is significantly lower than those resulting from other approaches.

For showing the daily variation of CN, computations for the modified SCS-CN method are shown in Table 7.6. These computations are similar to those shown in Table 7.6 (Example 7.4), except for P = 0, ET = drainage = 0.01 inch/day and F_c is neglected.

Table 7.6 Computation of CN-values using the modified SCS-CN method for $\Delta t = 1$ day (Example 7.6)

Day	P (inch)	I_a (inch)	RO (inch)	F_d (inch)	M (inch)	ET (inch)	S (inch)	CN
0					6.127		4.706	68.0
1	0.00	0.94	0	0	6.117	0.01	4.716	68.0
2	0.00	0.94	0	0	6.107	0.01	4.726	67.9
3	0.00	0.95	0	0	6.097	0.01	4.736	67.9
4	0.00	0.95	0	0	6.087	0.01	4.746	67.8
5	0.00	0.95	0	0	6.077	0.01	4.756	67.8

Example 7.7: How large a storm must fall on a watershed of CN = 80 to raise CN to 85. Assume evapotranspiration (including drainage) during the rain period has been nil, $f_c =$ 0, and $\lambda = 0.25$. Compare the resulting CN values using the Hawkins and modified Hawkins and modified SCS-CN models.

Solution: For the given ET = 0, M_t can be computed using the modified SCS-CN method (equation (7.37)) as:

$$F_{d\,(t,\,t+\Delta t)} = S_t - S_{t+\Delta t} + ET_{(t,\,t+\Delta t)}$$
$$= 1000/80 - 1000/85 + 0 = 0.735 \text{ inch}$$

From Table 2.7, $CN_I = 63$ corresponding to $CN_{II} = 80$. Therefore, at time t, M_t is computed as

$$M_t = 1000/63 - 1000/80 = 3.373 \text{ inches}$$

Since $f_c = 0$, $F_{c(t,\,t+\Delta t)} = 0$, rainfall P can be easily computed by trial and error using spread-sheet as shown in Table 7.7. From this table P = 3.07 inches and the corresponding F_d = 0.735 = 0.74 inch.

As shown in Example 7.3, the Hawkins model yields P = 0.95 inch, whereas the modified Hawkins model gives P = 0.76 inch. These values are much lower than those computed using the modified SCS-CN method. It is because of the accounting of antecedent moisture amount. Here, it is emphasized that the minimum amount of P required to raise the soil moisture by 0.735 inch should be equal to 0.735 plus 0.2 x 2.5 (according to the Hawkins model) which equals 1.235 or 1.24 inches. Therefore, a higher value of P yielded by the modified SCS-CN method is more appropriate.

Table 7.7 Determination of rainfall P for Example 7.7

Sl. No.	Trial value of P (inch)	S (inch)	$I_a = 0.25$ S (inch)	M (inch)	F_d (inch)
1	1	2.5	0.625	3.373	0.15
2	2	2.5	0.625	3.373	0.47
3	3	2.5	0.625	3.373	0.72
4	3.07	2.5	0.625	3.373	0.74
5	4	2.5	0.625	3.373	0.91
6	5	2.5	0.625	3.373	1.07
7	6	2.5	0.625	3.373	1.20

Example 7.8. Using the modified SCS-CN method, compute runoff for the given rainfall-runoff-evaporation series of Hemavati watershed (beginning date: June 1, 1974) (Table 7.8). Use a single linear reservoir routing for both rainfall-excess and baseflow. Assume S_{abs} = 3.39 mm, F_c = 21.36 mm/day, runoff storage coefficient K = 1.64 days, baseflow storage routing coefficient K_b = 30.93 days, S on the first day = S_{abs}, beginning baseflow = beginning outflow = 0, λ for the first day is equal to 0.25.

Solution: The computations for the first day are carried out in the following steps:

Step 1

Compute the initial abstraction, I_a, from equation (2.3) as

I_a = 0.25 x 3.39 = 0.85 mm

Check for $I_a \le P$. Apparently, the computed I_a is less than the first day rainfall.

Step 2

Check for the effective rainfall to be greater than the minimum infiltration for producing the rainfall-excess as

$P - I_a - F_c = 27.63 - 0.85 - 21.36 = 5.42$ mm

which is greater than 0.

Step 3

Since $P - I_a - F_c > 0$, RO can be computed using equation (7.34) as

$$RO_1 = \frac{(P_1 - I_{a(1)} - F_{c(1)})(P_{(1)} - I_{a(1)} - F_{c(1)} + M_1)}{P_{(1)} - I_{a(1)} - F_{c(1)} + M_1 + S_1}$$

Table 7.8 Rainfall series of Hemavati watershed for the months June-July, 1974

Day	Rainfall (mm)	Runoff (mm)	Evaporation (mm)	Day	Rainfall (mm)	Runoff (mm)	Evaporation (mm)
1	27.63	0.81	4.6	31	16.34	24.24	3.1
2	41.34	3.72	3.7	32	14.34	17.18	4.8
3	40.23	7.66	0.5	33	11.06	14.03	1.9
4	0.20	5.56	4.0	34	17.69	12.30	6.0
5	0.43	1.79	5.0	35	9.11	10.51	4.0
6	0.52	0.95	3.8	36	20.67	9.62	1.9
7	0.45	0.76	3.8	37	11.45	10.17	1.4
8	2.41	0.63	5.0	38	31.56	11.51	3.3
9	1.57	0.60	4.0	39	39.11	17.58	3.4
10	0.14	0.53	5.0	40	40.20	23.85	2.0
11	0.22	0.52	4.2	41	72.51	45.99	2.8
12	0.64	0.50	4.4	42	32.43	36.99	2.2
13	2.33	0.50	5.8	43	14.83	22.58	3.2
14	7.95	0.49	4.7	44	35.46	17.37	2.5
15	22.13	0.56	5.4	45	15.86	15.28	2.7
16	9.62	0.76	4.4	46	7.93	12.59	4.0
17	3.69	0.98	5.0	47	6.91	10.56	4.7
18	10.05	0.75	6.0	48	7.94	8.86	3.7
19	33.40	1.09	4.0	49	5.76	8.52	3.0
20	69.92	11.65	4.0	50	3.85	9.04	6.3
21	69.60	31.98	6.0	51	7.13	8.52	3.3
22	105.31	47.35	3.0	52	26.84	9.99	6.6
23	113.91	46.44	2.8	53	5.01	12.04	3.3
24	92.31	47.38	1.4	54	19.34	11.59	2.6
25	58.70	39.10	1.7	55	9.00	11.92	3.8
26	42.35	28.30	3.8	56	15.34	12.87	2.6
27	80.81	34.66	1.7	57	7.09	9.63	3.0
28	68.08	53.63	2.6	58	7.97	9.85	3.1
29	38.73	39.2	4.0	59	9.83	9.91	3.6
30	44.47	30.15	2.5	60	5.90	8.64	1.4
				61	6.55	7.79	1.3

$$= \frac{5.42x(5.42 + 0)}{5.42 + 0 + 3.39} = 3.33 \text{ mm}$$

Step 4

Compute the dynamic infiltration, F_d, from the water balance equation (equation (2.1)) as

$F_d = P - I_a - F_c - RO$

$= 5.42 \quad 3.33 = 2.09$ mm

Step 5

Compute the total infiltration as the sum of F_c and F_d as

$F = F_d + F_c = 2.09 + 21.36 = 23.45$ mm

Step 6

The rainfall-excess (RO) is routed using equation (7.47). To this end, the routing coefficients d_1 and d_2 are computed as (Note: $\Delta t = 1$ day):

$$d_0 = \frac{\Delta t / K}{2 + \Delta t / K} = \frac{1/1.64}{2 + 1/1.64} = 0.234$$

$d_1 = d_0 = 0.234$

$$d_2 = \frac{2 - \Delta t / K}{2 + \Delta t / K} = \frac{2 - 1/1.64}{2 + 1/1.64} = 0.533$$

Therefore, the routed RO is computed for given $DO_1 = 0.0$ as

$DO_2 = d_0\, RO_1 + d_1\, RO_1 + d_2\, DO_1$

$= 0.234 \times 3.33 + 0.234 \times 3.33 + 0.533 \times 0 = 1.56$ mm

Step 7

Similar to step 6, compute baseflow, O_b, as

$O_{b(2)} = g_0\, F_{c(1)} + g_1\, F_{c(1)} + g_2\, O_{b(1)}$

$= 0.016 \times 21.36 + 0.016 \times 21.36 + 0.968 \times 0$

$= 0.67$ mm

Step 8

The total outflow for the second day is the sum of DO_2 and $O_{b(2)}$, computed as

$O_2 = DO_2 + O_{b(2)} = 1.56 + 0.67 = 2.23$ mm

It is noted that for the first day, the given outflow is equal to zero.

Step 9

Revise the following parameters for the next (second) day as:

Compute PANC using equation (7.42) as

$$PANC = \frac{S_1}{S_{abs}} = \frac{3.39}{3.39} = 1.00$$

which is equal to the initial abstraction coefficient, λ, for the second day.

Compute evapotranspiration, ET, for the first day as:

$ET_1 = E_1[1 - (PANC)^2]$

$= 4.6x[1 - (1)^2] = 0.0$

Compute S for the next day using equation (7.37):

$S_{(2)} = S_{(1)} - F_{d(1)} + ET_{(1)}$

$= 3.39 - 2.09 + 0.0 = 1.30$ mm

which is within the range (0, 3.39) mm.

Step 10

Compute M_2 from equation (4.4) as

$M_2 = S_{abs} - S_2 = 3.39 - 1.30 = 2.09$ mm

Step 11

Repeat all the above steps for computing runoff series, as shown in Table 7.9.

Table 7.9 Runoff computation using modified SCS-CN method (Example 7.8)

Day	P (mm)	S (mm)	RO (mm)	F (mm)	O_b (mm)	O (c) (mm)	O (o) (mm)	ET (mm)	M (mm)	PANC
1	27.63	3.39	3.33	23.45	0	0	0.81	0	0	1
2	41.34	1.31	17.56	22.47	0.68	2.23	3.72	3.15	2.09	0.39
3	40.23	3.35	14.77	24.17	1.34	10.36	7.66	0.01	0.04	0.99
4	0.2	0.55	0	0	1.98	13.68	5.56	3.89	2.84	0.16
5	0.43	3.39	0	0	1.91	8.16	1.79	0	0	1
6	0.52	3.39	0	0	1.85	5.18	0.95	0	0	1
7	0.45	3.39	0	0	1.79	3.57	0.76	0	0	1
8	2.41	3.39	0	0	1.74	2.68	0.63	0	0	1
9	1.57	3.39	0	0	1.68	2.19	0.6	0	0	1
10	0.14	3.39	0	0	1.63	1.9	0.53	0	0	1
11	0.22	3.39	0	0	1.57	1.72	0.52	0	0	1
12	0.64	3.39	0	0	1.52	1.6	0.5	0	0	1
13	2.33	3.39	0	0	1.48	1.52	0.5	0	0	1
14	7.95	3.39	0	4.56	1.43	1.45	0.49	0	0	1
15	22.13	3.39	0	18.74	1.53	1.54	0.56	0	0	1
16	9.62	3.39	0	6.23	2.08	2.08	0.76	0	0	1
17	3.69	3.39	0	0.3	2.21	2.21	0.98	0	0	1
18	10.05	3.39	0	6.66	2.15	2.15	0.75	0	0	1
19	33.4	3.39	6.21	23.8	2.29	2.29	1.09	0	0	1
20	69.92	0.96	46.71	22.26	2.9	5.79	11.65	3.68	2.44	0.28
21	69.6	3.39	44.11	24.53	3.49	26.82	31.98	0	0	1
22	105.31	0.23	83.5	21.58	4.05	37.09	47.35	2.99	3.17	0.07
23	113.91	3	89.46	24.25	4.6	61.18	46.44	0.62	0.4	0.88
24	92.31	0.73	69.61	22.05	5.14	77.06	47.38	1.34	2.67	0.21
25	58.7	1.37	35.79	22.62	5.65	76.5	39.1	1.42	2.03	0.4
26	42.35	1.54	19.05	22.68	6.15	60.64	28.3	3.02	1.86	0.45
27	80.81	3.24	54.92	24.42	6.64	44.59	34.66	0.15	0.16	0.95
28	68.08	0.33	46.09	21.67	7.11	52.98	53.63	2.58	3.06	0.1
29	38.73	2.6	14.95	23.53	7.56	53.53	39.2	1.66	0.8	0.77
30	44.47	2.09	19.71	23.17	8	39.5	30.15	1.55	1.31	0.61
31	16.34	1.84	0	15.21	8.42	34.42	24.24	2.19	1.56	0.54
32	14.34	3.39	0	12.5	8.64	22.51	17.18	0	0	1
33	11.06	3.39	0	7.67	8.76	16.16	14.03	0	0	1
34	17.69	3.39	0	14.3	8.73	12.67	12.3	0	0	1
35	9.11	3.39	0	5.72	8.91	11.01	10.51	0	0	1
36	20.67	3.39	0	17.28	8.8	9.93	9.62	0	0	1
37	11.45	3.39	0	8.06	9.07	9.67	10.17	0	0	1
38	31.56	3.39	4.54	23.63	9.04	9.36	11.51	0	0	1
39	39.11	1.13	15.68	22.3	9.43	11.72	17.58	3.02	2.26	0.33
40	40.2	3.21	15.07	24.06	9.81	18.35	23.85	0.21	0.18	0.95
41	72.51	0.72	49.79	22.04	10.18	21.76	45.99	2.67	2.67	0.21
42	32.43	2.72	8.44	23.42	10.54	39.95	36.99	0.79	0.68	0.8
43	14.83	1.45	0	13.67	10.88	30.5	22.58	2.61	1.94	0.43
44	35.46	3.39	9.97	24.04	10.97	21.44	17.37	0	0	1
45	15.86	0.72	0	15.14	11.3	21.53	15.28	2.58	2.68	0.21
46	7.93	3.3	0	7.23	11.42	16.88	12.59	0.23	0.1	0.97
47	6.91	3.39	0	3.61	11.29	14.2	10.56	0	0	1
48	7.94	3.39	0	4.33	11.04	12.6	8.86	0	0	1
49	5.76	3.39	0	2.37	10.84	11.67	8.52	0	0	1
50	3.85	3.39	0	0.46	10.57	11.01	9.04	0	0	1
51	7.13	3.39	0	3.74	10.25	10.48	8.52	0	0	1

Table 7.9 Continued

Day	P (mm)	S (mm)	RO (mm)	F (mm)	O_b (mm)	O (c) (mm)	O (o) (mm)	ET (mm)	M (mm)	PANC
52	26.84	3.39	0.79	22.65	10.04	10.16	9.99	0	0	1
53	5.01	2.1	0	2.91	10.4	10.84	12.04	2.03	1.29	0.62
54	19.34	3.39	0	17.24	10.16	10.39	11.59	0	0	1
55	9	3.39	0	5.61	10.39	10.51	11.92	0	0	1
56	15.34	3.39	0	11.95	10.23	10.3	12.87	0	0	1
57	7.09	3.39	0	3.7	10.29	10.32	9.63	0	0	1
58	7.97	3.39	0	4.58	10.08	10.1	9.85	0	0	1
59	9.83	3.39	0	6.44	9.9	9.91	9.91	0	0	1
60	5.9	3.39	0	2.51	9.79	9.8	8.64	0	0	1
61	6.55	3.39	0	3.16	9.56	9.56	7.79	0	0	1

7.3 Application of the Modified SCS-CN Model

The following application illustrates the use of the modified SCS-CN application for long-term hydrologic simulation of the Hemavati watershed described in Chapter 3. The model performance is evaluated using the Nash and Sutcliffe (1970) efficiency criterion expressed as:

$$\text{Efficiency} = [1 - RV/IV] \times 100 \tag{7.51}$$

where

$$RV = \sum_{i=1}^{n} (O_i - \hat{O}_i)^2 \tag{7.52a}$$

and

$$IV = \sum_{i=1}^{n} (O_i - \bar{O}_i)^2 \tag{7.52b}$$

Here, RV is the remaining variance, IV is the initial variance, O_i is the observed runoff for ith day, $\hat{O}_i$ is the computed runoff for ith day, n is the total number of observations, and $\bar{O}$ is the overall mean daily observed runoff.

7.3.1 PARAMETER ESTIMATION

Model parameters, S_o [L], F_c [L], K [T], and K_b [T], are determined using the Marquardt algorithm of constrained least squares (Appendix B). Although these parameters were determined by trial-and-error for obtaining the maximum efficiency, it is also possible to derive these parameters physically or from rainfall-runoff data. S_o corresponds to the spatially averaged absolute potential maximum retention and approximates the S value derived for AMC I using the usual NEH-4 model utilizing various watershed characteristics (Chapter 2). F_c represents the spatially averaged minimum daily infiltration, which is derived from the given soil types of the watershed. Parameter K represents the storage coefficient of the direct surface runoff hydrograph and is analogous to the time lag of the watershed. It can also be derived from the rainfall-runoff data by plotting them on the semi-logarithmic scale. The slope of the fit represents K. The rainfall-runoff data set selected for derivation should correspond to high rainfall-runoff events excluding baseflow. Similarly, parameter K_b that represents the time lag of the baseflow to appear at the outlet of the watershed can be derived for low rainfall-runoff events or using other methods suggested in standard text books, for example, in the text books by Ponce (1989), Singh (1992), Singh and Singh (2001) among others.

For describing the range of variation of these parameters, the lower bound is taken as zero, for all the parameters are physically non-negative. The upper bound values are, however, decided from trial runs whether the estimated parameter values are well within the supplied range. If the estimated parameter value corresponds to the upper bound of the described range, the upper bound is increased to the extent the estimate falls in the prescribed range.

7.3.2 MODEL CALIBRATION AND VALIDATION

For model calibration and validation, the available 5-yr data set of the Hemavati watershed is split into two parts: one is used for calibration and the other for validation. Using these data, the model is calibrated and validated for 10 cases listed in Table 7.10. In this table, case 1 corresponds to the 3-year data (1974-75 to 1976-77; beginning date: June 1, 1974) for model calibration and validation is done on 2-year (1977-78 to 1978-79) data set. Thus, the data length shown in the table corresponds to these years. Case 2 stands for model calibration for 2-year data and validation for 3-year data. Similarly, cases 3 and 4 correspond to model calibration for 3 and 2 years of data, respectively, but for the data of only monsoon season (* marked), which represents the period of high flows from June 1 to October 31. Cases 2 and 4 are taken up for investigating into the consistent workability of the model. Cases 5 through 10 deal with the model calibration on 3 years and validation on 2 years data set, but for varying storm days ranging from 2 to 10 days. To this end, the available daily data were converted to 2 through 10 daily storm, leading to a reduction in the number of data points. These cases were considered for examining the impact of storm duration on model parameters.

For all the cases, the estimated parameters S_o, F_c, K, and K_b in calibration are given in the column 'calibration' of Table 7.10. It is apparent from this table that the resulting efficiencies in calibration vary from 73.31 to 82.89%, indicating a reasonably good performance of the model. The corresponding efficiencies in validation range from 71.86 to 82.28%, again showing a satisfactory model performance. The purpose of validation (or verification) is to check whether the calibrated model performs consistently during and beyond the calibration period. It is apparent from the table that in some cases, the resulting efficiencies in validation are higher than those in calibration. In general, the model should perform better in calibration than in validation, for the reason that the model parameters are optimized for calibration data set using the least square errors approach, implying the error between the observed and computed values are minimum. Thus, the estimated parameters when applied to another data set (for validation) should yield a lower efficiency than that resulting in calibration. However, the resulting higher efficiency-values in validation than in calibration may be attributed to discrepancies in hydrologic conditions during the periods of the two data sets (Schaake et al., 1996). Following this argument, the runoff factor was computed from the total annual rainfall-runoff records of both the calibration and validation periods as equal to 0.75 and 0.83. It implies that the model performs well for high runoff factor and poorly otherwise. It is supported from the comparison of the cases 1 and 2. For case 1, the efficiency in validation is higher than that in calibration and the reverse holds for case 2. The same inference can be drawn if cases 3 and 4 are compared. In an examination of the data, it was found that during 1977-78, the total rainfall was equal to 2941.87 mm and the corresponding runoff was equal to 2936.71 mm, yielding the runoff factor almost equal to 1. The results of both calibration and validation are shown in Figures 7.1 through 7.5. The figures show that the computed peaks of the hydrographs in general fairly match the observed ones. It is interesting to note that the computed baseflow using F_c and its routing appears to be natural, for its variation in these figures follows the trend more closely to that derived using sophisticated approaches (Ponce, 1989; Singh, 1992; Singh and Singh, 2001).

Table 7.10 Calibration and validation results for various duration storms of Hemavati watershed

Case	Storm days	Calibration						Validation	
		Data Length (years)	Parameters				Efficiency (%)	Data length (years)	Efficiency (%)
			S_o (mm)	F_c (mm)	K (days)	K_b (days)			
1	1	3	3.39	21.36	1.64	30.93	79.63	2	80.85
2	1	2	1.54	24.40	1.10	20.53	82.41	3	77.13
3	1	3*	3.04	20.65	1.63	29.32	74.85	2	78.22
4	1	2*	1.17	25.15	1.08	19.43	80.07	3	71.86
5	2	3	6.24	41.09	1.98	30.99	75.78	2	77.14
6	3	3	13.38	73.25	-	20.47	73.31	2	80.01
7	4	3	15.29	83.18	-	25.15	78.03	2	77.47
8	5	3	21.13	115.23	-	22.47	80.26	2	82.14
9	7	3	29.81	117.43	-	25.67	82.89	2	82.28
10	10	3	49.49	131.32	-	22.50	80.60	2	80.60

*Data of only monsoon season considered.

7.3.3 VOLUMETRIC STATISTIC

The monthly and yearly volumetric statistic is shown in Tables 7.11 and 7.12. For performance evaluation, the relative errors are computed analogous to equation (6.8) as

$$\text{Relative error (\%)} = (O_i - \hat{O}_i) \times 100 / O_i \tag{7.53}$$

where O_i and $\hat{O}_i$ correspond to the observed and computed monthly or annual runoff values, respectively. Thus, the relative errors may take on any value ranging from 0 to ∞, depending on the value of O_i. As O_i approaches zero, the relative error approaches infinity.

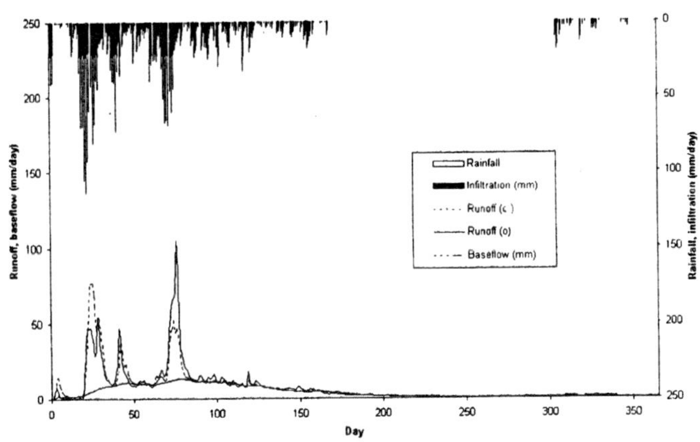

Figure 7.1. Model calibration for Hemavati data (year 1974-75).

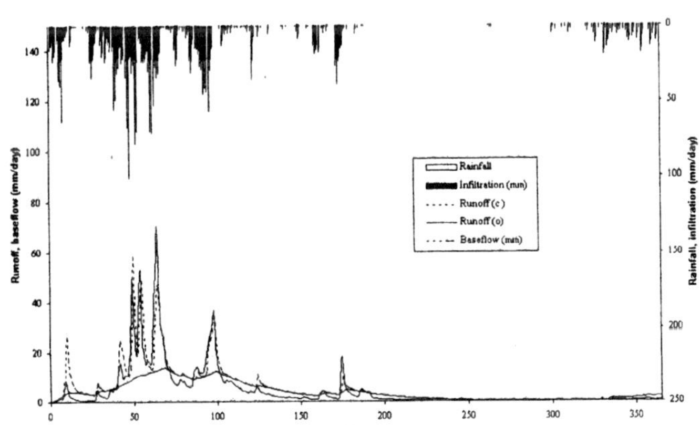

Figure 7.2. Model calibration for Hemavati data (year 1975-76).

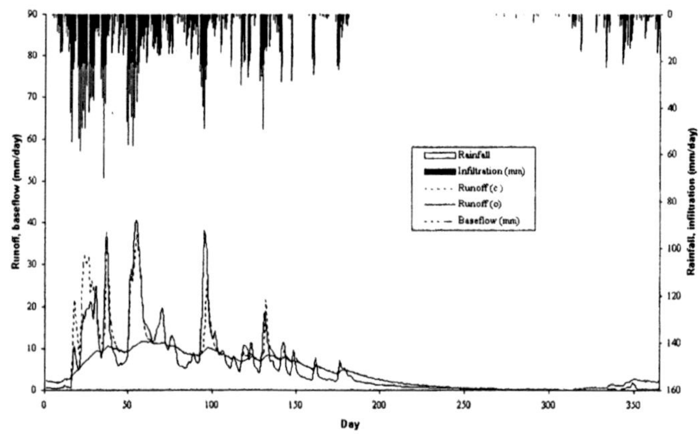

Figure 7.3. Model calibration for Hemavati data (year 1976-77).

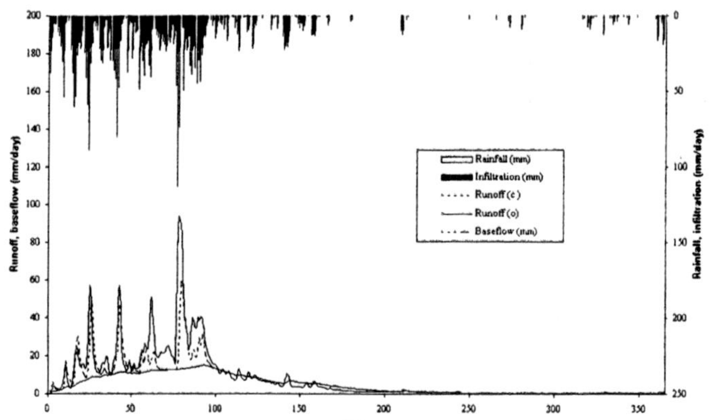

Figure 7.4. Model validation for Hemavati data (year 1977-78).

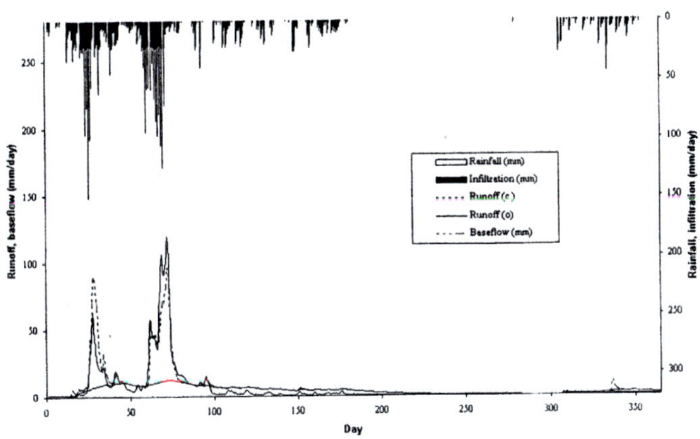

Figure 7.5. Model validation for Hemavati data (year 1978-79).

Table 7.11 Monthly volumetric statistics for calibration and validation for Case 1

Month	Rainfall (mm)	Rainfall-excess (mm)	F (mm)	O_b (mm)	ET (mm)	F_d (mm)	Runoff (comp.) (mm)	Runoff (obs.) (mm)	Rel. Error (%)
1974-75									
June	989.14	574.87	369.63	88.29	24.23	24.07	604.60	439.00	-37.7
July	535.01	112.44	346.43	299.26	17.26	13.02	470.25	451.52	-4.1
Aug	773.50	273.96	432.23	358.88	22.75	20.92	632.80	869.66	27.2
Sept	322.86	11.12	232.22	296.27	6.84	5.32	306.84	336.60	8.8
Oct	123.75	0.00	54.39	191.29	0.00	0.00	191.84	218.26	12.1
Nov	83.10	0.00	56.71	115.50	0.00	0.00	115.50	105.51	-9.5
Dec	0.00	0.00	0.00	49.65	0.00	0.00	49.65	50.47	1.6
Jan	0.00	0.00	0.00	18.45	0.00	0.00	18.45	26.12	29.4
Feb	0.00	0.00	0.00	6.43	0.00	0.00	6.43	14.87	56.8
Mar	0.10	0.00	0.00	2.82	0.00	0.00	2.82	13.50	79.1
Apr	99.58	0.00	52.12	24.80	0.00	0.00	24.80	15.04	-64.9
May	10.49	0.00	0.68	18.49	0.00	0.00	18.49	11.98	-54.3
1975-76									
June	374.42	74.50	239.97	94.91	11.72	8.30	168.14	53.99	-211.4
July	909.01	313.02	528.57	257.51	30.49	22.78	565.74	492.11	-15.0
Aug	526.50	149.94	311.39	353.08	17.99	16.52	507.48	536.13	5.3
Sept	310.12	102.95	161.60	290.35	11.46	11.30	394.90	343.13	-15.1
Oct	76.32	9.62	40.76	144.30	5.93	2.55	153.92	76.81	-100.4
Nov	202.87	15.22	150.14	104.39	4.57	4.46	119.53	101.37	-17.9
Dec	7.63	0.00	1.87	84.86	0.00	0.00	84.93	51.17	-66.0
Jan	0.00	0.00	0.00	31.59	0.00	0.00	31.59	16.15	-95.6
Feb	3.58	0.00	0.00	11.03	0.00	0.00	11.03	11.36	2.9
Mar	8.94	0.00	2.22	4.90	0.00	0.00	4.90	9.88	50.4

Table 7.11 Continued

Month	Rainfall (mm)	Rainfall-excess (mm)	F (mm)	O_b (mm)	ET (mm)	F_d (mm)	Runoff (comp.) (mm)	Runoff (obs.) (mm)	Rel. Error (%)
Apr	63.27	0.00	30.31	6.82	0.00	0.00	6.82	8.75	22.1
May	168.23	0.00	94.42	51.09	0.00	0.00	51.09	17.11	-198.6
1976-77									
June	671.51	246.26	376.10	125.67	21.55	21.32	342.89	211.76	-61.9
July	691.07	232.87	393.28	311.72	21.73	19.27	571.46	566.23	-0.9
Aug	308.81	0.00	216.82	311.87	0.00	0.00	314.01	319.89	1.8
Sept	358.53	56.46	239.07	250.42	13.95	9.99	304.84	366.29	16.8
Oct	273.39	37.60	187.66	231.01	12.50	9.90	270.56	233.72	-15.8
Nov	130.30	0.91	89.26	148.06	2.24	1.40	149.12	106.61	-39.9
Dec	0.00	0.00	0.00	83.95	0.00	0.00	83.95	45.39	-85.0
Jan	0.00	0.00	0.00	31.20	0.00	0.00	31.20	21.13	-47.7
Feb	0.43	0.00	0.00	10.90	0.00	0.00	10.90	6.82	-59.8
Mar	6.92	0.00	0.22	4.84	0.00	0.00	4.84	3.38	-43.2
Apr	71.42	0.00	44.35	10.13	0.00	0.00	10.13	2.32	-336.6
May	163.95	0.00	99.48	58.54	0.00	0.00	58.54	10.92	-436.1
1977-78									
June	752.12	255.2	437.81	141.39	19.05	19.26	391.72	418.65	6.4
July	729.17	211.53	467.6	330.3	22.74	20.34	534.61	639.78	16.4
Aug	874.27	289.77	520.97	408.61	24.63	23.26	687.66	1115.3 3	38.3
Sept	205.42	0	138.33	343.13	1.99	1.2	365.92	387.5	5.6
Oct	166.04	0	113.42	207.62	0	0	207.62	182.84	-13.6
Nov	55.52	0	32.33	130.11	0	0	130.11	86.5	-50.4
Dec	26.14	0	16.42	55.39	0	0	55.39	40.83	-35.7
Jan	0	0	0	29.23	0	0	29.23	26.13	-11.9
Feb	7.92	0	0.07	10.23	0	0	10.23	13.95	26.7
Mar	18.15	0	10.37	10.3	0	0	10.3	11.52	10.6
Apr	39.98	0	19.64	9.37	0	0	9.37	7.46	-25.6
May	67.14	0	37.06	16.57	0	0	16.57	6.22	-166.4
1978-79									
June	759.88	387.56	325.92	90.22	23.91	19.34	411.72	245.35	-67.8
July	566.18	146.22	340.88	285.07	11.95	9.83	420.66	362.66	-16.0
Aug	935.62	589.64	305.57	326.33	18.22	17.05	992.46	1180.8	15.9
Sept	216.55	14.55	148.78	208.12	5.52	2.72	222.88	116.68	-91.0
Oct	149.79	3.23	105.83	156.56	3.16	2.8	159.41	42.53	-274.8
Nov	167.08	0	103.15	130.88	0	0	131.26	39.27	-234.3
Dec	0	0	0	75	0	0	75	15.44	-385.8
Jan	0	0	0	27.87	0	0	27.87	10.9	-155.7
Feb	0	0	0	9.73	0	0	9.73	15.19	35.9
Mar	3.68	0	0.45	4.46	0	0	4.46	8.28	46.1
Apr	138.36	2.31	91.19	36.48	3.71	1.81	38.77	11.88	-226.3
May	127.21	17.12	72.3	71.82	4.8	2.78	88.93	12.79	-595.3

It is apparent from Table 7.11 that the relative errors computed for all 12 months during the whole 5-year period range from +1.6% to -595.3%. Here, '+' values indicate that the computed values are lower than the observed ones and vice versa for '–' values. It is also apparent from the table that the low runoff producing months (for example, January to May) yield relatively high relative errors. For the period of high flows (July 1 to September 30), the average '+' and '-' relative errors are of the order of 15.15% and

23.69%, respectively, which are well within the tolerance limit. On the other hand, these are computed respectively as 36.0% and 167.9% for the low runoff period (January to May), which are higher than those for the high flow period. Using monthly data, the efficiency (equation (7.51)) is computed as 91.02% for the calibration period (1974-75 to 1976-77) and 89.14% for the validation period (1977-78 to 1978-79). These efficiencies are quite high and indicate very good model performance during both periods. Similarly, the relative errors of the computed annual runoff volumes (Table 7.12) vary from +4.3% to −25.3%, which are within the acceptable limits of tolerance.

The model performance can be appreciated in view of (a) the limited number of model parameters (only four), (b) simplicity, and (c) no constraints imposed for matching the observed monthly or annual runoff volumes. In addition, there is little information available on baseflow and the lateral ground water interaction across the basin boundaries. The following discussion examines the impact of storm duration on model parameters.

Table 7.12 Annual volumetric statistic for calibration and validation periods

Year	Rainfall (mm)	Rainfall-excess (mm)	F (mm)	O_b (mm)	ET (mm)	F_d (mm)	Runoff (comp.) (mm)	Runoff (obs.) (mm)	Rel. error (%)
1974 -75	2937.53	972.39	1544.41	1470.13	71.08	63.33	2442	2553	4.3
1975 -76	2650.89	665.25	1561.25	1434.83	82.16	65.91	2100	1718	-22.2
1976 -77	2676.33	574.10	1646.24	1578.31	71.97	61.88	2152	1894	-13.6
1977 -78	2941.87	756.50	1794.02	1692.25	68.41	64.06	2449	2937	16.6
1978 -79	3064.35	1160.63	1494.07	1422.54	71.27	56.33	2583	2062	-25.3

7.3.4 EFFECT OF STORM DURATION ON MODEL PARAMETERS

It is apparent from Table 7.10 that all model parameters are greatly affected by the storm duration. The variation of these parameters with the storm duration is depicted in Figure 7.6. The figure shows a consistently increasing or decreasing trend with the storm duration, implying a distinct impact. The trends developed for S_o, F_c, and K_b are expressed, respectively, as

$$S_o = 5.073\,T_s - 3.39;\ r^2 = 0.99 \tag{7.54}$$

$$F_c = 52.295\,\text{Ln}(T_s) + 15.62;\ r^2 = 0.96 \tag{7.55}$$

$$K_b = -3.787\ \text{Ln}(T_s) + 30.40;\ r^2 = 0.48 \tag{7.56}$$

where T_s is the storm duration in days. Equations (7.54) and (7.55) are valid for T_s greater than or equal to 1 day. The first two trends exhibit a high value of r^2 indicating a strong dependence of S_o on T_s. A linear correlation of S_o with T_s infers S_o to be analogous to the ϕ-index for the watershed, consistent with the description of Schaake et al. (1996), Yu (1998), among others. The dependence of F_c on T_s is more convincing in a sense that it represents the volume of minimum infiltrated amount of water during the storm. For a given rainfall intensity, F_c will increase with the storm duration and vice versa. Parameter K_b, however, generally reduces with the increasing storm duration magnitude. The low value of r^2, however, reflects a poor correlation of K_b with T_s. Since $K \approx 2$ days for the Hemavati watershed, the impact of parameter K for the storm duration greater than 2 days is not evaluated, for its role apparently vanishes.

7.3.5 SENSITIVITY ANALYSIS

The sensitivity of the model parameters is evaluated by changing the parameter-values from their calibrated values. To this end, the parameters derived for case 1 were examined for investigating into the impact of the variation on the computed runoff values in the validation period (1978-79) in terms of efficiency (equation (7.51)). If efficiency increases, the computed values come closer to the observed ones and vice versa if the efficiency decreases with the varying parameter-values. However, the purpose of such an analysis lies in distinguishing parameters that are more sensitive for their cautious and judicious employment in the field.

The variation in efficiency with the variation in model parameter-values is depicted in Figure 7.7. It is apparent from this figure that the efficiency, in general, decreases if the parameter is either drastically increased or decreased from the calibrated one. Parameter S appears to be the least sensitive to variation. The variation in S_o-value from 10% to 400% of the calibrated value changes the efficiency in the range 80.85-76.42%. This parameter was further varied to 10 times the calibrated value and yielded an efficiency equal to 67.09%. It is because the modified model splits the total infiltration, F, into the one retained, F_d, and the other drained, F_c. The potential maximum retention, S, represents the potential volumetric space available for absorbing the dynamic infiltration amount whereas the static infiltration does not contribute to dynamic infiltration. It is again emphasized that the static portion actually contributes to percolation to meet the ground water table and finally appear as delayed flow or baseflow at the outlet of the basin.

It is seen from Figure 7.7 that parameter F_c is more sensitive than is S_o, for the efficiency varies rapidly with the change in the parameter value by the same extent. It can be seen that the variation of F_c from 10% to 300% leads to the efficiency in the range 81.15-64.45%, which exhibits a wider range than that due to S_o. However, these parameters are, as such, not much sensitive because the Hemavati watershed is a high runoff-producing watershed. As seen from Table 7.12, the runoff factor based on annual rainfall-runoff values can be computed in the range 0.65-1.00, signifying a high runoff factor. Thus, the losses due to evaporation, evapotranspiration, or dynamic infiltration are relatively minor compared to the surface runoff amount.

Parameter K is seen to be more sensitive than is S. The efficiency decreases faster than that due to any of the model parameters. Parameter K is, however, less sensitive than F_c when K increases beyond 75% of the calibrated value, and vice versa when K decreases from the calibrated value. Thus, it is more sensitive to reduction than increase beyond a 75%-value. A reduction or increase in K implies that the computed rainfall-excess is allowed to reach the outlet earlier or later than required and it results in the decrease in the model efficiency.

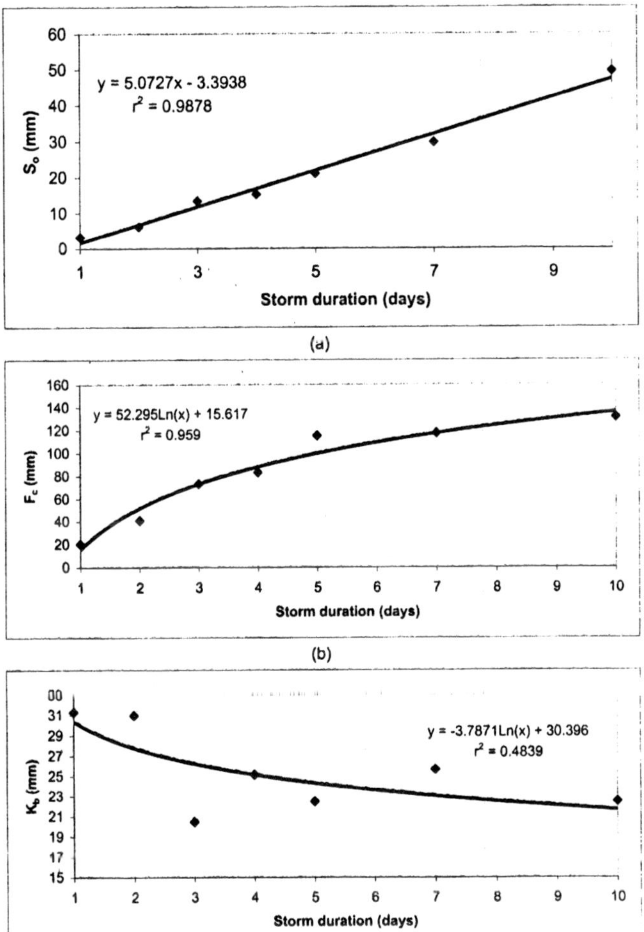

(c)

Figure 7.6. Impact of storm duration on (a) S_o, (b) F_c, and (c) K_b.

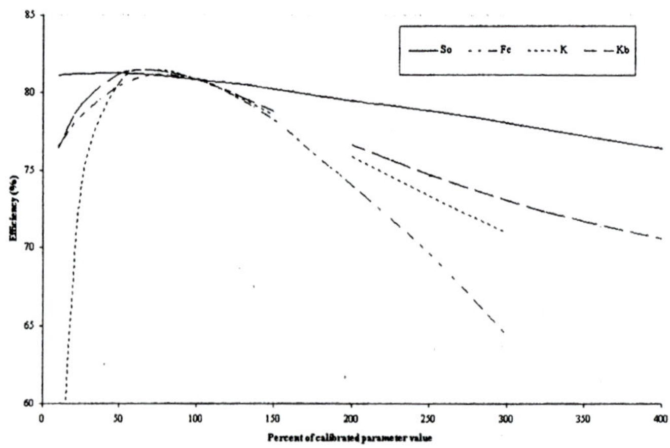

Figure 7.7. Sensitivity of model parameters for Case 1.

The model parameter K_b is apparently less sensitive than both F_c and K, but more than S_o. It is because the watershed, for the most part, is a hilly catchment and its outflow is largely governed by rainfall-excess that is, in general, influenced by F_c and its routing, where K influences most. The lesser amount of F_c than the direct surface runoff makes parameter K_b less sensitive than K.

7.4 Application of the Variations of the Modified SCS-CN Model

Here, it is appropriate to examine the applicability of the variants of the modified SCS-CN model, described in Chapter 4 (Table 4.2), to long-term hydrologic modelling. These models are described as Models 1 through 8 in the following discussion. The data applied to above case 1 (Table 7.10) is considered for calibration and validation. It is worth emphasizing that for the application of the models that exclude F_c (Models 1 through 4), the antecedent moisture, M, was taken equal to the total infiltration, F, of the previous day plus the antecedent moisture of the previous day less evapotranspiration, ET, computed as above in Section 7.2.3. The baseflow was taken equal to F less evapotranspiration. The results of calibration and validation are shown in Table 7.13. Since Model 1 does not account for M, F_c, and I_a, infiltration and evaporation losses are accounted for through S_o and, in turn, leads to the highest S_o-value equal to 458.98 mm. Furthermore, the calibrated model parameter K_b equal to 858.87 days (= 2.35 years) infers that the computed baseflow is not considered in simulation. Since the computed parameter K in all model applications is of the order of 0.623-2.295 days, it can be inferred that the application of Model 1 forces its parameters to assume their values such that the observed runoff is best simulated through the routed rainfall-excess. In other words, the evaporation or evapotranspiration losses are accommodated via these lost (or unaccounted) baseflows in modelling. The resulting calibration and validation efficiencies are, however, comparable with those derived using the modified SCS-CN model (equation (7.34)). Model 2 that accounts for M yields K_b equal to 29.71 days, which is a reasonable value, and reduces S_o to 109.03 mm from 458.98 mm. Model 3 which represents the existing SCS-CN method yields a low efficiency (= 73.22 %) in calibration. Inclusion of M in Model 3 to yield Model 4 further lowers the efficiency in calibration. Thus, based on efficiency, it can be inferred that the inclusion of I_a and M lowers the model efficiency.

Splitting F into F_c and F_d in Model 1 so that to yield Model 5 improves model efficiency significantly and it is equal to 84.60% in calibration and 85.61% in validation. However, parameter S_o assumes a value equal to 306.78 mm, which is comparable with that yielded by Model 1. Incorporation of I_a in Model 5 (to yield Model 6) leads to a significant reduction in model efficiency. The incorporation of M in Model 6 (to yield Model 7) improves model efficiency to 85.43 %. It implies that the problem basically lies with the incorporation of I_a rather than M. Therefore, the impact of λ, which governs I_a, is examined in what follows.

Table 7.13 Application results of the modified SCS-CN model and its seven variants

Model	Calibration					Validation
	Parameters				Efficiency	Efficiency
	S_o (mm)	F_c (mm)	K (days)	K_b (days)	(%)	(%)
1	458.98	-	1.928	858.87	79.47	82.16
2	109.03	-	0.623	29.71	78.16	73.77
3	50.34	-	2.295	0.20	73.22	80.56
4	49.95	-	1.200	0.20	71.95	80.31
5	306.78	9.84	1.184	23.64	84.60	85.61
6	3.09	20.26	1.659	31.62	79.64	80.84
7	333.30	11.35	1.049	20.49	85.43	86.35
8*	262.45	14.50	0.919	16.31	85.70	86.81

*Model 8 is applied with constant λ = 0.01

Since Model 4 also incorporates M besides I_a, it is used for evaluating the impact of λ on the model parameters and the resulting model efficiency. It is noted that in these applications, λ was taken to be constant, rather than varying according to PANC, as described earlier. The variation of the computed parameters and efficiencies with varying λ-values are depicted in Figure 7.8. It is apparent from Figure 7.8a that parameter S decreases consistently with the increasing value of λ, for the reason of maintaining a certain amount of I_a. The higher the λ-value, the lower will be the S_o-value and vice versa. It is apparent from Figure 7.8b that there has been only a marginal change in the computed K-values, for these are in the range of 1.603-1.712 days. However, K_b varies significantly with λ; as λ increases K_b reduces. Since S_o decreases with an increase in the λ-value, the computed infiltration F will also decrease with decreasing S_o and, in turn, will lead to a reduction in baseflow. The reduced baseflow will have less impact on the total computed runoff amount and, therefore, perhaps yield reduced K_b-values (Figure 7.8c) for

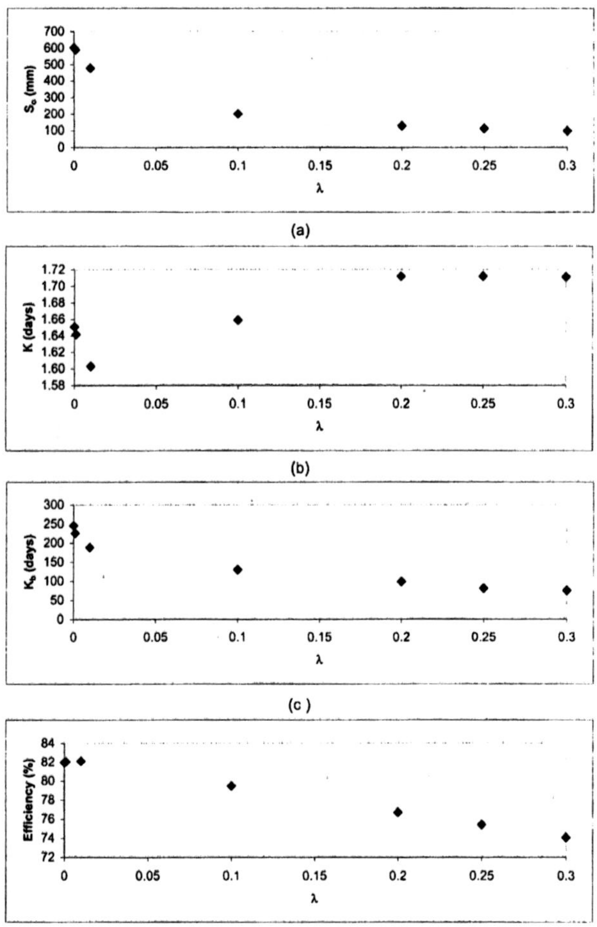

(d)

Figure 7.8. Variation of (a) S_o, (b) K, (c) K_b, and (d) efficiency with l of Model 4.

maximizing the efficiency. It implies that higher K_b values have less impact on the computed total runoff than in the otherwise situation. Figure 7.8d exhibits that as λ increases, the efficiency decreases and vice versa. The maximum efficiency is obtained when it is of the order of 0.01.

Since $\lambda = 0.01$ yielded maximum efficiency in the above trials, Model 8 (equation (7.34)) was applied again to the data employed in case 1 (Table 7.10), but for $\lambda = 0.01$, rather than varying λ with the above described PANC. The resulting efficiencies are shown in Table 7.13. These are 85.70% in calibration and 86.81% in validation, marginally improved over those from the application of Model 7. These applications, however, show that, for the Hemavati watershed, $S_o \approx 30$ cm (≈ 1 foot), $F_c \approx 10$ mm (per day) (≈ 0.016 inch/hr), $K \approx 1$ day, $K_b \approx 18$ days, and $\lambda \approx 0.01$. The rainfall-runoff data thus characterize the soils of an agricultural watershed to fall in the hydrologic group D (based on F_c-value), for which the absolute potential maximum retention (= S_o) is approximated as equal to 1 foot, which is equivalent to 3.33 feet of the soil depth for the soil porosity n = 0.3 (Singh and Yu, 1990). A depth of 3.33 feet is generally taken as the root zone depth that plays a significant part in evapotranspiration and, in turn, in the infiltration and soil moisture accounting. The computed values can be physically realized from the watershed that is characterised as: (a) land use: forests (12%), coffee plantations (29%), and agricultural lands (59%) and (b) soils: red loamy soil (67%) and red sandy soil (33%). Thus, the watershed is primarily characterised as an agricultural watershed of low infiltration capacity.

CHAPTER 8

TRANSPORT OF URBAN POLLUTANTS

The environmental pollution has been a matter of concern since long. It has affected almost all phases of the modern life the world over, especially more in the past two decades. Consequently, it has attracted globally the attention of scientists, engineers, hydrologists, planners, economists, sociologists, and environmentalists. Water pollution can occur above, on, or below the land surface in several forms. This chapter, however, deals specifically with the pollution of the surface water and its transport in urban areas during the processes of rainfall-runoff, river flow, and snowfall.

Urban nonpoint sources have been widely identified as major sources of pollution of surface water bodies. The 1988 Report of Congress (U.S. EPA, 1990) stated that urban storm water runoff is the fourth most extensive cause for the deterioration of the water quality of the lakes in the country. Typical urban land uses which contribute to pollution are residential land, commercial land use, industrial land use, open lands such as parks and golf courses, and land uses developed for transportation such as airports and highways (Novotony and Olem, 1994).

Nonpoint source pollution transport is generally modelled using the linear Fick's first law. In an attempt to develop mass rating curves similar to rating curves occurring in open channel hydraulics, an analysis is made for the occurrence of loops in mass rating curves, similar to looped rating curves occurring in open channel flows under dynamic wave conditions, for investigating into the efficacy of the Fick's law.

Since the book focuses on the SCS-CN method, it is appropriate to explore the possibilities for the employment of the method to pollution problems. The SCS-CN concept has not yet been employed in dealing with these problems. It has, however, been used in computation of the annual runoff from the annual rainfall using a procedure described in Chapter 7. These annual runoff values are used for determining the annual amount of pollutant load from urban areas using empirical runoff-pollutant load formulae. In this chapter, the SCS-CN concept is employed to compute event-based pollutant load from a micro urban watershed. A partitioning curve number (PCN) method analogous to the SCS-CN method is suggested for computation of the pollutant load generated by urban activities and transported by rainfall- or snowfall-generated runoff and river flows.

Thus, the objective of this chapter is three fold. First, (a) it presents the development of mass rating curves for computation of the mass of the dissolved, particulate, and (dissolved + particulate bound) metals transported by the rainfall-generated runoff from the pavement (or overland); (b) it describes the phenomenon of metal transport analogous to the wave propagation in wide open channels introducing a concept of equivalent mass depth of flow; and (c) it describes the hysteretic behaviour of metal transport using the wave analogy and extends the proportionality concept of the popular SCS-CN method for determining the potential maximum mass of heavy metals transported by a rain storm. Second, this chapter introduces (a) an SCS-CN inspired method for partitioning of heavy metals into dissolved and particulate bound forms and (b) an improved surrogate of the partitioning indices, frequently used in environmental hydrology and chemical engineering. Finally, a discussion is presented for the use of the SCS-CN method in determination of annual pollutant loads.

8.1 Heavy Metals

Urban highway storm water often contains elevated levels of heavy metals and particles (Foster and Charlesworth, 1996; Sansalone and Buchberger, 1997; Sansalone et al., 1998; and Li et al., 1999). Heavy metals like lead (Pb), cadmium (Cd), copper (Cu), nickel (Ni), zinc (Zn), and chromium (Cr) can pose acute and chronic threats to receiving water bodies and soils, for the dissolved fraction of these heavy metals has the potential for acute and long-term chronic toxicity for aquatic life. Unlike organic compounds, heavy metals are not degraded in the environment. In the United States of America, with recent signature of the Final Rule for Phase II Storm Water National Pollutant Discharge Elimination System (NPDES) regulations on 29 October 1999, issues related to heavy metals in roadway storm water have now come into direct focus at the national regulatory level. In the urban highway environment, heavy metals are generated primarily from the abrasion of metal-containing vehicular parts, including the abrasive interaction of tires against pavement, leaching of heavy metals from infrastructure, oil and grease leakage and industrial discharges (Armstrong, 1994; Ball et al., 1991; Lygren et al., 1984; Muschak, 1990). Pavement and vehicle part abrasion are sources of solids that can range from colloidal to gravel size particles abraded from pavement (Hamilton and Harrison, 1991). These solids wash off in lateral pavement sheet flow during rainfall-runoff events as part of heterogeneous mixture of heavy metals, organic compounds, oil and grease, and partition into particulate bound metals and dissolved metals (Sansalone and Buchberger, 1997).

Traffic and roadway maintenance activities are major sources of particulates and solids that are either initially abraded heavy metals that dissolve in the rainfall of low pH and low alkalinity or solids that act as a substrate to which dissolved heavy metals can partition (Sansalone and Buchberger, 1997). The sources of these solids and abraded heavy metals include construction/re-construction activities, maintenance operations including application of grit and salt (coated with anti-caking materials containing heavy metals) in northern climates, pavement degradation, littering, spillage, and vehicular-infrastructure abrasion. Within the category of vehicular-infrastructure abrasion, tire-pavement interaction generates a significant amount of particulate matter. Abraded pavement accounts for 40-50% and abraded tires account for 20-30% of the total particulate matter generated (Kobriger and Geinpolos, 1984). Most of these abraded solids and particulates contain heavy metals. Abraded tire-particles have a mean diameter of 20-μm, a density of 1.5-1.7 g/cm^3, and are the major sources of Zn and Cd (Sansalone and Tribouillard, 1999).

Annual heavy metal, total suspended solids (TSS), and chemical oxygen demand (COD) loadings from urban interstate highway pavements during rainwater flows have been shown to equal or exceed annual loadings and flows from untreated domestic wastewater for a given urban area (Sansalone et al., 1998). The urban interstate pavement alone typically constitutes less than 10% of the total pavement area in an urban area for generating a significantly disproportionate pollutant load, especially heavy metals, and the urban interstate combined with the major arterial pavement area constitute less than 10 to 15% of the total pavement area. It is reported that the storm water runoff from urban pavement carries greater pollutant loadings to receiving water bodies than point-source wastewater discharges from the same urban area (Klein et al., 1974).

Heavy metal concentrations in urban runoff can range from essentially background levels to milligram/liter levels during a single rainfall-runoff event. Since the concentration of hydrologically transported constituents can vary by orders of magnitude during a single runoff event, a single index designated as event mean concentration (EMC) is often used to characterize concentrations. EMC represents a flow average concentration for the event and, at a particular site, it can vary depending on the average daily traffic, vehicles during storm, rain intensity, previous dry days, and climate conditions (Sansalone and Buchberger, 1997; Sansalone et al., 1998; and Li et al., 1999). While the benefits of EMC data include simplicity and economics, drawbacks of EMC data include their inability to quantify the temporal and hysteretic variation of water quality parameters. Thus, one of the objectives of this chapter is to describe the hysteretic variation of the dissolved and particulate bound metals, develop normal rating curves, analogous to those used in channel hydraulics.

8.2 Metal Partitioning

In pavement rainfall-runoff or snowmelt, river flow, or any aqueous system, a temporal partitioning between heavy metals in solution and solids occurs irrespective of their being in suspension (total suspended solids (TSS) or volatile suspended solids (VSS)) or in the form of settleable solids, which eventually form a part of the fixed or mobile bed load. This partitioning includes specific mass transfer mechanisms of sorption, ion exchange, and surface complexation with both organic and inorganic sites on the solid matter. These partitioning reactions are generally non-reversible between the solid phase and soluble phase concentrations (Glenn et al., 2001).

The total concentration (c_T) of a heavy metal is the sum of the dissolved (c_d) and the particulate bound (c_p) concentrations. Expressed mathematically,

$$c_T = c_d + c_p \tag{8.1}$$

For practical purposes, the soluble or dissolved fraction is assumed to pass through the 0.45 μm membrane filter and, therefore, it contains both dissolved and part of the colloidal-bound heavy metals. The solid phase concentration c_p is defined as the product of the heavy metal concentration on the solid phase, c_s, expressed in mass/mass of solids and the concentration of the adsorbing solid material in the aqueous system, m, typically measured as either total solids (TS) or TSS in terms of mass/volume of aqueous solution. Expressed mathematically,

$$c_p = c_s\, m \tag{8.2}$$

Conceptually, when the rates of sorption and desorption are equal, the process is said to be in the concentration (instantaneous) equilibrium between the dissolved and solid phase concentrations of a heavy metal. The ratio of these phases is defined as the partitioning coefficient, K_d, for a particular heavy metal at a particular pH and Redox level (Glenn et al., 2001), expressed as

$$K_d = c_p/c_d \tag{8.3}$$

Coupling of equation (8.3) with equation (8.1) yields

$$f_d = \frac{D}{D+P} = \frac{c_d}{c_T} = \frac{1}{1+K_d} \tag{8.4}$$

and

$$f_p = \frac{P}{D+P} = \frac{c_p}{c_T} = \frac{K_d}{1+K_d} \tag{8.5}$$

where D is the dissolved mass of a heavy metal (mg), P is the particulate-bound mass of a heavy metal (mg), f_d and f_p are the dissolved and particulate-bound fractions, respectively. If $f_d > 0.5$, the metal is assumed to be mainly in the dissolved form. Otherwise, it is in the particulate bound form. For dimensional homogeneity, K_d is expressed as non-dimensional in equations (8.3)-(8.5) though it is usually expressed in l/kg. The larger the K_d value, the greater the partitioning of a heavy metal to the solid phase and vice versa. In pavement runoff, K_d–values generally range 10-10^5 l/kg in rainfall-runoff and these are in the typical range of 10^3-10^5 l/kg in snowmelt runoff. Much longer residence times lead to the dissolved fraction to be very low (Sansalone and Buchberger, 1997).

Since the values of K_d vary in a broad range, it is often difficult to physically realize the extent of partitioning for a given value of K_d or other partitioning indices, viz., f_p and f_d. Therefore, there exists a need to develop a parameter that is non-dimensional and varies in a relatively narrow range, such as the curve number (CN). CN varies from 0 to 100 and is widely used in characterizing watersheds in rainfall-runoff modelling. There also exists a need for an improved alternative to the exponentially decaying function describing the first order chemical process, such as the concept of the SCS-CN method (SCS, 1956), which is the generalization of the first-order rainfall-runoff process, as discussed in Chapter 3. Thus, the objective of this component of the chapter is to partition heavy metals into dissolved and particulate bound forms of heavy metals in rainfall-runoff, snowmelt, and river flow systems using the basic proportionality concept of the SCS-CN method; introduce a parameter partition curve number (PCN) analogous to the runoff curve number (CN) as an improved surrogate of partitioning indices K_d, f_d, and f_p; to correlate the parameter potential maximum desorption, ψ, with the ambient storm runoff alkalinity and pH characteristics in all the three aqueous systems; and to finally characterize these systems for their partitioning behavior using the PCN as an index.

8.3 Metal Transport

8.3.1 RATING CURVES IN OPEN CHANNEL HYDRAULICS

The rating curves in open channel hydraulics provide vital hydrologic information, the discharge, from measured stage for use in planning, design, operation, and management of water resource projects. The rating curves can be either single-valued (equation (1.81)) or double-valued depending on the flow behaviour. The single-valued rating curves are much more popular in hydrologic practice and research than the double-valued rating curves also known as looped or hysteretic rating curves. The hysteresis-based hydraulic literature reveals (Henderson, 1966; Cunge et al., 1980; Ponce, 1989; Mishra and Seth, 1994, 1996; Jain et al., 1996; Mishra et al. 1997; Mishra and Singh, 1999) that hysteresis is a manifestation of channel storage and/or channel roughness and directly relates to the flood wave attenuation. Since Henderson (1966), the hysteretic flow behaviour has been a topic of continued research (Cunge et al., 1980; French, 1985; Ponce, 1989; Perumal, 1994; Mishra and Seth, 1994, 1996; Jain et al., 1996; Mishra et al. 1997; Mishra and Singh, 1999; Ponce and Lugo, 2001). Using the non-dimensional hysteresis as an index, Mishra and Singh (1999) described flood waves (Table 8.1) in artificial as well as natural channels. In Table 8.1, F_0 is the Froude number and $\hat{\sigma}$ is the dimensionless wave number (Ponce and Simons, 1977). Hysteresis, η, is computed as (Mishra and Seth, 1994; 1996; Mishra et al., 1997; Jain et al., 1996; Mishra et al., 1996; and Mishra and Singh, 1999):

$$\eta = \frac{1}{2}\int_0^T \left(q\frac{dh}{dt} - h\frac{dq}{dt}\right)dt \tag{8.6}$$

where

$$h = \frac{H - H_{min}}{H_{max} - H_{min}} \tag{8.7}$$

and

$$q = \frac{Q - Q_{min}}{Q_{max} - Q_{min}} \tag{8.8}$$

Here T is the time period of the flood wave (= time of rise plus the time of recession); h is the dimensionless stage (a function of time, t); H_{max}, H_{min}, and H are, respectively, the maximum, minimum, and time varying computed depths at a site of interest. Similarly,

Q and q stand for discharge and dimensionless discharge, respectively. Using the analogy of Hagen's dam factor, Mishra and Seth (1996), among others, described hysteresis as an index of energy expended by the flood wave during its time period. This definition is identical to that given by Hayden et al. (1986) who described elastic hysteresis as the energy dissipated in a single cycle, which is equal to the area under the loading curve minus the area under the unloading curve. Mishra et al. (1997) described hysteresis-based studies to be useful for (i) identifying waves in natural channels for the application of a suitable simplified routing model; (ii) locating a fictitious downstream boundary for improving National Weather Service's Dam Break Flood Forecasting Model's (NWS DAMBRK) (Fread, 1984; 1985) results; (iii) computing the time base of the flood hydrograph; and (iv) processing the discharge hydrographs derived using an observed rating curve for phase difference. The process of metal transport is described utilizing this hysteresis concept.

Table 8.1 Criteria for wave types

Wave type	Hysteresis (η) (dimensionless)	Wave number, $\hat{\sigma}$, and $\hat{\sigma} F_o$ (dimensionless)	Phase difference (ϕ) (radians)
Kinematic	$\eta < 0.025$	$\hat{\sigma} \leq 0.03$	$\phi < 0.03$
Diffusion	$0.025 \leq \eta \leq 0.10$	$\hat{\sigma} F_o \leq 0.462$	$0.03 \leq \phi \leq 0.13$
Dynamic	$\eta > 0.10$	$\hat{\sigma} F_o > 0.462$	$\phi > 0.13$

8.3.2 GOVERNING FLOW AND METAL TRANSPORT EQUATIONS

The process of non-point source pollution is usually simulated using the following second-order partial-differential equation (Martin and McCutcheon, 1999):

$$\frac{\partial \dot{C}}{\partial t} + K\frac{\partial \dot{C}}{\partial x} = D'\frac{\partial^2 \dot{C}}{\partial x^2} \tag{8.9}$$

where $\dot{C}$ is the concentration; K and D′ are the diffusion (or chemical dispersion) parameters; and x and t are the space and time coordinates, respectively. Equation (8.9) represents the form of the diffusion wave equation, frequently used in open channel flow (Ponce, 1989; Singh, 1996; Rutschmann and Hager, 1996) and expressed as:

$$\frac{\partial Q}{\partial t} + c\frac{\partial Q}{\partial x} = D''\frac{\partial^2 Q}{\partial x^2} \tag{8.10}$$

where Q is the discharge (= uA), c is the diffusion wave celerity, D″ is the diffusion coefficient, u is the flow velocity, and A is the area of cross-section. Parameters c and D″ are frequently described as (Ponce, 1989; Mishra and Singh, 1999):

$$c = dQ/dA \tag{8.11}$$

and

$$D'' = Q_o/(2s_oW) \tag{8.12}$$

where Q_o is the reference flow, s_o is the bed slope, and W is the channel width that is equal to unity for a unit width rectangular channel. Equation (8.11) is the popular Seddon speed formula (Appendix C). Since equations (8.9) and (8.10) parallel each other, it is possible to describe the non-point source pollutant transport in overland flow (or impervious urban sheet flow), analogous to the flow in wide rectangular open channels, if $\dot{C}$ in equation (8.9) is replaced by a term equivalent to the product of flow depth and velocity. In a similar vein, the dynamic process of metal transport can be described using the equations analogous to the one-dimensional dynamic wave equations, described by the continuity and momentum equations, which respectively are:

$$\frac{\partial h}{\partial t} + u\frac{\partial h}{\partial x} + h\frac{\partial u}{\partial x} = 0 \tag{8.13}$$

and

$$\frac{\partial u}{\partial t} + u\frac{\partial u}{\partial x} + g\frac{\partial h}{\partial x} + g(s_f - s_o) = 0 \tag{8.14}$$

valid for a unit-width wide rectangular channel. Here, h is the depth of flow, g is the gravitational acceleration, s_o is the bed slope, and s_f is the friction slope described by Manning's equation (a variation of equation (1.79)):

$$u = \frac{1}{n}R^{2/3}s_f^{1/2} \tag{8.15}$$

where R is the hydraulic radius (= h for a wide rectangular channel) and n is Manning's roughness coefficient. Due to their highly non-linear nature, these equations do not yield an exact analytical solution and approximate methods (for example, linear and numerical methods) are, therefore, often resorted to. Ponce and Simons (1977) and Menendez and Norscini (1982) solved these equations for wide rectangular channels using a linear perturbation theory. Ponce and Simons (1977) derived propagation characteristics of kinematic, diffusion, steady-dynamic, gravity, and dynamic waves in open channel flow. These characteristics included wave celerity and attenuation defined by logarithmic decrement. Menendez and Norscini (1982) characterized these waves by wave celerity and phase difference and showed a link between logarithmic decrement and phase difference. A more general relationship between logarithmic decrement and phase speed was derived by Kundzewicz and Dooge (1989). Singh (1996) provided a good overview of these works. The work of Price (1973; 1985) is also notable in that he provided a more generalized solution of the St. Venant's equations as well as a relationship between attenuation and wave speed. Mishra and Singh (2001) described the wave characteristics, using the above hysteresis concept, as follows (see Appendix C):

$$\hat{c}_r = \pm\frac{1}{\sqrt{2}}\left\{\left[\left(\frac{1}{F_o^2}-\zeta^2\right)^2+\zeta^2\right]^{1/2}+\frac{1}{F_o^2}-\zeta^2\right\}^{1/2} \tag{8.16}$$

$$\delta = -2\pi\frac{\zeta \mp \frac{1}{\sqrt{2}}\left\{\left[\left(\frac{1}{F_o^2}-\zeta^2\right)^2+\zeta^2\right]^{1/2}-\frac{1}{F_o^2}+\zeta^2\right\}^{1/2}}{1 \pm \frac{1}{\sqrt{2}}\left\{\left[\left(\frac{1}{F_o^2}-\zeta^2\right)^2+\zeta^2\right]^{1/2}+\frac{1}{F_o^2}-\zeta^2\right\}^{1/2}} \tag{8.17}$$

$$\phi = \pi\ \tan^{-1}\frac{\zeta \mp \frac{1}{\sqrt{2}}\left\{\left[\left(\frac{1}{F_o^2}-\zeta^2\right)^2+\zeta^2\right]^{1/2}-\frac{1}{F_o^2}+\zeta^2\right\}^{1/2}}{\pm \frac{1}{\sqrt{2}}\left\{\left[\left(\frac{1}{F_o^2}-\zeta^2\right)^2+\zeta^2\right]^{1/2}+\frac{1}{F_o^2}-\zeta^2\right\}^{1/2}} \tag{8.18}$$

where $\hat{c} = c/u_o$ and $\hat{c}_r$ represents dimensionless relative celerity, ζ is the kinematic wave number ($= 1/\hat{\sigma}F_o^2$), δ is the logarithmic decrement, u_o is the normal flow velocity, and the dimensionless wave number is described as:

$$\hat{\sigma} = (2\pi/L)\,(h_o/s_o) \tag{8.19}$$

where h_o is the normal flow depth and L is the wave length. These derivations conform to those given by Ponce and Simons (1977) and Menendez and Norscini (1982) who have described in detail the wave characteristics employing the above derivations.

8.3.3 CONCEPT OF EQUIVALENT MASS DEPTH OF FLOW

The above dynamic wave equations (equations (8.13) and (8.14)) are the result of mass and momentum conservation. To describe the mechanism of metal transport analogous to the wave propagation in open channels, it is necessary that the flow depth, h, be replaced by a term describing the mass of the transported metal in such a way that the mass and momentum of flow in a channel remain conserved. To this end, a hypothetical flow in a real channel of equivalent mass depth of flow, h_e, is considered and described as

$$h_e(t) = \frac{m(t)}{\rho A_w} \tag{8.20}$$

where $h_e(t)$ is the equivalent mass depth of flow at time t, m(t) is the mass of the metal transported at time t, ρ is the metal density, and A_w is the area of the overland flow. In the present hypothetical flow, the flow velocity, u, is taken as equal to the velocity of the overland flow (direct surface runoff) $[L^3T^{-1}]$ divided by the area of the watershed $[L^2]$, for the flow velocity directly impacts the mass transport. Thus, u and h_e act exactly analogously to the two orthogonal components, u and h, respectively, of the full dynamic wave equations (equations (8.13) and (8.14)), leading to the description of characteristics of the metal transport phenomenon by equations (8.16)-(8.18), including the establishment of mass rating curves for m (or h_e) (equation (8.20)) and u. It is noted that in open channel hydraulics, rating curves are used to convert stage (h) to flow velocity (u) (or discharge) whereas in non-point source metal transport (for example, on an overland flow plane where the flow is described primarily by sheet flow), these can be of use in determination of the amount of mass transported by the known velocity of flow that is taken equivalent to the rate of rainfall-excess or direct runoff.

8.3.4 RELATION BETWEEN CONCENTRATION AND EQUIVALENT MASS DEPTH

The relation described by equation (8.20) can be shown equivalent to the concentration of flow that is more convincingly used in environmental engineering literature as below.

Concentration is defined as

$$\dot{C} = \frac{\text{Mass of the transported metal}}{\text{Volume of the solution}} \tag{8.21}$$

or, alternatively,

$$\dot{C} = \frac{\text{Mass of the transported metal}}{\text{Metal density x Volume of the solution}} \tag{8.22}$$

$\dot{C}$ in equation (8.21) has the dimension of $[ML^{-3}]$ whereas in equation (8.22), it is non-dimensional. The non-dimensional $\dot{C}$ is generally expressed in parts per million. Equation (8.22) can also be written in terms of the rates of mass and flow as:

$$\dot{C} = \frac{\text{Mass of the transported metal}}{\text{Metal density x Rate of direct surface runoff x } \Delta t} \tag{8.23}$$

where Δt is the time interval, the direct surface runoff is the solution, and the ratio of transported mass to Δt represents the rate of mass transported. Division of the numerator and denominator of the right-hand side of equation (8.23) by the area of the watershed yields

$$\dot{C} = [m / (\rho A_w)] / (u\Delta t) \tag{8.24}$$

or

$$\dot{C} = h_e / (u\Delta t) \tag{8.25}$$

which describes $\dot{C}$ as the ratio of the equivalent depth of mass [L] to the flow displacement [L], which, however, assumes the rate of rainfall-excess equivalent to the overland flow velocity.

8.4 SCS-CN Analogy for Metal Partitioning

In chemical and environmental engineering, the rate of metal desorption is taken proportional to the amount of solute adsorbed (O'Connor, 1985), which describes the first order reaction, activity, or process. Expressed mathematically,

$$\frac{dc_d}{dt} = -\beta c_d \tag{8.26}$$

where β is the decay factor. Equation (8.26) is solved for c_d as

$$c_d(t) = c_T e^{-\beta t} \tag{8.27}$$

where $c_d = c_{do} = c_T$ at $t = 0$, valid for the condition that the total mass is constant and does not decay with time. Equation (8.27) is analogous to the process of infiltration (equation (1.58) for $f_c = 0$):

$$f = f_o e^{-\alpha t} \tag{8.28}$$

where f is the infiltration rate, f_o is the initial infiltration rate, and α is the infiltration decay factor. The second equation describing the process is the law of universal mass conservation (equation (8.1)). Coupling equation (8.1) with equation (8.27) leads to

$$c_p = c_T(1 - e^{-\beta t}) \tag{8.29}$$

Equation (8.29) parallels the variant of the Mockus (1949) equation (equation (3.1) for initial abstraction $I_a = 0$):

$$Q = P(1 - e^{-\alpha t}) \tag{3.15}$$

c_T is analogous to the precipitation amount P, c_p to Q, and metal desorption parameter β to infiltration decay parameter α. Therefore, c_d is analogous to F and f_p to the runoff factor, C.

Equation (8.1) can be expressed in terms of the mass of dissolved, particulate-bound, and total mass or volume, as above. Similar to the SCS-CN proportionality hypothesis (equation (2.2) for $I_a = 0$):

$$\frac{Q}{P} = \frac{F}{S} \tag{2.5}$$

the following proportionality can be hypothesized as:

$$\frac{c_p}{c_T} = \frac{c_d}{\psi} \tag{8.30}$$

where ψ is a parameter, defined as the potential maximum desorption similar to the SCS-CN parameter, potential maximum retention (S). Equation (8.30) states that the ratio of the particulate-bound concentration to the total concentration is equal to the ratio of the dissolved concentration to the potential maximum desorption. This proportional equality is analogous to the SCS-CN's $C = S_r$ concept (Chapters 3 and 4), C corresponds to f_p that can be defined as the degree of adsorption and S_r to the degree of desorption. The implication of such an equality can be described as follows.

In a chemical activity, the instantaneous equilibrium condition (O'Connor, 1985) is attained when the rate of the formation of products equals the rate of the generation of reactants. Similarly, equation (8.30) is a kind of instantaneous equilibrium in terms of the equality of the degree of adsorption or sorption to the degree of desorption. Table 8.2 summarizes the analogous terms of rainfall-runoff and metal partitioning processes. Analogous to equation (2.5), as $c_p \to c_T$, $c_d \to \psi$ and as $c_p \to 0$, $c_d \to 0$. Combination of equations (8.1) and (8.30) leads to

$$c_p = \frac{c_T^2}{c_T + \psi} \tag{8.31}$$

Similar to the determination of S (Chapter 2), ψ can be derived from equation (8.31) as

$$\psi = \frac{c_T c_d}{c_p} = \frac{c_T}{K_d} \tag{8.32}$$

Equation (8.32) can also be expressed in terms of f_d and f_p, respectively, as:

$$\psi = \frac{c_T^2 f_d}{c_p} \tag{8.33a}$$

and

$$\psi = \frac{c_d}{f_p} \tag{8.33b}$$

It is noted that similar to the definitions of the SCS-CN variables, c_T, c_d, and c_p represent the cumulative concentration or masses or volumes of total, dissolved, and particulate-bound metals. Furthermore, following the method of generalization discussed in Chapter 3, it is possible to show that equation (8.31) is an approximate form of equation (8.29).

Table 8.2 Summary of analogous terms used in rainfall-runoff and metal partitioning processes

Sl.	Rainfall-runoff process		Metal partitioning process	
No.	Variable/parameter	Symbol/relation	Variable/parameter	Symbol/relation
1	Infiltration rate	f	Desorption rate	-
2	Uniform rainfall intensity	i_o	Uniform rate of input mass	c_o
3	Infiltration decay factor	α	Desorption decay factor	β
4	Cumulative rainfall	P	Total metal amount	c_T
5	Cumulative runoff	Q	Cumulative particulate bound metal amount	c_p
6	Cumulative Infiltration	F	Cumulative dissolved metal amount	c_d
7	Potential maximum retention	S	Potential maximum desorption	ψ
8	Initial abstraction	I_a	Initial flush	i_f
9	P - i_o relation	$P = i_o t$	c_T - c_o relation	$c_T = c_o t$
10	i_o-α-S relation	$i_o = \alpha S$	c_o-β-ψ relation	$c_o = \beta\psi$
11	Curve number	CN	Partition curve number	PCN
12	Runoff factor	C = Q/P	Particulate bound fraction or degree of adsorption	$f_p = c_p/c_T$
13	Degree of saturation	S_r	Degree of desorption	-
14	Mass conservation	P = F + Q	Mass conservation	$c_T = c_p + c_d$
15	Proportionality concept	Q/P = F/S	Proportionality concept	$c_p/c_T = c_d/\psi$
16	SCS-CN method	$Q = \frac{(P - I_a)^2}{P - I_a + S}$	PCN-method	$c_p = \frac{(c_T - i_f)^2}{c_T - i_f + \psi}$
17	I_a-S hypothesis	$I_a = \lambda S$	i_f-ψ hypothesis	$i_f = \lambda_f \psi$
18	Applicability of SCS-CN method	$P \geq I_a$	Applicability of PCN method	$c_T \geq i_f$
18	S-CN mapping	CN = 1000/(10+S)	ψ-PCN mapping	PCN = 1000/(10+ψ)

Similar to the derivation of equation (3.10):

$$\alpha = \frac{i_o}{S} \tag{3.10}$$

which is valid for $f_c = 0$, it is possible to derive an analogous relation:

$$\beta = \frac{c_o}{\psi} \tag{8.34}$$

Equation (8.34) assumes a linear growth of c_T with time, similar to P (= $i_o t$, where i_o is the uniform rainfall intensity), as

$$c_T = c_o t \tag{8.35}$$

where c_o is the uniform rate of input mass and is analogous to the uniform rainfall intensity, i_o.

The curve number CN is described as (see Chapter 3):

$$CN = \frac{F}{S} x100 \tag{3.137}$$

Similarly, a partition curve number (PCN) can be defined as

$$PCN = \frac{c_d}{\psi} x100 \tag{8.36}$$

Thus, PCN represents the percent degree of desorption or the percent degree of adsorption from equation (8.30). Thus, analogous to the runoff curve number (CN), PCN can also be defined as the adsorption curve number, for it indicates the potential of the medium for adsorption or formation of the particulate bound metal. The greater the PCN, the greater will be the production of particulate bound metal and vice versa. To bring PCN to a common base amount of input material (analogous to the base precipitation amount equal to 10 inches for the SCS-CN method (Chapter 3) it can be generalized as

$$PCN = \frac{c_d}{\psi} 100 = \frac{c_p}{c_T} 100 = \frac{100 c_T}{c_T + \psi} \tag{8.37}$$

For c_T = 10 units, equation (8.37) reduces to

$$\mathrm{PCN} = \frac{1000}{10+\psi} \tag{8.38}$$

Equation (8.38) can also be expressed in terms of the frequently used partitioning parameter, K_d, using equation (8.32) as

$$\mathrm{PCN} = \frac{100\,K_{d(10)}}{1+K_{d(10)}} \tag{8.39}$$

where $K_{d(10)}$ refers to K_d for 10 units of c_T. Thus, PCN is defined as the percent degree of adsorption of the 10-units of a metal. The advantage of developing such a curve number for metal partitioning is that the PCN defines the adsorption capacity of the 10 units of heavy metal and has the efficacy to distinguish one metal from the other on the basis of the adsorption capacity of solution media; the higher the PCN, the lesser will be the adsorption capacity and vice versa. K_d is distinguished from PCN in that it can take any value from 0 to ∞, analogous to the SCS-CN parameter S, whereas PCN can range from 0-100, analogous to CN.

An introduction of the concept of the first flush (Foster and Charlesworth, 1996; Sansalone and Buchberger, 1997; and others), designated as initial flush, i_f, analogous to the initial abstraction, I_a, leads to the modification of equation (8.31) as

$$c_p = \frac{(c_T - i_f)^2}{c_T - i_f + \psi} \tag{8.40}$$

which is analogous to the SCS-CN equation (equation (2.5)):

$$Q = \frac{(P - I_a)^2}{P - I_a + S} \tag{2.5}$$

Equation (8.40) is valid for $c_T \geq i_f$, $c_p = 0$ otherwise. The second hypothesis of the SCS-CN method is written as (equation (2.3)):

$$I_a = \lambda S \tag{2.3}$$

Parallel to equation (2.3), the initial flush, i_f, can be related to ψ as

$$i_f = \lambda_1 \psi \tag{8.41}$$

where λ_1 is the initial flush coefficient, similar to the initial abstraction coefficient, λ.

8.5 Application of Wave Analogy

8.5.1 EXPERIMENTAL WATERSHED

As mentioned in Chapter 6, an experimental urban watershed in Cincinnati (U.S.A.) of 15x20 m asphalt pavement of I-75 (Figures 6.1a,b) was selected. Rainfall-runoff data were collected during 1995-97 by Sansalone and Buchberger (1997). The storm water runoff diverted through the epoxy-coated converging slab, a 2.54 cm diameter Parshall flume, and a 2 m long 25.4 cm diameter PVC pipe to a 2000 liter storage tank was measured at a regular 1 min interval using an automated 24 bottle sampler with polypropylene bottles. The rainfall was recorded using a tipping bucket gauge in increments of 0.254 mm. Cincinnati receives an annual average 1020 mm of rainfall and 420 mm of snow. High rainfall occurs in March (average = 106 mm) and July (average = 104 mm) and highest snowfall occurs in January (average = 150 mm) (Soil, 1982). The average winter and summer temperatures are 0 ^{0}C and 23.4 ^{0}C, respectively. The water quality data at every 2 min interval were collected during rainfall-runoff events at the experimental site and samples were analyzed for dissolved and particulate bound metals for the rainfall-runoff events of June 17, July 7, and August 8, 1996 (Sansalone and Buchberger, 1997; Sansalone et al., 1998; and Li et al., 1999). An example set of hydrologic and water quality data for the event of August 8, 1996 is shown in Table 8.3 with the linearly interpolated values of water quality data. It is worth emphasizing that for each discrete sample, dissolved and particulate heavy metal concentrations were obtained after sample preparation and digestion through ICP-AES analysis. Each dissolved or particulate sample concentration at time i, c_i, was multiplied by the runoff volume during 1-min interval, v_i, to determine solid mass (m_i) as

$$m_i = c_i\, v_i \tag{8.42}$$

8.5.2 DEVELOPMENT OF LOOPED MASS RATING CURVES

Looped mass rating curves for dissolved and particulate bound metals were developed for 12 metal elements, viz., zinc (Zn), cadmium (Cd), lead (Pb), nickel (Ni), manganese (Mn), iron (Fe), chromium (Cr), magnesium (Mg), aluminum (Al), calcium (Ca), copper (Cu), and sodium (Na), as illustrated below.

Example 8.1: Compute the equivalent mass depth for the dissolved metal Zn using the rainfall-runoff-water quality data of the Cincinnati I-75 highway for the August 8, 1996 rainstorm event, shown in Table 8.3. Assume the density of Zn at 20 ^{0}C equal to 7.14 gm/cc. Watershed area = 300 sq. m.

Solution: Computations are shown in Table 8.3. In this table, the first col. shows the time (min), the second the rainfall intensity (mm/hr), the third the runoff rate (mm/hr), and the fourth the equivalent mass depth, h_e, (10^{-9} mm). Cols. fifth to ninth stand, respectively, for the cols. first to fourth. The dissolved mass in micro grams was computed using equation (8.42) for known concentration of runoff for each element and the volume of runoff during the given time interval of 1 min. The equivalent mass depth, h_e, can be computed using equation (8.20), for example, at the time 10 min as below:

Given are the dissolved mass of Zn at 10 min, m(10), = 15824 micro gram; metal density, ρ, = 7.14 gm/cc; and the area of the watershed, A_w, = 300 sq. m. The equivalent mass depth, h_e, is computed as:

Table 8.3 Example data for the dissolved solids of Zn of event Aug. 8, 1996.

Time	Rainfall intensity	Runoff rate	Dissolved Mass	Equivalent mass depth, h_e	Time	Rainfall intensity	Runoff rate	Dissolved Mass	Equivalent mass depth, h_e
(min)	(mm/hr)	(mm/hr)	(μg)	(10^{-9} mm)	(min)	(mm/hr)	(mm/hr)	(μg)	(10^{-9} mm)
0	2.04	0.00	0.00	0.00	24	18.36	21.81	18261	8.53
1	2.04	0.00	0.00	0.00	25	18.36	21.09	18094	8.45
2	2.04	0.00	0.00	0.00	26	18.36	19.09	17927	8.37
3	2.04	0.00	0.00	0.00	27	18.36	18.43	23084	10.78
4	2.04	0.00	0.00	0.00	28	9.18	20.97	28241	13.18
5	2.04	0.00	0.00	0.00	29	9.18	18.05	30945	14.45
6	2.04	0.00	0.00	0.00	30	9.18	15.25	33650	15.71
7	2.04	0.01	0.00	0.00	31	9.18	12.72	31879	14.88
8	2.04	0.08	0.00	0.00	32	9.18	13.04	30108	14.06
9	18.36	0.18	0.00	0.00	33	9.18	11.29	30347	14.17
10	18.36	0.59	15824	7.39	34	6.12	10.92	30587	14.28
11	18.36	5.80	92783	43.32	35	6.12	7.97	29863	13.94
12	18.36	12.26	169743	79.25	36	6.12	7.76	29138	13.60
13	18.36	13.32	125357	58.52	37	0.00	5.74	24149	11.27
14	73.45	19.79	80971	37.80	38	0.00	5.49	19159	8.94
15	73.45	36.47	179588	83.84	39	0.00	3.42	15695	7.33
16	55.08	56.26	278204	129.88	40	0.00	2.07	12230	5.71
17	91.81	56.06	212691	99.30	41	0.00	2.27	13764	6.43
18	110.17	68.56	147179	68.71	42	0.00	1.93	15297	7.14
19	91.81	78.29	93083	43.46	43	0.00	1.79	16830	7.86
20	18.36	75.49	38987	18.20	44	0.00	1.93	14575	6.80
21	36.72	56.08	31658	14.78	45	0.00	1.25	12320	5.75
22	18.36	39.74	24329	11.36	46	0.00	1.36	10384	4.85
23	18.36	28.86	21295	9.94	47	0.00	0.62	8448	3.94

$$h_e(10) = \frac{15824\text{x}10^{-6}(\text{gm})}{7.14(\text{gm/cm}^3)\text{x}300\text{x}10^4(\text{cm}^2)}$$

$$= 7.387\text{x}10^{-10}\,\text{cm} = 7.387\text{x}10^{-9}\,\text{mm}$$

Similarly, $h_e(t)$ for any time t can be computed.

The runoff rate in Table 8.3 represents the direct surface runoff observed at the outlet of the 15 m x 20 m micro urban watershed. It can be computed using a rainfall-runoff model, as shown in Chapter 6. Thus, the advantage of developing rating curves for pollutants lies in the fact that for a given amount of rainfall intensity, the mass of either dissolved or particulate bound metals can be easily determined.

With the above assumption of the direct surface runoff rate observed at the outlet to be equivalent to the velocity of overland flow, rating curves analogous to those in open channels were developed for dissolved as well as particulate bound metal elements described above and these are shown in Figures 8.1 and 8.2, respectively, for the event of August 8, 1996. Here, it is noted that the y-axes of the rating curves for various elements shown in Figures 8.1 and 8.2 represent the mass of flow (Table 8.3), rather than h_e. It is apparent from these figures that several metals (for example, Pb, Mn, Mg, Cu, and Na in Figure 8.1) exhibit a relation tantamount to steady-state flow rating curves observed in open channels. All other dissolved and particulate bound elements show mild to strong loops in the rating curves. Data points that deviate from the looped shape in a few plots in Figure 8.1 (for example, the plots for Zn, Cd, Mn among others) are attributed to the multiplicity of peaks of the runoff hydrograph. It is shown in Figures 8.3 and 8.4 using, respectively, the data of Table 8.3 and the event of June 18, 1996 for the dissolved Zn. Figure 8.4 exhibits formation of three loops in the rating curves in accordance with the three runoff hydrograph peaks, discussed later. Such a phenomenon of mixed waves is typical in open channel flows and the hysteretic behavior (or loop formation) greatly depends on the initial (input) flow and channel storage conditions (Mishra and Singh, 1999). The underlying difference, however, is that the natural flow waves normally travel downstream, dissipate during their travel, and show a counter-clockwise loop in the rating curve. These waves are known as primary waves. On the other hand, secondary waves travel upstream, amplify during their travel, and exhibit clock-wise loop rating curves, as shown in Figures 8.3 and 8.4. Mishra and Singh (1999) explained such a phenomenon using hysteresis, η; the value of η represents the type of wave (Table 8.1), counter-clock-wise loop rating curve exhibits a primary wave whereas clock-wise rating curve shows a secondary wave. The '+' sign in equations (8.16)-(8.18) correspond to primary waves and the '-' sign to secondary waves (Ponce and Simons, 1977). Here, the physical process of the present metal transport on the paved surface is explained in what follows.

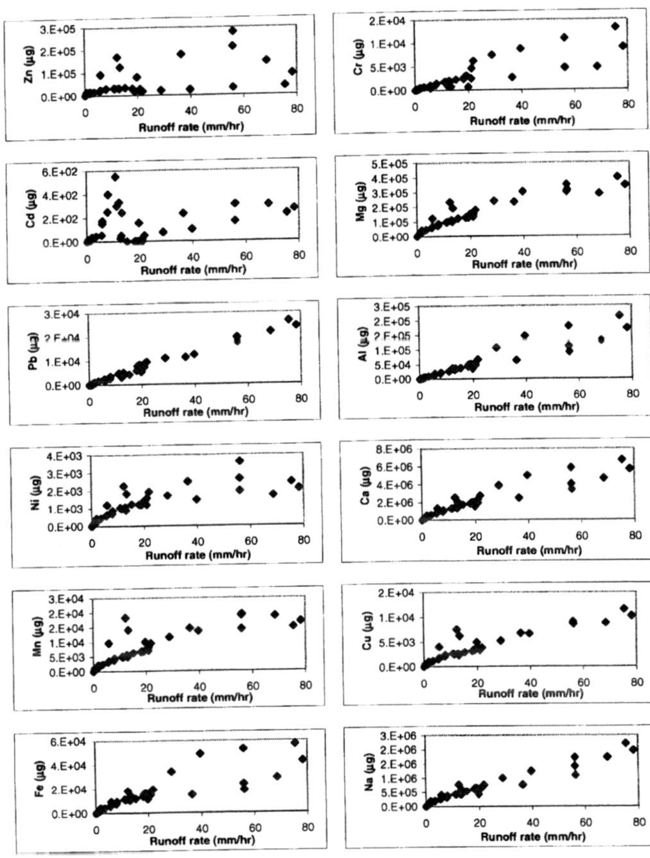

Figure 8.1. Relation between runoff rate (mm/hr) and mass of dissolved metals (µg) for Aug. 8, 1996 event.

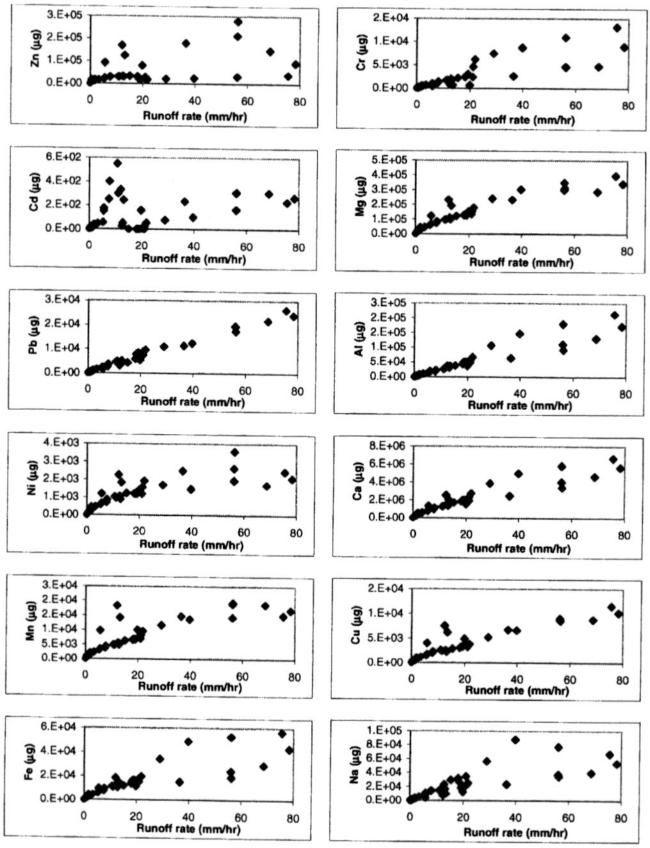

Figure 8.2. Relation between runoff rate (mm/hr) and mass of particulate bound metals (µg) for Aug. 8, 1996 event.

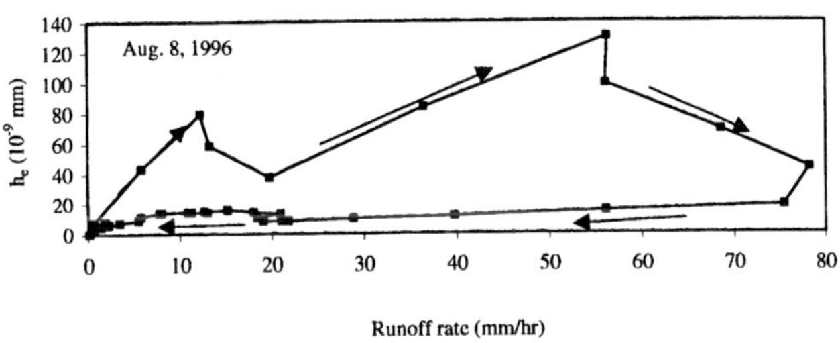

Figure 8.3. Relation between runoff rate (mm/hr) and equivalent Zn dissolved mass depth of flow (10^{-9} mm).

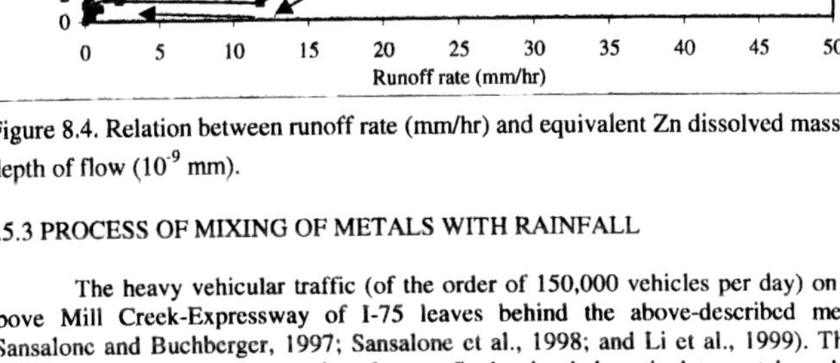

Figure 8.4. Relation between runoff rate (mm/hr) and equivalent Zn dissolved mass depth of flow (10^{-9} mm).

8.5.3 PROCESS OF MIXING OF METALS WITH RAINFALL

The heavy vehicular traffic (of the order of 150,000 vehicles per day) on the above Mill Creek-Expressway of I-75 leaves behind the above-described metals (Sansalone and Buchberger, 1997; Sansalone et al., 1998; and Li et al., 1999). These metals are deposited on the road surface as finely abraded particulate metals and the metal layer so formed represents the layer of the highest mass concentration. The rainfall occurring over the paved road surface generates the following processes:

a) The rainfall impacts the road surface to break the metal layer.
b) The rainfall of low pH (< 5) and low alkalinity (Sansalone et al., 1998) causes finely abraded heavy metals to mix with the rain water.
c) The rainfall converts to rainfall-excess, for the land surface allows infiltration, alhough it is insignificant on the paved watershed.
d) The rainfall-excess builds up storage on the road surface before the start of the direct surface runoff; higher the storage, the greater will be the rate of direct surface runoff and vice versa.
e) The low pH and poorly buffered water storage or pondage on the road surface allows dissolution (or diffusion) of metals.

The breaking up of the metal layer leads to the mixing of particulate bound and dissolved metals with the rainfall-excess.

Both dissolved and particulate bound metals are carried by the direct surface runoff to the outlet. The process of dissolution (or molecular diffusion or dispersion) is usually described by the linear Fick's first law that describes the phenomenon of molecular diffusion as to be directly proportional to the concentration gradient; the higher the gradient, the higher will be the rate of dissolution and vice versa. It is expressed mathematically as:

$$\vartheta = -D' \frac{\partial \dot{C}}{\partial x} \tag{8.43}$$

where ϑ is the net flux $[ML^{-2}T^{-1}]$, D' is the coefficient of molecular diffusion $[L^2T^{-1}]$, $\dot{C}$ is the concentration $[ML^{-3}]$, and ∂x is the mixing length [L].

Martin and McCutcheon (1999) among many others have described the process of molecular diffusion at length. At the outset of a rainfall event, the low pH rainfall-excess leading to surface pondage allows the highest rate of dissolution of metals in a given time interval because of the maximum concentration gradient existing between the metal layer and the rainfall-excess. It is driven by pH and alkalinity conditions. Such a phenomenon is seen in Figure 8.5, depicting the rainfall-runoff-metal transport event of August 8, 1996, which is a relatively intense event of short duration compared to other events. There are as many peaks in the hydrograph of equivalent mass depth of flow, h_e, as in the direct surface runoff hydrograph or in the rainfall hyetograph. Depending on the amount of rainfall intensity, the storage on the pavement surface further increases with the passage of time and because of the ongoing process of dissolution, the concentration gradient decreases with time, which, in turn, decreases the rate of dissolution of metal mass or the mass of the metal during the given time interval [= m(t)]. The difference between the rate of runoff and the rate of dissolution produces a time lag (or phase lag) between them. A phenomenon of positive time lag is apparent in Figure 8.6 depicting the event of June 18, 1996. The slow rising trend of the rainfall hyetograph results in a slow piling up of water on the paved surface, allowing more time for water to stay on the surface, leading to a slow rate of dissolution with the passage of time and, consequently, a positive phase lag between the rate of runoff and h_e, as seen in Figures 8.6 and 8.7, in which the peaks of h_e occur later than do the peaks of the direct surface runoff. It is noted that the event of August 8, 1996 exhibits the case of a negative time lag.

8.5.4 DEVELOPMENT OF NORMAL MASS RATING CURVES

The use of normal flow rating curves is more common in practice than the looped rating curves. On similar grounds, it is useful to develop normal mass rating curves for water quality analyses. To this end, the best-fit curves for all the above-described three events and for all the dissolved and particulate bound metal elements were derived from the data, as shown in Figures 8.1 and 8.2 and illustrated in Figure 8.8 for the event of August 8, 1996. These rating curves are described using the following power relation:

$$u = a\,m^{b} \tag{8.44}$$

where 'u' is the rate of direct surface runoff, 'm' is the mass of the dissolved or particulate bound metals, and 'a' and 'b' are the mass rating coefficient and exponent, respectively. Such a relationship (equation (8.44)) was adopted for the reason of its resemblance with Manning's (equation (1.79)) or Chezy's (equation (1.80)) friction laws Chapter 1) expressed as:

$$u_o = \frac{1}{n} R^{2/3} s_o^{1/2} \tag{8.45}$$

or

$$u_o = \chi\sqrt{R s_o} \tag{8.46}$$

where R is the hydraulic radius (= h_o for wide rectangular channels such as overland flows) and χ is the Chezy coefficient.

Equations (8.45) and (8.46) can, in general, be expressed in the form of flow rating equation as

$$u = \dot{\alpha} h^{\dot{\beta}} \tag{8.47}$$

where u and h represent the channel flow velocity and depth, respectively, and parameters $\dot{\alpha}$ and $\dot{\beta}$ represent the flow rating coefficient and exponent, respectively, dependent on the channel and flow characteristics, as follows:

$$\dot{\alpha} = \frac{1}{n} s_o^{1/2} \tag{8.48}$$

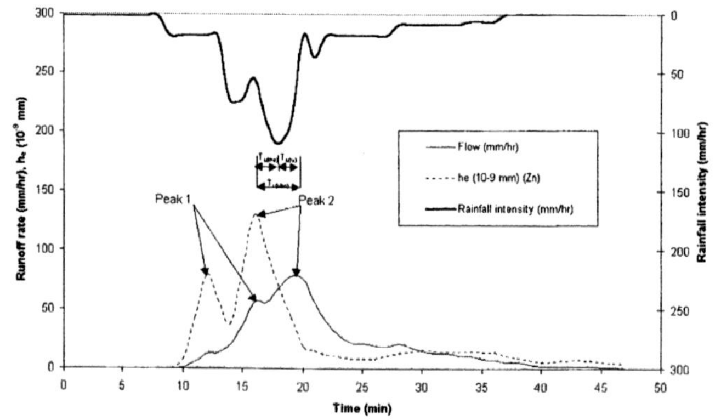

Figure 8.5. Temporal variation of rainfall intensity, runoff, and equivalent mass depth, h_e, (Event: Aug. 8, 1996).

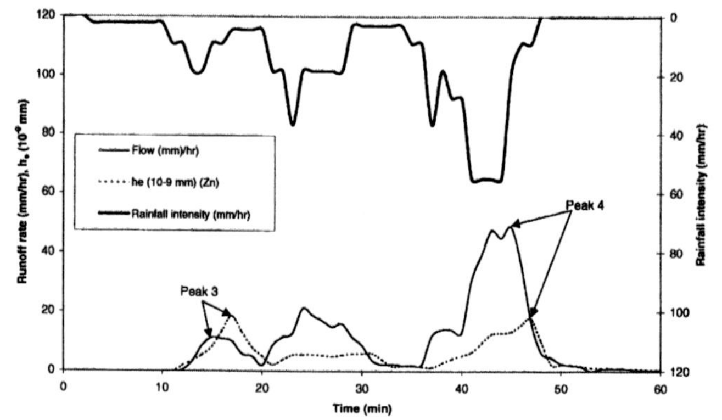

Figure 8.6. Temporal variation of rainfall intensity, runoff, and equivalent mass depth, h_e, (Event: June 18, 1996).

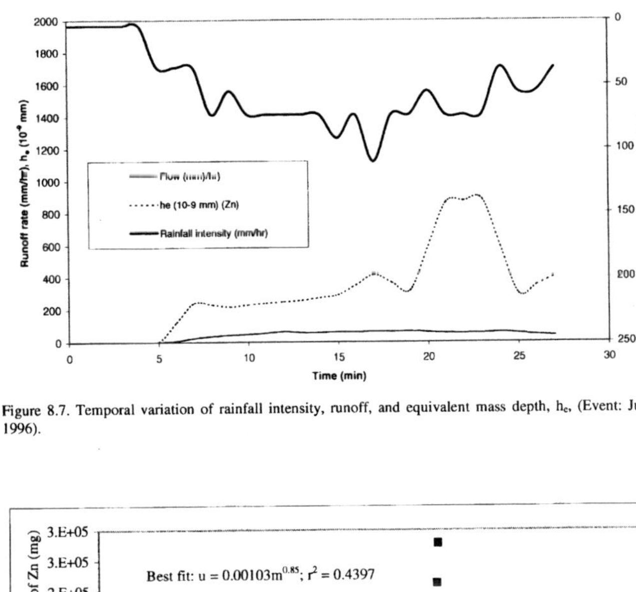

Figure 8.7. Temporal variation of rainfall intensity, runoff, and equivalent mass depth, h_e, (Event: July 7, 1996).

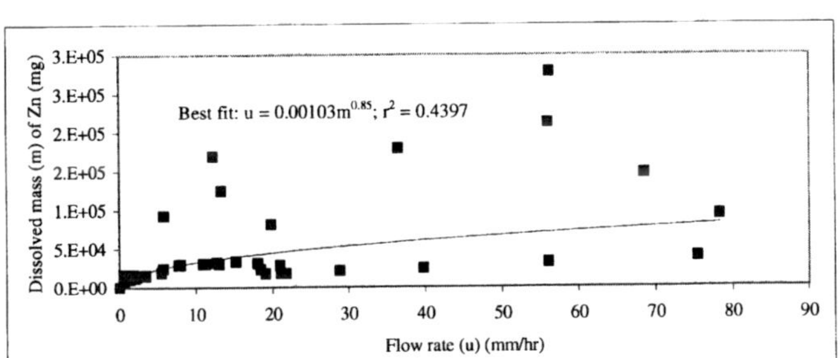

Figure 8.8. Single (best-fit) and loop rating curves for the dissolved mass of Zn for the event of Aug. 8, 1996.

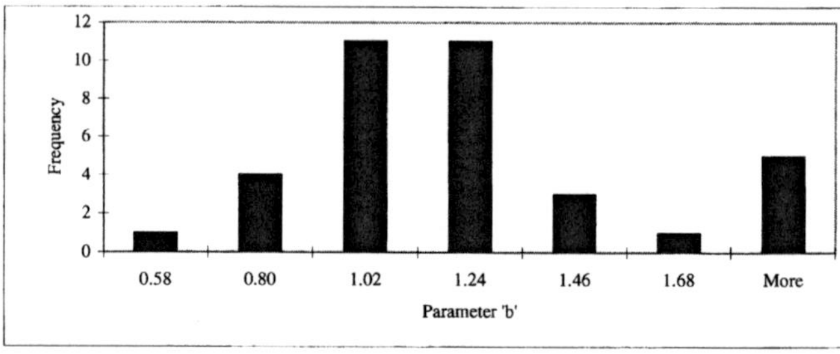

Figure 8.9. Histogram showing the variation of parameter b.

and $\dot{\beta}$ = 2/3 (Manning's friction) or 1/2 (Chezy's friction) the type of flow (laminar, turbulent, or transition). Here, it is noted that the dimensionless relative celerity, $\hat{c}_r$, (equations (8.16)-(8.18)) of the open channel flow is equal to $\dot{\beta}$ (see Appendix C). In a similar vein, the exponent 'b' of equation (8.44) describes the kinematic wave celerity of mass transport of both dissolved and particulate bound metals.

Furthermore, the term D′ in equation (8.43) describing the process of dissolution (or molecular diffusion) reflects the resistance to molecular diffusion (Martin and McCutcheon, 1999), and is thus analogous to the friction in open channel flows. Therefore, a concept of equivalent mass roughness, analogous to Manning's roughness, representing the dissolution of metals (dissolved) or mixing (particulate bound) metals is considered. It is expressed as:

$$n_e = s_o^{1/2} / a\rho^b \tag{8.49}$$

which can be derived from equations (8.20) and (8.44) and a relation analogous to equation (8.45). From equation (8.49) the product of n_e and ρ^b should form a constant for a given watershed. The computed parameters 'a' and 'b'; the equivalent mass roughness, n_e; and the product of n_e and ρ^b for all events and all dissolved and particulate bound metals are given in Tables 8.4 and 8.5. The average values of these parameters are shown in Table 8.6. In these computations, the bed slope is taken as 0.02, the area of watershed equal to 300 sq. m, and the equivalent depth of mass h_e is taken in 10^{-9} mm. It is apparent from Table 8.4 that parameter 'b' varies from 0.52 to 1.91. As shown in the histogram of parameter 'b' (Figure 8.9) developed using the data of Table 8.4, most values fall in a close range (1.02, 1.24). It is noted that b = 1.0 represents a straight-line relationship between the rate of surface runoff, u, and the equivalent mass depth of flow, h_e. In surface water flows, or overland flow in particular, the range of variation of parameter 'b' is equivalent to 50-75% turbulent flows, described by Manning's friction (Ponce, 1989). Parameter 'a' (Table 8.4), however, shows a wider range of variation than does parameter 'b' and exhibits a link with

Table 8.4 Computation of mass rating parameters ($u = a\,m^b$)

Event	Dissolved solids											
	Zn		Cd		Pb		Ni		Mn		Fe	
	b	a	b	a	b	a	b	a	b	a	b	a
Aug896	0.85	1.03E-03	1.20	3.14E-02	1.16	6.83E-04	1.78	5.04E-05	1.91	5.74E-07	1.36	3.48E-05
June1896	1.29	4.83E-05	1.01	1.52E-01	0.98	8.12E-03	1.01	1.29E-02	1.06	1.04E-03	0.61	3.19E-02
Jul796	1.12	1.12E-05	0.92	3.49E-02	0.66	4.65E-02	0.78	1.49E-02	1.16	1.90E-04	0.97	8.54E-04
	Cr		Mg		Al		Ca		Cu		Na	
	b	a	b	a	b	a	b	a	b	a	b	a
Aug896	1.15	2.15E-03	1.80	8.15E-09	1.27	2.06E-05	1.62	1.14E-09	1.86	3.24E-06	1.81	5.25E-10
June1896	1.22	2.78E-03	1.07	6.57E-05	0.93	1.03E-03	1.11	3.70E-06	1.10	3.49E-03	1.09	8.55E-06
Jul796	0.58	4.33E-01	0.90	1.48E-04	0.71	4.14E-03	1.02	3.75E-06	1.01	9.27E-04	0.89	6.76E-05
	Particulate bound solids											
	Zn		Cd		Pb		Ni		Mn		Fe	
	b	a	b	a	b	a	b	a	b	a	b	a
Aug896	1.16	7.80E-04	0.96	7.19E-01	1.27	3.06E-03	1.12	5.86E-02	1.17	4.04E-03	1.18	4.68E-05
June1896	1.11	2.69E-03	1.14	6.12E-01	1.05	2.21E-02	1.01	1.26E-01	0.85	6.90E-02	0.96	9.04E-04
Jul796	0.64	3.93E-02	0.64	1.07E+00	0.70	7.30E-02	0.52	5.31E-01	1.00	1.90E-02	0.99	4.25E-04
	Cr		Mg		Al		Ca		Cu		Na	
	b	a	b	a	b	a	b	a	b	a	b	a
Aug896	1.02	1.13E-01	1.14	1.32E-04	1.09	4.17E-04	1.05	8.50E-05	1.37	1.79E-03	1.07	4.19E-04
June1896	0.98	2.40E-01	0.77	7.08E-03	1.01	1.97E-03	0.85	9.06E-04	1.00	4.51E-02	1.02	1.19E-03
Jul796	0.82	4.40E-01	0.83	1.83E-03	0.79	6.11E-03	0.74	1.33E-03	0.77	6.16E-02	0.59	2.37E-02

Note: a and b are rating curve coefficient and exponents, respectively; u in mm/hr; and m in microgram.

Table 8.5 Computation of rating parameters

Event	Zn	Cd	Pb	Ni	Mn	Fe	Cr	Mg	Al	Ca	Cu	Na
	Dissolved Solids											
	n_e											
Aug. 8, 96	3.22E+03	3.94E+02	1.58E+04	2.18E+06	3.04E+08	6.62E+05	4.85E+03	1.46E+10	7.96E+05	5.21E+10	4.63E+07	2.33E+11
June 18, 96	3.60E+05	4.00E+01	6.80E+02	4.74E+02	7.29E+03	4.36E+01	4.90E+03	1.19E+05	4.38E+03	2.39E+06	2.48E+03	9.55E+05
Jul. 7, 96	8.50E+05	1.25E+02	3.56E+01	1.78E+02	5.75E+04	6.13E+03	2.88E+00	2.75E+04	4.94E+02	1.71E+06	6.60E+03	5.81E+04
	$n_e{*}\rho^b$											
Aug. 8, 96	1.70E+04	4.14E+03	1.54E+05	7.22E+07	1.29E+10	9.64E+06	4.67E+04	5.08E+11	9.68E+06	1.25E+12	1.80E+09	8.09E+12
June 18, 96	4.52E+06	2.91E+02	4.67E+03	3.44E+03	5.90E+04	1.45E+02	5.41E+04	9.77E+05	2.71E+04	2.09E+07	2.15E+04	8.05E+06
Jul. 7, 96	7.78E+06	7.54E+02	1.29E+02	8.32E+02	5.66E+05	4.09E+04	9.03E+00	1.61E+05	2.01E+03	1.27E+07	4.79E+04	3.33E+05
	Particulate Bound Metals											
	n_e											
Aug. 8, 96	1.40E+04	7.02E+00	5.32E+03	1.62E+02	2.75E+03	2.46E+05	5.59E+01	7.67E+04	1.99E+04	8.43E+04	1.31E+04	1.84E+04
June 18, 96	3.33E+03	1.63E+01	3.22E+02	4.92E+01	4.94E+01	5.66E+03	2.28E+01	3.64E+02	3.11E+03	3.76E+03	1.31E+02	5.46E+03
Jul. 7, 96	3.98E+01	1.45E+00	2.62E+01	1.86E+00	3.11E+02	1.32E+04	6.95E+00	1.74E+03	4.47E+02	1.71E+03	4.00E+01	5.41E+01
	$n_e{*}\rho^b$											
Aug. 8, 96	1.37E+05	4.61E+01	6.44E+04	1.48E+03	2.73E+04	2.50E+06	4.14E+02	7.24E+05	1.69E+05	6.64E+05	1.93E+05	1.51E+05
June 18, 96	2.95E+04	1.52E+02	2.55E+03	3.60E+02	2.64E+02	3.73E+04	1.56E+02	1.67E+03	2.26E+04	2.01E+04	9.30E+02	4.07E+04
Jul. 7, 96	1.41E+02	5.14E+00	1.03E+02	5.17E+00	2.20E+03	9.15E+04	3.49E+01	8.92E+03	2.12E+03	7.33E+03	1.79E+02	1.72E+02

Note: ρ is the mass density; h_e is the equivalent mass depth (10^{-9} mm); n_e is the equivalent roughness; and a and b are rating curve coefficient and exponents, respectively.

Table 8.6 Computation of average values of rating parameters for dissolved and particulate bound solids

Parameter	Zn	Cd	Pb	Ni	Mn	Fe	Cr	Mg	Al	Ca	Cu	Na
	Dissolved solids											
'b'	1.09	1.04	0.93	1.19	1.38	0.98	0.98	1.26	0.97	1.25	1.32	1.26
'a'	3.63E-04	7.28E-02	1.84E-02	9.28E-03	4.10E-04	1.09E-02	1.46E-01	7.12E-05	1.73E-03	2.48E-06	1.47E-03	2.54E-05
n_e	4.05E+05	1.86E+02	5.50E+03	7.27E+05	1.01E+08	2.23E+05	3.24E+03	4.87E+09	2.66E+05	1.74E+10	1.54E+07	7.76E+10
$n_e \rho^b$	4.11E+06	1.73E+03	5.28E+04	2.41E+07	4.29E+09	3.22E+06	3.35E+04	1.69E+11	3.24E+06	4.16E+11	6.02E+08	2.71E+12
	Particulate bound solids											
'b'	0.97	0.91	1.01	0.88	1.01	1.04	0.94	0.91	0.96	0.88	1.05	0.89
'a'	1.43E-02	8.00E-01	3.27E-02	2.39E-01	3.07E-02	4.59E-04	2.64E-01	3.01E-03	2.83E-03	7.74E-04	3.62E-02	8.44E-03
n_e	5.79E+03	8.25E+00	1.89E+03	7.09E+01	1.04E+03	8.83E+04	2.86E+01	2.62E+04	7.83E+03	3.00E+04	4.45E+03	7.98E+03
$n_e \rho^b$	5.57E+04	6.80E+01	2.24E+04	6.15E+02	9.91E+03	8.79E+05	2.01E+02	2.46E+05	6.46E+04	2.30E+05	6.48E+04	6.40E+04

Note: ρ is the mass density; n_e is the equivalent roughness; and a and b are rating curve coefficient and exponents, respectively.

Table 8.7 Characteristics of the selected wave events for metal transport

Wave No.	Time to peak rainfall intensity, $t_{p(i)}$	Time to peak runoff, $t_{p(u)}$	Time to peak h_e, $t_{p(he)}$	Time period, T	Rainfall-runoff process		Molecular diffusion			Rainfall-runoff-molecular diffusion		
					$T_{L(iu)}$	Average wave celerity	$T_{L(ihe)}$	ϕ	Wave type	$T_{L(uhe)}$	ϕ	Wave type
	(min)	(min)	(min)	(min)	(min)	(m/s)	(min)	(radian)		(min)	(radian)	
1	15	17	12	7	2	1.7952	-3	-2.6928	SDYW	-5	-4.4880	SDYW
2	18	19	16	8	1	0.7854	-2	-1.5708	SDYW	-3	-2.3562	SDYW
3	14	15	17	7	1	0.8976	3	2.6928	PDYW	2	1.7952	PDYW
4	44	45	47	5	1	1.2566	3	3.7699	PDYW	2	2.5133	PDYW

Notations: $T_{L(iu)}$ is the time difference between rainfall-peak and direct surface runoff peak; $T_{L(ihe)}$ is the time difference between peak rainfall and peak equivalent mass depth of flow; $T_{L(uhe)}$ is the time difference between direct surface runoff and equivalent mass depth of flow, h_e; T is the time period; ϕ is the phase difference; PDYW stands for primary dynamic wave for metal transport; and SDYW stands for secondary dynamic wave for metal transport.

parameter 'b', as shown in Figure 8.10. Apparently, parameter 'a' varies inversely with parameter 'b'. Thus, parameter 'a' is also sensitive to the variation in the type of flows, as described above and, therefore, the product of n_e and ρ^b (equation (8.49)) will also vary, as seen in Table 8.5, with the change in parameter 'b'.

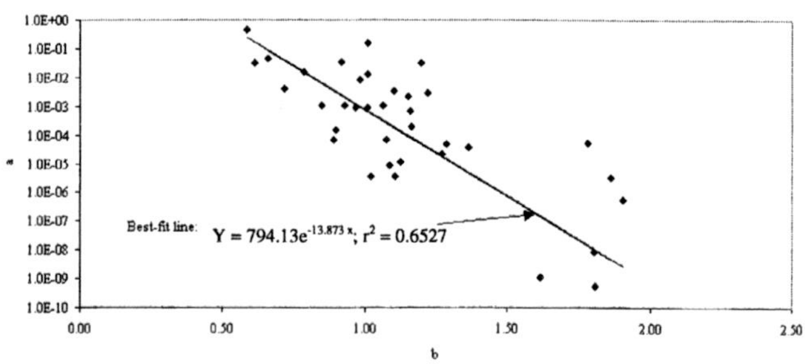

Figure 8.10. Variation of the rating coefficient, a, with the rating exponent, b.

Example 8.2: Determine the mass of a metal dissolved by a rainfall of intensity equal to 73 mm/hr. The mass rating curve is given by $u = a\,m^b$, where u = rainfall-excess rate, m = mass of the metal, a = 0.001, and b = 0.85. Assume: φ-index = 36 mm/hr.

Solution: Given: a = 0.001, b = 0.85, rainfall intensity = 73 mm/hr and φ-index = 36 mm/hr. Therefore, u = rainfall intensity - φ-index = 73.0 – 36.0 = 37 mm/hr. The dissolved mass of the metal can be computed from equation (8.44) as

$$m = \left(\frac{1}{a}u\right)^{1/b} = (37/0.001)^{1/0.85} = 236785 \text{ micro-gram} = 0.24 \text{ gram}$$

The actual amount of the metal transported in a given time interval also depends on other factors, such as the actual number of vehicles that have passed through the stretch before the occurrence of the rainstorm. During heavy rainy days, it is more likely that the number of vehicles will be less than that during dry days and vice versa. It can be realized from Figure 8.7 depicting the event of July 7, 1996. Apparently, for the same order of peak rainfall intensity, the peak of the h_e-hydrograph (Figure 8.7) is several times larger than that of August 8, 1996 (Figure 8.5). Thus, the number of antecedent dry days significantly influences the number of vehicles that have passed through the road stretch before the start of the rain and, in turn, the potential depth of the metal layer.

8.5.5 WAVE ANALYSIS

Similar to the waves in open channels, the wave characteristics of the events shown in Figures 8.5–8.7 are determined using the phase difference and non-dimensional hysteresis, η. The phase difference, ϕ, is computed as (Mishra and Seth, 1996):

$$\phi = \frac{2\pi}{T}[t_{p(h_e)} - t_{p(u)}] \tag{8.50}$$

where ϕ is in radian; T is in hr; $t_{p(he)}$ is the time of rise of the h_e-wave (min) and $t_{p(u)}$ is the time of rise of the u-wave (min). Menendez and Norscini (1983) described ϕ as a kinematic parameter that drives attenuation.

Mishra and Singh (1999) derived a mathematical expression linking the phase difference, ϕ, with the non-dimensional hysteresis, η. The computation of ϕ is illustrated for the selected four discrete peaks shown in Figures 8.5 and 8.6 for the dissolved Zn metal. The wave phenomenon of molecular diffusion pertinent to the present case study can be described in three components: (1) Rainfall-runoff process, (2) molecular diffusion process, and (3) rainfall-runoff-molecular diffusion process. To this end, three time lags are considered as shown in Figure 8.5 for peak no. 2: a) Time lag between peaks of rainfall and runoff, designated as $T_{L(iu)}$; b) time lag between rainfall and equivalent mass depth of flow, h_e, designated as $T_{L(ihe)}$; and c) time lag between the runoff and equivalent mass depth of flow, designated as $T_{L(uhe)}$. $T_{L(iu)}$ describes the lag between peak rainfall intensity and the peak runoff rate and represents the storage effect and helps describe the average wave celerity of the dynamic rainfall-runoff process.

The time lag between the peak rainfall intensity and h_e (= $T_{L(ihe)}$) is taken for the reason that if the direct surface runoff observed at the outlet of the watershed is accounted for the time lag $T_{L(iu)}$, it will represent the rainfall-excess over the watershed, where the dissolution or mixing of metals with water takes place. Thus, $T_{L(ihe)}$ indirectly represents the lag between the rates of rainfall-excess and molecular diffusion; the greater the lag, more dynamic will be the process than in the otherwise situation. The time lag between runoff and h_e (= $T_{L(uhe)}$) considers both the rainfall-runoff and molecular diffusion processes and is the end result observed at the outlet of the watershed. The peaks are chosen for the reason that they are easily discernible. For convenience of description, the time to peak rainfall intensity is designated as $t_{p(i)}$, time to peak runoff as $t_{p(u)}$, and time to peak h_e as $t_{p(he)}$. Thus,

$$T_{L(iu)} = t_{p(u)} - t_{p(i)} \tag{8.51}$$

$$T_{L(ihe)} = t_{p(he)} - t_{p(i)} \tag{8.52}$$

$$T_{L(uhe)} = t_{p(he)} - t_{p(u)} \tag{8.53}$$

For the example discrete peaks of Figures 8.5 and 8.6, the computed time lags are given in Table 8.7. In this table, the time period corresponds to the time of rise plus the time of recession of h_e-wave. Using these time lags, the above processes are described below.

Using $T_{L(iu)}$, the average wave celerity is computed as (Mishra and Seth, 1996):

$$\text{Average wave celerity} = \Delta x/\Delta T_t \tag{8.54}$$

where Δx is the reach length and ΔT_t is the time of travel [$= T_{L(iu)}$], which is equivalent to the storage routing coefficient of reservoir routing used in catchment routing (Chapters 1, 6, and 7). For selected events, the average wave celerity varies from 0.79 to 1.80 m/s. In Table 8.7, negative time lags indicate that the h_e-wave precedes rainfall intensity, i-, or runoff u-wave, leading to a negative phase difference describing the analogous process of secondary waves in open channels. Based on the phase difference criteria (Table 8.1), the waves of molecular diffusion and the rainfall-runoff-molecular diffusion are described to be either primary dynamic wave (PDYW) or secondary dynamic wave (SDYW).

Example 8.3: In a storm event on a micro urban watershed, measurements were made at discrete times for rainfall intensity, runoff hydrograph, and pollutant load. Peaks of rainfall intensity, runoff hydrograph, and equivalent mass depth were observed at 18, 19, and 16 minutes, respectively. The time base of the runoff hydrograph was of the order of 8 minutes. Using the wave concept, characterize this event for molecular diffusion and rainfall-runoff-molecular diffusion processes.

Solution: Given $t_{p(i)} = 18$ min, $t_{p(u)} = 19$ min, $t_{p(he)} = 16$ min, and time period T = 8 min. The time lags to describe the processes of rainfall-runoff ($=T_{L(iu)}$), molecular diffusion ($=T_{L(ihe)}$), and rainfall-runoff-molecular diffusion ($=T_{L(uhe)}$) are determined using equations (8.51) through (8.53), respectively, as:

$T_{L(iu)} = t_{p(u)} - t_{p(i)} = 19 - 18 = 1$ min

$T_{L(ihe)} = t_{p(he)} - t_{p(i)} = 16 - 18 = -2$ min

$T_{L(uhe)} = t_{p(he)} - t_{p(u)} = 16 - 19 = -3$ min

Here '+' sign represents the occurrence of primary wave and '-' sign stands for secondary wave.

Using equation (8.50), compute the phase lag (ϕ) for the molecular diffusion and rainfall-runoff-molecular diffusion processes, respectively, as

Table 8.8 Description of rainfall-runoff events

Event	T* (min)	$T_{L(iu)}$ (min)	Average Wave Celerity ($=\Delta x/ T_{L(iu)}$) (m/s)	Wave length = Average Wave celerity x T (m)
Aug. 8, 96	12	1	0.333	239.76
June 18, 96	15	7	0.048	43.20
Jul. 7, 96	28	1	0.333	559.44

*Time period is based on the direct surface runoff hydrograph.

Table 8.9 Times to peak rainfall intensity, runoff, and equivalent mass depth, h_e

Event	$t_{p(i)}$ (min)	$t_{p(u)}$ (min)	$t_{p(he)}$ (min) Zn	Cd	Pb	Ni
			Dissolved solids			
August 8, 96	18	19	16	16	20	20
June 18, 96	23	30	25	25	26	25
July 7, 96	17	18	23	25	25	27
			Particulate bound metals			
August 8, 96	18	19	22	20	22	22
June 18, 96	23	30	25	25	25	25
July 7, 96	17	18	27	27	21	27

Table 8.9 Continued

Event	$t_{p(he)}$ (min) Mn	Fe	Cr	Mg	Al	Ca	Cu	Na
	Dissolved solids							
August 8, 96	16	20	20	20	20	20	20	20
June 18, 96	25	31	27	25	25	25	25	25
July 7, 96	27	25	23	27	27	27	23	23
	Particulate bound metals							
August 8, 96	22	22	22	22	22	22	22	22
June 18, 96	25	25	25	25	25	25	25	25
July 7, 96	27	25	25	27	27	27	27	27

$$\phi = \frac{2\pi}{T}[t_{p(h_e)} - t_{p(u)}] = \frac{2x3.14}{8}[16-18] = -1.57 \text{ radian}$$

$$\phi = \frac{2\pi}{T}[t_{p(h_e)} - t_{p(u)}] = \frac{2x3.14}{8}[16-19] = -2.355 \text{ radian}$$

From Table 8.1, both the above processes are characterized by the secondary dynamic wave. It implies that this event cannot be modeled by a second-order partial differential equation such as equation (8.9) which represents a diffusion wave equation.

Similar to the above, computations for all 12 dissolved and particulate bound metals were made for all the above three rainfall-runoff events and these are presented in Table 8.8. The corresponding times to peak of rainfall, runoff, and h_e, are given in Table 8.9. In Table 8.8, the time period T is computed from the direct surface runoff hydrograph corresponding to the discrete highest rainfall intensity of the event. The computed times to peak (Table 8.9) correspond to this discrete event. The wave description for these events is given in Table 8.10. It is seen that the processes of molecular diffusion of Zn, Cd, and Mn elements are analogous to secondary dynamic waves in open channels and the other elements exhibit primary dynamic wave phenomena. Based on the overall process described by the rainfall-runoff-molecular diffusion process, the process is, however, described by the secondary dynamic wave for all metals for the event of June 18, 1996 and metals Zn, Cd, and Mn for the event of August 8, 1996. The hysteresis, η, values are computed for complete events, as shown in Figures 8.1 and 8.2 for dissolved and particulate bound metals, respectively. These

Table 8.10 Wave description for dissolved solids

Event	Zn	Cd	Pb	Ni	Mn	Fe	Cr	Mg	Al	Ca	Cu	Na
Molecular Diffusion Process												
	$T_{L(tbe)}$											
Aug. 8, 96	-2	-2	2	2	-2	2	2	2	2	2	2	2
June 18, 96	2	2	3	2	2	8	4	2	2	2	2	2
Jul. 7, 96	6	8	8	10	10	8	6	10	10	10	6	6
	ϕ											
Aug. 8, 96	-1.0472	-1.0472	1.0472	1.0472	-1.0472	1.0472	1.0472	1.0472	1.0472	1.0472	1.0472	1.0472
June 18, 96	0.8378	0.8378	1.2566	0.8378	0.8378	3.3510	1.6755	0.8378	0.8378	0.8378	0.8378	0.8378
Jul. 7, 96	1.3464	1.7952	1.7952	2.2440	2.2440	1.7952	1.3464	2.2440	2.2440	2.2440	1.3464	1.3464
	Wave type											
Aug. 8, 96	SDYW	SDYW	PDYW	PDYW	SDYW	PDYW	PDYW	PDYW	PDYW	PDYW	PDYW	PDYW
June 18, 96	PDYW	PDYW	PDYW	PDYW	PDYW	PDYW	PDYW	PDYW	PDYW	PDYW	PDYW	PDYW
Jul. 7, 96	PDYW	PDYW	PDYW	PDYW	PDYW	PDYW	PDYW	PDYW	PDYW	PDYW	PDYW	PDYW
	Rainfall-Runoff-Molecular Diffusion Process											
	$T_{L(tbe)}$											
Aug. 8, 96	-3	-3	1	1	-3	1	1	1	1	1	1	1
June 18, 96	-5	-5	-4	-5	-5	1	-3	-5	-5	-5	-5	-5
Jul. 7, 96	5	7	7	9	9	7	5	9	9	9	5	5
	ϕ											
Aug. 8, 96	-1.5708	-1.5708	0.5236	0.5236	-1.5708	0.5236	0.5236	0.5236	0.5236	0.5236	0.5236	0.5236
June 18, 96	-2.0944	-2.0944	-1.6755	-2.0944	-2.0944	0.4189	-1.2566	-2.0944	-2.0944	-2.0944	-2.0944	-2.0944
Jul. 7, 96	1.1220	1.5708	1.5708	2.0196	2.0196	1.5708	1.1220	2.0196	2.0196	2.0196	1.1220	1.1220

Table 8.10 Continued

Event	Zn	Cd	Pb	Ni	Mn	Fe	Cr	Mg	Al	Ca	Cu	Na
	Wave type											
Aug. 8, 96	SDYW	SDYW	PDYW	PDYW	SDYW	PDYW	PDYW	PDYW	PDYW	PDYW	PDYW	PDYW
June 18, 96	SDYW	SDYW	SDYW	SDYW	SDYW	PDYW	SDYW	SDYW	SDYW	SDYW	SDYW	SDYW
Jul. 7, 96	PDYW	PDYW	PDYW	PDYW	PDYW	PDYW	PDYW	PDYW	PDYW	PDYW	PDYW	PDYW
	η											
Aug. 8, 96	0.4488	0.1907	0.0518	0.1345	0.1905	0.3224	0.3553	0.0609	0.2265	0.1922	0.0401	0.1198
June 18, 96	0.5065	0.4618	0.4153	0.3132	0.2309	0.1391	0.6301	0.2861	0.3705	0.3218	0.3605	0.2730
Jul. 7, 96	0.1505	0.3972	0.4007	0.4541	0.0225	0.2325	0.2725	0.4055	0.5000	0.2920	0.2377	0.4003

Table 8.11 Wave description for particulate bound metals

Event	Zn	Cd	Pb	Ni	Mn	Fe	Cr	Mg	Al	Ca	Cu	Na
	Molecular Diffusion Process											
	$T_{L(ibe)}$											
Aug. 8, 96	4	2	4	4	4	4	4	4	4	4	4	4
June 18, 96	2	2	2	2	2	2	2	2	2	2	2	2
Jul. 7, 96	10	10	4	10	10	8	8	10	10	10	10	10
	ϕ											
Aug. 8, 96	2.0944	1.0472	2.0944	2.0944	2.0944	2.0944	2.0944	2.0944	2.0944	2.0944	2.0944	2.0944
June 18, 96	0.8378	0.8378	0.8378	0.8378	0.8378	0.8378	0.8378	0.8378	0.8378	0.8378	0.8378	0.8378
Jul. 7, 96	2.2440	2.2440	0.8976	2.2440	2.2440	1.7952	1.7952	2.2440	2.2440	2.2440	2.2440	2.2440
	Wave type											
Aug. 8, 96	PDYW	PDYW	PDYW	PDYW	PDYW	PDYW	PDYW	PDYW	PDYW	PDYW	PDYW	PDYW
June 18, 96	PDYW	PDYW	PDYW	PDYW	PDYW	PDYW	PDYW	PDYW	PDYW	PDYW	PDYW	PDYW
Jul. 7, 96	PDYW	PDYW	PDYW	PDYW	PDYW	PDYW	PDYW	PDYW	PDYW	PDYW	PDYW	PDYW

Table 8.11 Continued

Event	Zn	Cd	Pb	Ni	Mn	Fe	Cr	Mg	Al	Ca	Cu	Na
	Rainfall-Runoff-Molecular Diffusion Process											
	$T_{L(uhe)}$											
Aug. 8, 96	3	1	3	3	3	3	3	3	3	3	3	3
June 18, 96	-5	-5	-5	-5	-5	-5	-5	-5	-5	-5	-5	-5
Jul. 7, 96	9	9	3	9	9	7	7	9	9	9	9	9
	ϕ											
Aug. 8, 96	1.5708	0.5236	1.5708	1.5708	1.5708	1.5708	1.5708	1.5708	1.5708	1.5708	1.5708	1.5708
June 18, 96	-2.0944	-2.0944	-2.0944	-2.0944	-2.0944	-2.0944	-2.0944	-2.0944	-2.0944	-2.0944	-2.0944	-2.0944
Jul. 7, 96	2.0196	2.0196	0.6732	2.0196	2.0196	1.5708	1.5708	2.0196	2.0196	2.0196	2.0196	2.0196
	Wave type											
Aug. 8, 96	PDYW	PDYW	PDYW	PDYW	PDYW	PDYW	PDYW	PDYW	PDYW	PDYW	PDYW	PDYW
June 18, 96	SDYW	SDYW	SDYW	SDYW	SDYW	SDYW	SDYW	SDYW	SDYW	SDYW	SDYW	SDYW
Jul. 7, 96	PDYW	PDYW	PDYW	PDYW	PDYW	PDYW	PDYW	PDYW	PDYW	PDYW	PDYW	PDYW
	η											
Aug. 8, 96	0.3752	0.5499	0.3419	0.4111	0.4356	0.4077	0.3715	0.3975	0.3989	0.4050	0.2773	0.3530
June 18, 96	0.0613	0.3172	0.2268	0.0255	0.1354	0.0345	0.0647	0.2589	0.0743	0.2067	0.0966	0.2834
Jul. 7, 96	0.4773	0.4564	0.5484	0.5186	0.3112	0.3015	0.3530	0.4173	0.4568	0.4574	0.4666	0.5023

are computed using equations (8.6)-(8.8) replacing 'H' by 'h_e' and 'Q' by 'u'. Since the three events shown in Figures 8.5-8.7 are multi-peaked events and the η-values stand for the complete events, these are not comparable with the computed ϕ-values that stand for a single discrete event. The computed η-values show the average wave behavior of the complete events.

For example, in Table 8.10, the ϕ-criteria describe all waves for all metals as dynamic waves whereas the dissolution (or molecular diffusion) process of Mn is equivalent to kinematic wave and of metals Pb, Mg, and Cu equivalent to diffusion waves in open channels. The difference between the two descriptions can be explained as follows. At low flows, the waves are usually kinematic or diffusion waves in nature and as the flow magnitude increases, the wave behavior tends towards dynamic wave. High magnitude dam break flood waves (Fread, 1985) are a classical example of dynamic waves. Thus, the mixed event described by the non-dimensional hysteresis, η, will be less dynamic than the discrete peak events described by the phase difference, ϕ.

Similar to the above wave description, the process of the particulate bound metals can be explained from Table 8.11. It is seen that all the waves identified correspond to the primary dynamic wave of open channel flow. Thus, the process of building of h_e follows the process of rainfall-excess (u) generation, as shown in Figure 8.6. On the other hand, as apparent from the table, the overall process of rainfall-runoff-molecular diffusion process is described by both primary and secondary dynamic waves. The mixing of all the metals during the June 18, 96 event is described by secondary dynamic waves. It implies that the h_e-wave precedes the u-wave and the former follows the rainfall intensity i-wave. It is because the mixing primarily occurs over the paved surface and the direct surface runoff is further lagged because of storage effects in routing, leading to the precedence of h_e-wave.

Parallel to the work of Ponce et al. (1996), the implication of such an analysis is that the waves identified as kinematic and diffusion waves can be simulated by equation (8.9) whereas for simulating dynamic waves, higher order terms than the present ones in equation (8.9), identical to those described by Ferrick (1985) for full dynamic wave (equations (8.13) and (8.14)), will be needed. Here, it is worth emphasizing that since these characteristics are valid for a point measurement at the outlet of the paved watershed, the described wave phenomena hold for the same point. Like the measurements of open channel flow, measurements at several points in space would be necessary to describe the wave propagation. Furthermore, the wave kinds described in Table 8.7 for metal transport may be entirely different from the overland flow waves. For establishing the kind of flow waves over the land, measurements of depth and velocity on the paved surface would be required.

8.5.6 DETERMINATION OF POTENTIAL MASS DEPTH OF FLOW

The proportionality concept of the SCS-CN method (equation (2.2)) is extended to determination of the potential mass of dissolved, particulate bound, and (dissolved + particulate bound) metals. The hypothesis states that the ratio of the actual amount of direct runoff to the maximum potential runoff is equal to the ratio of the amount of actual mass of dissolved or particulate bound metals to the potential maximum amount of the metals. Expressed mathematically,

$$\frac{Q}{P}=\frac{M_a}{M_p} \tag{8.55}$$

where M_a and M_p correspond to the total actual and potential mass of the metal element in an event, respectively. M_p can be computed for a known Q (direct surface runoff), P (rainfall), and actual mass, M_a, of the metal under study. The determination of M_p is illustrated below.

Example 8.4: Determine the potential mass of a metal element that is available for transport on a micro urban watershed. A rainfall of 50 mm on this watershed generated a 30 mm runoff which transported 1 gram mass of this metal.

Solution: Given: P = 50 mm, Q = 30 mm, and M_a = 1 gram. Using equation (8.55), the potential mass of the metal, which can be transported by a rainstorm is determined as:

$$\frac{Q}{P}=\frac{M_a}{M_p} \text{ or } M_p=\frac{M_a P}{Q}=\frac{1\text{x}50}{30}=1.67 \text{ gram}$$

The values of M_p were computed for all events utilizing the data reported by Sansalone and Buchberger (1997), Sansalone et al. (1998), and Li et al. (1999) and their variation with the antecedent dry period (ADP) is shown in Figures 8.11-8.13 for dissolved, particulate bound, and (dissolved + particulate bound) metals, respectively, for the event of August 8, 1996. Apparently, except for Zn, all other metals generally exhibit a similar trend of variation; M_p first increases with ADP, reaches a peak, and then approaches zero with the increasing dry period. On the other hand, the Zn metal, a major constituent, exhibits a distinguished and convincing (though apparently hysteretic) behaviour with ADP in all its three forms of dissolved, particulate bound, and (dissolved + particulate bound). For this metal, M_p is found to generally increase with ADP. It is worth mentioning that M_p is analogous to the SCS-CN parameter, S, and ADP to the antecedent moisture condition (AMC). The hysteretic relation is mild for dissolved (Figure 8.11) and strong for the particulate bound (Figure 8.12) and (dissolved + particulate bound) (Figure 8.13) Zn metal.

8.5.7 LIMITATIONS OF WAVE ANALOGY

The above described wave analogy suffers from the following limitations:

i) It does not describe the physical process of molecular diffusion and is, therefore, a conceptualisation of the process of metal transport.

ii) The waves in open channels are governed by the continuity and momentum equations. The wave phenomenon of metal transport as such does not involve momentum. The molecular diffusion in metal transport is independent of the gravitational force, a governing factor in open channel flow.

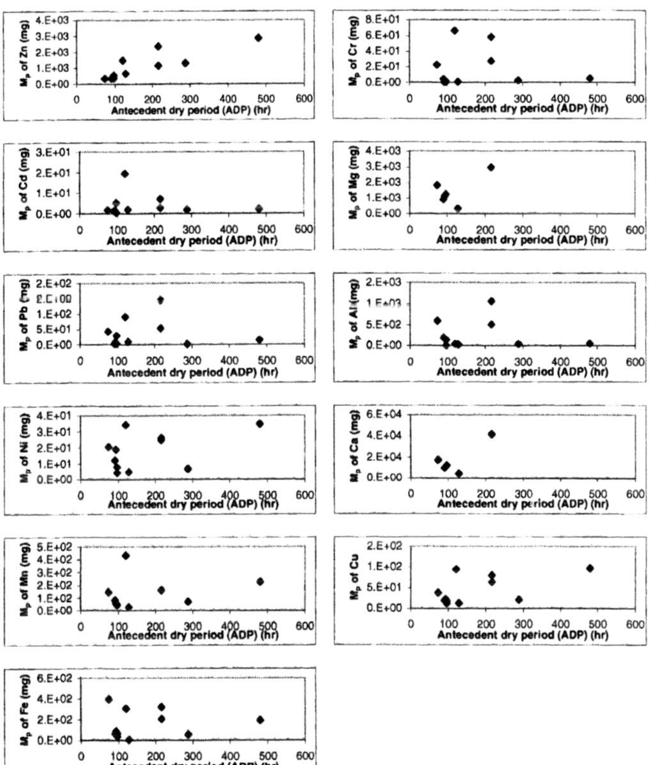

Figure 8.11. Behaviour of the potential maximum mass (M_p) of dissolved solids (μg) with the antecedent dry period (ADP) (hr).

iii) The assumption of the rate of rainfall-excess equivalent to the overland flow velocity is not realizable in practice, and thus, is a forced assumption.

iv) The process of metal deposition on the road stretch under study is independent of the rainfall-runoff process and the time lag (equation (8.51)) also depends on the rainfall intensity and duration.

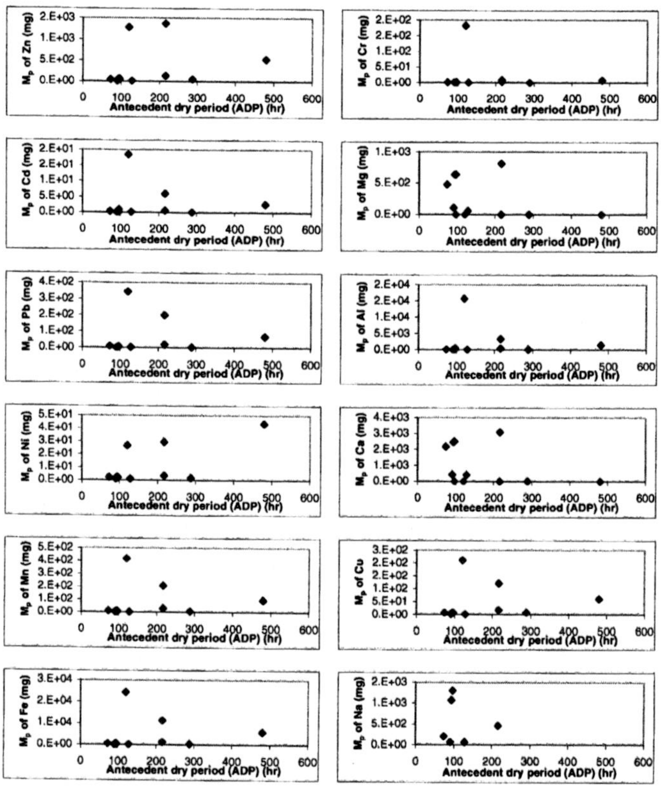

Figure 8.12. Behaviour of the potential maximum mass (M_p) of particulate bound solids (mg) with the antecedent dry period (ADP) (hr).

Despite the above limitations, the above wave analysis of non-point source pollution presents a simple alternative for computing the pollutant load for various metals from rainfall-runoff events and reveals the existence of strong non-linearity (or high ϕ- or η-values) in the process of molecular diffusion of various metals. It indicates the inefficacy of the governing second order partial differential equation (equation (8.9)) to simulate events exhibiting strong loops in the mass rating curves, and therefore, needs to be re-investigated for non-linearity, such as revision of the linear Fick's first law for non-linearity as a viable alternative.

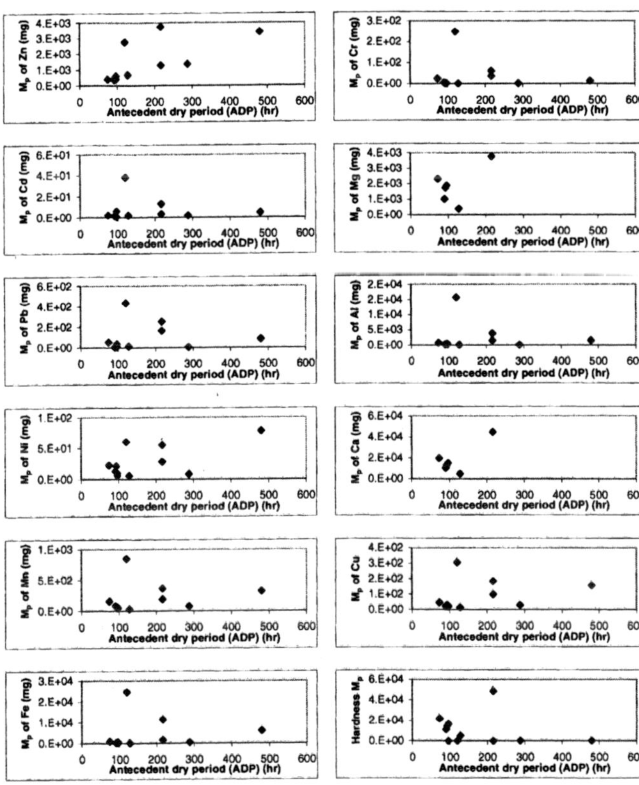

Figure 8.13. Behaviour of the potential maximum mass (M_p) of total (dissolved + particulate bound) solids (mg) with the antecedent dry period (ADP) (hr).

Table 8.12 Characteristics of the rainfall-runoff events

Sl. No.	Event	Previous dry hours	Rain duration (min)	Runoff duration (min)	Rain depth (mm)
1	Aug. 8, 1996	216	51	52	14.1
2	June 18, 1996	73	63	76	11.3
3	July 7, 1996	117	50	60	40.4
4	Oct. 17, 1996	182	616	609	29.1

Table 8.12 Continued

Sl. No.	Runoff volume (L)	Runoff factor, C	CN	IPRT (min)	APRT (min)
1	3877	0.91	99.6	7	3
2	2779	0.82	99.2	5	2
3	9643	0.79	96.9	4	1.5
4	3693	0.42	91.4	5	7

Note: IPRT and APRT stand for the initial and average pavement residence times, respectively.

8.6 Application of the SCS-CN Analogy to Metal Partitioning in the Rainfall-Runoff Environment

The above-described PCN-based method, designated as PCN method, was applied to analyze the partitioning behavior of the above 12 metals during the four rainfall-runoff events summarized in Table 8.12. Apparently, the July 8, 1996 event was the most intense storm event whereas the Oct. 17, 1996 event was the least intense. The Aug. 8, 1996 event yielded the highest runoff coefficient (C = 0.91) and CN = 99.6 and the Oct. 17, 1996 event yielded the lowest C (= 0.42) value and CN = 91.4. These events are designated as events 1 through 4 according to Sl. No. of Table 8.12. Since it is a paved urban watershed, the expected high CN-values for all the events lie in a narrow range (91.4, 99.6), exhibiting a high runoff potential of the watershed. On the other hand, C-values range from 0.42-0.91, which is a quite broad range of C-variation, not generally exhibiting the watershed characteristic. In Table 8.12, IPRT and APRT stand for initial and average pavement residence times, respectively. These are the indicators of the residence times of the rainwater on the pavement. APRT is computed using the method of moments (Sansalone et al., 1998) and IPRT is used for computing the amount of first or initial flush, i_f. A higher magnitude of the initial flush refers to the higher amount of metal abstraction by rainwater from the paved watershed during IPRT. Similar to the deduction of I_a from the total precipitation amount, the initial flush is also deducted from c_T for the completeness of the SCS-CN analogy. However, the underlying difference between i_f and I_a is that I_a is initially abstracted from the rainwater through evaporation, and, thus represents a loss whereas i_f appears at the outlet and is not abstracted.

8.6.1 DERIVATION OF K_d AND PCN

Similar to the SCS-CN method, the quantities of c_p, c_d, and c_T were first computed in the unit of micrograms in each time interval of the rainstorm, as above, and then cumulative values were determined for employing the PCN method. The computation of K_d values is shown below.

Example 8.5: During a rainstorm, a metal of 0.7 gram partitioned into 0.5 gram of particulate-bound form and 0.2 gram of dissolved form. Using these data, characterize the solution medium for partitioning of the metal using partitioning coefficient, dissolved fraction, and particulate-bound fraction.

Solution: Here, c_p = 0.5 gram, c_d = 0.2 gram, and c_T = 0.7 gram. These values can also be represented in terms of concentration by assuming a constant volume of the rainfall-excess. The partitioning coefficient K_d, dissolved fraction f_d, and particulate-bound fraction f_p, can be computed, respectively, using equations (8.3)-(8.5) as:

$$K_d = c_p/c_d = 0.5/0.2 = 2.5$$

$$f_d = \frac{c_d}{c_T} = \frac{0.2}{0.7} = 0.29$$

and

$$f_p = \frac{c_p}{c_T} = \frac{0.5}{0.7} = 0.71$$

Since $f_p > 0.5$, the metal is mainly in the particulate-bound form, and $f_d < 0.5$ indicates that the metal is not mainly in dissolved form. Alternatively, the solution has a significant adsorption potential. The K_d-value of 2.5, however, does not indicate a significant potential of the medium for partitioning.

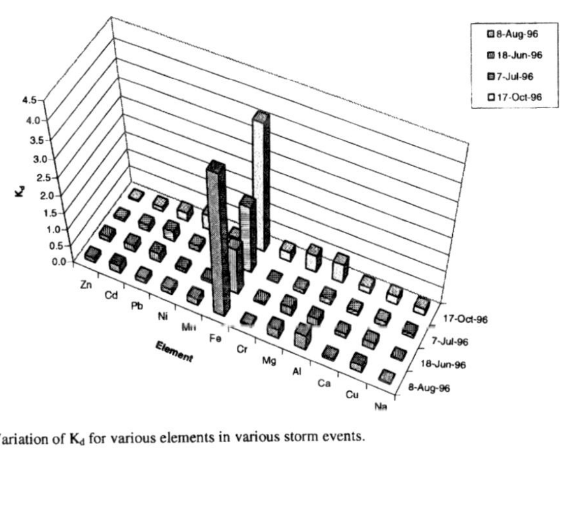

Figure 8.14. Variation of K_d for various elements in various storm events.

The K_d-values derived from equation (8.3) and shown in Figure 8.14 are based on the cumulative quantities of c_p and c_d. The K_d-values ranged 0.04-4.09, 0.02-1.36, 0.01-2.0, and 0.04-3.65 for events 1 through 4, respectively. Apparently, these values vary greatly from one event to the other and, therefore, it is difficult to distinguish the partitioning capability of various elements in these storms utilizing these values. However, the partitioning of Fe is clearly distinguished from other metals, for its K_d-values have far exceeded all others in all the storm events. Here, a short discussion on the variation of K_d during the storm is in order.

Variation of K_d for the Zn metal during the Aug. 8, 1996 storm (event 1) is depicted in Figure 8.15. It is noted that the low K_d-values were multiplied by 1000 to show their variation with time during the storm. Apparently, K_d increases with the setting up of the storm, continues to increase with increasing rainfall, attains a temporary constant value at about 10 minutes, then further increases with increasing rainfall intensities, and attains approximately an equilibrium condition at about 22 minutes. After 22 minutes, a further small rise and then a continuous decrease is also apparent. An increase in the K_d values exhibits an increase in the particulate bound metal formation with respect to the dissolved metal amount. Since K_d represents the ratio of the amounts of particulate and dissolved amounts, it assumes a decreasing trend at the fag end of the storm. It is because as one component of c_T (c_p or c_d) increases, the other decreases to conserve the mass. Therefore, the ratio of c_p to c_d (= K_d) either shows a consistently increasing or decreasing trend and, in turn, a fluctuating trend. The K_d-values plotted in Figure 8.14 represent the final K_d-value attained in a storm for a metal.

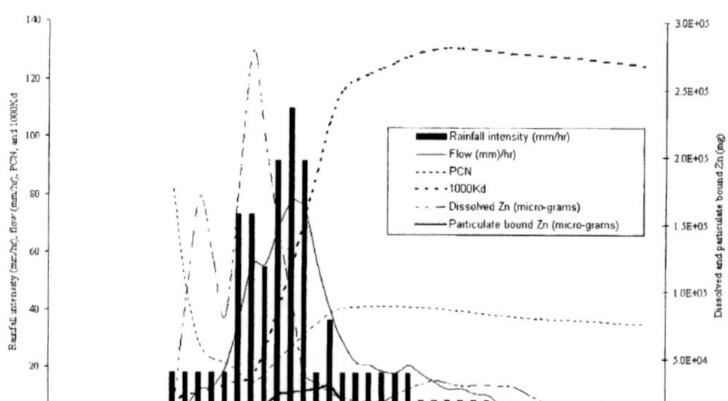

Figure 8.15. Variation of dissolved and particulate bound Zn, non-dimensional PCN, and K_d during the Aug. 8, 1996 storm.

For computing PCN, the ψ-values were first determined using equation (8.32) and then from equation (8.38) the PCN-values were computed for all 12 metals and for all 4 storms using the base amount of the input mass equal to 10 grams. The following example illustrates the determination of PCN from ψ values.

Example 8.6: Using the partitioning data of Example 8.5, compute the potential maximum desorption and the partitioning curve number and characterize the medium.

Solution: The potential maximum desorption, ψ, can be computed from equation (8.32) as:

$$\psi = \frac{c_T c_d}{c_p} = \frac{0.7x0.2}{0.5} = 0.28 \text{ gram}$$

The partitioning curve number PCN can be estimated from equation (8.38) as

$$PCN = \frac{1000}{10+\psi} = \frac{1000}{10+0.28} = 97$$

Similar to the SCS curve number (Chapter 2), the PCN-value of 97 indicates a very high adsorption potential of the medium. It is noted that a watershed having a CN-value of 97 is considered to be of a very high runoff potential.

It is apparent from Figure 8.15 that PCN assumes a very high value at the outset of the storm, implying that Zn has the greatest tendency for adsorption or particulation and it reduces with dissolution of the metal as rainfall continues, reaches a minimum value, and again increases to attain a final constant value, which represents the PCN value for the complete storm.

Table 8.13 PCN-based ranking of metals

Element	8-Aug-96	18-Jun-96	7-Jul-96	17-Oct-96	Av. PCN	Ranking
Zn	35	68	1	0	26	IV
Cd	100	100	85	93	94	XII
Pb	82	97	32	62	68	VIII
Ni	97	97	36	82	78	XI
Mn	84	76	16	13	47	VI
Fe	93	90	46	63	73	X
Cr	82	94	4	96	69	IX
Mg	28	40	1	5	18	III
Al	61	66	2	27	39	V
Ca	1	4	0	0	1	II
Cu	93	96	26	52	67	VII
Na	1	1	0	0	1	I

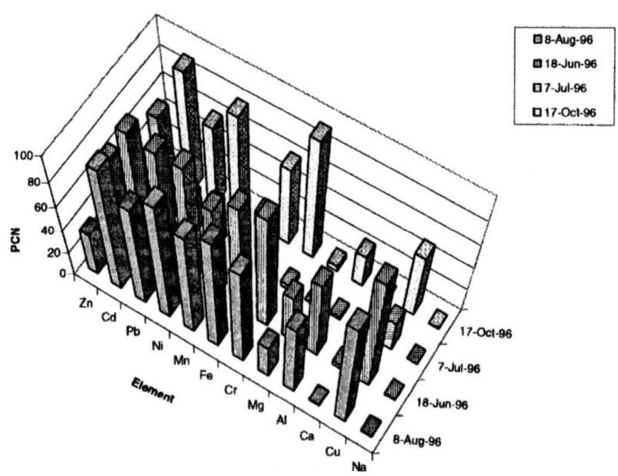

Figure 8.16. Variation of PCN for various elements in various storm events.

Similar to the derivation of the K_d-values, the computed PCN-values for a metal also correspond to the final value of PCN (or ψ). The computed PCN-values are given in Table 8.13 and depicted in Figure 8.16. As apparent from the table and the figure, metals Na and Ca exhibit low PCN-values and are clearly distinguishable from others in that they are least adsorbed or most dissolved compared to others during the above storms. On the other hand, elements Cd, Ni, and Fe are most adsorbed or least dissolved. On the other hand, the K_d values signify only Fe to be adsorbed most. Thus, based on these PCN values, it is possible to rank different metals in the order of their adsorption characteristics. To this end, the derived PCN-values for individual storms were averaged to obtain a representative average value, based on which the derived ranking is given in Table 8.13. It can be appreciated in view of the ranking of the metals based on their chemical characteristics, shown in Table 8.14 and described below.

The covalent theory holds for divalent metal ions (McBride, 1994). The predicted order of preference using this theory differs from that based on the concept of electrostatics, according to which the bonding preference depends on the charge-to-radius ratio. These produce a different order of preference for divalent metals, as shown in the table. It is noted that the trivalent ions, such as Cr^{+3}, have a greater bonding preference to all divalent metal ions. Another concept of hydrolysis is based on the tendency of metals to hydrolyze. Accordingly, the bonding preference of metals is decided based on the ratio of metal ions to iron oxide (McBride, 1994). It is apparent from Table 8.14 that there is no standard theory yet available to yield a consistent ranking of metals.

Table 8.14 Order of preference of metals

Sl. No.	Concept	Order of preference of metals	Reference
1	Covalent theory	Cu > Ni > Co > Pb > Cd > Zn > Mg > Sr	(McBride, 1994)
2	Electrostatics	Ni > Mg > Cu > Co > Zn > Cd > Sr > Pb	(McBride, 1994)
3	Hydrolysis	Pb > Cu > Zn > Co > Ni > Cd > Mn	(McBride, 1994)

According to the theory of hydrolysis (Table 8.14), the Ni, Cd, and Mn metals have the tendency to hydrolyze or dissolve least of the above 12 metals. This is consistent with the PCN-based ranking (Table 8.13), according to which Ni and Cd are ranked as XI and XII, respectively. Metal Mn is, however, an exception, for it is ranked as VI in contrast with the concept of hydrolysis (Table 8.14). It is emphasized here that a high PCN-value indicates a great potential for adsorption and weak for dissolution. Similarly, Pb and Cu are ranked as VIII and VII, respectively, according to the above average values of PCN. These are, however, close to each other in ranking. The least adsorbing element Cr (of highest bonding preference) is placed at IX according to the average PCN-value. Given a large set of data, it is, thus, possible to rank metals in order of their behavior in the partitioning process using PCN as an index, similar to the characterization of watersheds using CN. The characterization of the partitioning behavior in snowmelt, rainfall-runoff, and river flow systems is discussed later in this chapter.

Example 8.7: During a rainfall-runoff event, the masses of three metals X, Y, Z transported by runoff from a highway stretch were measured. The observations yielded the partitioning curve number (PCN) values as 94, 55, and 72 for the respective elements. Rank these elements for their adsorption potential in the considered medium.

Solution: A high PCN indicates a higher adsorption potential in the medium and vice versa. Thus, X > Z > Y is the order of their adsorption potential.

8.6.2 RELATIONS BETWEEN ψ AND CHEMICAL CHARACTERISTICS OF RAINFALL

The intensity of the process of adsorption or metal partitioning is generally attributed to rainfall pH and its alkalinity, traffic levels, and residence times (Sansalone and Glenn, 2000; Glenn et al., 2001). As shown in Figure 8.15, PCN varies with time during the storm rainfall, of which the pH and alkalinity characteristics were also measured during the experimental work. The ψ-values computed at each time step were correlated with these characteristics, leading to the following general linear relationship:

$$Y = a_1X_1 + a_2X_2 + b \tag{8.56}$$

where Y represents the ψ value on a natural logarithmic scale (= ln ψ) (microgram); X_1 alkalinity (milligram $CaCO_3$/litre); X_2 is the pH value of the rainwater at the scale of 0-14; and a_1, a_2, and b represent the parameters. The ψ-values were taken on a logarithmic scale to reduce high magnitudes and to enhance the coefficient of determination, r^2, for which the computed values of ψ using equation (8.32) are in perfect agreement with those computed using equation (8.56). The computed values of the parameters a_1, a_2, and b and r^2 for all metals and all storms are shown in Table 8.15. It is seen from this table that parameter a_1, in general, assumes a negative value for all the metals and for all the events, except the Oct. 17, 1996 event (event 4). It implies that, in general, the magnitude of ψ at an instant during the rainstorm is inversely related to the alkalinity of the rainwater; the higher the alkalinity, the lower will be the ψ or higher the CN and vice versa. Except for the June 18, 1996 event (event 2), all other events for all the metals exhibit a positive value of parameter a_2, implying that ψ increases with the increasing pH of the rainwater and vice versa.

Parameter b, in general, assumes a positive high value compared to the parameters a_1 and a_2. Since ψ is taken on the logarithmic scale, the non-zero values of parameter b indicate the existence of ψ because of its dependency on other unknown factors besides the pH and alkalinity. Except for the metal Zn in event 1, for which r^2 = 0.81, a high value, all metals show r^2 values less than 0.62, exhibiting a poor correlation, and in event 2, r^2 values for all metals are less than 0.5, which also show a poor correlation. However, the r^2-values for events 3 and 4 range from 0.82-0.89 and from 0.73-0.91 (except for Na for which $r^2 = 0.42$, which is very low), respectively. The high values of r^2 exhibit good correlations. Thus, it can, in general, be inferred that ψ (or PCN) generally depends on the pH and alkalinity characteristics of the rainwater. In an attempt to identify which of these characteristics affect more, the computed ψ-values were further correlated with pH and alkalinity separately for event 1 for element Zn. pH correlated with ψ with r^2 = 0.001 (exhibiting virtually no correlation) and the correlation with the alkalinity yielded $r^2 = 0.76$, showing a reasonably good correlation, implying that ψ relies more on alkalinity than on the pH. It might be because of the abundance of $CaCO_3$ in the asphalt-mix pavement material. The rainwater during storms mixes with the exposed material that is rich in calcium contents.

8.6.3 RELATION BETWEEN i_f AND ψ

Similar to the relationship between the initial abstraction and the potential maximum retention, S, of the SCS-CN method (equation (2.3)), it is appropriate to relate the initial flush with the PCN parameter potential maximum desorption, for deriving the parameter λ_1. According to NEH-4 (SCS, 1971), the derivation of λ is based on the assumption that S includes I_a (Chen, 1982) whereas it excludes it (Chapter 3). Therefore, the ψ-values computed using equation (8.33a) were related to the initial flush, i_f, values for the events of Aug. 8, 96; Dec. 16, 96; June 12, 95; June 18, 96; May 21, 96; Nov. 25, 96; and Oct. 17, 96 events published by Sansalone et al. (1998). The relations for all 12 metals are depicted in Figure 8.17 and the computed λ_1 values along with the r^2-values are given in Table 8.16. It is apparent from this figure and the table that the relations for metals Zn, Cd, Mn, and Cu yield r^2 values equal to 0.992, 0.879, 0.785, and 0.621, respectively, which exhibit reasonable fits. The relations for other metals are poorly correlated because of the low or even negative r^2-values. In rainfall-runoff modeling, the relations between I_a and S are no better than those derived for these metals in the present study. The relationship of NEH-4 (SCS, 1971) exemplifies it. Based on the elements exhibiting a good correlation, a value of λ_1 equal to 0.02 appears to be an appropriate value.

Table 8.15 Relations for desorption capacity

Element	Aug. 8, 1996				June 18, 1996			
	a_1	a_2	b	r^2	a_1	a_2	b	r^2
Zn	-0.0850	0.1835	16.83	0.81	-0.0211	-0.0654	15.98	0.29
Cd	-0.0902	0.2854	9.45	0.62	-0.0353	-0.1873	11.59	0.46
Pb	-0.0757	1.9085	1.01	0.44	-0.0612	-0.3541	16.12	0.47
Ni	-0.0737	0.5028	9.97	0.43	-0.0418	-0.1685	14.71	0.45
Mn	-0.0764	0.4581	12.16	0.57	-0.0334	-0.1932	17.00	0.45
Fe	-0.0413	1.1804	9.73	0.32	-0.0738	-0.4110	18.39	0.49
Cr	-0.0278	2.1235	-1.72	0.40	-0.0684	-0.3473	16.39	0.42
Mg	-0.0565	0.7420	11.98	0.32	-0.0408	-0.2312	18.95	0.45
Al	-0.0547	1.2667	6.59	0.38	-0.0720	-0.3742	19.30	0.49
Ca	-0.0464	0.8580	14.68	0.31	-0.0369	-0.2063	21.60	0.46
Cu	-0.0410	0.7179	8.51	0.33	-0.0395	-0.2128	15.04	0.46
Na	-0.0348	0.9265	13.57	0.35	-0.0370	-0.2113	22.94	0.50

Table 8.15 Continued

Element	July 7, 1996				Oct. 17, 1996			
	a_1	a_2	b	r^2	a_1	a_2	b	r^2
Zn	-0.0486	1.8159	7.99	0.86	0.0341	0.4008	18.20	0.73
Cd	-0.0820	1.8717	1.53	0.83	0.0295	0.4523	9.81	0.84
Pb	-0.1898	2.8714	-2.21	0.85	-0.0658	1.1933	7.05	0.76
Ni	-0.0869	2.0406	2.77	0.82	-0.0095	0.7573	8.96	0.91
Mn	-0.0417	1.5406	7.11	0.88	0.0139	0.5568	13.77	0.91
Fe	-0.0827	2.2605	0.92	0.88	-0.0365	0.9636	8.57	0.83
Cr	-0.0385	7.5433	-33.20	0.86	0.0143	0.5616	8.41	0.91
Mg	-0.1242	2.6651	2.96	0.88	0.0181	0.5202	15.10	0.89
Al	-0.1885	2.9354	0.33	0.86	-0.0339	0.9519	10.13	0.82
Ca	-0.0378	2.3035	6.76	0.86	0.0250	0.4666	18.49	0.84
Cu	-0.0797	1.7879	5.03	0.89	0.0177	0.5230	11.95	0.89
Na	-0.0481	2.4092	5.47	0.85	0.0623	0.1649	19.79	0.42

Table 8.16 Computation of λ_1-values

Element	λ_1	r^2	Element	λ_1	r^2
Zn	0.017	0.992	Cr	0.0	-0.219
Cd	0.023	0.879	Mg	0.049	0.447
Pb	0.002	-0.030	Al	0.012	0.116
Ni	0.027	0.304	Ca	0.011	0.078
Mn	0.017	0.785	Cu	0.046	0.621
Fe	0.182	0.252	Na	0.004	-0.161

8.6.4 RELATION BETWEEN ADP AND ψ

Similar to the existence of the relations between the antecedent moisture condition (AMC) and the potential maximum retention (Chapter 3), it is possible to relate the potential maximum desorption with the antecedent dry period (ADP). ADP actually affects the total quantity of the metals deposited on the pavement due to the vehicular traffic, which, in turn, affects ψ (equation (8.32)). The developed relations are depicted in Figure 8.18 for all 12 metals. In this figure, the y-axes of the linear trends correspond to ψ (μg) and the x-axes to ADP (hours). The apparent coefficient of determination, r^2, ranges from –0.17 to 0.48, implying that ψ not only depends on ADP but also on other factors, such as leaching of metal elements from infrastructure, oil and grease leakages, and so on.

Example 8.8: Determine the partitioning curve number for the Zn metal using the characteristics of the rainfall measured in Cincinnati on August 8, 1996 as: pH = 8.73 and alkalinity = 18 milligram $CaCO_3$/litre. Also characterize the rainwater (medium) for the partitioning of Zn metal.

Solution: For determination of PCN, the potential maximum adsorption, ψ, is computed using equation (8.56) as

$$Y = a_1X_1 + a_2X_2 + b$$

where Y = ln ψ. Given: X_1 = 18 milligram and X_2 = 8.73. From Table 8.15, the parameters are: a_1 = -0.085, a_2 = 0.1835, and b = 16.83. Thus,

$$\ln \psi = -0.085 \times 18 + 0.1835 \times 8.73 + 16.83 = 16.91$$

Therefore, $\psi = 21.90 \times 10^6$ micrograms = 21.9 grams.

PCN can be computed from equation (8.38) as:

$$PCN = \frac{1000}{10+\psi} = \frac{1000}{10+21.9} = 31$$

The low value of PCN (= 31) indicates that the medium with the given alkalinity and pH levels has a low adsorption potential for the Zn metal.

8.7 Application of the SCS-CN Analogy to Metal Partitioning in the Snowmelt Environment

8.7.1 SNOWMELT WATER QUALITY DATA

Four primary highway sites in Greater Cincinnati, Ohio, U.S.A., were selected for collecting the snowmelt water quality data. Three sites were located on Inter State 75 (I-75), a major north-south freeway between Michigan and Florida and another on U.S. Route 126, 1 km east of I-75. An additional site located in the residential area of urban Cincinnati was taken as a control site, where the snowfall was unaffected by traffic activities. Figure 8.19 illustrates the locations of each site and the detailed plan views of the four primary sampling sites designated as 1SW, 1SE, 2ES, and 3SW are given in Figure 8.20. The first number of these designations corresponds to the site number, the first letter indicates the direction of traffic movement, and the last letter refers to the highway side of sampling.

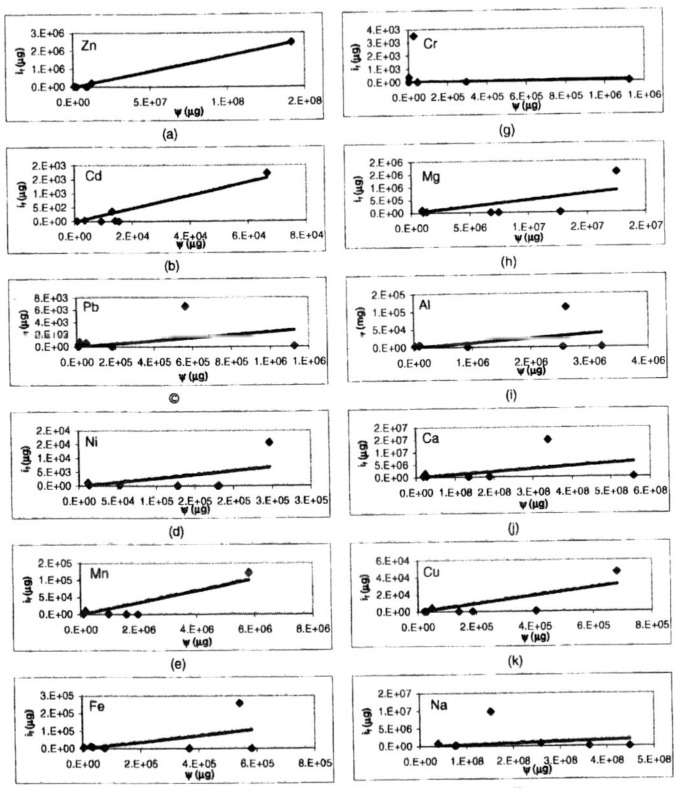

Figure 8.17. Relation between initial flush, i_f, (μg) and the potential maximum desorption, ψ, (μg)

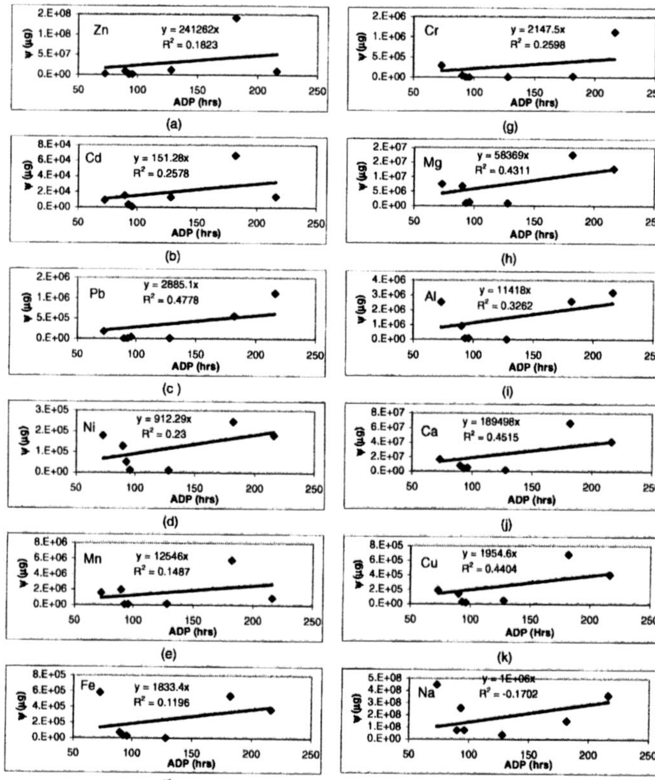

Figure 8.18. Relation between antecedent dry period (ADP) (hours) and the potential maximum desorption, ψ, (μg)

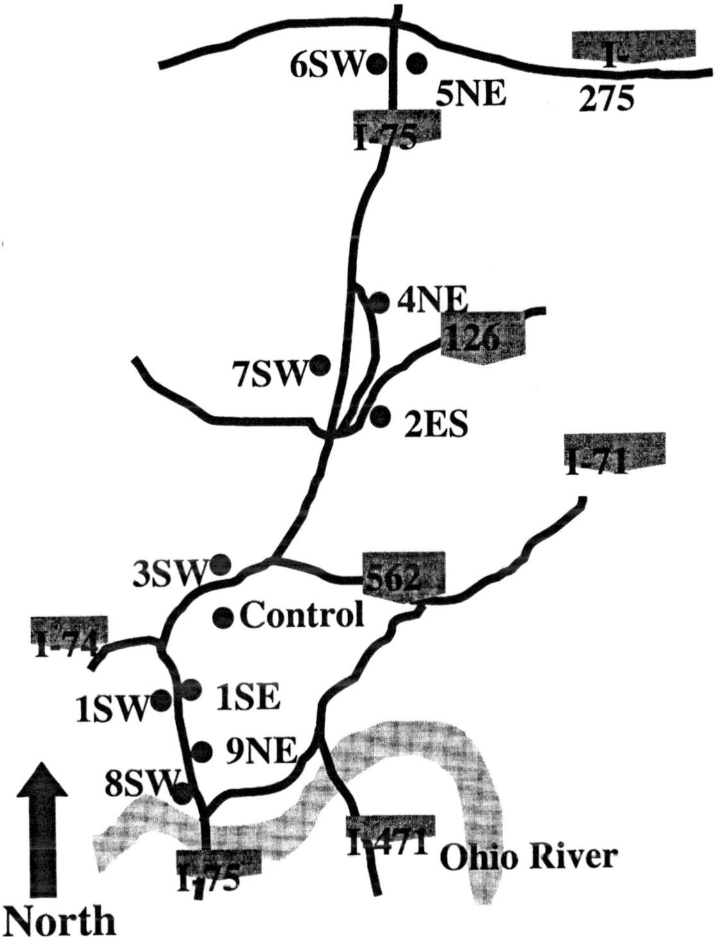

Figure 8.19. Location of four primary snow sampling sites and the control for metropolitan Cincinnati study.

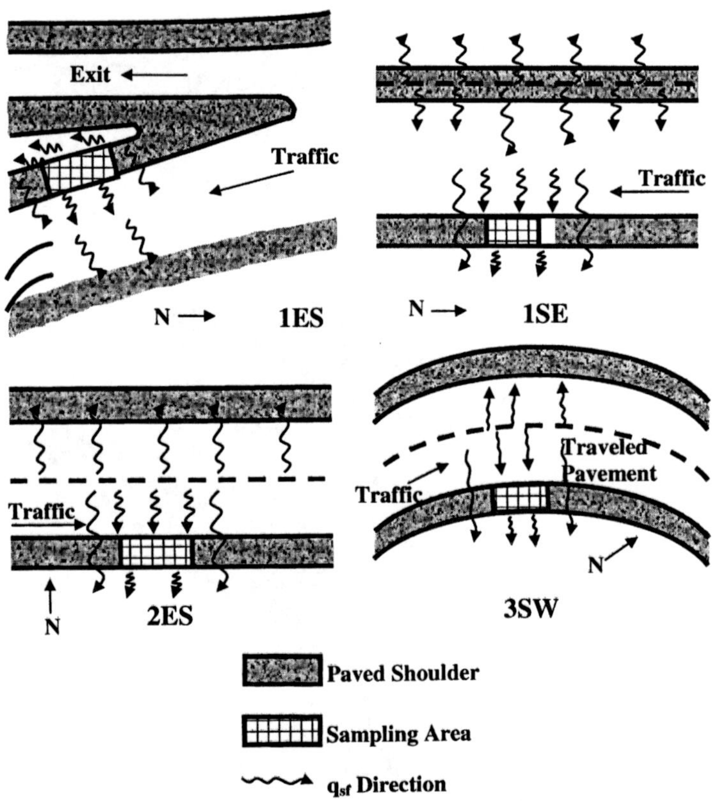

Figure 8.20. Plan view schematics of four primary snow sampling sites for metropolitan Cincinnati study.

Snow samples were collected from all the above sites at every 4 hrs during snowfall and then at times varying from 6-24 hrs. Typically, 18 samples were collected at each of the sites for the 102-hr duration until the snow banks melted fully. From each site 4-litre rectangle-shaped snow samples were collected from the leading edge of the snow bank exposed to traffic. After sealing, these samples were transported to the laboratory, where these were melted and placed in 2-litre polypropylene containers. Using a 0.45-µm membrane filter and vacuum filtration apparatus, aliquots were separated from each sample for the acid-preserved of the dissolved and particulate-bound fractions. Water quality parameters were derived following standard methods (Standard, 1995).

8.7.2 METAL PARTITIONING IN SNOWMELT MEDIUM

The above PCN method was applied to analyze the partitioning behavior of traffic-generated snow-bound metals, viz., Al, Ca, Cd, Cu, Fe, Mg, Na, and Zn. Of the available partitioning indices, viz., K_d, f_d, and f_p, the most versatile K_d (O'Connor, 1985; Sansalone and Glenn III, 2000; Sansalone and Buchberger, 1997; Sansalone et al., 1998; Sansalone and Tribouillard, 1999) was taken for comparison with the present metal partitioning parameter, PCN, for evaluating the performance of each in describing their partitioning behavior in a system. To this end, the variation of K_d and PCN with time is illustrated for metal Al at 1SW in Figure 8.21. It is apparent from this figure that as the snowfall increases, parameter K_d also generally increases. and it continues to rise even beyond the cessation of snowfall. The fluctuations in the K_d-curve after the end of snowfall appear to be a function of the variation in temperature. As the temperature increases, the rate of increase of K_d (with time) also increases and vice versa. For example, when temperature rises sharply at 72 hrs, K_d also undergoes a sudden rise. On the other hand, the variation of PCN is, for the most part, unaffected by the changes in temperature. PCN first increases with the outset of snowfall and then continuously decreases, leading to a constant PCN-value well within the duration of snowfall. Such a constant value can be taken to be representative of an event, similar to the curve number, CN, in the rainfall-runoff process. The first increase of PCN with the setting up of snowfall implies that Al has a greater tendency for adsorption or particulation in the beginning of snowfall than in its later part and PCN reduces with dissolution of the metal with the snowmelt and finally reaches a minimum value, which represents the PCN value for the complete storm. Here, it is noted that the fundamental difference between K_d and PCN is that PCN represents the adsorption potential of a 10-unit metal concentration in a certain environment, whereas K_d simply represents the proportion of the adsorbed material to the dissolved, irrespective of the concentration amount of the metal under study. Thus, the K_d and PCN-values at the end of 102-hr, when the snow melts completely, are taken to be representative of the particular event in the following investigations.

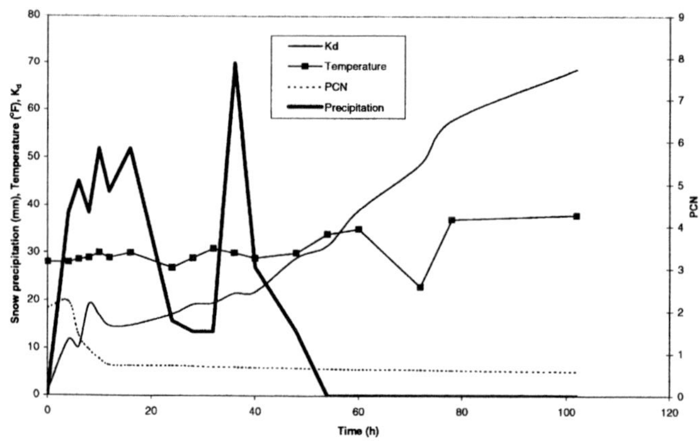

Figure 8.21. Variation of snowfall (water equivalent), K_d, and PCN with time for metal Al at site 1SW.

8.7.3 RELATION OF PCN AND K_d WITH THE MEDIUM CHARACTERISTICS

In an attempt to investigate the dependence of PCN and K_d on the physical and chemical characteristics of the medium during the process of snowmelt, these were linearly correlated with temperature, relative humidity, pH, and Cl- content as:

$$Y = a_1 X_1 + a_2 X_2 + a_3 X_3 + a_4 X_4 + b \tag{8.57}$$

where Y is the PCN or K_d; and X_1 through X_4 represent respectively the temperature (°F), relative humidity (%), pH of snowmelt at the scale of 0-14, and the cumulative Cl- content (mg/l); and a_1-a_4 and b are the parameters. For all the above metal elements observed at all the four sites, the computed parameters for PCN and K_d are summarized in Table 8.17. As apparent from this table, PCN correlates with the above characteristics in the range of the coefficient of determination, r^2, as: (0.28, 0.75) at site 1SW, (0.45, 0.84) at 1SE, (0.64, 0.94) at 2ES, and (0.21, 0.78) at 3SW. Similarly, K_d correlates with r^2 ranging from 0.24-0.80 for the site 1SW, from 0.03-0.93 for 1SE, from 0.30-0.98 for 2ES, and from 0.16-0.86 for 3SW. Thus, both the indices exhibit almost the same kind of behavior, as also shown in Figure 8.22, which compares the coefficient of correlation, r^2, (Table 8.17) for PCN and K_d. As seen in Figure 8.22, for some metals, K_d exhibits a higher r^2-value than the PCN and for other metals, the reverse holds. Furthermore, for several metals at various sites, the r^2-values are not generally very high to expect a strong relationship of PCN and K_d with physical and

Table 8.17 Correlation of PCN and K_d with the medium characteristics at four sites

Element	ISW						ISE					
	a_1	a_2	a_3	a_4	b	r^2	a_1	a_2	a_3	a_4	b	r^2
PCN												
Al	0.0180	-0.0154	-0.4482	0.0	6.95	0.6762	-0.0023	-0.0109	-1.7076	0.0	14.59	0.6093
Ca	-0.0014	-0.0016	0.0169	0.0	0.04	0.3162	0.0002	0.0	-0.0248	0.0	0.18	0.5529
Cd	-1.0624	-0.5916	3.0901	-0.0013	107.84	0.7475	-0.5672	0.6342	-43.3842	-0.0004	306.92	0.8367
Cu	-1.5221	-0.2618	5.5819	-0.0010	56.42	0.4801	-0.2565	0.7327	-55.5511	0.0	354.38	0.6848
Fe	-0.8462	-0.5746	10.2972	-0.0010	26.26	0.3492	0.4040	0.0331	-38.1876	0.0	271.20	0.4671
Mg	-0.0076	-0.0054	0.0371	0.0	0.77	0.5601	0.0009	-0.0006	-0.2221	0.0	1.79	0.7720
Na	0.0	0.0	0.0008	0.0	0.0	0.2776	0.0003	0.0	-0.0023	0.0	0.0	0.4547
Zn	-0.0272	-0.0201	0.0890	0.0	3.43	0.5396	-0.0018	-0.0005	0.1415	0.0	-0.79	0.7254
K_d												
Al	-0.0797	-0.8038	0.0525	0.0005	87.20	0.7913	0.1750	-0.5293	-5.6127	0.0005	90.60	0.8546
Ca	0.0057	-0.0168	0.0851	0.0	1.69	0.5115	-0.0107	-0.0212	0.5482	0.0	-0.82	0.8440
Cd	0.0711	0.0320	-1.5352	0.0	8.60	0.2717	-0.0283	0.0370	-1.2260	0.0	9.55	0.2171
Cu	0.0980	-0.0245	1.6676	0.0	-3.95	0.4915	-0.2690	-0.2149	2.2341	0.0001	21.55	0.8879
Fe	3.3650	2.8408	64.2009	0.0025	-696.28	0.2397	-0.0605	0.0646	-7.7573	0.0	92.87	0.0301
Mg	0.0247	-0.0231	0.8602	0.0002	-0.56	0.7125	-0.1361	-0.0347	7.3858	0.0	-41.59	0.9294
Na	-0.0148	-0.0030	0.0681	0.0	0.25	0.4014	0.0017	0.0002	-0.0374	0.0	0.21	0.3856
Zn	0.0334	-0.0941	0.0732	0.0001	9.84	0.7965	0.0797	-0.0542	-0.5020	0.0	6.93	0.8850

Table 8.17 Continued

Element	2ES						3SW					
	a_1	a_2	a_3	a_4	b	r^2	a_1	a_2	a_3	a_4	b	r^2
PCN												
Al	-0.0414	-0.0117	-1.2227	0.0	12.91	0.8546	-0.0197	-0.0216	-0.7447	0.0	9.88	0.7462
Ca	-0.0018	-0.0002	-0.0824	0.0	0.70	0.6364	-0.0010	-0.0009	-0.0148	0.0	0.24	0.3345
Cd	-0.4327	-0.3311	-27.0139	-0.0003	281.56	0.9414	-2.2440	-0.0332	-16.0173	-0.0007	237.78	0.7813
Cu	-0.7768	-0.4678	-48.1436	0.0	450.65	0.8640	0.9048	0.5185	-64.8522	-0.0003	423.64	0.7416
Fe	-0.7816	-0.1041	-32.9119	0.0	288.56	0.7073	-0.7683	-0.7049	-12.6742	-0.0005	197.18	0.2558
Mg	-0.0217	-0.0035	-0.8619	0.0	7.76	0.7721	-0.0094	-0.0081	-0.1205	0.0	2.19	0.4862
Na	0.0	0.0	-0.0004	0.0	0.0	0.7654	0.0008	0.0	-0.0059	0.0	0.02	0.3968
Zn	-0.0153	-0.0030	-0.6081	0.0	6.36	0.8455	0.0073	0.0074	0.1027	0.0	-0.94	0.2136
K_d												
Al	1.9585	-0.1846	19.6698	0.0	-170.12	0.6241	4.5535	1.4640	-26.2308	0.0030	-68.16	0.8611
Ca	0.0185	-0.0034	-0.0831	0.0	1.01	0.2926	0.0823	0.0417	1.4540	0.0	-14.82	0.4548
Cd	0.0316	0.0039	-0.6789	0.0	4.66	0.9758	0.9215	-0.9228	-0.0700	-0.0006	111.16	0.1576
Cu	-0.0613	-0.0159	-7.1159	0.0	67.76	0.7385	6.0266	1.0660	-24.0299	0.0014	-83.49	0.7574
Fe	6.4645	-0.7804	60.1465	-0.0005	-446.06	0.4880	2.1093	0.4252	46.5360	-0.0004	-380.87	0.3990
Mg	0.2059	-0.0564	2.8203	0.0	-17.01	0.6310	0.1258	0.1551	5.7211	0.0	-50.75	0.3859
Na	-0.0002	0.0	-0.0101	0.0	0.09	0.7093	-0.0017	-0.0013	-0.0271	0.0	0.40	0.2966
Zn	0.3128	-0.0153	3.3230	0.0	-28.88	0.7057	0.6749	0.1013	10.0502	0.0004	-100.57	0.8532

chemical characteristics of the medium. Within the premise of the available extent of correlation (Table 8.17), the behavior of metals adsorption is analyzed below.

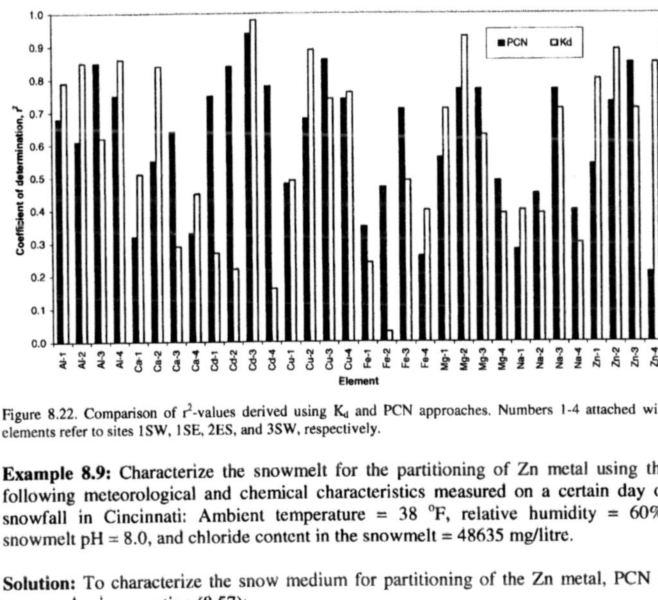

Figure 8.22. Comparison of r^2-values derived using K_d and PCN approaches. Numbers 1-4 attached with elements refer to sites 1SW, 1SE, 2ES, and 3SW, respectively.

Example 8.9: Characterize the snowmelt for the partitioning of Zn metal using the following meteorological and chemical characteristics measured on a certain day of snowfall in Cincinnati: Ambient temperature = 38 °F, relative humidity = 60%, snowmelt pH = 8.0, and chloride content in the snowmelt = 48635 mg/litre.

Solution: To characterize the snow medium for partitioning of the Zn metal, PCN is computed using equation (8.57):

$Y = a_1 X_1 + a_2 X_2 + a_3 X_3 + a_4 X_4 + b$

where Y = PCN, X_1 = temperature = 38 °F, X_2 = relative humidity = 60%, X_3 = pH of snowmelt = 8.0, X_4 = Cl-content = 48635 mg/litre. From Table 8.17, parameter

$a_1 = -0.0272$, $a_2 = -0.0201$, $a_3 = 0.089$, $a_4 = 0.0$, and $b = 3.43$. Thus,

$PCN = -0.0272 \times 38 - 0.0201 \times 60 + 0.089 \times 8 + 0.0 \times 48635 + 3.43 = 1.9 \approx 2$

A very low value of PCN characterizes the snowmelt medium as a poor medium for partitioning of the Zn metal. Alternatively, the medium has very poor adsorption potential for the Zn metal. It is noted that the chloride content does not play any role in the description of PCN.

According to Stumm and Morgan (1981), the extent of sorption varies positively with temperature, implying that sorption decreases with the decrease in temperature and vice versa. As a corollary, the adsorption is inversely related to the temperature. Analogous to CN that represents the runoff potential of a watershed, PCN represents the adsorption potential of a metal in a system. Thus, for PCN (Table 8.17), the inverse behavior of temperature to metal adsorption is supported by all the metals except Al at site 1SW; by Al, Cd, Cu, and Zn at site 1SE; by all the metals at site 2ES; and by all the metals except Cu and Zn at site 3SW. The other metals at sites exhibit a contrasting behavior. However, K_d, in general, shows a reverse trend; it generally increases with the rise in the temperature and vice versa. Similarly, the effect of relative humidity, pH, and Cl- content can be explained from Table 8.17.

8.7.4 PCN- AND K_d-BASED RANKING OF METALS

PCN and K_d can also be used for ranking of the metals for their adsorption potential in a system, as shown in Figure 8.23. It is apparent from this figure that PCN ranks the Cd metal to exhibit the highest adsorption potential at all the sites. However, K_d ranks the metal Fe to be of the highest adsorption potential at the sites 1SW and 2ES, and Al at the sites 1SE and 3SW. Analogous to the CN that has the efficacy to distinguish two watersheds from each other for their hydrologic (runoff producing potential) behavior, PCN also generally describes the consistent adsorption potential of the metal at various sites. However, both parameters PCN and K_d describe Na as the least adsorbing metal. The ranking by PCN at sites 1SW and 1SE exactly matches with each other, reflecting the superior efficacy of PCN to that of K_d. The other sites, however, exhibit a little discrepancy in the ranking of Zn, Cu, Al, and Fe. Based on the results of Figure 8.23 and judgment, the following overall ranking of the metals in the snowmelt environment can be recommended:

$$Cd > Cu > Zn > Al > Fe > Mg > Ca > Na$$

8.8 Application of the SCS-CN Analogy to Metal Partitioning in the Riverflow Environment

8.8.1 DON RIVER FLOW AND WATER QUALITY DATA

The Don River (latitude: 43°42′ and Longitude: 79°25′) is about 32 km long and drains a greatly urbanized watershed, in which the city of Toronto is situated. The river drains a number of sewage disposals including the treated sewage effluent and runoff from a snowmelt yard. A complete description of the Don River system and its water quality parameters is given by Warren and Zimmerman (1993, 1994). This analysis utilizes the data published by Warren and Zimmerman (1994).

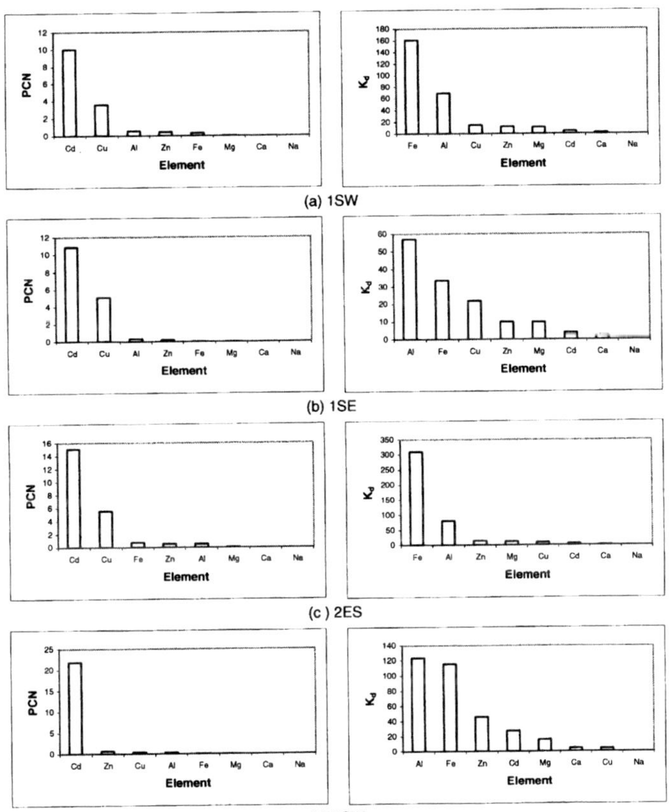

Figure 8.23. PCN- and K_d-based ranking of metal elements at the four sites.

8.8.2 METAL PARTITIONING IN RIVER FLOW SYSTEM

Similar to the above, the computed PCN- and K_d-values for 10 samples of each Cd, Cu, and Zn metal in the Don River flows are shown in Table 8.18. The average of the PCN- and K_d-values, shown in the last two columns of this table, are

Table 8.18 PCN- and K_d-based partitioning analysis of Don river metals

Element	Sample	[a]Leachable (µg/g)	[a]Reducible (µg/g)	[a]Oxidizable (µg/g)	[a]Residual (µg/g)	Total (µg/g)	[a]TSS (mg/l)
Cd	1	3.02	1.37	0	0	4.39	15.4
	2	5.26	1.93	0	0	7.19	8.8
	3	2.49	0.26	0	0	2.75	4.8
	4	2.79	0.47	0.81	0	4.07	21.6
	5	4.31	1.12	0	0	5.43	5.9
	6	2.61	1.34	0	0	3.95	7.6
	7	4.47	0	0	0	4.47	9.4
	8	2.43	1.68	0	0.52	4.63	79.3
	9	3.12	0	1.11	0	4.23	7.8
	10	1.34	4.08	0.4	0	5.82	160.7
Cu	1	43.19	28.02	76.06	0	147.27	15.4
	2	26.77	19.01	0	115.34	161.12	8.8
	3	10.45	2.32	13.93	34.91	61.61	4.8
	4	56.57	20.99	176.2	18.86	272.62	21.6
	5	54.77	31.42	57.9	30.46	174.55	5.9
	6	29.48	74.78	29.42	168.05	301.73	7.6
	7	56.59	0	27.69	63.09	147.37	9.4
	8	23.19	3.81	79.74	42.42	149.16	79.3
	9	18.49	26.06	28.59	32.98	106.12	7.8
	10	24.67	3.94	119.5	25.77	173.88	160.7
Zn	1	322.7	130.98	85.47	295.52	834.67	15.4
	2	519	155.37	25.27	4454.4	5154.04	8.8
	3	179.03	0	74.27	1012.68	1265.98	4.8
	4	486.08	298.61	0	661.23	1445.92	21.6
	5	763.78	576.78	0	159	1499.56	5.9
	6	307.81	249.7	0	478.48	1035.99	7.6
	7	374.12	0	0	229.29	603.41	9.4
	8	245.52	264.66	108.3	267.75	886.23	79.3
	9	365.17	163.199	0	308.24	836.609	7.8
	10	124.39	755.62	25.78	45.36	951.15	160.7

[a]Source: Warren and Zimmerman (1994)

Table 8.18 Continued

Element	Sample	c_p (mg/l)	[a]c_d (mg/l)	c_T (mg/l)	$K_d = (c_p/c_d)$	ψ (mg/l)	PCN	Average K_d	Average PCN
Cd	1	0.000068	0.000230	0.000298	0.293939	0.001012	99.990	0.232	99.925
	2	0.000063	0.000610	0.000673	0.103725	0.006491	99.935		
	3	0.000013	0.000190	0.000203	0.069474	0.002925	99.971		
	4	0.000088	0.000360	0.000448	0.244200	0.001834	99.982		
	5	0.000032	0.000040	0.000072	0.800925	0.000090	99.999		
	6	0.000030	0.000470	0.000500	0.063872	0.007828	99.922		
	7	0.000042	0.000540	0.000582	0.077811	0.007480	99.925		
	8	0.000367	0.001600	0.001967	0.229474	0.008572	99.914		
	9	0.000033	0.001000	0.001033	0.032994	0.031309	99.688		
	10	0.000935	0.002300	0.003235	0.406641	0.007956	99.921		
Cu	1	0.002268	0.005000	0.007268	0.453592	0.016023	99.840	0.463	99.483
	2	0.001418	0.006000	0.007418	0.236309	0.031390	99.687		
	3	0.000296	0.005000	0.005296	0.059146	0.089537	99.113		
	4	0.005889	0.009000	0.014889	0.654288	0.022755	99.773		
	5	0.001030	0.006000	0.007030	0.171641	0.040957	99.592		
	6	0.002293	0.011000	0.013293	0.208468	0.063766	99.366		
	7	0.001385	0.011000	0.012385	0.125934	0.098347	99.026		
	8	0.011828	0.011000	0.022828	1.075308	0.021230	99.788		
	9	0.000828	0.009000	0.009828	0.091971	0.106857	98.943		
	10	0.027943	0.018000	0.045943	1.552362	0.029595	99.705		
Zn	1	0.012854	0.004000	0.016854	3.213480	0.005245	99.948	2.006	99.780
	2	0.045356	0.015000	0.060356	3.023703	0.019961	99.801		
	3	0.006077	0.007000	0.013077	0.868101	0.015064	99.850		
	4	0.031232	0.012000	0.043232	2.602656	0.016611	99.834		
	5	0.008847	0.008000	0.016847	1.105926	0.015234	99.848		
	6	0.007874	0.007000	0.014874	1.124789	0.013223	99.868		
	7	0.005672	0.005000	0.010672	1.134411	0.009408	99.906		
	8	0.070278	0.024000	0.094278	2.928252	0.032196	99.679		
	9	0.006526	0.013000	0.019526	0.501965	0.038898	99.613		
	10	0.152850	0.043000	0.195850	3.554647	0.055097	99.452		

[a]Source: Warren and Zimmerman (1994)

taken to be representative of the metal's adsorption potential. It is apparent that the PCN values of these metals are of the order of 99(+), indicating a very high adsorption potential in the Don River flows. The average values of K_d, however, range from 0.232 to 2.006. According to K_d, the metals are ranked for adsorption potential as Zn > Cu > Cd, whereas PCN describes the ranking as: Cd > Zn > Cu. The PCN-based ranking generally agrees with that in the other two systems, except for Cu > Zn. It is worth emphasizing that the difference in the PCN-values of these metals in the Don River system is quite marginal to adhere strictly to the PCN-described ranking for Zn and Cu. Thus, following the ranking of the metals in the other two systems, the ranking in this river flow system should be as: Cd > Cu > Zn. Consequently, an inference can be drawn that in different systems, the ordering of metals for their partitioning potential does not vary with the solution media which, however, affect the extent of partitioning.

8.8.3 RELATION BETWEEN PARTITIONING PARAMETERS AND MEDIUM CHARACTERISTICS

Similar to the above, PCN and K_d (Table 8.18) are correlated with the physical and chemical characteristics of the Don River flows as follows:

$$Y = a_1X_1 + a_2X_2 + a_3X_3 + b \qquad (8.58)$$

where Y is PCN or K_d, X_1-X_3 correspond to temperature (°C), pH of river flow, and Cl^- content of the flow (mg/liter), and a_1-a_3 represent the parameters. The computed values of parameters along with the coefficient of determination, r^2, are shown in Table 8.19. It is seen that the r^2-values are not too high to describe a good correlation. However, it is apparent from this table that both PCN and K_d varies positively with the temperature, implying that the adsorption potential of metals increases with the temperature and vice versa. It is consistent with the results of Warren and Zimmerman (1994) but contrasts the above-described conclusion of Stumm and Morgan (1981). Similarly, the effect of the other medium characteristics on the PCN- or K_d-variation can be described.

Table 8.19 Correlation of PCN and K_d with the Don river flow characteristics

Element	PCN				
	a_1	a_2	a_3	b	r^2
Cd	0.0312	-0.0874	0.0001	100.43	0.60
Cu	0.1314	-0.7915	0.0005	100.91	0.66
Zn	0.0182	0.1771	-0.0001	98.36	0.50

Table 8.19 Continued

Element	K_d				
	a_1	a_2	a_3	b	r^2
Cd	0.0666	-0.5868	0.0003	4.48	0.38
Cu	0.0635	-0.1089	0.0008	0.56	0.48
Zn	0.3100	-1.0623	0.0018	8.04	0.43

Example 8.10: Characterize the flow medium of the Don River for partitioning of the Zn metal using the following information: temperature = 10 °C, pH of river flow = 8.3, and Cl^- content = 163.8 mg/liter.

Solution: To characterize the river flow medium for the partitioning potential of the Zn metal, PCN is calculated from equation (8.58):

$$Y = a_1X_1 + a_2X_2 + a_3X_3 + b$$

where Y = PCN, X_1 = temperature = 10 °C, X_2 = pH of river flow = 8.3, and X_3 = Cl^- content of the flow = 163.8 mg/liter. From Table 8.19, parameters are: a_1 = 0.0182, a_2 = 0.1771, a_3 = -0.0001, and b = 98.36. Thus,

$$PCN = 0.0182 \times 10 + 0.1771 \times 8.3 - 0.0001 \times 163.8 + 98.36 = 99.99$$

This PCN value which is very close to 100 indicates that the metal Zn has an extremely high potential for adsorption in the given river flow medium.

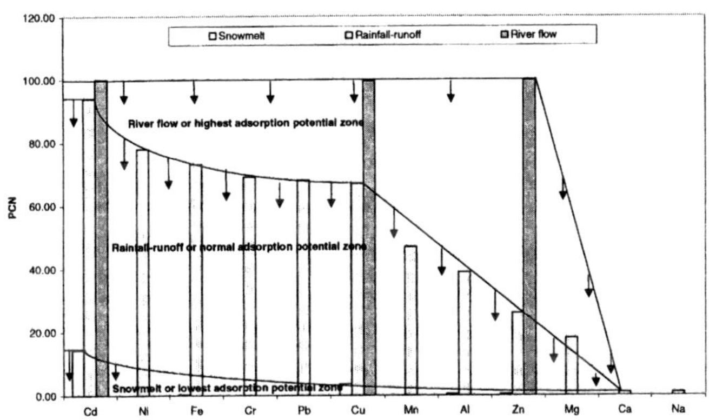

Figure 8.24. PCN-based characterization of partitioning media.

8.9 PCN-Based Characterization of Media

Similar to the characterization of watersheds for the runoff producing potential using the curve number CN, PCN can be used for characterization of the systems (solution media) based on the adsorption potential of metals. To this end, the above computed average PCN-values for all the three systems are plotted against the elements in their approximate descending order of adsorption potential (or PCN), as shown in Figure 8.24. Apparently, the three systems are distinguished from each other in that these indicate three distinct zones for the adsorption potential of the metal elements under study. The river flow zone can be characterized as the highest adsorption potential zone, the rainfall-runoff zone as the normal adsorption potential zone, and the snowmelt zone as the lowest adsorption potential zone. For metal adsorption, the river flow zone is analogous to the urban watersheds of the highest runoff potential, the rainfall-runoff zone to the agricultural watersheds of the normal runoff potential, and the snowmelt zone to the rangelands of the lowest runoff potential. It can, therefore, be inferred that PCN can be taken as a partitioning parameter for describing the adsorption potential of the media.

Example 8.11: Based on the data of Examples 8.8-8.10, compare the partitioning potential of the three types of media, rainfall-runoff, snowfall-snowmelt, and river flow, for the Zn metal.

Solution: The PCN values determined in Examples 8.8-8.10 for the three media rainfall-runoff, snowfall-snowmelt, and river flow are: 37, 2, and 99.99, respectively. These values show that the river flow has the highest adsorption potential (or partitioning potential) for the Zn metal, the snowmelt has the minimum potential, and the medium of rainfall-runoff lies in the middle.

8.10 Determination of Annual Pollutant Loads

While most studies including the one described earlier in this chapter have focused on estimating pollutant loads from individual storm events, the forthcoming discussion, based on the work of Pandit and Gopalakrishnan (1996, 1997), is devoted to the computation of annual pollutant loads required to obtain necessary permits. It can also be used to study the effect of watershed urbanization by comparing the current annual pollutant loads from a watershed to the projected future loads. Furthermore, the best management practices (BMP) for a watershed can be designed on the basis of keeping the projected future annual loads to the levels of current annual loads or past (historical) annual loads.

8.10.1 NPDES PERMIT

As described earlier, the Environmental Protection Agency (EPA) published in 1990 regulations that established requirements for municipalities to obtain the NPDES permit prior to discharging storm water to the waters of the United States. One of the application requirements was the estimation of annual pollutant load of the cumulative discharges from all identified municipal outfalls under average climatic conditions (Smoley, 1993). The estimation of annual pollutant loads is necessary to formulate and fulfill planning and management goals.

8.10.2 DRY- AND WET-WEATHER CONDITIONS

Annual pollutant loads can be separately estimated under dry- and wet-weather conditions. Generally, wet-weather conditions are defined as those periods when storm water or direct runoff occurs due to rainfall events. The remaining time, during which there is either no rainfall or no storm water is generated as a result of rainfall, is defined as a dry-weather condition. The direct runoff, which is the accumulated surface runoff (or overland flow) and interflow, causes a fairly rapid rise in the flow rate and water levels. During dry conditions, water discharges into the storm water system as baseflow. In rural basins, the baseflow primarily occurs due to ground water flow or agricultural discharge and, in urban basins, it can occur due to several reasons, such as discharges from air conditioners, groundwater leakage into storm water pipes, industrial discharge, or residential discharge (for example, sprinkler systems). Baseflow is also termed as dry-weather flow or non-storm flow and it occurs at a fairly uniform rate unlike storm flow (or storm runoff). In other words, baseflow does not cause sharp rises in flow rates or water levels.

8.10.3 METHODOLOGY FOR ESTIMATION OF ANNUAL LOADS

One of the methods widely used for estimating annual pollutant loads is known as the concentration-times-flow method or the constant concentration method. It is expressed as:

$$AL = (AL)_w + (AL)_d \tag{8.59}$$

where AL is the annual load, $(AL)_w$ is the annual load under wet-weather conditions, and $(AL)_d$ is the annual load under dry-weather conditions. The annual loads under wet- and dry-weather conditions can be calculated, respectively, as:

$$(AL)_w = (EMC)_w (AARV)_w \tag{8.60}$$

and

$$(AL)_d = (EMC)_d (AARV)_d \tag{8.61}$$

in which $(EMC)_w$ and $(EMC)_d$ are the average annual event mean concentration (mass/volume) under wet and dry conditions, respectively; and $(AARV)_w$ and $(AARV)_d$ are the average annual runoff volume under wet and dry conditions, respectively. Representative EMC values for wet- and dry-weather conditions can be obtained by direct measurements or from literature sources. For example, Sansalone and Buchberger (1997) provided the total EMC for some elements generated by expressways of Cincinnati and Milwaukee along with the EPA discharge criteria, as shown in Table 8.20. It is apparent from this table that EMC varies significantly with the time of the year at I-75 highway of Cincinnati, and it varies from one site to the other in Milwaukee.

Table 8.20a Event mean concentration (EMC) data versus EPA discharge criteria (Total metal EMC = dissolved + particulate bound)

Data / Event	Total EMC (µg/l), Cincinnati (I-75)					Average total EMC (µg/l), Milwaukee			Discharge criteria (µg/l)[a]
	April 8	April 30	July 15	Sept. 8	Oct. 3	I-94	I-794	Highway 45	
Zn	459	628	15,244	3,616	1,427	465	336	371	120.0
Cd	5	6	11	5	8	11	32	29	5.6
Cu	43	70	325	166	71	155	88	75	18.0
Ni	9	23	91	83	11	---[c]	---	---	1,700.0
Pb	62	31	44	88	97	817	1,457	738	82.0
Cr	35	14	29	14	14	---	---	---	1,400.0
Mn[b]	120	175	820	337	166	---	---	---	None
Fe[b]	3477	932	4,676	6,415	5,178	---	---	---	None
Al[b]	2,224	1,859	270	1,621	5,496	---	---	---	None
Hardness	52	70	92	68	56	---	---	---	---
Number of events	1	1	1	1	1	107	35	29	---
Average daily traffic/1,000	150	150	150	150	150	116	53	85	---
Pavement type	Asphalt	Asphalt	Asphalt	Asphalt	Asphalt	Asphalt	Concrete	Concrete	---

Note: [a]Ohio EPA criteria for discharge to modified warmwater surface waters.
[b]Not U.S. EPA priority pollutant
[c]No data from Federal Highway Administration (Driscoll et al., 1990)

Table 8.20b Dissolved event mean concentration (EMC) versus EPA discharge criteria

Data / Event	Dissolved EMC (μg/l), Cincinnati (I-75)					Average dissolved EMC (μg/l), Milwaukee			Discharge criteria (μg/l)[a]
	April 8	April 30	July 15	Sept. 8	Oct. 3	I-94	I-794	Highway 45	
Zn	209.0	1,322.0	14,786.0	3,051.0	904.0	---[c]	---	---	110.0
Cd	2.0	4.0	9.0	2.0	3.0	---	---	---	3.7
Cu	13.0	44.0	279.0	103.0	25.0	---	---	---	17.0
Ni	5.0	16.0	76.0	37.0	10.0	---	---	---	1,400.0
Pb	13.0	14.0	16.0	16.0	21.0	---	---	---	65.0
Cr	9.0	6.0	28.0	6.0	10.5	---	---	---	550.0
Mn[b]	61.0	161.0	744.0	238.0	72.0	---	---	---	None
Fe[b]	43.0	137.0	583.0	204.0	93.0	---	---	---	None
Al[b]	6.0	22.0	84.0	58.0	229.0	---	---	---	None
Number of events	1	1	1	1	1	---	---	---	---

Note: [a]U.S. EPA criteria for discharge to modified warmwater surface waters.

[b]Not U.S. EPA priority pollutant

[c]No data from Federal Highway Administration (Driscoll et al., 1990)

For the purpose of obtaining an NPDES permit, the average EMC value for a site is generally determined by defining the mean EMC of three or more representative storm events (Smoley, 1993). A "representative" storm event under wet conditions must meet the following three criteria: 1) the depth of the storm must be greater than 0.25 cm (0.1 inch); 2) the storm must be preceded by at least 72 hours of dry weather; and 3) where feasible, the depth and duration of the storm event should not vary by more than 50% from the depth and duration of the average storm event of the area.

(a) *Estimation of AARV Values Under Dry Conditions*

Under dry conditions, the average annual runoff volume can be calculated as

$$(AARV)_d = 365\ (Q_d) \tag{8.62}$$

where Q_d is the average daily dry flow or baseflow. Since there is usually less variation in baseflow, as compared to flow under wet conditions, a relatively good estimation of baseflow can be obtained by just a few sample measurements during the year.

(b) *Estimation of AARV Values Under Wet Conditions*

For the computation of the annual runoff from urban watersheds can be determined using any of the expressions given in Table 2.3 (Chapter 2). Because of the large variation in runoff volumes under wet conditions, Pandit and Gopalakrishnan (1996) computed AARV values for a wet condition, $(AARV)_w$, as:

$$(AARV)_w = (P_a)\ (ASRC)(A) \tag{8.63}$$

where P_a = annual rainfall depth, A = watershed area, and ASRC = annual storm runoff coefficient. The ASRC is the annual storm runoff coefficient and defined as the ratio of the annual runoff and annual rainfall depth. It is determined using the SCS-CN-based continuous simulation model (Pandit and Gopalakrishnan, 1996) described in Chapter 7. Based on the daily runoff values computed using the SCS-CN method, the annual runoff, Q_a, corresponding to the annual rainfall, P_a, can be derived. Thus,

$$ASRC = \frac{Q_a}{P_a} \tag{8.64}$$

8.10.4 APPLICATION RESULTS

(a) *Variation of ASRCs with Pervious Curve Number for Orlando, Florida*

Using daily rainfall-runoff data for 13 years between 1981 and 1993, Pandit and Gopalakrishnan (1996) performed spreadsheet calculations for computing ASRC values for the Orlando, Florida watershed. Computations were made for 0, 20, 38, 65, 85, and 100 percent impervious watersheds that had pervious curve numbers of 50, 60, 65, 70, 75, 80, 85, 90, and 95 under AMC II. All impervious areas were assumed to be directly connected (Chapter 2) with CN = 98. The variation of average ASRC values computed for the 13-year period with the pervious curve number is shown in Figure

8.25. As expected, the average ASRC values are higher in more impervious (or lower weighted CN) watersheds than in less pervious (or higher weighted CN) areas.

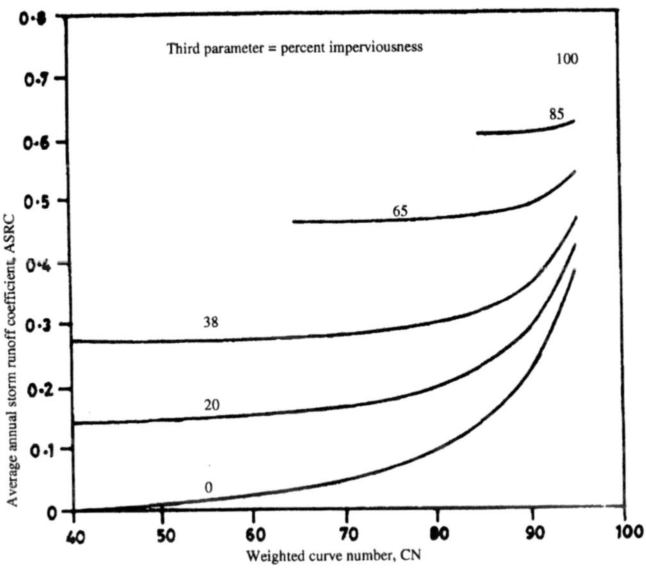

Figure 8.25. Relation between the average annual storm runoff coefficient and weighted curve numbers for 0-100 percent impervious watersheds for Ontario, Florida. Data of 13 years (1981-1993) used.

(b) *Iso-Mean-ASRC Graphs for Florida*

Using a continuous simulation model, Pandit and Gopalakrishnan (1997) calculated ASRC values for urban watersheds in 10 cities of Florida: Daytona Beach, Gainesville, Jacksonville, Miami, Oasis Ranger Station, Orlando, Punta Gorda, Tallahassee, Tampa, and Vero Beach. In the calculations, the assumptions were: 1) The watersheds had either soil A (least runoff producing soil) or soil D (highest runoff producing soil) to determine the effect of soil type on ASRC values. 2) The watersheds were of 0 percent, 20 percent, 38 percent, 65 percent, 85 percent, or 100 percent imperviousness. The values for imperviousness were selected to make the analysis consistent with that of SCS (1986), characterizing urban watersheds as low, medium, and high density residential watersheds based on the respective average percent imperviousness of 20 percent, 38 percent, and 65 percent. The commercial watersheds are described by the average percent imperviousness of 85.3. The cover in the pervious part of the watersheds was assumed to be a grassy area in good hydrologic condition.

From NEH-4 table (Table 2.1), the CN value for the pervious area in these simulations was 39 for soil A and 80 for soil D.

The computed 10-year average ASRC values are given in Table 8.21. The soil type is not indicated in 100 percent impervious watershed in the table, for it does not affect the ASRC computations. It is seen from the table that ASRC values increase as the percent imperviousness of watersheds increases; the more impervious is the watershed, the higher will be the runoff potential. The difference in the ASRC values for soils A and D is fairly large. For example, there is a 17.8% difference between the respective ASRC values for 38 percent impervious watersheds located in Tampa. Furthermore, this difference decreases with the increase in the imperviousness of the watershed. It implies that the soil type plays an important role in the description of the runoff producing potential of rural watersheds, which have a percent imperviousness close to zero, and residential watersheds, which usually have a percent imperviousness between 20 percent and 60 percent. On the other hand, the soil type is less important for highly commercial or industrial watersheds, which normally have a percent imperviousness greater than 70 percent.

It is convenient (Pandit and Gopalakrishnan, 1997) to depict ASRC values graphically in the form of iso-ASRC, as shown in Figures 8.26 and 8.27. The development of these graphs was based on the ASRC values calculated for ten cities of Florida using 20 percent and 65 percent impervious watersheds with soil type D. Similar graphs could be constructed for watersheds of different percent imperviousness and soil types. These graphs are useful in field for computation of annual pollutant loadings given the event mean concentration (EMC) values, as illustrated later.

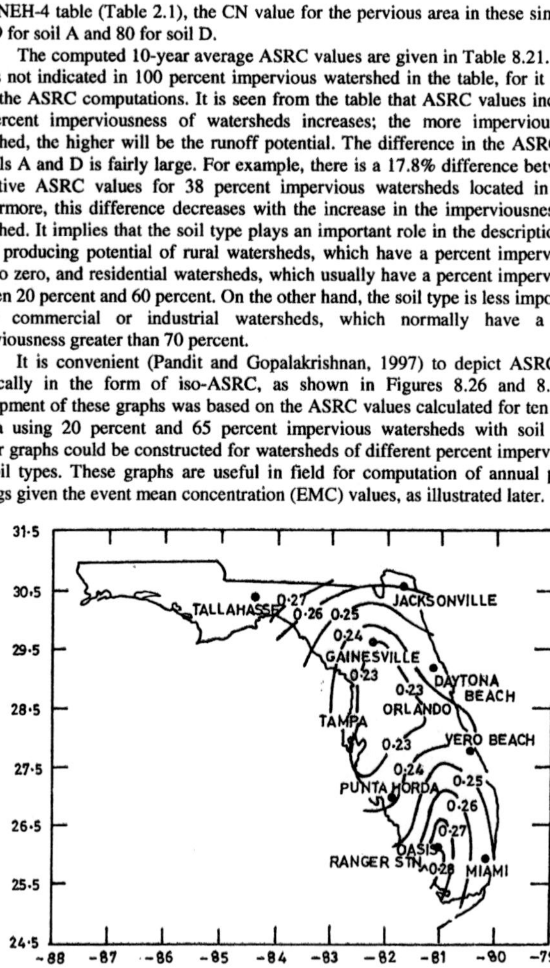

Figure 8.26. Iso-ASRC lines for Florida for 20 percent impervious watersheds of soil type D.

Table 8.21 Mean ASRC values computed from continuous simulation method for 10 cities of Florida using 10 years daily rainfall data (from 1984-1993)

City	Mean annual rainfall (inch)	Percent imperviousness										
		0 %		20 %		38 %		65 %		85 %		100 %
		Soil A	Soil D	Soil A	Soil D	Soil A	Soil D	Soil A	Soil D	Soil A	Soil D	
Daytona Beach	116.03	0.005	0.109	0.143	0.238	0.269	0.323	0.457	0.485	0.599	0.612	0.702
Gainesville	122.61	0.009	0.097	0.147	0.228	0.271	0.317	0.458	0.482	0.599	0.610	0.702
Jacksonville	134.90	0.012	0.128	0.154	0.260	0.282	0.344	0.476	0.507	0.622	0.636	0.728
Miami	108.94	0.016	0.124	0.151	0.249	0.274	0.331	0.457	0.488	0.597	0.611	0.699
Oasis Ranger Station	123.57	0.009	0.136	0.144	0.226	0.272	0.318	0.462	0.484	0.604	0.615	0.715
Orlando	145.47	0.011	0.149	0.152	0.278	0.281	0.353	0.474	0.511	0.620	0.637	0.726
Punta Gorda	120.65	0.008	0.106	0.149	0.238	0.276	0.327	0.467	0.541	0.611	0.623	0.715
Tallahassee	153.26	0.010	0.145	0.158	0.281	0.292	0.363	0.494	0.531	0.647	0.664	0.757
Tampa	105.03	0.008	0.100	0.146	0.230	0.271	0.319	0.458	0.483	0.599	0.610	0.702
Vero Beach	123.02	0.010	0.114	0.149	0.244	0.275	0.329	0.464	0.492	0.607	0.620	0.710

Note: mean annual rainfall values are for the period 1984 to 1993.

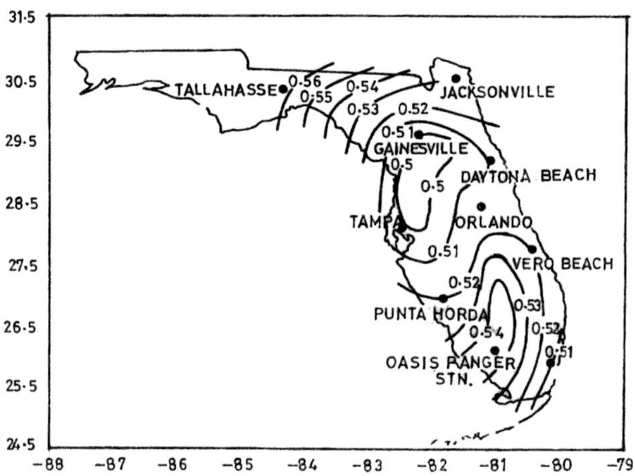

Figure 8.27. Iso-ASRC lines for Florida for 65 percent impervious watersheds of soil type D.

(c) Comparison of Mean ASRC Values Computed for Different Cities

Since ASRC is an important constituent in the computation of annual pollutant loadings, it is appropriate to compare ASRC values statistically. Figures 8.26 and 8.27 indicate the existence of some regional variability in the iso-ASRC values. Figure 8.26 shows a range of (0.23, 0.28) indicating a difference of 21.7%. On the other hand, Figure 8.27 shows a range of (0.50, 0.56) exhibiting a difference of 12%. The largest ASRC values occur in the northwest, and the smallest values in central Florida.

To show that the ASRC values for 100 percent impervious watersheds located in various cities were significantly different from each other, Pandit and Gopalakrishnan (1997) conducted the non-parametric Wilcoxon rank-sum test (Lapin, 1983). The comparison was made in pairs by testing the following null hypothesis: average ASRC value of city A was equal to the average ASRC value of city B. The null hypothesis was tested for a total of 45 times. Rejection of the null hypothesis leads to a strong conclusion that the ASRC means are statistically different. The conclusion that the means are statistically equal is usually weak if the null hypothesis is not rejected although it can be strengthened if the number of data points being compared is increased (say, 30 years instead of 10 years). The null hypothesis was rejected 24 times ($\alpha = 0.1$) and rejected 21 times. Jacksonville, Miami, Tallahassee, and Tampa, which respectively represent the northeast, southeast, northwest, and midwest regions of Florida had statistically different mean ASRC values from each other. Orlando which is in central Florida also had statistically different mean ASRC values from Miami,

Tallahassee, and Tampa. The null hypothesis was, however, not rejected for Orlando and Jacksonville. Since the watersheds were 100 percent impervious, the soil type or land use did not contribute to these differences. Therefore, it is clear that the differences in average ASRC values were due to differences in regional rainfall patterns.

(d) *Confidence Limits of ASRC Values*

It is also important to ascertain the confidence limits of ASRC values, for describing the variation of ASRC values with the change in the data length. Pandit and Gopalakrishnan (1997) determined the 95% confidence limits for the average ASRC values corresponding to 85 percent impervious watersheds with A and D soils as 0.577-0.619 (mean = 0.599) and 0.584-0.637 (mean = 0.610), respectively. The implication of these limits, as described by Lapin (1983), is that there is a 95% confidence that this interval contains the true 10-year average ASRC value. Employment of these limits in equations (8.59)-(8.61) enables determination of the confidence interval for the predicted average annual loads.

Thus, the continuous simulation model also lends a degree of confidence in the predicted ASRC values. The fraction of uncertainty in the average ASRC prediction due to the use of one 10-year record versus another 10-year record can be determined, to some extent, by calculating the difference between the upper and lower bounds of the confidence interval. Pandit and Gopalakrishnan (1997) showed that this difference for 95% confidence limits was as much as 70% for 10 percent impervious watersheds and it was approximately 9% for 85 percent impervious watersheds. It implies that the knowledge of confidence limits is more crucial in relatively pervious watersheds. It is noted that the confidence interval will be narrow if a long simulation period (say, 30 years) is used and vice versa.

Example 8.12: A 3-acre residential watershed in Orlando, Florida, consists of 40% directly connected impervious area. The average annual rainfall for the city is 50 inches and baseflow is equal to zero. Rainfall-runoff data collected during certain five events yielded an average pervious curve number CN = 74. The event mean concentration for total suspended solids during wet-weather conditions was 40 mg/l. Estimate the annual pollutant loading from the residential watershed.

Solution: For computing the annual pollutant load (= total suspended load) from the residential watershed, the average ASRC value for the watershed can be derived as 0.33. For given watershed area A = 3 acre and average annual rainfall = 50 inches, the average annual runoff volume (AARV) can be computed from equation (8.63) as

$(AARV)_w = (P_a)(ASRC)(A) = 50 \times 0.33 \times 3 = 49.5$ acre-inches (= 5092 m^3)

Then, using equation (8.60), the annual total suspended load can be computed as

$(AL)_w = (EMC)_w (AARV)_w = 40$ mg/l x 5092 m^3 = 203.68 kg

Example 8.13: A 45-acre commercial-residential watershed in Melbourne, Florida, consists of 65 percent directly connected pervious area. The major soil in this watershed is classified as D type soil. Melbourne is located approximately half way between Daytona Beach and Vero Beach on the east coast of Florida and has an average annual rainfall of 48 inches. To meet the requirement of NPDES permit, runoffs were measured during a dry period when there was no antecedent rainfall for approximately two weeks. Four discrete grab samples were taken (two samples per day for two consecutive days) with a greater than 4-hr interval between each grab. The average dry-weather flow rate from this watershed was 35 gallons per minute. The average event mean concentrations for total suspended solids during dry- and wet-weather conditions were 5 and 25 mg/l, respectively. Determine the annual amount of the total suspended solids from the Melbourne watershed.

Solution: The dry-weather average annual runoff volume can be computed from equation (8.62) as

$(AARV)_d = 365 \times (Q_d) = 365 \times 35$ (gallon per minute)-day

$= 365 \times 24 \times 60 \times 3.789 \times 10^{-3} \times 35\ m^3 = 69695\ m^3$

Using equation (8.61), the dry-weather average annual load can be computed as

$(AL)_d = (EMC)_d\,(AARV)_d = 5$ mg/l x $69695\ m^3 = 348$ kg

From Figure 8.27, the average annual runoff coefficient (ASRC) can be derived as 0.51. Thus, the average annual runoff volume for wet-weather conditions can be computed from equation (8.63) as

$(AARV)_w = (P_a)\,(ASRC)(A) = 45$ acres x 0.51 x 48 inches = 1101.6 acre-inches

In metric units, $(AARV)_w = 113321\ m^3$. The average annual total suspended for wet weather can be determined from equation (8.60):

$(AL)_w = (EMC)_w\,(AARV)_w = 25$ mg/l x $113321\ m^3 = 2833$ kg

Thus, the total annual suspended load can be computed from equation (8.59) as

$AL = (AL)_w + (AL)_d = 348 + 2833 = 3181$ kg

8.11 Summary

In this chapter, the mechanism of dissolved and particulate bound metal transport for 12 metal including Zn, Cd, Pb, Ni, Mn, Fe, Cr, Mg, Al, Ca, Cu, and Na elements was described to be analogous to that of wave propagation in open channels using the data of 300 sq. m stretch of Mill Creek-Expressway of I-75 in Cincinnati, U.S.A. The occurrence of strong loops (or hysteresis) in the rating curves developed for all 12 dissolved and particulate bound metals suggested a need for the development of a metal transport equation similar to the dynamic wave equation valid for open channel flows. Introducing a concept of equivalent mass depth, normal mass rating curves of the form of Manning's equation were developed for easy computation of the mass of dissolved and particulate bound metals. The proportionality concept of the SCS-CN method was extended for determining the potential maximum metal mass, M_p, of all the 12 elements transported by a storm. M_p was found to be related to the antecedent dry period for the major constituent of the pollutant load.

In the second component of this chapter, a PCN method analogous to the SCS-CN method was introduced for partitioning of all the 12 metals into dissolved and particulate bound forms using the data of an instrumented pavement of an interstate highway in Cincinnati. Similar to the SCS-CN method, using the available data, empirical relations were developed for characterizing the PCN parameter potential maximum desorption, ψ, on the basis of pH and alkalinity of rainfall and the antecedent dry period. Parameter PCN was described as a more reliable and versatile index than the frequently used partitioning indices, viz., K_d, f_p, and f_d, for PCN has the efficacy of distinguishing one metal from another on the basis of the adsorption capacity in metal partitioning.

Finally, an SCS-CN-based approach available in literature for the computation of annual pollutant loads was described and illustrated with examples.

CHAPTER 9

SEDIMENT YIELD

Soil erosion is the result of soil exposure to erosive energy of rainfall and flowing waters. Although extensive efforts have been made in the past to model the processes of erosion and sediment transport, the understanding on the subject is still less than complete. Consequently, there exists lack of a universally accepted formula for determination of the sediment yield from a watershed (Shen and Julien, 1992). Recent advancements have employed approximate solutions of the full dynamic equations for determination of sediment yield. Erosion, degradation, and sediment yield from watersheds are related to a complex interaction between topography, geology, climate, soil, vegetation, land use, and man-induced influences. Water, wind, and ice are the primary agents of soil erosion, with water being the most prominent of them.

Mathematical modeling of sediment yield started about a half century ago. The works of Cook (1936), Zingg (1940), Smith (1941), Ellison (1944, 1947) among others are noteworthy. The analysis of Ellison (1944) forms the basis for most of the recent studies on soil erosion modeling. His concepts of soil erosion due to detachment and transport by both rainfall and runoff were further developed and modeled by Meyer and Wischmeier (1969). An addition of these concepts in the Stanford Watershed Model (Negev, 1967) proved further impetus to soil erosion modeling. Clark et al. (1985) presented a treatise on the impact of erosion and sedimentation on environment in general and water quality in particular.

Methods for estimation of sediment yield and its temporal variation can be classified as empirical methods and process-based methods. Empirical methods basically combine all soil erosion governing processes into one equation which utilize empirical coefficients/factors for rainfall characteristics, soil properties, ground cover conditions, etc. Such methods include the universal soil loss equation (USLE) (Musgrave, 1947; Wischmeier and Smith, 1978), the modified universal soil loss equation (MUSLE) (Williams, 1978), and the revised universal soil loss equation (RUSLE) (Renard et al., 1991). These methods are simple in application and, therefore, frequently used in different parts of the world.

The process-based methods attempt to solve the fundamental equations of flow of water and transport of sediment. Some of the process-based models for soil erosion include Aerial Non-point Source Watershed Environment Response (ANSWERS) model (Beasley et al., 1980), Water Erosion Prediction Project (WEPP) model (Nearing et al., 1989), Agricultural Non-Point Source (AGNPS) pollution model (Young et al., 1987), KINEROS (Smith et al., 1995; Woolhier et al., 1990), and Systeme Hydrologique Europeen Sediment (SHESED) model (Wicks and Bathrust, 1996). These models more realistically simulate the process of rainfall-runoff-soil erosion. However, due to temporal variations in rainfall inputs and pronounced spatial heterogeneity prevalent in catchment areas, the process-based models may also produce unsatisfactory results (Wu et al., 1993). The effect of temporal variations in rainfall on sediment yield can be approximately simulated by analysing the isolated rain-storm events (Kothyari et al., 1997) whereas a Geographical Information System (GIS)-based technique is better suited for quantification of the spatial variation in rainfall, topographic, and drainage features of a catchment (Shultz, 1994; Jain and Kothyari, 2000).

Similar to the importance of the SCS-CN method in rainfall-runoff modeling, the universal soil loss equation (USLE) (Appendix D) is viewed as a tool in hydrology and environmental engineering for computing the potential soil erosion from small watersheds. A great deal of published material on this method along with its applications is available in hydrologic literature including those in Ponce (1989), Singh (1992), Novotny and Olem (1994). It is noted that USLE was developed for estimation of the annual soil loss from small plots of an average length of 22 meters. Therefore, its application to individual storm events and large areas is subject to error. Novotny and Olem (1994) found its accuracy to increase if it was coupled with a hydrologic rainfall-excess model. According to the current practice, the hydrologic information derived from a rainfall-runoff model is utilized in computation of the potential erosion using USLE for determination of the sediment yield (Knisel, 1980; Leonard et al., 1987; Young et al., 1987; Rode and Frede, 1997), which is important in watershed management (Tideman, 1996).

In this chapter, a sediment yield model is developed by coupling the basic proportionality concept of the SCS-CN method (Chapter 2) with the universal soil loss equation (USLE). The coupling leads to an analytical method that uses a $C = S_r = DR$ concept, where C is the runoff coefficient, S_r is the degree of saturation, and DR is the sediment delivery ratio, and requires data of storm rainfall and watershed characteristics. It also permits an interpretation of the SCS-CN parameter potential maximum retention (S) from USLE and describes the ratio of the potential erosion (A) to S as a constant for the watershed under dry conditions. The method is applied to rainfall-runoff-sediment data of four watersheds of urban, agricultural, and rangeland land uses, and its results are discussed.

9.1 Computation of Sediment Yield

The universal soil loss equation (USLE), described in Appendix D, estimates the potential soil loss from upland (sheet and rill) erosion as:

$$A = (R)(K)(L_s)(V_c)(M_p) \tag{D.1}$$

where A is the potential soil loss for a given storm or period (tonnes/ha), R is the rainfall factor, K is the soil erodibility factor, L_s is the slope-length factor, V_c is the vegetative cover (or cropping management) factor, M_p is the erosion control practice factor. Equation (D.1) determines the soil loss per unit area due to erosion by rainfall-runoff, excluding wind erosion, and does not yield direct sediment yield estimates.

The potential erosion, computed as above, is used to determine the sediment yield using the sediment delivery ratio, DR. Erosion is distinguished from the sediment yield in that the former represents the potential erosion that is taken equal to the sum of sheet (upland) and channel erosion and the latter refers to the sediment measured in the receiving body of water in a given time period. Vegetation dissipates rainfall energy, binds the soil, increases porosity by its root system, and reduces soil moisture by evapotranspiration to affect the sediment yield. The delivery ratio (DR) is a dimensionless ratio of the sediment yield (Y) to the total potential erosion (A) in the contributing watershed. Expressed mathematically,

$$DR = \frac{Y}{A} \tag{9.1}$$

Equation (9.1) is used to compute the sediment yield for a given 'A' from USLE (equation (D.1)). Thus, the delivery ratio (varying from 0-1) acts as a scaling parameter. It generally decreases with the size of the basin (Roehl, 1962).

Example 9.1: Determine the sediment yield for the two alternative watershed conditions described in Example D.1 if the sediment delivery ratio is equal to 0.6 for alternative (a) and 0.9 for alternative (b).

Solution: The potential maximum erosion (A) computed for alternative (a) in Example D.1 = 230 tonnes, and A = 2.3 tonnes for alternative (b). Thus, the sediment yield (Y) for the two alternatives can be computed from equation (9.1) using the given sediment delivery ratio (DR) as follows. For alternative (a), sediment yield Y = DR x A = 0.6 x 230 = 138 tonnes and for alternative (b), Y = DR x A = 0.9 x 2.3 = 2.07 tonnes.

Novotny and Olem (1994) equated the nonpoint pollution with soil loss and calibrated for DR by minimizing the difference between the USLE-computed upland erosion and measured sediment yield estimates. According to Wolman (1977), the DR concept conceals a number of processes that contribute to temporal or permanent deposition of sediments in an eroding watershed, for these are highly variable, intermittent, and describable only statistically. The correlation of DR with the runoff coefficient (Novotny and Olem, 1994) indicates a significant effect of infiltration and other hydrologic losses on the magnitude of DR. DR is affected by rainfall impact, overland flow energy, vegetation, infiltration, depression and ponding storage, change of slope of overland flow, drainage, and so on (Novotny et al., 1979; Novotny, 1980; Novotny et al., 1986; Novotny and Chesters, 1989). Since these factors vary with time throughout the year, the sediment yield also varies with time. Walling and Kane (1983) showed that certain minerals of eroded sediment, such as Si, Al, Fe, Ti and K, do not vary with a certain range of hydrological conditions, while others, particularly those associated with organic fractions, may vary significantly.

Water primarily drives erosion from watersheds (Langbein and Schumm, 1958; Gottschalk, 1964; and Leopold et al., 1964; Walling and Webb, 1983; Singh, 1985), if wind effects are ignored. It is evidenced by the existence of a direct relation between annual precipitation and erosion (Langbein and Schumm, 1958). The amount of soil erosion is, however, governed by the soil texture which determines both permeability and erodibility of soils. Permeability describes infiltration, which, in turn, determines hydrologic activeness of the soil surface in terms of both runoff generation and soil erosion. This kind of relationship is found useful for undisturbed watersheds of continental climates (Walling and Webb, 1983). Most of the sediment carried by streams in humid areas originates from sheet erosion and appears in the form of wash load or suspended load (Gottschalk, 1964; and Leopold et al., 1964).

According to the SCS-CN method, the runoff generation is closely linked with infiltration. Thus, the processes of runoff generation and soil erosion are closely interrelated, consistent with the description of the potential erosion 'A' as a function of the density of soil particles and ψ-θ characteristics of the soil. However, the SCS-CN method and USLE have not yet been investigated for their interrelationship. The SCS-CN proportionality concept can be extended for computing the delivery ratio. The following text integrates these equations for (a) better interpretation of the parameters of both the methods and (b) derivation of an SCS-CN-based sediment yield model.

9.2 Analytical Derivation

The SCS-CN method can be coupled with USLE analytically using the $C = S_r$ concept (equation (3.133)) described in Chapter 3. In brief, for $I_a = 0$ (i.e. immediate ponding situation), the SCS-CN proportionality hypothesis (equation (2.2)) equates the runoff factor C (= Q/P) to the degree of saturation, S_r, that is defined as (Figure 1.18):

$$S_r = \frac{F}{S} = \frac{V_w}{V_v} \tag{3.130}$$

which is valid for the completely dry antecedent moisture condition (AMC) of the soil, for which $V_a = V_v$. The SCS-CN parameter S can also be signified from USLE, in what follows.

Equation (D.1) computes the potential soil loss from a watershed in tonnes per hectare of the watershed. Since A is the potential mass of soil per unit watershed area which can be observed at the outlet of the watershed, it is equal to the product of the density of solids, ρ_s, and the volume of solids per unit surface area, V_s (Figure 1.18). Expressed mathematically,

$$A = V_s \rho_s \tag{9.2}$$

Equation (9.2) can also be expressed in terms of V and V_v as:

$$A = (V - V_v) \rho_s \tag{9.3}$$

Its division by S yields

$$\frac{A}{S} = \frac{(V - V_v)\rho_s}{V_a} \tag{9.4}$$

From Figure 1.18,

$$V_a = V_v - V_w = V_v(1 - S_{ro}) \tag{9.5}$$

where V_w is the initial moisture and S_{ro} is the initial degree of saturation. Thus, substitution of equation (9.5) into equation (9.4) yields

$$\frac{A}{S} = \frac{(V - V_v)\rho_s}{V_v(1 - S_{ro})} = \frac{(1 - n)}{n(1 - S_{ro})}\rho_s \tag{9.6}$$

where n is the soil porosity (dimensionless). It follows that

$$S = \frac{n(1 - S_{ro})}{(1 - n)\rho_s} A \tag{9.7}$$

Coupling of equation (9.7) with equation (9.2) leads to a definition of S as

$$S = \frac{n(1 - S_{ro})}{(1 - n)\rho_s}(R)(K)(L_s)(V_c)(M_p) \tag{9.8}$$

Thus, S depends on A, n, ρ_s, and S_{ro}. It also implies that S depends on all the factors affecting the potential erosion of the watershed; the higher the potential erosion from a watershed, the higher will be the value of S and vice versa. It is of common knowledge that highly porous soils (of high n), such as sands, yield high S-values, whereas nonporous soils (of low n), such as loams and clays, yield low S-values. S decreases with an increase in the antecedent moisture and vice versa. According to equation (9.7), heavy sediments (of high mass density) will yield low S-values and vice versa.

Equation (9.8) can also be written in terms of the potentially erodible soil depth, V, as:

$$V = (1/\rho_s)(R)(K)(L_s)(V_c)(M_p) + S_I \tag{9.9}$$

where S_I is equal to S of a completely dry soil or AMC I (Chapter 4). Thus, it is possible to determine the potential erodible depth of a watershed using USLE and NEH-4 tables (Tables 2.1 and 2.7). Since equation (9.7) also describes 'A' in terms of the watershed characteristic explained by S and the material of the soil, 'A' can be better interpreted in terms of the potential maximum erodible depth of the watershed, as discussed in Chapter 5.

9.2.1 COUPLING OF SCS-CN METHOD WITH USLE

Similar to the SCS-CN proportional equality (or $C = S_r$) concept, it is possible to extend it for sediment yield as

$$C = S_r = DR \tag{9.10}$$

in which all variables range (0, 1). Using the usual definitions and $I_a = 0$, equation (9.10) can be expanded as:

$$\frac{Q}{P} = \frac{F}{S} = \frac{P}{P+S} = \frac{Y}{A} \qquad (9.11)$$

Thus, equation (9.11) defines Y as

$$Y = C\,A \qquad (9.12)$$

Equation (9.12) implies that the sediment yield is directly proportional to the potential maximum erosion 'A' and the runoff factor C is the proportionality constant. Alternatively,

$$Y = \frac{AP}{P+S} \qquad (9.13)$$

For given watershed characteristics or A and S, the actual sediment yield Y increases with the rainfall amount, which is in conformity with the general notion that the higher the rainfall, the higher will be the sediment erosion and its transport and hence the higher the sediment yield, and vice versa. As $S \rightarrow 0$ (or $CN \rightarrow 100$), $Y \rightarrow A$ since $Q \rightarrow P$. Similarly, as $S \rightarrow \infty$ (or $CN \rightarrow 0$), $Y \rightarrow 0$ since $Q \rightarrow 0$. It is consistent with the general notion that the direct surface runoff (or surface water) primarily drives sediment erosion. Equation (9.13) integrates the SCS-CN method with USLE and can be modified for various elements of the rainfall-runoff-erosion process.

Example 9.2: Determine the sediment yield from a watershed having experienced a rainfall of 10 cm, and generated the direct surface runoff amount equal to 7 cm. The potential soil erosion was computed from USLE and it was equal to 10 tonnes. Assume initial abstractions are nil.

Solution: To determine the sediment yield, compute, first, the potential maximum retention from equation (2.5) as follows:

$S = P(P-Q)/Q$

which is valid for an initial abstraction, $I_a = 0$. This equation enables a determination of S as:

$S = 10 \times (10 - 7)/7 = 4.29$ cm

Then, the sediment yield Y can be computed from equation (9.13) as

$$Y = \frac{AP}{P+S} = \frac{10x10}{10+4.29} = 7 \text{ tonnes.}$$

(i) *Incorporation of I_a*

The initial abstraction I_a can be incorporated in equation (9.13) as:

$$Y = \frac{(P - I_a)A}{P - I_a + S} \tag{9.14}$$

Taking $I_a = 0.2S$ which is a standard practice, equation (9.14) can be recast as

$$Y = \frac{[P - 0.2S]A}{P + 0.8S} \tag{9.15}$$

Equation (9.15) infers that the sediment yield reduces with increasing initial abstraction and vice versa.

Example 9.3: Determine the sediment yield for the data of Example 9.2 if initial abstractions are not nil. Assume an initial abstraction coefficient λ equal to 0.2. Also, analyze the impact of initial abstraction on the sediment yield.

Solution: Since $I_a \neq 0$, the SCS-CN parameter S (for λ = 0.2) is computed using equation (2.14) as

$$S = 5[P + 2Q - \sqrt{Q(4Q + 5P)}]$$

$$= 5 \text{ x } [10 + 2\text{x } 7 - \sqrt{7\text{x}(4\text{x}7 + 5\text{x}10)}\ ] = 3.17 \text{ cm}$$

A substitution of this value in equation (9.15) gives the sediment yield Y as:

$$Y = \frac{[P - 0.2S]A}{P + 0.8S} = \frac{[10 - 0.2\text{x}3.17]\text{x}10}{10 + 0.8\text{x}3.17} = 7.47 \text{ tonnes.}$$

which is greater than that of Example 9.2. It implies that the incorporation of initial abstraction reduces the potential maximum retention for a given set of P-Q data and, therefore, increases the sediment yield. In reality, the existence of initial abstractions will, however, reduce the direct surface runoff and, consequently, the sediment yield for a given amount of rainfall.

(ii) *Incorporation of Antecedent Moisture*

For incorporation of the antecedent moisture M which represents the amount of moisture in the soil profile before the start of a storm, the SCS-CN proportionality (equation (2.2)) can be modified according to $C = S_r$ concept as:

$$\frac{Q}{P - I_a} = \frac{F + M}{S + M} \tag{9.16}$$

Its development is discussed in Chapter 4. Combination of equation (9.16) with the water balance equation (equation (2.1)) leads to

$$\frac{Q}{P - I_a} = \frac{P - I_a + M}{P - I_a + M + S} \tag{9.17}$$

which is the same as equation (4.29).

Thus, similar to the derivation of equation (9.14), Y can be derived using equation (9.17) as:

$$Y = \frac{(P - I_a + M)A}{P - I_a + S + M} \tag{9.18}$$

It is noted that, according to the above volumetric concept, $S + M = V_a + V_w = V_v$. Thus, equation (9.8) is revised for M as

$$S = \frac{n}{(1-n)\rho_s}(R)(K)(L_s)(V_c)(M_p) - M \tag{9.19}$$

Similar to equation (9.15), equation (9.19) can also be expressed as:

$$Y = \frac{(P - 0.2S + M)A}{P + 0.8S + M} \tag{9.20}$$

Equation (9.20) infers that the sediment yield will increase with the increase in the antecedent moisture amount, M, and vice versa. Furthermore, for AMC I for which $S_{ro} = 0$ or $M = 0$, equation (9.6) describes A/S_I ratio depending on the soil type and the density of its particles, which are constant for a watershed. Thus, the basic hypothesis of the rainfall-runoff-sediment yield model can be described from equation (9.11) as a proportional equality, which makes the ratio of the potential erosion to S_I (or S for AMC I) equal to the ratio of the actual erosion (yield) to the actual infiltration. Expressed mathematically,

$$\frac{A}{S_I} = \frac{Y}{F} = \text{constant} \tag{9.21}$$

which relates the sediment yield to infiltration amount, as discussed earlier. In equation (9.21), the "constant" depends on the material of the watershed soil. It is noted that an availability of the values of this constant for various watersheds will enable a determination of A from NEH-4 tables (Tables 2.1 and 2.7), as shown later. Furthermore, since S_I is constant for a watershed, the potential erosion is also constant. As stated above, equation (9.21) describes the dependence of sediment yield on the amount of infiltration.

Example 9.4: Determine the sediment yield for the data of Example 9.3 with an initial moisture amount of 1 cm in the soil profile. What happens if the initial moisture is increased. Also find out the A/S_I-ratio for this watershed. Assume initial abstraction coefficient λ equal to 0.2.

Solution: Since both I_a and M are non-zero, S is computed from equation (4.33) as:

$$\frac{S}{P}=\frac{1}{2\lambda^2}[(2+M^*)\lambda+(1-\lambda)C-\sqrt{(\lambda M^*)^2+(1-\lambda)^2C^2+2\lambda C\{2+(1+\lambda)M^*\}}]$$

Substituting $M^* = 1/10 = 0.1$, $\lambda = 0.2$, and $C = 7/10 = 0.7$, one obtains:

$$\frac{S}{P}=\frac{1}{2\lambda^2}[(2+M^*)\lambda+(1-\lambda)C-\sqrt{(\lambda M^*)^2+(1-\lambda)^2C^2+2\lambda C\{2+(1+\lambda)M^*\}}]$$

$$=\frac{1}{2x(0.2)^2}[(2+0.1)x0.2+(1-0.2)x0.7$$

$$-\sqrt{(0.2x0.1)^2+(1-0.2)^2x(0.7)^2+2x0.2x0.7x\{2+(1+0.2)x0.1\}}]$$

$$=12.5x[2.1x0.2+0.8x0.7-\sqrt{0.0004+0.64x0.49+0.28x2.12}]$$

$$=12.5x[0.98-\sqrt{0.9076}]=0.342$$

Thus, for given P = 10 cm, S = 0.342 x 10 = 3.42 cm. Its substitution in equation (9.20) yields the sediment yield:

$$Y=\frac{(P-0.2S+M)A}{P+0.8S+M}=\frac{(10-0.2x3.42+1)x10}{10+0.8x3.42+1}=7.51 \text{ tonnes.}$$

The above computed sediment yield of 7.51 tonnes is greater than that (= 7.47 tonnes) computed in Example 9.3 with M = 0. It implies that sediment yield increases with the increase in the initial soil moisture amount, and vice versa.

For computing A/S_I ratio, it is necessary to first determine the potential maximum retention S for completely dry initial condition or M = 0. It corresponds to that computed in Example 9.3 for M = 0, where it is equal to 3.17 cm. Since A = 10 tonnes, $A/S_I = 10/3.17 = 3.15$ tonnes/cm.

(iii) *Incorporation of Initial Flush*

Similar to the concept of initial abstraction, the concept of initial flush, I_f, is quite popular in environmental engineering (Foster and Charlesworth, 1996; Sansalone and Buchberger, 1997). The initial abstraction is a loss of water primarily due to evaporation and it does not contribute to runoff. On the other hand, the initial flush is not a loss of sediments, rather it appears at the outlet of the watershed. For the most part, it is contributed by the initial runoff generated at the start of rainfall, after meeting the initial abstraction requirements. To incorporate I_f in equation (9.20) it is necessary to redefine the sediment delivery ratio, DR, as

$$DR=\frac{Y-I_f}{A-I_f} \tag{9.22}$$

For simplicity, following the second SCS-CN hypothesis (equation (2.3)), I_f can also be related to the potential erosion, A, as

$$I_f=\lambda_1 A \tag{9.23}$$

where λ_1 is the initial flush coefficient, similar to the initial abstraction coefficient. Combination of equation (9.23) with equation (9.22) yields

$$DR = \frac{Y - \lambda_1 A}{(1-\lambda_1)A} \tag{9.24}$$

Thus, as above, equation (9.20) can be further modified for the initial flush as

$$Y = \left[\frac{(1-\lambda_1)[P - 0.2S + M]}{P + 0.8S + M} + \lambda_1\right]A \tag{9.25}$$

Here, it is possible to show that, for M = 0, as $S \rightarrow 0$, $Y \rightarrow A$ and as $S \rightarrow \infty$, $Y \rightarrow 0$, for $Y \geq 0$.

Example 9.5: Determine the sediment yield for the data of Example 9.4 considering the initial flush and find out its impact on the sediment yield. Assume the initial flush coefficient λ equal to 0.01.

Solution: Parameter S computed using P, Q, M, and I_a is equal to 3.42 cm. Its substitution into equation (9.25) yields the sediment yield:

$$Y = \left[\frac{(1-\lambda_1)[P - 0.2S + M]}{P + 0.8S + M} + \lambda_1\right]A$$

$$= \left[\frac{(1-0.01)[10 - 0.2x3.42 + 1.0]}{10 + 0.8x3.42 + 1} + 0.01\right]x10 = 7.54 \text{ tonnes.}$$

which is greater than that (= 7.51 tonnes) computed in Example 9.4. It implies that the sediment yield increases with the increase in initial flush.

As shown by Novotny and Olem (1994) using field data, there exists a power relation between C and DR:

$$DR = \alpha C^{\beta} \tag{9.26}$$

where α and β are respectively the coefficient and exponent of the power relation. The relationship depicted by Novotny and Olem (1994) for a particular data set yields α = 0.97 and β = 1.03. These values suggest that the above derived relations assuming C = DR should generally hold, for both parameters are approximately equal to 1. However, for completeness, a relationship for determining the sediment yield, Y, using the power relation (equation (9.26)) is given as:

$$Y = \left\{(1-\lambda_1)\alpha\left[\frac{P - 0.2S + M}{P + 0.8S + M}\right]^{\beta} + \lambda_1\right\}A \tag{9.27}$$

This is the general form of the model based on the existing SCS-CN method, for computation of the sediment yield from the rainfall amount and catchment characteristics.

Example 9.6: Determine the sediment yield for the data of Example 9.5 considering a power relation between the sediment delivery ratio and runoff coefficient as: DR = $0.97C^{1.03}$. All other assumptions are the same as in Example 9.5.

Solution: Parameter S computed in Example 9.5 is equal to 3.42 cm. Its substitution into equation (9.27) yields the sediment yield:

$$Y = \left\{(1-\lambda_1)\alpha\left[\frac{P-0.2S+M}{P+0.8S+M}\right]^{\beta} + \lambda_1\right\}A$$

$$= \left\{(1-0.01)x0.97\left[\frac{10-0.2x3.42+1}{10+0.8x3.42+1}\right]^{1.03} + 0.01\right\}x10 = 7.15 \text{ tonnes}$$

9.3 Application

9.3.1 STUDY AREAS

Cincinnati Watershed

Rainfall-runoff-water quality data for 13 events of the paved Cincinnati watershed (Figures 6.1a,b) were collected during 1995-97 by Sansalone and Buchberger (1997). The total solids (TS) which is the sum of dissolved (DS) and suspended (TSS) solids are taken as sediment yield data in the following application.

Nelson Farm

Nelson Farm (watershed no. 2) is an agricultural watershed located near Como, MS, U.S.A. Its physiological details are available elsewhere (Dabney et al., 1997). The rainfall-runoff-sediment yield data include 14 events of 1997 used by Huntington et al. (1998).

Lucky Hills Watershed

The Lucky Hills watershed is a 3-acre rangeland watershed near Tombstone, Arizona (Agriculture Research Service (ARS), U.S. Department of Agriculture, location 63.101). In his rainfall-runoff-erosion study, Smith (1977) used seven events of 1975, and these are used here.

Nagwan Watershed

The Upper Siwane River drains the Nagwan watershed which is 92.46 sq. km in size and is located in the Upper Damodar Valley of Hazaribagh district in Bihar (India). The watershed lies between 85.25^0 and 85.43^0 E longitudes and between 23.99^0 and 24.12^0 N latitudes. It is undulating in nature and its slope varies from 2.1 to 9.1 %

(average = 2.3 %). Its major soil is silty clay but sandy loam, clay loam, loam, and loamy sand soils are also found in the watershed which falls in the sub-humid tropical region of India, receiving an annual rainfall of 1256 mm. Major crops grown in the watershed are paddy (rice), maize, minor millet in Kharif (monsoon season), and mustard in Rabi (winter season). The Soil Conservation Department of Damodar Valley Corporation, Hazaribagh, and Indo-German Bilateral Project (IGBP) on 'Watershed Management', New Delhi, monitor the data. Daily rainfall-runoff-sediment yield data were available (Tripathi, 1999) for years 1991, 1996, and 1997 and these were used to derive 47 events of storm duration varying from 1 to 12 days.

Table 9.1 Formulation of rainfall-sediment yield and rainfall-runoff models

Model	Rainfall-sediment yield models	Model	Rainfall-runoff models
S1	$Y=\frac{AP}{P+S}$	R1	$Q=\frac{P^2}{P+S}$
S2	$Y=\frac{A(P-0.2S)}{P+0.8S}$	R2	$Q=\frac{(P-0.2S)^2}{P+0.8S}$
S3	$Y=\frac{A(P-\lambda S)}{P+(1-\lambda)S}$	R3	$Q=\frac{(P-\lambda S)^2}{P+(1-\lambda)S}$
S4	$Y=\frac{A[P-\lambda S+M]}{P+(1-\lambda)S+M}$	R4	$Q=\frac{(P-\lambda S)(P-\lambda S+M)}{P+(1-\lambda)S+M}$
S5	$Y=\frac{A[P-0.2S+M]}{P+0.8S+M}$	R5	$Q=\frac{(P-0.2S)(P-0.2S+M)}{P+0.8S+M}$
S6	$Y=\left[\frac{(1-\lambda_1)[P-0.2S+M]}{P+0.8S+M}+\lambda_1\right]A$	R6	$Q=\frac{(P-0.2S)(P-0.2S+M)}{P+0.8S+M}$
S7	$Y=\left[\frac{(1-\lambda_1)[P-\lambda S+M]}{P+(1-\lambda)S+M}+\lambda_1\right]A$	R7	$Q=\frac{(P-\lambda S)(P-\lambda S+M)}{P+(1-\lambda)S+M}$
S8	$Y=\left\{(1-\lambda_1)\alpha\left[\frac{[P-0.2S+M]}{P+0.8S+M}\right]^{\beta}+\lambda_1\right\}A$	R8	$Q=\frac{(P-0.2S)(P-0.2S+M)}{P+0.8S+M}$
S9	$Y=\left\{(1-\lambda_1)\alpha\left[\frac{[P-\lambda S+M]}{P+(1-\lambda)S+M}\right]^{\beta}+\lambda_1\right\}A$	R9	$Q=\frac{(P-\lambda S)(P-\lambda S+M)}{P+(1-\lambda)S+M}$

9.3.2 DISCUSSION OF RESULTS

Based on the above analytical development (equation (9.27)) for determination of the sediment yield, nine models can be formulated as shown in Table 9.1. Model S1 excludes the initial abstraction I_a, antecedent moisture M, and initial flush I_f components. Model S2 accounts only for initial abstraction with $\lambda = 0.2$ (a standard value) and in Model S3, λ is allowed to vary. Model S4 accounts for both initial abstraction and antecedent moisture but allows the variation of λ. Model S5 is distinguished from Model S4 for $\lambda = 0.2$. Models S6 and S7 include all I_a, M, and I_f but the former assumes $\lambda = 0.2$. Models S8 and S9 correspond to equation (9.26) with the difference that the former assumes a constant value of λ (= 0.2) and the latter allows it to vary. These models were applied to the data of the above described four watersheds and their performance is evaluated using the Nash and Sutcliffe (1970) efficiency computed using equation (7.51).

Parameters of the aforementioned models were estimated using the non-linear Marquardt algorithm of the least squares procedure (Appendix B). These parameters were initially set as zero in all cases and the lower and upper limits were decided by trial and error. If the computed value of a parameter in a run did not fall in the prescribed range, the limit was extended accordingly for the next run. If the subsequent runs produced the estimate within the prescribed range, the parameter estimate was assumed to be optimal globally. Approximately, three to five runs or a few more in some cases were required to compute the final estimates of model parameters.

The estimated parameters and the efficiencies resulting from computation of the sediment yield are shown in the third last column of Tables 9.2-9.5. In these tables, superscripts 'a' and 'b' stand for the efficiency resulting from the computed sediment yield and runoff, respectively. Here, it is noted that Models S1-S9 determine the sediment yield using rainfall and ignore a direct involvement of the observed runoff. In application of these models, variables A and M were also taken as parameters due to the lack of their observations. It is seen from Table 9.2 that parameter S for the Cincinnati watershed varies from 2.43 to 15.10 mm or CN (equation (2.7)) varies from 94 to 99, λ from 0.23 to 0.68, λ_1 from 0.01 to 0.07, A from 909 to 1090 grams, M from 0.0 to 0.51 mm, α from 1.01 to 1.02, and β from 2 to 3. The ratio of A/S_1 varies from 0.024 to 0.116 tonnes/ha-cm. Clearly, the parameters vary in a narrow range and, therefore, support their credibility. The estimated high values of CN exhibit a high runoff potential of the watershed, which is consistent with the paved nature of the Cincinnati watershed. The resulting efficiency (with superscript 'a') varies from 75.79 to 79.51%, which is quite high to indicate a satisfactory performance of the models in computing the sediment yield (from the Cincinnati watershed.

Similarly, the parameters shown in Tables 9.3-9.5 for Nelson Farm, Lucky Hills, and Nagwan watersheds, respectively, can also be generally described to vary in a close range. For the S-values derived from the use of Model S2 that is based on the existing SCS-CN method, these watersheds yield curve numbers equal to 62, 71, and 42, respectively. It implies that the Lucky Hills watershed has a fairly good runoff potential, the Nagwan watershed has the lowest runoff potential, and the Nelson Farm falls in the middle.

To examine their validity, the above curve numbers are correlated with those derived from S-values (equation (2.7)) computed using P-Q data as (Chapter 2):

$$S = 5[P + 2Q - \sqrt{Q(4Q + 5P)}] \tag{2.14}$$

The resulting median curve numbers for the Cincinnati, Nelson Farm, Lucky Hills, and Nagwan watersheds are as 98, 89, 93, and 76, respectively. These curve numbers when plotted against each other (Figure 9.1) exhibit a quadratic relation:

$$y = -0.007\ x^2 + 1.394\ x + 30.05;\ r^2 = 1 \tag{9.28}$$

where y is the CN-value corresponding to P-Q data, x is the CN derived using the rainfall-sediment yield model (Model S2), and r^2 is the coefficient of determination. Here, r^2 comes out to be 1 due to only one degree of freedom. Therefore, a large set of data is required for a more reliable relationship. Figure 9.1 shows that the sediment yield model underestimates the value of CN. It is because of the deposition of sediment particles during their transport by runoff, yielding a lesser amount of sediment yield at the outlet of the watershed than actually eroded in the watershed. It is also evident from the results of the above watersheds. The Cincinnati watershed, which is of high runoff potential and small in size, does not allow the process of deposition to occur whereas this process is quite dominant in other agricultural watersheds.

Model S1 is the simplest of all the above 9 models and the resulting efficiency of 75.79% in Cincinnati application (Table 9.2) is indicative of (a) the dependence of the sediment yield on the SCS-CN-generated runoff and (b) the applicability of $C = S_r = DR$ concept (equation (9.10)). It is further supported by the efficiency (= 77.98%) of Model S2 that is directly based on the existing SCS-CN method (equation (2.6)). If λ is allowed to vary as in Model S3, the efficiency improves to 78.03%. It is noted that λ in Model S3 is estimated as 0.23 which is well within the range of (0.1, 0.3) (Chen, 1982; SCD, 1972). The incorporation of the antecedent moisture M, as in Models S4 and S5, marginally reduces the resulting efficiency. It is because of its variation from one rainfall event to another. The assumption of M as a constant is, therefore, not appropriate. The inclusion of the concept of initial flush (Models S6 and S7) and allowing the variation of λ in optimization (Model S7) slightly reduces the efficiency to 77.92%. This reduction actually undermines the validity of equation (9.23) and, therefore, requires further improvement in future. However, a high magnitude of efficiency supports the general applicability of the concept. The efficiency is improved to 78.56% (Model S8) and 79.51% (Model S9) if DR is allowed to vary nonlinearly with the runoff factor C (equation (9.26)).

Table 9.2 Results of various models application to Cincinnati data

Sl. No.	Parameters of rainfall-sediment yield models							Eff.[a] (%)	Eff.[b] (%)	A/S_l ratio (t/ha-cm)
	S (mm)	λ	λ_1	A (g)	M (mm)	α	β			
1	15.10	-	-	1090	-	-	-	75.79	88.93	0.024
2	8.81	-	-	961	-	-	-	77.98	88.26	0.036
3	8.15	0.23	-	942	-	-	-	78.03	88.14	0.039
4	8.23	0.25	-	944	0.21	-	-	78.02	88.12	0.037
5	8.72	-	-	961	0.0	-	-	77.98	88.27	0.037
6	8.64	-	0.01	947	0.0	-	-	77.80	88.27	0.037
7	7.45	0.37	0.05	909	0.51	-	-	77.92	87.70	0.038
8	4.05	-	0.03	929	0.0	1.02	2.0	78.56	82.85	0.076
9	2.43	0.08	0.07	913	0.19	1.01	3.0	79.51	81.10	0.118

Similar to the above, the results of models application to other watersheds can also be explained from Tables 9.3-9.5. Apparently, the SCS-CN-based sediment yield models perform best on Lucky hills watershed with efficiencies ranging from 84.56 to 99.47% but these perform less satisfactorily on the data of Nagwan watershed, which might be attributed to (a) the derivation of rainstorms from daily data and (b) the less satisfactory applicability of the SCS-CN method to watersheds of low curve numbers, such as forest watersheds (Chapter 2 and 3). The efficiencies resulting from application to the Nelson Farm range from 59.40 to 86.96% indicating a satisfactory performance of the model.

Table 9.3 Results of various models application to data of Nelson farm

Sl. No.	Parameters of rainfall-sediment yield models							Eff.[a] (%)	Eff.[b] (%)	A/S_I ratio (t/ha-cm)
	S (cm)	λ	λ_I	A (t/ha)	M (cm)	α	β			
1	98.3	-	-	22.10	-	-	-	59.40	29.86	0.22
2	15.5	-	-	8.78	-	-	-	79.49	53.50	0.57
3	4.71	0.72	-	4.63	-	-	-	80.47	77.10	0.98
4	28.6	0.17	-	8.03	9.12	-	-	86.97	43.52	0.21
5	25.3	-	-	7.71	8.54	-	-	86.94	43.48	0.23
6	25.2	-	0.001	7.62	8.61	-	-	86.95	43.84	0.23
7	26.2	0.19	0.0	7.66	8.90	-	-	86.96	44.02	0.22
8	24.1	-	0.076	2.56	7.75	3.0	1.0	86.93	45.44	0.08
9	26.5	0.18	0.009	2.57	8.45	3.03	1.0	86.96	44.95	0.07

Table 9.4 results of various models application to data of Lucky Hills watershed

Sl. No.	Parameters of rainfall-sediment yield models							Eff.[a] (%)	Eff.[b] (%)	A/S_I ratio (t/ha-cm)
	S (inch)	λ	λ_I	A (t)	M (inch)	α	β			
1	97.60	-	-	36.70	-	-	-	84.56	-33.96	0.12
2	3.99	-	-	3.67	-	-	-	99.47	30.50	0.30
3	4.02	0.198	-	3.69	-	-	-	99.47	30.31	0.30
4	4.05	0.206	-	3.70	0.037	-	-	99.47	28.83	0.29
5	4.07	-	-	3.70	0.018	-	-	99.47	29.27	0.29
6	4.07	-	0.0	3.71	0.018	-	-	99.47	29.27	0.29
7	4.04	0.212	0.002	3.69	0.052	-	-	99.47	28.11	0.29

Table 9.5 Results of various models application to data of Nagwan watershed

Sl. No.	Parameters of rainfall-sediment yield models							Eff.[a] (%)	Eff.[b] (%)	A/S_I ratio t/ha-cm
	S mm	λ	λ_I	A t/ha	M Mm	α	β			
1	955	-	-	41.5	-	-	-	57.89	24.03	0.43
2	351	-	-	3.92	-	-	-	64.25	8.22	0.11
3	477	0.137	-	4.76	-	-	-	64.57	0.64	0.10
4	356	0.196	-	3.94	0.03	-	-	64.27	8.19	0.11
5	356	-	-	3.94	1.52	-	-	64.27	7.33	0.11
6	373	-	0.002	4.05	4.99	-	-	64.32	3.49	0.11
7	569	0.157	0.001	5.72	18.80	-	-	64.82	-20.11	0.10
8	326	-	0.028	7.89	0.0	1.27	2.0	66.25	16.90	0.24
9	326	0.190	0.027	7.85	0.11	1.27	2.0	66.49	20.74	0.24

Note: Superscript 'a' and 'b' in Tables 9.2-9.5 stand for efficiency in computation of sediment yield and direct runoff, respectively.

The above evaluation, however, excludes the observed runoff, Q, in estimation of the sediment yield. In practice, the sediment yield from upland areas is generally correlated with the observed runoff (Singh and Chen, 1983), rather than the rainfall. Therefore, to further support the validity of the coupling of the SCS-CN method with

USLE, it is appropriate to check the credibility of the estimated values of the parameters by computing the direct surface runoff using these values and comparing the computed runoff with the observed runoff.

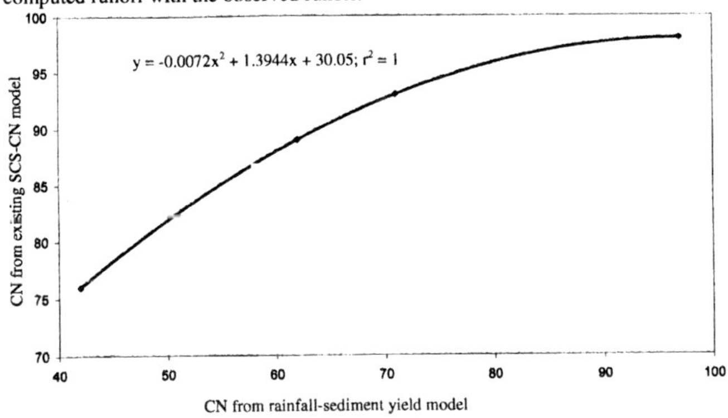

Figure 9.1. Relationship between curve numbers derived from (a) the rainfall-sediment yield model and (b) the existing SCS-CN method using rainfall-runoff data.

For computation of the direct runoff, SCS-CN-based models, designated as Models R1-R9 (Table 9.1) corresponding to Models S1-S9, respectively, were used. Models R1-R3 are equivalent to equation (2.5) with $I_a = 0$, $I_a = 0.2S$, and $I_a = \lambda S$, respectively. Models R4 and R5 are equivalent to equation (9.17) with $I_a = \lambda S$ and $I_a = 0.2S$, respectively. Models R4, R7, and R9 are the same in formulation, and so are the Models R5, R6, and R8.

Using the parameters of Tables 9.2-9.5, the direct runoff was computed and the resulting model efficiencies are shown in the second last column of these tables. It is seen that Model R1 that is the simplest in form exhibits the highest efficiency (= 88.93%) and the most complicated Model R9 shows the lowest efficiency (= 81.18%) in the Cincinnati watershed. These values are reasonably high to suggest a satisfactory model performance. The efficiencies are, however, quite low in application to other watersheds, indicating a poor performance of the model. It is because of the under-estimation of CN-values (or over-estimation of S-values) as described above. Therefore, it is necessary to transform the S-values of Tables 9.2-9.5 to those corresponding to the rainfall-runoff data using equation (9.28). To this end, it is appropriate to evaluate first the models for adoption for field use utilizing the results of the Cincinnati watershed (Table 9.2).

Although the resulting efficiencies are reasonably high to adopt any model formulation for determination of the sediment yield and direct runoff, a more appropriate model based on the computed efficiencies can be chosen. As a trade-off, Models S2-S7 for determination of the sediment yield and the corresponding runoff

Models R2-R7 appear to be more suitable for adoption, for their efficiencies in computation of sediment yield vary from 77.80 to 78.03%, and these vary from 88.12 to 88.27% in runoff computation. Since the efficiency from application of Model S2 (or Model R2) deviates only marginally from application of other models, it is appropriate to adopt the existing SCS-CN method for the reasons of simplicity and possible use of the SCS-CN tables in determination of the sediment yield in future. Such an inference can also, in general, be derived from the results of other watersheds.

The use of equation (9.28) in the existing SCS-CN application (Model R2) modifies the efficiency of runoff computations (Tables 9.2-9.5) to 85.45 % from 88.26 % in Cincinnati application, 63.64 % from 53.50 % in Nelson Farm application, 99.72% from 30.50 % in Lucky Hills application, and 44.44 % from 8.22 % in Nagwan application, showing a general improvement in the resulting efficiencies. Except for Nagwan application, where the model shows a less than satisfactory performance, other watersheds indicate satisfactory performance. The results of sediment yield and runoff computations are depicted in Figures 9.2-9.5 for Cincinnati, Nelson Farm, Lucky Hills, and Nagwan watersheds, respectively. The closeness of data points to the line of perfect fit (LPF) indicates a satisfactory model performance.

It is also appropriate to investigate the estimated parameter values for A/S_I ratio which, according to equation (9.6), should be a constant value for a watershed. The computed values of this ratio are given in the last column of Tables 9.2-9.5. Apparently, Models S2-S7 yield A/S_I ratio in the range of (0.036, 0.039), (0.22, 0.98), (0.29, 0.30), and (0.10, 0.11) tonnes/ha-cm for the watersheds of Cincinnati, Nelson Farm, Lucky Hills, and for Nagwan, respectively. These values in general vary in a narrow range to assume a constant value of this ratio for Model S2 as 0.036, 0.57, 0.30, and 0.11 tonnes/ha-cm for the respective watersheds. Apparently, the A/S_I ratio increases with the reduction in the runoff potential (or low CN value) of the watershed. Such a determination of A/S_I ratio for various watersheds is important in estimation of the sediment yield using NEH-4 tables as follows.

1. Determine the AMC II CN value for a watershed from NEH-4 table.
2. Convert the AMC II CN value to AMC I using the NEH-4 table.
3. Compute S_I using equation (2.7) and transform it to obtain S_T-value relevant for sediment yield model using equation (9.28).
4. Multiply the S_T-value of step 3 with the available A/S_I ratio to obtain A in tonnes/ha-cm.

Such a procedure may help in future in determination of the sediment yield of ungauged watersheds using the concept of homogeneous watersheds (Singh et al., 2001). Furthermore, it may also help in the revision of the empirically derived components of USLE, for the SCS-CN rational is better understood, accepted, and established than USLE. The application of the procedure is illustrated by the following example.

Example 9.7: Compute the sediment yield using the following data:

Soil type = Hydrologic group B

Land use = Row crop with crop residue cover and straight row plough

A/S_I ratio = 0.57 tonnes/ha-cm

Sediment delivery ratio = 0.73

Solution: For the given land use and soil type, find the AMC II curve number (CN) from Table 2.1 as equal to 80, which corresponds to Sl. No. 9 of this table. Convert AMC II CN to AMC I CN from Table 2.7, which yield CN for AMC I = 63. Further conversion of this CN from equation (2.7) provides S_I = 5.87 inches (= 14.92 cm). Thus A can be computed from the given A/S_I ratio of 0.57. It follows that A = 8.50 tonnes/ha. The sediment yield can be computed using equation (9.1) as: Y = 0.73 x 8.50 = 6.21 tonnes/ha.

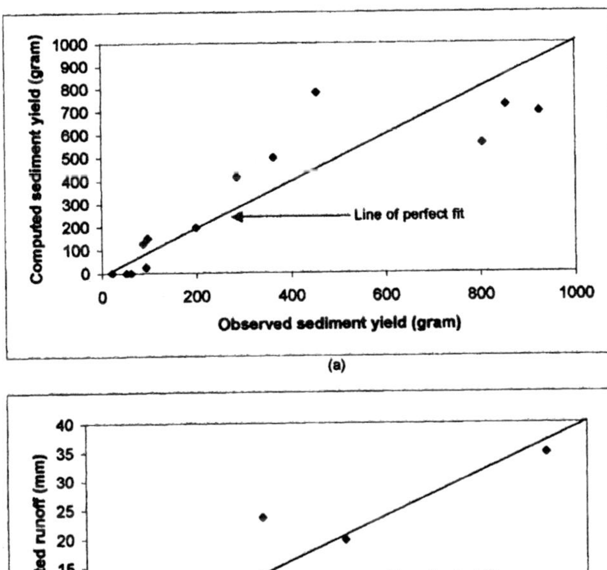

(a)

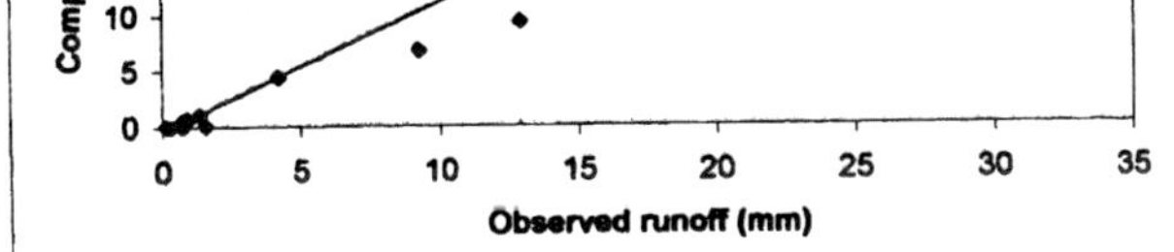

(b)

Figure 9.2. Comparison between observed and computed (a) sediment yield and (b) runoff using Models S2 and R2, respectively, for Cincinnati watershed.

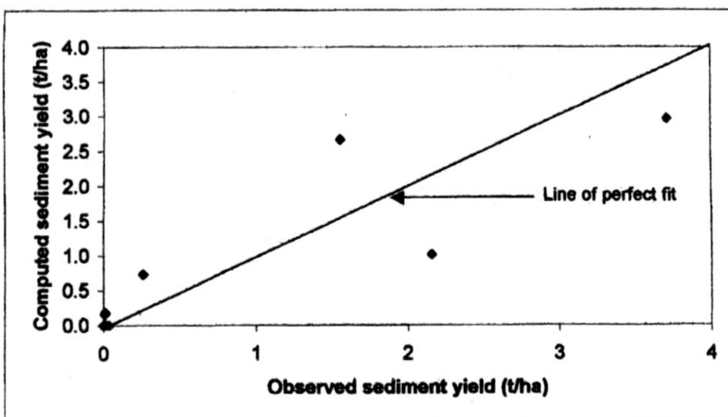

(a)

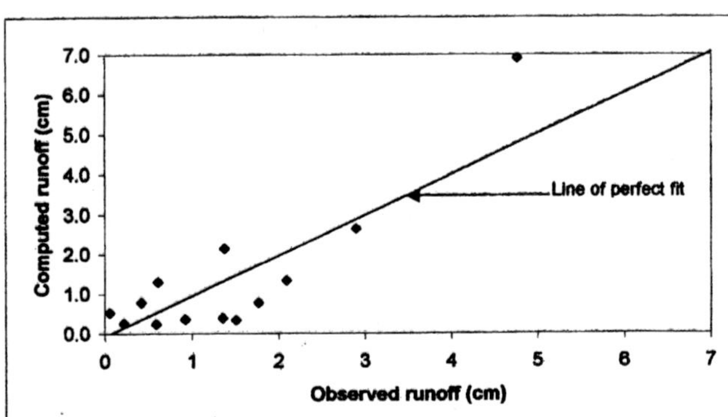

(a)

Figure 9.3. Comparison between observed and computed (a) sediment yield and (b) runoff using Models S2 and R2, respectively, for Nelson Farm.

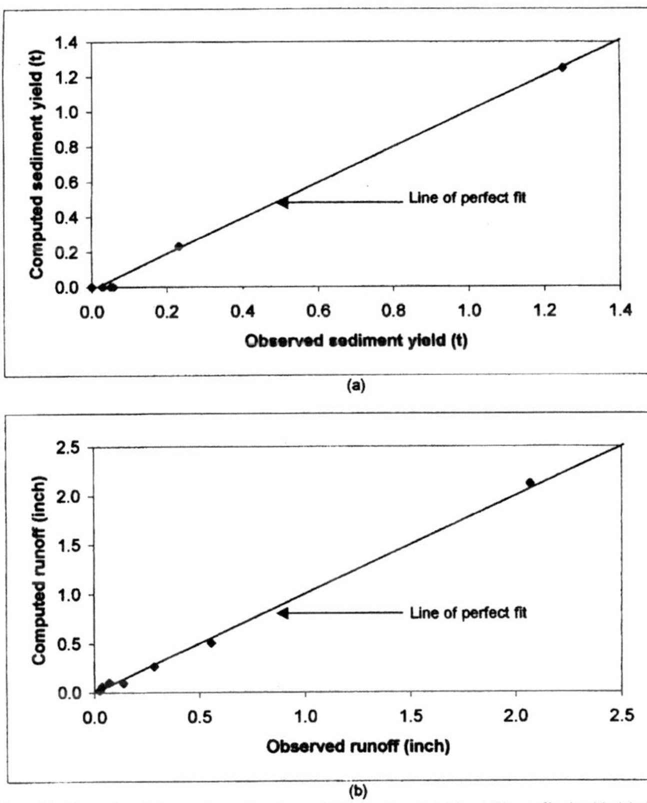

Figure 9.4. Comparison between observed and computed (a) sediment yield and (b) runoff using Models S2 and R2, respectively, for Lucky Hills watershed.

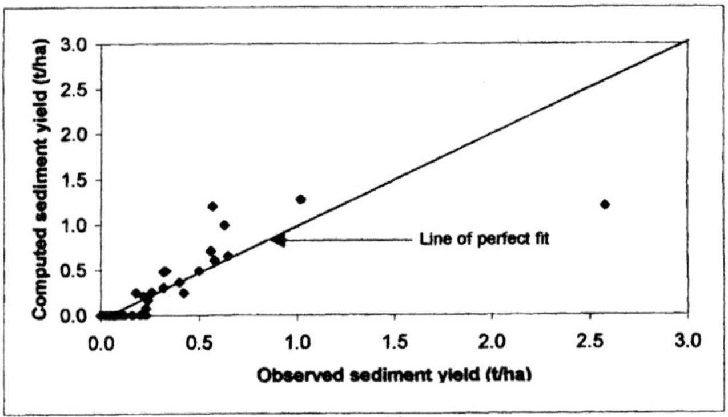

(a)

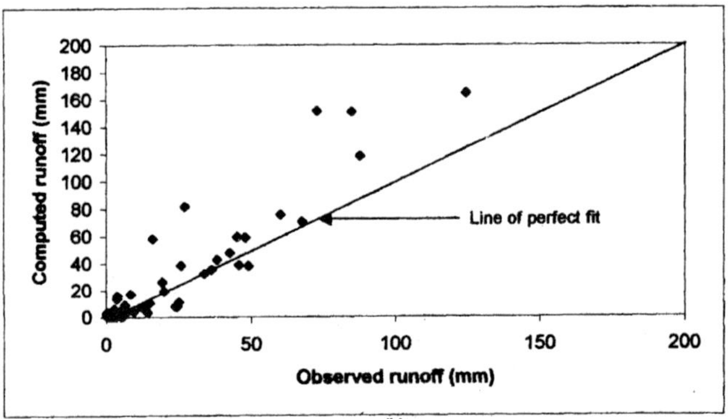

(b)

Figure 9.5. Comparison between observed and computed (a) sediment yield and (b) runoff using Models S2 and R2, respectively, for Nagwan watershed.

APPENDIX A

A.1 SCS-CN Theory for 'S' Including I_a

For a limiting case F = S and for S to include I_a, equations (2.1) and (2.2) reduce, respectively, to

$$P = S + Q \tag{A.1}$$

and

$$\frac{P - S}{P - I_a} = \frac{S - I_a}{S} \tag{A.2}$$

For a given single storm, I_a may vary from 0 to S. Thus, the right-hand side of equation (A.2) varies from 1 to 0 while the left-hand side varies from (1-S/P) to 1. Inequality of both sides at the extreme value of I_a thus contrasts the common belief that equation (2.2) is valid for all values of P. A further examination of equation (A.2) reveals the existence of a limiting P (say P_s) beyond which the available pores in the soil column are completely depleted for the applicability of equation (A.1). Thus, P_s can be expressed from equation (A.2) as:

$$P_s = \frac{S^2 - SI_a + I_a^2}{I_a} \tag{A.3}$$

The magnitude of P_s is greater than or equal to S for all values of I_a ranging from 0 to S. Here, it can be shown that the SCS-CN method is valid only for $I_a \leq P \leq P_s$. Specifically, when $I_a = 0$, P_s in equation (A.3) becomes infinity, making equation (2.2) valid for all values of P. The solution of equations (2.1) and (2.2) for Q and F, respectively, yields

$$Q = 0 \qquad \text{for } P \leq I_a \tag{A.4a}$$

$$Q = \frac{(P - I_a)^2}{P - I_a + S} \qquad \text{for } I_a \leq P \leq P_s \tag{A.4b}$$

$$Q = P - S \qquad \text{for } P \geq P_s \tag{A.4c}$$

and

$$F = P - I_a \qquad \text{for } P \leq I_a \tag{A.5a}$$

$$F = \frac{(P - I_a)S}{P - I_a + S} \qquad \text{for } I_a \leq P \leq P_s \tag{A.5b}$$

$$F = S - I_a \qquad \text{for } P \geq P_s \tag{A.5c}$$

Equation (A.5a) is equivalent to $F = 0$ because I_a is always equal to P before the time to ponding. The substitution of equation (A.3) into equation (A.4b) or equation (A.4c) for $P = P_s$, leads to

$$Q = \frac{(S - I_a)^2}{I_a} \tag{A.6}$$

whereas substitution of equation (A.3) into equation (A.5b) leads to equation (A.5c).

Because the magnitude of P_s varies with I_a (equation (A.3)), the role of I_a and hence P_s in the runoff and infiltration computation can be demonstrated by expanding and rearranging equations (A.4b) and (A.5b), respectively, as

$$Q = P - S - \left[I_a - \frac{S^2}{P - I_a + S} \right] \tag{A.7}$$

and

$$F = S - I_a + \left[I_a - \frac{S^2}{P - I_a + S} \right] \tag{A.8}$$

At $P = P_s$, the substitution of equation (A.3) makes the term within the parentheses in equations (A.7) and (A.8) equivalent to zero. Equations (A.7) and (A.8) are not valid for $P > P_s$ and it can be demonstrated by considering their limits. As $P \to \infty$, equations (A.7) and (A.8) converge, respectively, to

$$Q = P - S - I_a \tag{A.9}$$

and

$$F = S \tag{A.10}$$

Hawkins (1978a; 1978b; 1979) among others obtained equation (A.9) as a mathematical limit of the SCS-CN method. However, equation (A.9) is physically not realizable for mass balance reasons under the fully saturated condition. Similarly, equation (A.10) is invalid because F will approach the potential maximum retention that is equal to S excluding I_a under the fully saturated condition. This explains why equations (A.4) and (A.5) need three expressions to cover the whole range of P.

A.1.1 DETERMINATION OF RUNOFF AND INFILTRATION RATES

Since the existing SCS-CN method does not contain any expression for time t, an additional unknown such as surface detention is required. It, however, cannot be determined without the help of an additional hydrodynamic equation. The instantaneous water balance equation corresponding to equation (2.1) can be written as

$$P(t) = I_a(t) \qquad \text{for } P(t) \leq I_a \qquad \text{(A.11a)}$$

$$P(t) = I_a + F(t) + Q(t) + D(t) \qquad \text{for } P(t) \geq I_a \qquad \text{(A.11b)}$$

where P(t), I(t), F(t), Q(t), and D(t) are the cumulative precipitation, initial abstraction, infiltration, runoff, and surface detention, respectively, at any time 't' since rainfall begins. Note that $I(t_p) = I_a$, where t_p is the time to ponding. The magnitude of D(t) varies, depending mainly on the roughness characteristics and the slope of the soil surface on which rainwater moves. According to field observations during rainfall, D(t) rises rapidly in the early stage and then converges very slowly to a maximum value if the rainfall of uniform intensity continues indefinitely. In reality, there may be no maximum D(t) or more than one maximum D(t), depending on the variation of rainfall intensity, dP/dt, with t. The state of flow with no change in D(t) corresponds to the equilibrium stage when input equals the output. Immediately after the cessation of rainfall, D(t) drops rapidly and then diminishes very slowly to zero. Because most part of D(t) eventually converts to Q(t) in the rainfall-runoff process, equation (2.2) can be extended to a general form as

$$\frac{F(t)}{S} = \frac{Q(t) + D(t)}{P(t) - I_a} \qquad \text{(A.12)}$$

Mathematically, equation (2.2) may be regarded as a special case of equation (A.12) because $F(t) \rightarrow F$, $Q(t) + D(t) \rightarrow Q$, and $P(t) \rightarrow P$ as $t \rightarrow \infty$. Thus, similar to the derivation of equation (A.4) and (A.5), the following expressions for Q and F, respectively, can be derived from equations (A.11) and (A.12) as

$$Q(t) + D(t) = 0 \qquad \text{for } P(t) \leq I_a \qquad \text{(A.13a)}$$

$$Q(t) + D(t) = \frac{[P(t) - I_a]^2}{P(t) - I_a + S} \qquad \text{for } I_a \leq P(t) \leq P_s \qquad \text{(A.13b)}$$

$$Q(t) + D(t) = P(t) - S \qquad \text{for } P(t) \geq P_s \qquad \text{(A.13c)}$$

and

$$F(t) = 0 \text{ or } P(t) - I_a \qquad \text{for } P(t) \leq I_a \qquad \text{(A.14a)}$$

$$F(t) = \frac{[P(t) - I_a)S}{P(t) - I_a + S} \qquad \text{for } I_a \leq P(t) \leq P_s \qquad \text{(A.14b)}$$

$$F(t) = S - I_a \qquad \text{for } P(t) \geq P_s \qquad \text{(A.14c)}$$

Despite the apparent generality of the combined expression for Q(t) + D(t) (equation (A.13)), it is not possible to separate Q(t) from D(t) unless an additional hydrodynamic equation is incorporated with equations (A.11) and (A.12) for obtaining a solution. However, equation (A.14) containing only one unknown can be solved explicitly for F(t) or dF/dt.

The mathematical expressions of the rates of change in Q(t), D(t), I(t), and F(t) are designated herein as dQ/dt, dD/dt, dI/dt, and dF(t), respectively. Given the rainfall intensity dP/dt, expressions for runoff and infiltration rates can be derived, respectively, by differentiating equations (A.13) and (A.14) with respect to t as follows:

$$\frac{dQ}{dt}+\frac{dD}{dt}=0 \qquad \text{for } P(t)\le I_a \qquad (A.15a)$$

$$\frac{dQ}{dt}+\frac{dD}{dt}=\frac{[P(t)-I_a][P(t)-I_a+2S]}{[P(t)-I_a+S]^2}\frac{dP}{dt} \qquad \text{for } I_a\le P(t)\le P_s \qquad (A.15b)$$

$$\frac{dQ}{dt}+\frac{dD}{dt}=\frac{dP}{dt} \qquad \text{for } P(t)\ge P_s \qquad (A.15c)$$

and

$$\frac{dF}{dt}=0 \text{ or } \frac{dP}{dt}-\frac{dI_a}{dt} \qquad \text{for } P(t)\le I_a \qquad (A.16a)$$

$$\frac{dF}{dt}=\frac{S^2}{[P(t)-I_a+S]^2}\frac{dP}{dt} \qquad \text{for } I_a\le P(t)\le P_s \qquad (A.16b)$$

$$\frac{dF}{dt}=0 \qquad \text{for } P(t)\ge P_s \qquad (A.16c)$$

Thus, the instantaneous water balance equation (equation (A.11)) can be written in differential form as

$$\frac{dP}{dt}=\frac{dI_a}{dt} \qquad \text{for } P(t)\le I_a \qquad (A.17a)$$

$$\frac{dP}{dt}=\frac{dF}{dt}+\frac{dQ}{dt}+\frac{dD}{dt} \qquad \text{for } P(t)\ge I_a \qquad (A.17b)$$

Equations (A.15) and (A.16), however, reveal the existence of a discontinuity at $P(t) = P_s$.

For $P \ge I_a$, equations (A.15) and (A.16) can be recast, respectively as:

$$\frac{dQ}{dt}+\frac{dD}{dt}=\left[1-\left(\frac{S-F(t)}{S}\right)^2\right]\frac{dP}{dt} \qquad (A.18)$$

and

$$\frac{dF}{dt} = \left(\frac{S - F(t)}{S}\right)^2 \frac{dP}{dt} \tag{A.19}$$

In the state (variable) formulation of the soil-cover-moisture complex, equation (A.19) helps remove the discontinuity of dF/dt at $P(t) = P_s$ from equation (A.16).

It can be readily shown that the integration of equation (A.19) with respect to t with the help of an appropriately prescribed condition at $P = I_a$ yields equation (A.14). The rainfall intensity, dP/dt, is not necessarily constant during rainfall, except when the rainfall is linearly varying. The infiltration capacity decay curve derived from equation (A.19) may be subject to such a uniform rainfall condition, but its application can be extended beyond this condition, provided that the rainfall intensity is larger than or equal to the infiltration capacity. The following initial (f_o) and final (f_c) infiltration rates under uniform rainfall are obtained from equation (A.19). When $P(t) = I_a$, $F(t) = 0$. Therefore, substituting $F(t) = 0$ into equation (A.19) leads to

$$f_o = \frac{dP}{dt} \tag{A.20}$$

For $F(t) = S - I_a$, coupling of equation (A.19) with equation (A.20) yields

$$f_c = \left(\frac{I_a}{S}\right)^2 f_o \tag{A.21}$$

Since $F(t) = S - I_a$ at $P(t) = P_s$, the f_c-derivation from equation (A.21) suggests its occurrence at $P(t) = P_s$ or larger. Therefore, such an f_c-expression does not yield an asymptotic value. In terms of the available soil moisture M(t), infiltration rates can be derived as follows.

By definition,

$$M(t) = S - I_a - F(t) \tag{A.22}$$

It thus follows from equation (A.19) that

$$\frac{dF}{dt} = \left(\frac{M(t) + I_a}{S}\right)^2 \frac{dP}{dt} \tag{A.23}$$

Under the assumption of uniform rainfall, equation (A.23) can be shown to be analogous to the Holtan-Overton equation. To this end, equation (A.23) can be expanded with the help of equations (A.20) and (A.21) as

$$f = f_c + 2\sqrt{f_o f_c}\left(\frac{M(t)}{S}\right) + f_o\left(\frac{M(t)}{S}\right)^2 \tag{A.24}$$

where $f = dF/dt$. Since $M(t)$ is subject to the same constraints of $F(t)$ as specified in equation (A.14), f_c in equation (A.24) does not correspond to an asymptotic value valid for $t \rightarrow \infty$. It, however, is merely a constant for $t > t_s$, where t_s is the time at which $P = P_s$.

A.1.2 SCS-CN PARAMETERS

SCS-CN model has two parameters, S and I_a, that need to be determined before it can be applied. Strictly speaking, I_a includes interception and surface and subsurface storage before runoff begins. Therefore, for a given soil-vegetation-land use complex, even a rough estimation of I_a is extremely difficult. To avoid its estimation for various complexes, SCS (1971) developed the following approximate relationship between I_a and S using rainfall-runoff data for small experimental watersheds less than 10 acres in size:

$$I_a = 0.2\,S \tag{A.25}$$

Although SCS (1971) admitted that equation (A.25) has a large standard error of estimate, it was assumed valid for the lack of another better relationship. Furthermore, the introduction of equation (A.25) in the SCS-CN equations greatly simplifies the SCS-CN procedure from a two-parameter to one-parameter model. However, the general applicability of equation (A.25) still remains in question. For example, the field data analyzed by many investigators such as Hawkins (1976) and Aron et al. (1977) for various small watersheds do not strongly support the validity of equation (A.25). This may necessitate the use of a more general two-parameter concept by postulating that

$$I_a = \lambda S \tag{A.26}$$

where λ is the initial abstraction coefficient. Obviously, λ ranges from 0 to 1, for I_a can vary from 0 to S. Substitution of equation (A.26) into equation (A.21) yields

$$\lambda = \sqrt{\frac{f_c}{f_o}} \tag{A.27}$$

If the time to ponding, t_p, can be accurately estimated from soil and rainfall properties, such as initial and saturated hydraulic conductivities, initial and saturated moisture contents, and mean rainfall intensity, the corresponding value of f_o can then be read from any applicable infiltration decay curve or simply as the rainfall intensity in the case of uniform rainfall. For a given soil-cover complex, f_c is equivalent to the saturated hydraulic conductivity that is fixed. Therefore, known or given the values of f_o, f_c, and S, the λ value can be determined from equation (A.27) and, therefore, I_a can be estimated using equation (A.26).

APPENDIX B

B.1 Levenberg-Marquardt Algorithm

Marquardt (1963) provided an elegant and improved version of the non-linear optimization method originally proposed by Levenberg (1944). The method primarily provides a smooth variation between the two extremes of the inverse-Hessian method and the steepest descent method. The latter method is used when the trial solution is far from the minimum and it tends continuously towards the former as the minimum is approached. This Levenberg-Marquardt method is also called as the Marquardt method, which works well in practice and has become the standard of non-linear least-squares routines. Before describing the methodology, it is appropriate to introduce the fitting of non-linear models for a better understanding of the methodology. A complete treatise of optimization procedures is however available elsewhere (Press et al., 1998).

B.1.1 FITTING OF NON-LINEAR MODELS

Consider that a non-linear model has a set of M unknown parameters a_k, k = 1,2,3,...M. Its best-fit parameters are determined by minimizing the chi-square (χ^2) merit function. Because of the non-linear dependences, the minimization is supposed to proceed iteratively. For given trial values of the parameters, Marquardt developed a procedure to improve the trial solution. This procedure is then repeated until χ^2 stops (or effectively stops) decreasing. The χ^2 is expected to assume a quadratic form:

$$\chi^2(a) \approx \gamma - d \cdot a + (1/2)\, a \cdot D \cdot a \tag{B.1}$$

which can be derived using the Taylor series expansion of a given function and ignoring the third and higher order terms. In equation (B.1), d is an M-vector, D is an M x M matrix, and γ is a constant. If the approximation is fairly reasonable, the current value of parameter 'a', designated as a_{cur}, can be improved in a single leap by the parameters that correspond to the minimum χ^2-value, designated as a_{min} and expressed as:

$$a_{min} = a_{cur}\, D^{-1} \cdot [-\nabla \chi^2 (a_{cur})] \tag{B.2}$$

where ∇ stands for gradient. Equation (B.2), may be a poor local approximation to the shape of the function that is tried to minimize at a_{cur}. In such a case, the steepest-descent method can be resorted to. In other words,

$$a_{next} = a_{cur} - \text{constant} \times \nabla \chi^2 (a_{cur}) \tag{B.3}$$

where constant is small enough not to exhaust the downhill direction.

To use equation (B.2) or equation (B.3) it is necessary to compute the gradient of the χ^2 function at any set of parameters 'a'. For using equation (B.2), a determination of the second derivative matrix (or Hessian matrix) of the χ^2 function, matrix D, at any 'a' is also required.

Since the form of χ^2 function that represents an error criterion similar to the coefficient of determination, the Nash and Sutcliffe (1970) efficiency, or the root mean square error (RMS), the Hessian matrix is known a priori. It is noted that equation (B.3) is resorted to only when equation (B.2) fails to improve the fit. A failure of equation (B.2) is also an indication of a local optimum.

B.1.2 DETERMINATION OF THE GRADIENT AND HESSIAN MATRIX

Suppose the model to be fitted is represented as:

$$y = y(x;a) \tag{B.4}$$

where y is the dependent variable of independent variable x and 'a' represents the model parameters. Assume, for example, that the χ^2 represents a merit function, expressed as:

$$\chi^2(a) = \sum_{i=1}^{N}\left[\frac{y_i - y(x_i;a)}{\sigma_i}\right]^2 \tag{B.5}$$

where i is an integer that varies from 1 to N, N is the number of data points, y_i is the i_{th} observed value, σ_i is a measurement error (standard deviation) of the ith data point, and $y(x_i;a)$ is the ith computed value for given parameter values of 'a'. The gradient of χ^2 with respect to the parameters 'a', which will be zero at the minimum value of χ^2, is given as:

$$\frac{\partial \chi^2}{\partial a_k} = -2\sum_{i=1}^{N}\frac{[y_i - y(x_i;a)]}{\sigma_i^2}\frac{\partial y(x_i;a)}{\partial a_k} \tag{B.6}$$

The second partial derivative of equation (B.6) yields

$$\frac{\partial^2 \chi^2}{\partial a_k \partial a_l} = 2\sum_{i=1}^{N}\frac{1}{\sigma_i^2}\left[\frac{\partial y(x_i;a)}{\partial a_k}\frac{\partial y(x_i;a)}{\partial a_l}\right] - [y_i - y(x_i;a)]\frac{\partial^2 y(x_i;a)}{\partial a_k \partial a_l} \tag{B.7}$$

By convention, the factor of 2 is removed by defining

$$\beta_k \equiv -\frac{1}{2}\frac{\partial^2 \chi^2}{\partial a_k} \qquad \alpha_{kl} \equiv \frac{1}{2}\frac{\partial^2 \chi^2}{\partial a_k \partial a_l} \tag{B.8}$$

Equation (B.8) makes $[\alpha] = (1/2)D$ in equation (B.2). It leads to re-writing of equation (B.2) as a set of linear equations:

$$\sum_{l=1}^{M} \alpha_{kl}\delta a_l = \beta_k \tag{B.9}$$

Equation (B.9) is solved for the increments δa_l, which, upon addition to the current approximation, yields the next approximation. In a least squares sense, the matrix $[\alpha]$, which is equal to one-half times the Hessian matrix, is normally called the curvature matrix. Thus, the steepest descent formula (equation (B.3)) finally converts to

$$\delta a_l = \text{constant} \times \beta_k \tag{B.10}$$

It is noted that the components α_{kl} of the Hessian matrix (equation (B.7)) depend on both the first and second derivatives of the basis functions with respect to their parameters. The second derivative can be ignored for the following reasons.

Second derivatives occur because of the dependence of the gradient (equation (B.6)) on $\partial y/\partial a_k$. Therefore, the next derivative simply contains those terms involving $\partial^2 y/\partial a_l \partial a_k$. The second derivative can be ignored if it is equal to zero (as in the linear case) or it is small enough to be negligible compared to the term involving the second derivative. It also has an additional possibility of being significantly small in practice. The term multiplying the second derivative in equation (B.7) is $[y_i - y(x_i;a)]$. For a successful model, this term should just be the random measurement error of each point, which can be of either sign and should, in general, be uncorrelated with the model. Therefore, the second derivative terms tend to compensate (or cancel out) when they are summed over i.

It is worth emphasizing that the second derivative term can be destabilizing if the model fits badly or is contaminated by outlier points that are unlikely to be offset by compensating points of opposite sign. Thus, a coupling of equation (B.7) with equation (B.8) yields a formula for α_{kl}:

$$\alpha_{kl} = \sum_{i=1}^{N} \frac{1}{\sigma_i^2}\left[\frac{\partial y(x_i;a)}{\partial a_k}\frac{\partial y(x_i;a)}{\partial a_l}\right] \tag{B.11}$$

Here it is noted that a minor or even major fiddling with $[\alpha]$ has no effect on the final estimate of parameters; rather it affects only the iterative route followed for reaching the optimum set of parameter values.

B.1.3 LEVENBERG MARQUARDT METHOD

This method is based on two important observations. According to the first, the components of the Hessian matrix, even if they are not usable in any precise manner, yield some information about the order-of-magnitude scale of the problem. This insight decides the scale of magnitude of the constant in equation (B.10) in what follows.

As seen in equation (B.5), the quantity χ^2 is non-dimensional. However, β_k has the dimensions of $1/\alpha_k$ which may be dimensional. It is noted that each component of β_k can have different dimension. Thus, the constant of proportionality in equation (B.10) must have the dimensions of α_k^2. A scanning of the components of $[\alpha_k]$ reveals that there is only one obvious quantity bearing these dimensions and it is $1/\alpha_{kk}$, the reciprocal of the diagonal element. It, in turn, sets the scale of the constant. Since this scale may be too large, it is appropriate to divide the constant by some (non-dimensional) fudge factor λ (different from SCS-CN's initial abstraction coefficient), with a possibility to reduce the number of iterations. Alternatively, equation (B.10) can be replaced by

$$\delta a_l = \frac{1}{\lambda\alpha_{ll}}\beta_l \qquad \text{or} \qquad \lambda\alpha_{ll}\delta a_l = \beta_l \tag{B.12}$$

Here, it is necessary that α_{ll} be positive, which, however, is guaranteed by the definition given by equation (B.11).

The second observation consists of two components: (i) equations (B.12) and equation (B.9) are combined for defining a new matrix α' as:

$$\begin{aligned} \alpha'_{jj} &\equiv \alpha_{jj}(1+\lambda) \\ \alpha'_{jk} &\equiv \alpha_{jk} \qquad (j \neq k) \end{aligned} \tag{B.13}$$

and (ii) equations (B.12) and (B.9) are replaced by

$$\sum_{l=1}^{M} \alpha'_{kl}\,\delta a_l = \beta_k \tag{B.14}$$

When λ is very large, the matrix α' is forced to become diagonally dominant, which makes equation (B.14) become identical to equation (B.12). On the other hand, as λ approaches zero, equation (B.14) switches over to equation (B.9).

(a) *Application Steps of the Method*

(i) Employ an initial guess for the set of fitted parameters 'a'.
(ii) Compute $\chi^2(a)$ from equation (B.5).
(iii) Supply a modest value for λ, say $\lambda = 0.001$.
(iv) Solve the linear equations (equation (B.14)) for δa and evaluate χ^2 (a+δa).
(v) If χ^2 (a+δa) $\geq \chi^2$ (a), increase λ by a factor 10 (or any other substantial factor) and go to step (iv).
(vi) If χ^2 (a+δa) $< \chi^2$ (a), decrease λ by a factor of 10, update the trial solution a $\leftarrow$ a + δa, and go to step (iv).
(vii) Stop iteration if the following described convergence criterion is met.

(b) *Convergence Criterion*

It is necessary to provide a condition for stopping the procedure at a minimum required convergence level. Iterating to convergence to machine accuracy is generally of no use and unnecessary since the minimum is at best only a statistical estimate of parameters 'a'. It is known that a change in the parameters that change χ^2 by an amount much less than one is never statistically meaningful.

Furthermore, it is not uncommon to find the parameters wandering around near the minimum in a flat valley of complicated topography. The reason is that the Marquardt method generalizes the method of normal equations and, therefore, has the same problem as the method with regard to near-degeneracy of the minimum. Outright failure by a zero pivot is possible, but it is unlikely. More often, a small pivot will generate a large correction, which is rejected for increasing the value of λ. For sufficiently large λ-value, the matrix $[\alpha']$ is positive definite and can have no small pivots. Thus, the method tends to stay away from zero pivots, but at the expense of a tendency to wander around doing steepest descent in very un-steep degenerate valleys.

The above considerations suggest that, in practice, iteration may be stopped on the first or second occasion when χ^2 decreases by a negligible amount, say either less than 0.01 absolutely or some fractional amount like 0.001. However, after a step where χ^2 increases, it is imperative not to stop iterations because it shows that λ has not yet adjusted itself optimally.

Once the acceptable minimum is reached, the covariance matrix of standard errors in the fitted parameters can be computed as:

$$[C] \equiv [\alpha']^{-1} \tag{B.15}$$

where [C] is the covariance matrix.

APPENDIX C

C.1 Analytical Derivation for Wave Characteristics

Equations (8.13) and (8.14) can be linearized as below (Mahmood and Yevjevich, 1975; Ponce and Simons, 1977):

$$\frac{\partial h'}{\partial t} + u_o \frac{\partial h'}{\partial x} + h_o \frac{\partial u'}{\partial x} = 0 \tag{C.1}$$

$$\underset{\text{I}}{\frac{\partial u'}{\partial t}} + \underset{\text{II}}{u_o \frac{\partial u'}{\partial x}} + \underset{\text{III}}{g \frac{\partial h'}{\partial x}} + \underset{\text{IV}}{g\, s_o \left(\frac{h'}{h_o} - 2\frac{u'}{u_o}\right)} = 0 \tag{C.2}$$

where u' and h' are small velocity and depth perturbations about the normal velocity of flow u_o and the normal depth h_o, respectively. Expressed mathematically,

$$u' = u - u_o \tag{C.3}$$

$$h' = h - h_o \tag{C.4}$$

After Menendez and Norscini (1982), the solution of equations (C.1) and (C.2) follows:

$$\frac{u'}{u_o} = \hat{u}\ \exp[i(\hat{\sigma}\hat{x} - \hat{\omega}\hat{t})] \tag{C.5}$$

$$\frac{h'}{h_o} = \hat{h}\ \exp[i(\hat{\sigma}\hat{x} - \hat{\omega}\hat{t} - \gamma)] \tag{C.6}$$

where

$$\hat{\sigma} = (2\pi/L)\,(h_o/s_o) \tag{C.7}$$

$$\hat{\omega} = (2\pi/T)\,(h_o/(u_o s_o)) \tag{C.8}$$

$$\hat{t} = (t\, u_o s_o)/h_o \tag{C.9}$$

$$\hat{x} = (x\, s_o)/h_o \tag{C.10}$$

$$\gamma = \phi/\pi \tag{C.11}$$

$\hat{u}$ and $\hat{h}$ are, respectively, the dimensionless amplitudes of u-wave and h-wave; T is the time period of the wave; and L is the wave length ($L = c\, T$). γ ranges between 0 and 1. $\gamma = 0$ corresponds to $\phi = 0$ and $\gamma = 1$ to $\phi = \pi$. Substitution of equations (C.3) and (C.4) into equations (C.5) and (C.6), respectively, yields

$$(u - u_o)/u_o = \hat{u}\ \exp[i(\hat{\sigma}\hat{x} - \hat{\omega}\hat{t})] \tag{C.12}$$

$$(h - h_o)/h_o = \hat{h}\ \exp[i(\hat{\sigma}\hat{x} - \hat{\omega}\hat{t} - \gamma)] \tag{C.13}$$

From equations (C.12) and (C.13), the following can be derived:

$$\frac{(u - u_o)/u_o}{(h - h_o)/h_o} = \frac{\hat{u}}{\hat{h}} e^{i\gamma} = \Omega e^{i\gamma} \tag{C.14}$$

where

$$\Omega = \hat{u}/\hat{h} \tag{C.15}$$

For a unit-width rectangular channel, the Seddon (1900) speed formula can be written as

$$c = \frac{dQ}{dA} = \frac{d(uh)}{dh} \tag{C.16}$$

where Q is the discharge and A is the cross-sectional area of the channel. Thus, the Seddon speed is derived from the stage-discharge relation or rating curve. Equation (C.16) can, alternatively, be written as

$$c = u + h\frac{du}{dh} \tag{C.17}$$

or alternatively,

$$c = (1 + \frac{du/u}{dh/h})u = (1 + \dot{\beta})u \tag{C.18}$$

where $\dot{\beta}$ is the exponent of the rating curve. Under the assumption of a small perturbation analysis, equations (C.18) infers the following:

$$\frac{(u - u_o)/u_o}{(h - h_o)/h_o} = \dot{\beta} \tag{C.19}$$

Therefore, from equations (C.14) and (C.19) the following can be derived:

$$\dot{\beta} = \Omega\ e^{i\gamma} \tag{C.20}$$

Equation (C.20) can also be written as

$$\dot{\beta} = \Omega \cos\gamma + i\ \Omega \sin\gamma \tag{C.21a}$$

or

$$\dot{\beta} = \dot{\beta}_R + \dot{\beta}_I \tag{C.21b}$$

where $\dot{\beta}_R = \Omega \cos\gamma$ and $\dot{\beta}_I = \Omega \sin\gamma$. Therefore,

$$\frac{dQ}{dA} = (1 + \dot{\beta}_R)\ u_o + \dot{\beta}_I\ u_o \tag{C.22}$$

The first part on the right-hand side of the Seddon speed formula (equation (C.22)) accounts for translation and the second the attenuation. It is also consistent with the work of Price (1973; 1985), for wave celerity is a function of attenuation. Substitution of equations (C.12) and (C.13) into equation (C.1) (the continuity equation) and employment of equations (C.21a) and (C.21b) yield

$$\dot{\beta} = \dot{\beta}_R = \frac{\hat{\omega}}{\hat{\sigma}} - 1 \tag{C.23}$$

or

$$\dot{\beta}_R = \hat{c} - 1 = \hat{c}_r \tag{C.24}$$

where $\hat{c} = c/u_o$ and $\hat{c}_r$ represents the dimensionless relative celerity. Equation (C.24) establishes a link between the dimensionless relative celerity from the Seddon speed formula and the dimensionless relative speed from the linear solution of the St. Venant equations. The continuity equation represents a kinematical condition for which the phase shift is equal to zero. It is worth noting that equation (C.24) only equates the real component of parameter '$\dot{\beta}$' to $\hat{c}_r$, and thus, explicitly describes their equivalence. The value of '$\dot{\beta}$' or $\hat{c}_r$ will differ from one wave type to the other, as shown below.

C.1.1 KINEMATIC WAVE

The kinematic wave formulation consists of equation (C.1) and the IV term of equation (C.2). Substitution of equations (C.12) and (C.13) into equations (C.1) and (C.2) (IV term) and employment of equation (C.21a) yield

$$\dot{\beta} = \frac{1}{2} \tag{C.25}$$

from which

$$\dot{\beta}_R = \hat{c}_r = \frac{1}{2}; \text{ and } \dot{\beta}_I = 0 \qquad (C.26)$$

Employment of equations (C.11) and (C.21a) in equation (C.26) yields

$$\phi = \pi \tan^{-1}(0) = 0 \qquad (C.27)$$

The logarithmic decrement δ is related with φ as below (Singh, 1996):

$$\delta = -2\pi \frac{\Omega \sin \phi}{1 + \Omega \cos \phi} \qquad (C.28)$$

or alternatively,

$$\delta = -2\pi \frac{\dot{\beta}_I}{1 + \dot{\beta}_R} \qquad (C.29)$$

Employment of equation (C.26) to equation (C.29) yields

$$\delta = 0 \qquad (C.30)$$

Thus, kinematic waves do not exhibit attenuation and, in turn, the change in storage or wave type. According to the Rankine (1870) hypothesis, such a wave is of permanent nature and propagates without any energy loss. Similarly, the characteristics of other waves are derived. It is worth mentioning that equation (C.29) derived from the continuity equation representing kinematic wave (Singh, 1996) is used for computing δ for other waves using the same reasoning as of the relation $\dot{\beta} = \hat{c}_r$ (equation (C.24)) and the derived results conform to those of Ponce and Simons (1977) as shown below.

C.1.2 DIFFUSION WAVE

Diffusion wave is composed of equation (C.1) and the III and IV terms of equation (C.2). Usual substitution yields

$$\dot{\beta} = \frac{1}{2} + i(\frac{\hat{\sigma}}{2}) \qquad (C.31)$$

from which

$$\dot{\beta}_R = \hat{c}_r = \frac{1}{2}; \text{ and } \dot{\beta}_I = \frac{\hat{\sigma}}{2} \qquad (C.32)$$

From equations (C.29) and (C.32), δ can be computed as

$$\delta = -2\pi(\frac{\hat{\sigma}}{3}) \tag{C.33}$$

and the employment of equations (C.11) and (C.21a) in equation (C.32) yields

$$\phi = \pi \tan^{-1}(\frac{\hat{\sigma}}{2}) \tag{C.34}$$

It is evident from equations (C.33) and (C.34) that both δ and ϕ are non-zero for a non-zero $\hat{\sigma}$. If $\hat{\sigma} = 0$, the wave becomes kinematic for which both ϕ and δ are equal to zero. Thus, a diffusion wave ($\hat{\sigma} \neq 0$) exhibits a certain amount of phase lag. According to the Rankine hypothesis, diffusion waves are of non-permanent nature and will exhibit attenuation in propagation until a stable condition of kinematic wave ($\hat{\sigma} = 0$) is attained.

C.1.3 STEADY-DYNAMIC WAVE

The steady-dynamic wave formulation consists of equation (C.1) and the II, III, and IV terms of equation (C.2). The derived characteristic equation is

$$\dot{\beta}(i\hat{\sigma} F_o^2 - 2) + i\hat{\sigma} + 1 = 0 \tag{C.35}$$

where F_o is the Froude number [$= u_o/\sqrt{(gh_o)}$]. From equation (C.35) the real and imaginary parts can be separated as

$$\dot{\beta} = \frac{2 - \hat{\sigma}^2 F_o^2}{4 + \hat{\sigma}^2 F_o^4} + i\frac{2\hat{\sigma} + \hat{\sigma} F_o^2}{4 + \hat{\sigma}^2 F_o^4} \tag{C.36}$$

Equation (C.36) yields the following characteristics:

$$\hat{c}_r = \frac{2 - \hat{\sigma}^2 F_o^2}{4 + \hat{\sigma}^2 F_o^4} \tag{C.37}$$

$$\delta = -2\pi \frac{\hat{\sigma}(2 + F_o^2)}{6 - \hat{\sigma}^2 F_o^2 (1 - F_o^2)} \tag{C.38}$$

$$\phi = \pi \tan^{-1} \frac{2\hat{\sigma} + \hat{\sigma} F_o^2}{2 - \hat{\sigma}^2 F_o^2} \tag{C.39}$$

For $F_o = 0$, equations (C.37) through (C.39) reduce to equations (C.32) through (C.34), respectively, describing the characteristics of diffusion waves. $F_o = 0$ implies that either $u_o = 0$ or $h_o = \infty$. Physically, in open channel flow, either of the situations is not feasible for propagation of kinematic and diffusion waves. By making the second term of equation (C.2) describing the rate of change in the magnitude of velocity perturbations with space equal to zero, the mechanism of the above waves can be described, following equation (8.14). Under the diffusion wave, velocity does not change along the flow direction whereas the depth of flow changes. Under the kinematic wave, both the velocity and depth remain unchanged during the course of propagation, whereas a steady-dynamic wave is characterized by a change in both velocity and depth with space. The change in velocity with space in steady-dynamic wave eventually introduces Froude number to account for the inertial force. Here, it is important to note that the continuity equation (equation (8.13)) does account for variations in stage and velocity with space and variation in stage with time, whereas approximate formulations ignore their impact on momentum. Consequently, these approximations exhibit lesser capacity to lose energy (or attenuation or damping capacity) than the dynamic wave (Mishra and Singh, 1999, 2001). According to Rankine's hypothesis, these waves are of non-permanent nature and will attenuate during propagation until a stable condition of kinematic wave ($\hat{\sigma}$ = 0) is attained.

C.1.4 GRAVITY WAVE

Equation (C.1) and I, II, and III terms of equation (C.2) represent the gravity wave formulation. The characteristic equation is

$$\dot{\beta}[(-\hat{\omega}+\hat{\sigma})F_o^2] + \hat{\sigma} = 0 \tag{C.40}$$

Substituting equation (C.24) into equation (C.40) yields

$$\dot{\beta} = \dot{\beta}_R = \hat{c}_r = \pm\frac{1}{F_o} \tag{C.41}$$

where + sign represents primary wave propagating downstream and – sign corresponds to secondary wave propagating upstream. According to equation (C.2), these waves occur when both s_f and s_o are equal to zero, and thus, no energy is lost. Consequently, the imaginary component of '$\dot{\beta}$' becomes zero making both δ and ϕ equal to zero.

C.1.5 DYNAMIC WAVE

Equations (C.1) and (C.2) represent the dynamic wave. The desired characteristic equation is derived as

$$\dot{\beta}^2 i\hat{\sigma}F_o^2 + 2\dot{\beta} - (i\hat{\sigma}+1) = 0 \tag{C.42}$$

Let

$$K = 1/\hat{\sigma}F_o^2 \tag{C.43}$$

which represents the kinematic wave number. Coupling of equation (C.43) with equation (C.42) leads to the following solution for β:

$$\dot{\beta} = iK \pm \sqrt{\hat{\sigma}K - K^2 - iK} \tag{C.44}$$

Assume that

$$X\cos\theta = \hat{\sigma}K - K^2 \tag{C.45}$$

and

$$X\sin\theta = K \tag{C.46}$$

Equations (C.45) and (C.46) lead to

$$X = \sqrt{K^2 + (\hat{\sigma}K - K^2)^2} \tag{C.47}$$

and

$$\cos\theta = \frac{\hat{\sigma}K - K^2}{\sqrt{K^2 + (\hat{\sigma}K - K^2)^2}} \tag{C.48}$$

Substitution of equations (C.45) and (C.46) into equation (C.44) leads to

$$\beta_R = \pm\left[\frac{X + X\cos\theta}{2}\right]^{1/2} \tag{C.49}$$

and

$$\beta_I = K \mp \left[\frac{X - X\cos\theta}{2}\right]^{1/2} \tag{C.50}$$

A usual substitution of equations (C.49) and (C.50) for deriving $\hat{c}_r$, δ, and ϕ, as above, finally leads to equations (8.16) through (8.18), respectively.

C.1.6 PHYSICAL INTERPRETATION OF RATING EXPONENT

The rating exponent '$\dot{\beta}$' can be physically interpreted in terms of the laws of thermodynamics as follows. The amount of energy expended to produce a wave is (Rankine, 1870):

$$E = p\,u \tag{C.51}$$

where E = energy expended in unit time, p = pressure, and u = flow velocity. A part of this energy (E_c) contributes to translating a wave with a certain speed c and the remaining part (E_δ) contributes to altering the channel storage. Expressed mathematically,

$$E = E_c + E_\delta \tag{C.52}$$

For a unit mass, $E_c + E_\delta$ can be determined, respectively, as

$$E_c = (1/2)\,c^2 = (1/2)\,(1+\dot{\beta}_R)^2\,u_o^2 \tag{C.53}$$

and

$$E_\delta = (1/2)\,(p_1 + p_2)\,(S_2 - S_1) \tag{C.54}$$

In Eq. (C.54), p is the pressure, S is the storage (different from SCS-CN parameter S), and subscripts 1 and 2 correspond to undisturbed and disturbed states, respectively. The work done in altering the channel storage is attributed to the imaginary component of '$\dot{\beta}$' as described earlier. For a unit mass, E_δ is equivalent to:

$$E_\delta = \frac{1}{2}\dot{\beta}_I^2\,u_o^2 \tag{C.55}$$

Therefore, from equations (C.53) and (C.55)

$$\frac{E_\delta}{E_c} = \left[\frac{\dot{\beta}_I}{1+\dot{\beta}_R}\right]^2 \tag{C.56}$$

Substitution of equation (C.29) into equation (C.56) yields

$$\frac{E_\delta}{E_o} = \frac{\delta^2}{4\pi^2} \tag{C.57}$$

or alternatively,

$$\delta = 2\pi\sqrt{\frac{E_\delta}{E_c}} \tag{C.58}$$

Therefore, δ represents the ratio of the magnitude of the work done in altering the channel storage or phase to the energy expended in translation. In a looped rating curve (Figure C.1), E_δ corresponds to the shaded area. Thus, the real component of '$\dot{\beta}$' signifies the energy expenditure in translation and the imaginary component describes the energy expenditure (or loss) in attenuation. Physically, '$\dot{\beta}$' can be described using equation (C.19) as the ratio of two longitudinal orthogonal strains. The strain in the numerator can be interpreted as a longitudinal strain in the horizontal direction and that in the denominator as a longitudinal strain in the vertical direction. Equation (C.19) can be re-written as

$$\dot{\beta} = \frac{h\,du}{u\,dh} \tag{C.59}$$

The numerator of equation (C.59) represents the energy expended in a unit time at constant pressure and the denominator represents the energy expended in a unit time at constant storage (or volume). From the concepts of thermodynamics, these are analogous to the specific heats c_p and c_v which are produced at constant pressure and constant volume, respectively. Here it is necessary to mention that in a closed gaseous system, the work done for altering the volume of a gas converts to heat to increase the kinetic energy of the molecules of the gas or indirectly increase its temperature. In open channel flow, the heat produced in the work done on altering channel storage is conducted through water into the atmosphere and eventually dissipated through space.

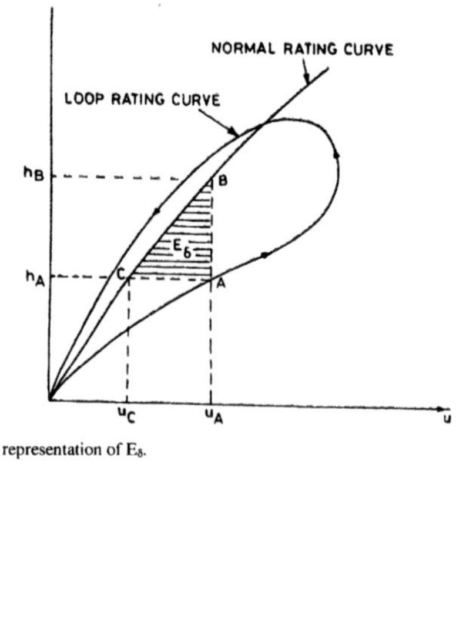

Figure C.1. Graphical representation of E_δ.

Parameter '$\dot{\beta}$' can also be described using the imaginary component of equation (C.21a) as

$$\dot{\beta}_I = \Omega \sin\gamma \tag{C.60}$$

where Ω is the amplitude ratio of velocity and depth of flow (equation (C.15)). Assume U and H are velocity and depth amplitudes of u-wave and h-wave, respectively. Their normalization using u_o and h_o, respectively, leads to re-writing of equation (C.15) as

$$\Omega = \frac{\hat{u}}{\hat{h}} = \frac{Uh_o}{Hu_o} \tag{C.61}$$

Since U and H represent the amplitudes about u_o and h_o, respectively, the last part of equation (C.61) represents the ratio of the maximum possible amount of c_p and c_v and can be designated as $\dot{\beta}_{max}$, analogous to '$\dot{\beta}$'. Thus,

$$\Omega = \dot{\beta}_{max} \tag{C.62}$$

Therefore, equation (C.60) can be re-written as

$$\frac{\dot{\beta}_I}{\dot{\beta}_{max}} = \sin\gamma \tag{C.63}$$

The left-hand side of equation (C.63) represents the ratio of energy loss (non-dimensional) in a unit time to the potential maximum energy (non-dimensional) available for being lost in a unit time or, indirectly, the potential maximum energy (per unit time) of the input wave. This ratio is identical to the definition of damping capacity (DC) (Murphy, 1957). Thus, DC is solely dependent on γ, which is a function of phase lag, ϕ, an attenuation driving parameter (Menendez and Norscini, 1983). For $\gamma = 0$, DC = 0. Thus, the damping capacity of a kinematic wave is zero. It can vary between -1 and $+1$. Here, '-' sign corresponds to the situation when h-wave precedes u-wave and the '+' sign shows that the h-wave follows the u-wave. Therefore, it describes the magnitude of the damping effect. To distinguish DC from hysteresis, η is re-derived from equation (8.6) using the linear perturbation theory as

$$\eta = \pi \hat{u}\hat{h} \sin\gamma \tag{C.64}$$

which is valid for simple cosine waves. In equation (8.6), σ was replaced by h, ε by u, and the total derivative by partial ones. In equation (C.64), η is made non-dimensional using u_0h_0. The term $\hat{u}\hat{h}$ on the right-hand side of this equation represents the power (non-dimensional) (or energy per unit time) of the input wave.

Murphy (1957) described hysteresis as an index of energy expenditure that describes how quickly vibrations will damp out. Similarly, from equation (C.64) η can be defined as an index of energy expenditure representing the magnitude and quickness of wave subsidence in unit time period of wave propagation. The larger power of the input wave will yield larger η and more quickly the wave will subside or damp out and vice versa. Such a definition also infers that the waves of larger η-values (for example, dynamic waves) will exhibit shorter time periods than those of smaller η-values (for example, diffusion waves) for a given power of the wave. Thus, the damping capacity describes the magnitude of subsidence, whereas the hysteresis indicates both the magnitude and quickness of subsidence.

APPENDIX D

D.1 Universal Soil Loss Equation

The universal soil loss equation (USLE) estimates the potential soil loss from long-term upland (sheet and rill) erosion. Mathematically, it is expressed as (Wischmeier and Smith, 1965):

$$A = R\ K\ L_s\ V_c\ M_p \tag{D.1}$$

where A is the potential soil loss for a given storm or period (tonnes/ha), R is the rainfall factor, K is the soil erodibility factor, L_s is the slope-length factor, V_c is the vegetative cover (or cropping management) factor, M_p is the erosion control practice factor. Equation (D.1) expresses soil loss per unit area due to erosion by rainfall-runoff, excludes wind erosion, and does not yield direct sediment yield estimates. It is, however, more versatile than the earliest, most successful equation after Musgrave (1947).

The rainfall factor, R is computed as (Foster et al., 1977):

$$R = 0.5R_r + 7.5Qq^{1/3} \tag{D.2}$$

where R_r is the rainfall energy factor, Q is the runoff volume (cm), and q is the peak runoff rate (cm/hr). Since q is more related to detachment than Q (Williams and Berndt, 1977), a reduction in peak discharge by the vegetation cover will also reduce the sediment transport. Alternatively, the rainfall factor, R, can be computed as

$$R = \Sigma[(2.29 + 1.15 \ln X_i)\ D_i]\ I \tag{D.3}$$

where i is the time interval of rainfall hyetograph, D_i is the rainfall volume during i^{th} time interval (cm), I is the maximum 30-min rainfall intensity of the storm (cm/hr), and X_i is the rainfall intensity (cm/hr).

The soil erodibility factor, K, is a measure of the potential erodibility of soil. It is a function of soil texture. The slope-length factor, L_s, accounts for the overland runoff length and slope. For slopes > 4%, it can be determined as:

$$L_s = L^{1/2}\ (0.0138 + 0.00974\ Y + 0.00138\ Y^2) \tag{D.4}$$

where Y is the percent slope over the runoff length and L is the length in meters from the point of origin of the overland flow to the point where the slope decreases to the extent that sedimentation begins. The vegetation cover factor or cropping management factor, V_c, estimates the effect of ground cover conditions, soil conditions, and general management practices on erosion rates. The erosion control practice factor, M_p, takes into account the effectiveness of erosion control practices, such as land treatment, by contouring, compacting, establishing sedimentation basins, and other control structures. Generally, V_c reflects the protection of the soil surface against the impact of rain droplets and subsequent loss of soil particles, whereas M_p includes treatments that retain eroded particles and prevent them from further transport. The experimentally derived values of the above factors for various soil-vegetation-land use complexes are available elsewhere (Singh, 1992; Ponce, 1989; Novotny and Olem, 1994; Haan et al., 1994; Singh and Singh, 2001).

Example D.1: A 50-ha land of square shape is to be developed as a residential area. It has a drainage ditch at the center, which is to be replaced by a storm sewer. The average slope of the plot towards the ditch is 2.4%. The composition of soils for the plot suggests soil erodibility factor to be equal to 0.33. For the following rainstorm, determine the potential soil erosion for two conditions: (a) when all vegetation is stripped from the soil surface (100% pervious) and (b) subsequent to construction when 25% area is impermeable due to the construction of streets, roofs, driveways, etc.:

Time (min)	0-30	30-60	60-90	90-120	120-150	150-180
Rainfall intensity (cm/hr)	0.30	2.50	1.25	0.70	0.20	0.10

Solution: First of all, determine the rainfall factor using equation (D.3) utilizing the given rainfall values. The available data show that the maximum 30-min rainfall intensity is 2.5 cm/hr. Thus,

$$R = \Sigma[(2.29 + 1.15 \ln X_i) D_i]\, I$$
$$= [(2.29 + 1.15 \ln 0.30) \times 0.15$$
$$+ (2.29 + 1.15 \ln 2.50) \times 1.25$$
$$+ (2.29 + 1.15 \ln 1.25) \times 0.6175$$
$$+ (2.29 + 1.15 \ln 0.70) \times 0.35$$
$$+ (2.29 + 1.15 \ln 0.2) \times 0.10$$
$$+ (2.29 + 1.15 \ln 0.1) \times 0.05] \times 2.5 = 16.4$$

For computing the L_s factor for the given plot, determine the overland flow depth L as:

$$L = 0.5\,(50 \times 100 \times 100)^{1/2} = 353.5 \text{ m}$$

Here, percent slope Y = 2.4. Use equation (D.4) for computing L_s as:

$$L_s = L^{1/2}\,(0.0138 + 0.00974\,Y + 0.001138\,Y^2)$$
$$= (353.5)^{1/2} \times \{0.0138 + 0.00974 \times 2.4 + 0.00138\,(2.4)^2\} = 0.85$$

Since factors R, K, and L_s will be same for both the given alternatives, V_c and M_p need to be evaluated for each alternative. Only M_p will be determined if erosion control practices are implemented during construction.

For alternative (a) that refers to the period during construction, V_c is estimated assuming no vegetative protective cover. In this case, $V_c = 1$. In the absence of erosion control practices, $M_p = 1$. Thus, the potential soil loss, A, is estimated as:

$$A = 16.4 \times 0.33 \times 0.85 \times 1 \times 1 = 4.60 \text{ tonnes/ha}$$

Therefore, for 50-ha plot, A = 50 x 4.60 = 230 tonnes.

For alternative (b) that represents the period after construction, assume that the pervious areas are covered by lawns, leading to a reduction of V_c to 0.01. Thus, the potential soil loss, A, is computed as:

$$A = 16.4 \times 0.33 \times 0.85 \times 0.01 \times 1 = 0.046 \text{ tonnes/ha}$$

Therefore, total potential soil loss = 50 x 0.046 = 2.3 tonnes.

REFERENCES

Chapter 1

1. ASCE (1996),'Hydrology Handbook,' ASCE Manuals and Reports on Engineering Practice No. 28, American Society of Civil Engineers, New York.
2. Bowen, I.S. (1926),'The Ratio of Heat Losses by Conduction and by Evaporation from Any Water Surface,' Physical Review 27, pp. 779-787.
3. Branson, F.A., G.F. Gifford, K.G. Renard, and R.F. Hadley (1981),'Rangeland Hydrology, Society for Range Management, Range Science Series No. 1, Kendall/Hunt Publ. Co., Dubuque, Iowa.
4. Bras, R.L. (1990),'Hydrology: An Introduction to Hydrologic Science,' Addison-Wesley Publishing Company, New York.
5. Bruce, J.P. and G.K. Rodgers (1962),'Water balance in the Great Lakes System, Great lakes basin,' Am. Assoc. for the Advancement of Science, Pub. No. 71, Washington D.C.
6. Chen Cheng-lung (1982), An evaluation of the mathematics and physical significance of the soil conservation service curve number procedure for estimating runoff volume, in Rainfall-Runoff Relationship, edited by V.P. Singh, Water Resour. Pub., Littleton, Colo. 80161.
7. Chorley, R.J. (1978),'Chapter 1: The hillslope hydrologic cycle,' Hillslope Hydrology, M.J. Kirkby, ed., John Wiley & Sons, New York, N.Y.
8. Chow, V.T. (1964),'Handbook of Applied Hydrology,' McGraw Hill, New York.
9. Collis-George, N. (1977),'Infiltration equations for simple soil systems,' J. Water Resources Research, Vol. 13, No. 2, pp. 395-403.
10. Dickinson, T.W. and H.D. Ayers (1965),'The effect of storm characteristics on the unit hydrograph,' Transactions of the Engineering Institute of Canada 8(A-15), pp. 3-7.
11. Dooge, J.C.I. (1973),'Linear theory of hydrologic systems,' Tech. Bull. No. 1468, U.S. Dept. of Agriculture, Agriculture Research Service, Washington D.C.
12. Dunin, F.X. (1969),'The infiltration component of a pastoral experimental catchment,' J. Hydrology, 7, pp. 121-131.
13. Dunne, T. and R.D. Black (1970),'An experimental investigation of runoff production in permeable soils,' Water Resources Research, Vol. 6, No. 2, pp. 478-490.
14. Dunne, T. and L.B. Leopold (1978),'Water in Environmental Planning,' Freeman and Co., San Francisco.
15. Feddes, R.A., P. Kabat, P.J.T. Van Bakel, J.J.B. Bronswijk, and J. Halbertsma (1988),'Modeling soil water dynamics in the unsaturated zone-State of the art,' J. Hydrology, 100, 69-111.
16. Foster, I.D.L. and S.M. Charlesworth (1996),'Heavy metals in the hydrological cycle: Trends and explanation,' Hydrological Processes, Vol. 10, pp. 227-261.
17. Gayle, G.A. and R.W. Skaggs (1978),'Surface storage on bedded cultivated lands, Trans. ASAE, 21(1), pp. 101-104, 109.
18. Green, W.H. and C.A. Ampt (1911),'Studies on soil physics, I. Flow of air and water through soils,' J. Agric. Sci., 4, pp. 1-24.
19. Gray, D.D., P.G. Katz, S.M. deMonsabert, and N.P. Cogo (1982),'Antecedent moisture condition probabilities,' J. Irrig. and Draing. Div., ASCE, Vol. 108, No. IR2.
20. Haithaway, G.A. (1945),'Design of drainage facilities,' Trans., ASCE, Vol. 110, pp. 697-730.
21. Hall, L.S. (1934),'Evaporation from water surfaces,' Trans. ASCE, Vol. 99, pp. 715-724.
22. Hall, M.J. (1984),'Urban Hydrology,' Elsevier Applied Science Publishers, London.
23. Hamilton, R.S. and R.M. Harrison (1991),'Highway Pollution,' Elsevier Science Publishing Co., Inc., New York.
24. Hargreaves, G.H. and Z.A. Samani (1982),'Estimating potential evapotranspiration,' Tech. Note, J. Irrig. And Drainage Engrg., ASCE, Vol. 108, No. 3, pp. 225-230.
25. Hargreaves, G.H. and Z.A. Samani (1985),'Reference crop evapotranspiration from temperature,' Applied Engineering in Agriculture, Vol. 1, No. 2, pp. 96-99.
26. Hawkins, R.H. (1973),'Improved prediction of storm runoff from mountain watersheds, J. Irrig. and Drainage Div., ASCE, 99(4), 519-523.
27. Hawkins, R.H. (1981),'Interpretations of source area variability in rainfall-runoff relations,' Proc., Int. Symp. on Rainfall Runoff Modeling, Water Resources Pub., Littleton, Colo., 303-324.

28. HEC (1981),'HEC-1 Flood Hydrograph Package- Users Manual,' Hydrologic Engineering Center, U.S. Army Corps of Engineers, Davis, California.
29. Hicks, W.I. (1944),'A method of computing urban runoff,' Trans., ASCE, Vol. 109, pp. 1217-1233.
30. Holtan, H.N. (1961),'A concept of infiltration estimates in watershed engineering,' ARS41-51, U.S. Dept. of Agric. Service, Washington, D.C.
31. Horton, R.E. (1919),'Rainfall interception,' Monthly Weather Review, Vol. 47, September, pp. 603-623.
32. Horton, R.I. (1933),'The role of infiltration in the hydrologic cycle,' Trans., Am. Geophysical Union, 14, 446-460.
33. Horton, R.I. (1938),'The interpretation and application of runoff plot experiments with reference to soil erosion problems,' Proc., Soil Science Society of America, 3, pp. 340-349.
34. Jensen, M.E. (1966),'Empirical methods of estimating or predicting evapotranspiration using radiation,' Proc. Conference on Evapotranspiration, A.S.C., Chicago, Dec., pp. 65.
35. Jensen, M.E. and H.R. Haise (1963),'Estimating evapotranspiration from solar radiation,' J. Irrig. and Drainage Div., A.S.C.E., Vol. 89, pp. 15-41.
36. Jensen, M.E., D.C.N. Robb, and C.E. Franzoy (1970),'Scheduling irrigations using climate, crop and soil data,' Trans. Am. Soc. Ag. Engr., Vol. 96, pp. 25-28.
37. Johnstone, D. and W.P. Cross (1949),'Elements of Applied Hydrology,' Ronald Press, New York.
38. Kalinin, G.P. and P.I. Milyukov (1958),'On the computation of unsteady flow in open channels,' Meteorologiya Gidrologiya Zhuzurnal, Leningrad, Vol. 10, pp. 10-18.
39. Kirkby, M.J., and R.J. Chorley (1967),'Throughflow, overland flow, and erosion,' Bull., Int. Assoc. on Eci. Hydrol., 12, pp. 5-21.
40. Kirpich, Z.P. (1940),'Time of concentration of small agricultural watersheds,' Civil Engineering, Vol. 10, June, p. 362.
41. Kostiakov, A.N. (1932),'On the dynamics of the coefficients of water percolation in soils,' Sixth Commission, Int. Soc. Soil Science, Part A, pp. 15-21.
42. Le Bissonnais, Y. and M.J. Singer (1993),'Seal formation, runoff, and interrill erosion from seventeen California soils,' Soil Sci. Soc. of Am. J., 57, 224-229.
43. Li, Y., S.G. Buchberger, and J.J. Sansalone (1999),'Variably saturated flow in storm-water partial exfiltration trench,' J. Environmental Engineering, A.S.C.E., Vol. 125, No. 6, pp. 556-565.
44. Linacre, E.T. (1969), Closing discussion 'Climate and evaporation from crops,' J. Irrig. Drainage Div., A.S.C.E., Vol. 95, pp. 348-352.
45. Linsley, R.K. (1982),'Rainfall-runoff models-An overview,' In: Proc., Int. Symp. on Rainfall-Runoff Relationship (ed. V.P. Singh), Mississippi, May 18-21, 1981, Water Resources Publications.
46. Linsley, R.K., M.A. Kohler, and L.H. Paulhus (1982),'Hydrology for Engineers,' 3rd Ed., McGraw Hill, New York.
47. Mahmood, K. and V. Yevjevich (1975),'Unsteady Flow in Open Channels,' Vol. I, Water Resources Pub., PO Box 303, Fort Collins, Colorado 80522.
48. McCarthy, G.T. (1938),'The unit hydrograph and flood routing,' unpublished manuscript, presented at a conference of the North Atlantic Division, U.S. Army Corps of Engineers, June 24.
49. Mein, R.G. and C.L. Larson (1971),'Modeling the infiltration component of the rainfall-runoff process,' WRRC Bull. 43, Water Resources Research Center, University of Minnesota, Minneapolis, Minnesota.
50. Meyer, A.F. (1944),'Evaporation from lakes and reservoirs,' Minnesota Resources Commission, St. Paul, Minnesota, June 14, Monteith, J.L (1959),'The reflection of short wave radiation by vegetation,' Quarterly Journal of the Royal Meteorological Society, Vol. 85, pp. 586-592.
51. Mishra, S.K., and V. P. Singh (1999),'Hysteresis-based flood wave analysis,' J. Hydrologic Engrg., A.S.C.E., Vol. 4, No. 4, pp. 358-365.
52. Mishra, S.K. and V.P. Singh (2002), Chapter 13, 'SCS-CN-based hydrologic simulation package,' Mathematical Models in Small Watershed Hydrology, (eds.) V.P. Singh and D.K. Frevert, Water Resources Publications, P.O. Box 2841, Littleton, Colorado 80161.
53. Mishra, S.K., S.R. Kumar, and Vijay P. Singh (1999),'Calibration of a general infiltration model,' J. Hydrological Processes, Vol. 13, pp. 1691-1718.
54. Monteith, J.L. (1965),'Evaporation and the environment,' In: The State and Movement of Water in Living Organisms, XIXth Symposium, The Soc. for Expt. Biol., Swansea, Cambridge University Press, 205-234.
55. Morel-Seytoux H.J. and J. Khanji (1975),'Equation of infiltration with compression and counterflow effects,' Hydrol. Sci. Bull., 20(4), 505-517.

56. Nash, J.E. (1957),'The form of the instantaneous unit hydrograph,' Publication 42, Wallingford, England, International Association Scientific Hydrology, pp. 114-112.
57. Nash, J.E. (1958),'Determining runoff from rainfall,' Proc. Institution of Civil Engineers, Ireland, Vol. 10, pp. 163-184.
58. Nash, J.E. (1959),'Systematic determination of unit hydrograph parameters,' Journal of Geophysical Research, Vol. 64, No. 1, pp. 111-115.
59. Nash, J.E. (1960),'A unit hydrograph study, with particular reference to British catchments,' Proc. Institution of Civil Engineers, London, Vol. 17, pp. 249-282.
60. NWS (1973),'Atlas 2: Precipitation atlas of the Western United States,' NOAA National Weather Service.
61. NWS (1977),'Five- to sixty-minute precipitation frequency for the Eastern and Central United States,' Technical Memorandum NWS-HYDRO-35, National Weather Service.
62. Onstad, C.A. (1984),'Depression storage on tilled soil surfaces,' Trans., A.S.A.E., 27(3), pp. 729-732.
63. Overton, D.E. (1964),'Mathematical refinement of an infiltration equation for watershed engineering,' ARS 41-99, U.S. Dept. of Agric. Service, Washington, D.C.
64. Papadakis, C.N. and M.N. Kazan (1987),'Time of concentration in small rural watersheds,' Proc. Engineering Hydrology Symp., A.S.C.E., Williamsburg, Virginia, Aug. 3-7, pp. 633-638.
65. Parlange, J.-Y. (1971),'Theory of water movement is soils, 2, One-dimensional infiltration,' Soil Sci., 111(3), pp. 170 174.
66. Penman, H.L. (1948),'Natural evaporation from open water, bare soil, and grass,' Proc. Roy. Soc. London, Vol. 193, pp. 120-133.
67. Perreault, P. (1674),'De l'origine des fontaines,' Paris, In: Proc., Int. Symp. on Rainfall-Runoff Relationship (ed. V.P. Singh), Mississippi, May 18-21, 1981, Water Resources Publications.
68. Philip, J.R. (1957),'Theory of Infiltration, Chapter 1 and Chapter 4,' Soil Sci., 83(5), pp. 345-357.
69. Ponce, V.M. (1989),'Engineering Hydrology: Principles and Practice,' Prentice Hall, Englewood Cliffs, New Jersey 07632.
70. Ponce, V.M., and R.H. Hawkins (1996),'Runoff curve number: Has it reached maturity?,' J. Hydrolog. Engrg., A.S.C.E., Vol. 1, No. 1, pp. 11-19.
71. Preistley, C.H.B. and R.J. Taylor (1972),'On the assessment of surface heat flux and evaporation using large scale parameters,' Monthly Weather Review, Vol. 100, pp. 81-92.
72. Rode, A.A. (1965),'Theory of soil moisture,' Vol. 1, (Translated from Russian). Published for U.S.D.A. and N.S.F. by the Israel Program for Scientific Translation, Jerusalem, 1969.
73. Rohwer, C. (1931),'Evaporation from free water surface,' U.S. Dept. of Agri. Tech. Bull. 271, Washington, D.C.
74. Rohwer, C. (1934),'Evaporation from different types of pans,' Trans. A.S.C.E., Vol. 99, pp. 673-677.
75. Sansalone, J.J and S.G. Buchberger (1997),'Partitioning and first flush of metals in urban roadway storm water', J. Environmental Engineering, A.S.C.E.,Vol.123, No. 2, pp. 134-143.
76. Sansalone, J.J., J.M. Koran, J.A. Smithson, and S.G. Buchberger (1998),'Physical characteristics of urban roadway solids transported during rain events,' J. Environmental Engineering, A.S.C.E., Vol. 125, No. 6, pp. 556-565.
77. Sansalone, J.J. and T. Triboullard (1999),'Variation in characteristics of abraded roadway particles as a function of particle size- Implications for water quality and drainage,' In: Transporation Research Record 1690, TRB, National Research Council, Washington D.C., pp. 153-163.
78. Schaake, J.C, V.I. Koren, Qing-Yun Duan, K. Mitchell, and F. Chen (1996),'Simple water balance model for estimating runoff at different spatial and temporal scales,' J. Geophysical Research, Vol. 101, No. D3, pp. 7461-7475.
79. SCS (1956, 1964, 1969, 1971, 1972, 1985, 1993),'Hydrology,' National Engineering Handbook, Supplement A, Section 4, Chapter 10, Soil Conservation Service, USDA, Washington, D.C.
80. Sherman, L.K. (1932),'Streamflow from rainfall by unit-graph method,' Engineering News Record, Vol. 108, April 7, pp. 501-505.
81. Singh, P. and V.P. Singh (2001),'Snow and Glacier Hydrology,' Kluwer Academic Publishers, P.O. Box 322, Dordrecht, The Netherlands.
82. Singh, V.P. (1988),'Hydrologic Systems: Vol. I. Rainfall-Runoff Modeling,' Prentice-Hall, Englewood Cliffs, New Jersey.
83. Singh, V.P. (1992),'Elementary Hydrology,' Prentice Hall, Englewood Cliffs, New Jersey 07632.
84. Singh, V.P. (1996),'Kinematic Wave Modeling in Water Resources: Surface-Water Hydrology,' John Wiley & Sons, Inc., New York.

85. Singh, V.P. and F.X. Yu (1990),'Derivation of infiltration equation using systems approach,' J. Irrig. and Drain. Engrg., A.S.C.E., Vol. 116, No. 6, pp. 837-857.
86. Sleight, R.B. (1917),'Evaporation from the surfaces of water and river-bed materials,' J. Agri. Res., July, pp. 209-261.
87. Smith, R.E. (1972),'The infiltration envelope: Results from a theoretical infiltrometer,' J. Hydrology, 17, pp. 1-21.
88. Smith, R.E. and J.-Y. Parlange (1978),'A parameter-efficient hydrologic infiltration model,' J. Water Resources Res., Vol.14, No. 3, pp. 533-538.
89. Snyder, F.F. (1938),'Synthetic unit-graphs,' Trans., American Geophysical Union, Vol. 19, pp. 447-454.
90. Snyder, W.M. (1971),'A proposed watershed retention function,' J. Irrig. and Draing. Div., A.S.C.E., 97, pp. 193-201.
91. Stephens, J.C. (1965), Discussion of 'Estimating evaporation from isolation", J. Hydaul. Div., A.S.C.E., Vol. 91 (HY5), pp. 171-182.
92. Subcommittee (1934),'Standard equipment for evaporation studies report,' Subcommittee on Irrigation and Hydraulics, Trans. A.S.C.E., Vol. 99, pp. 716-718.
93. Terstriep, M.L. and J.B. Stall (1974),'The Illinois Urban Drainage Area Simulator, ILLUDAS,' Bull. 58, Illinois State Water Survey, Urbana, Ill.
94. Tholin, A.L. and C. J. Keifer (1960),'The hydrology of urban runoff,' Trans., A.S.C.E., Vol. 125, pp. 1308-1379.
95. Thornthwaite, C.W., H.G. Wilm, et al. (1944),'Report of the committee on transpiration and evaporation,' 1943-44,' Transactions, American Geophysical Union, Vol. 25, part V, pp. 683-693.
96. USAE (1986),'Program description and user manual for SSARR model, streamflow synthesis and reservoir regulation', U.S. Army Corps of Engineers, Portland, Oregon.
97. USDA (1940),'Influences of vegetation and watershed on runoff, silting, and streamflow,' Misc. Publ. No. 397, U.S.D.A., Washington, D.C.
98. USDC (1955),'Rainfall intensity-duration-frequency curves,' Technical Paper No. 25, Weather Bureau, U.S. Department of Commerce.
99. USDC (1960),'Generalized estimates of probable maximum precipitation for the United States West of the 105th Meridian, Technical Paper No. 38, Weather Bureau, U.S. Department of Commerce.
100. USDC (1962),'Generalized estimates of probable maximum precipitation and rainfall frequency data for Puerto Rico and Virgin Islands for areas to 400 square miles, durations to 24 hours, and return periods from 1 to 100 years,' Technical Paper No. 42, Weather Bureau, U.S. Department of Commerce.
101. U.S. Dept. of Interior (1952),'Water loss investigations: Volume 1 Lake Hefner studies,' Technical Report, Geological Survey Circular 229, Washington.
102. USGS (1974),'The water budget control,' In: Water Loss Investigations, Lake Hefner Studies, U.S. Geological Survey, Professional Paper No. 269.
103. van Bavel, C.H.M. (1968), Discussion of "Climate and evaporation of crops,' J. Irrig. and Drainage Div., A.S.C.E., Vol. 68, pp. 533-535.
104. Viessman, W., J.W. Knapp, G.L. Lewis, and T.E. Harbaugh (1977),'Introduction to Hydrology,' 2nd edition, Harper and Row, New York.
105. White, W.N. (1932),'A method of estimating ground water supplies based on discharge by plants and evaporation from soil: Results of investigation in Escalante Valley, Utah,' U.S. Geological Water Supply Paper 659-A.
106. Young, A.A. (1947),'Evaporation from water surface in California: Summary of pan records and pan coefficients,' 1881-1946, Bull. Public Works Dept., Calif. 54.
107. Zhao, D. (1981),'A semi-linear model in infiltration theory,' Hydrological Res. Report, 2, 246, pp. 246-255, P.R. China.

Chapter 2

1. Anderson, J.R., E.E. Hardy, J.T. Roach, and R.E. Witmer (1976),'A land use and land cover classification system for use with remote sensor data,' U.S. Geological Survey Professional Paper, 964, 28 pp.
2. Andrews, R.G. (1954),'The use of relative infiltration indices for computing runoff (unpublished),' Soil Conservation Service, Fort Worth, Texas, 6p.

3. Bondelid, T.R., R.H. McCuen, and T.J. Jackson (1982),'Sensitivity of SCS models to curve number variation,' Water Resources Bull., 18(1), pp. 111-116.
4. Bosznay, M. (1989),'Generalisation of SCS curve number method,' J. Irrigation and Drainage Engineering, A.S.C.E., 115(1), 139-144.
5. Brakensiek, D.L. and W.J. Rawls (1988),'Effects of agricultural and rangeland management systems on infiltration,' in Modeling Agricultural, Forest, and Rangeland Hydrology, Am. Soc. of Ag. Engrs., St. Joseph, Mich, p. 247.
6. Cazier, D.J and R.H. Hawkins (1984),'Regional application of the curve number method,' Water Today and Tomorrow, Proc., A.S.C.E., Irrigation and Drainage Division Special Conf., A.S.C.E., NewYork, N.Y.
7. Chen Cheng-lung (1982),'An evaluation of the mathematics and physical significance of the Soil Conservation Service curve number procedure for estimating runoff volume,' in Rainfall-Runoff Relationship, edited by V.P. Singh, Water Resour. Pub., Littleton, Colo. 80161.
8. Cowan, W.L. (1957) (In: Rallison and Miller (1982)), Personal communications, Letter to H.O. Ogrosky dated Oct. 15, 1957, 7 pp.
9. Finnerty, B., M. Smith, D.J. Seo, V. Koren, and G.E. Moglen (1997),'Space-time scale sensitivity of the Sacramento model to radar-gauge precipitation inputs,' J. Hydrology, Vol. 203, Nos. 1-4, pp. 21-38.
10. Freebairn, D.M., S.C. Gupta, C.A. Onstad, and W.J. Rawls (1989),'Antecedent rainfall and tillage effects upon infiltration,' Soil Sci. Soc. Am. Jour., Vol. 53, pp. 1183-1189.
11. Grove, M., J. Harbor, and B. Engel (1998),'Composite vs. distributed curve numbers: Effects on estimates of storm runoff depths,' J. Am. Water Resources Association, Vol. 34, No. 5, pp. 1015-1033.
12. Hawkins, R.H. (1975),'The importance of accurate curve numbers in the estimation of storm runoff,' Water Resources Bull., 11(5), pp. 887-891.
13. Hawkins, R.H. (1978),'Runoff curve numbers with varying site moisture,' J. Irrig. and Draing. Div., A.S.C.E., Vol. 104 (IR4), pp.389-398.
14. Hawkins, R.H. (1984),'A comparison of predicted and observed runoff curve numbers,' Proc., Spec. Conf., Irrig. and Drain. Div., A.S.C.E., New York, N.Y., pp. 702-709.
15. Hawkins, R.H. (1993),'Asymptotic determination of runoff curve numbers from data, J. Irrig. and Drain. Engrg., A.S.C.E., Vol. 119, No. 2, pp. 334-345.
16. Heany, J.P., W.C. Huber, and S.J. Nix (1976),'Storm water management model: level I-preliminary screening procedures,' EPA Rep. No. 66/2-76-275, U.S. Environmental Protection Agency, Cincinnati, Ohio.
17. Hjelmfelt, A.T. Jr. (1980),'Empirical investigation of curve number technique,' J. Hydraul. Div., A.S.C.E., Vol. 106, No. Hy9, pp. 1471-1476.
18. Hjelmfelt, A.T. (1991),'Investigation of curve number procedure,' J. Hydr. Engrg., A.S.C.E., 117(6), 725-737.
19. Hjelmfelt, A.T. Jr., K.A. Kramer, R.E. Burwell (1982),'Curve numbers as random variables,' Proc. Int. Symp. on Rainfall-Runoff Modeling, (Ed. V.P. Singh), Water Resources Publication, Littleton, Colo., 365-373.
20. Horner, W.W. (1940),'The analysis of hydrologic data for small watersheds,' Tech. Paper 30, Soil Conservation Service, U.S. Dept. of Agri., Washington D.C., p. 30.
21. Horton, R.I. (1933),'The role of infiltration in the hydrologic cycle,' Trans., Am. Geophysical Union, 14, 446-460.
22. Jenson, S.K. and J.O. Domingue (1988),'Extracting topographic structure from digital elevation data for geographic information system analysis,' Photogrammetric Engineering and Remote Sensing, Vol. 54, No. 11, pp. 1593-1600.
23. McCuen, R.H. (1982),'A Guide to Hydrologic Analysis using SCS Methods,' Prentice Hall, Englewood Cliffs, New Jersey 07632.
24. McCuen, R.H. (2002),'Approach to confidence interval estimation for curve numbers,' J. Hydrologic Engrg., Vol. 7, No. 1, 43-48.
25. Mein, R.G. and C.L. Larson (1971),'Modeling the infiltration component of the rainfall-runoff process,' Water Resources Research Centre, University of Minnesota, Graduate School, Minneapolis, Minnesota.
26. Miller, N. and R.C. Cronshey (1989),'Runoff curve numbers, the next step,'Proc., Int. Conf.,Channel Flow and Catchment Runoff, Uni. of Virginia, Charlottesville, Va.
27. Mishra, S.K., N.K. Goel, S.M. Seth, and D.K. Srivastava (1998),'An SCS-CN-based long-term daily flow simulation model for a hilly catchment,' Int. Symp. 'Hydrology of Ungauged Streams in Hilly Regions for Small Hydro Power Development,' Mar. 9-10, New Delhi, India.

28. Mockus, V. (1949),'Estimation of total (peak rates of) surface runoff for individual storms,' Exhibit A of Appendix B, Interim Survey Report Grand (Neosho) River Watershed, U.S.D.A., Dec. 1.
29. Mockus, V. (1964), Letter to Orrin Ferris, March 5, 6p, In: Rallison, R.E, Origin and evolution of the SCS runoff equation, Proc., A.S.C.E. Symp. Watershed Management, Boise, Idaho, July, 1980.
30. Moglen, G.E. (2000),'Effect of orientation of spatially distributed curve numbers in runoff calculations,' J. Am. Water Resources Association, Vol. 36, No. 6.
31. Moglen, G.E. and M.J. Casey (1998),'A perspective on the use of GIS in hydrologic and environmental analysis in Maryland,' Infrastructure, 3(4), pp. 15-25.
32. Moore, R.J. (1983),'The probability-distributed approach to spatial conceptual rainfall-runoff modeling,' Report to Flood Protection Commission, Ministry of Agriculture, Fisheries and Food, Institute of Hydrology, Wallingford, Oxfordshire.
33. Moore, R.J. (1985),'The probability-distributed principle and runoff production at point and basin scales,' J. Hydrological Sciences, I.A.H.S., 30, 2, 6, pp. 273-297.
34. Moore, R.J. and R.T. Clarke (1981),'A distribution function approach to rainfall-runoff modeling,' Water Resources Res., 17 (5), pp. 1367-1382.
35. Moore, R.J. and R.T. Clarke (1982), A distribution function approach to modeling basin soil moisture deficit and streamflow, In: Statistical Analysis of Rainfall and Runoff (ed. V.P. Singh), Water Resources Publications, 173-190.
36. Moore, R.J. and R.T. Clarke (1983),'A distribution function approach to modeling basin sediment yield,' J. Hydrology, 65, 239-257.
37. NRCS (2000),'USDA-NRCS soils data,' Natural Resources Conservation Service, http://www.ftw.nrcs.usda.gov/soils_data.html.
38. O'Callaghan, J.F. and D.M. Mark (1984),'The extraction of drainage networks from digital elevation data,' Computer Vision, Graphics and Image Processing, Vol. 28, pp. 323-344.
39. Ogrosky, H.O. (1956),'Service objectives in the field of hydrology,' Unpublished, Soil Conservation Service, Lincoln, Nebraska, 5pp.
40. Pandit, A. and G. Gopalakrishnan (1996),'Estimation of annual storm runoff coefficients by continuous simulation,' J. Irrig. and Drainage Engineering, Vol. 122, No. 4, pp. 211-220.
41. Pandit A. and J. Regan (1997),'What is the impervious curve number?,' Proc. Stormwater and Water Quality Management Modeling Conference, Feb. 20-21, Toronto, Canada.
42. Ponce, V.M. (1989),'Engineering Hydrology: Principles and Practices,' Prentice Hall, Englewood Cliffs, New Jersey.
43. Ponce, V.M., and R.H. Hawkins (1996),'Runoff curve number: Has it reached maturity?,' J. Hydrol. Engrg., A.S.C.E., Vol. 1, No. 1, pp. 11-19.
44. Ragan, R.M. (1991),'A geographic information system to support state-wide hydrologic and non-point pollution modelling,' Technical Report FHWA/MD-91/02, College Park, Maryland.
45. Rallison, R.E. (1980),'Origin and Evolution of the SCS runoff equation,' Proc., A.S.C.E. Irrig. and Drain. Symp. on Watershed Management, A.S.C.E., New York, N.Y., Vol. II, 912-924.
46. Rallison, R.E. and R.C. Cronshey (1979), Discussion to 'Runoff curve numbers with varying soil moisture,' J. Irrig. and Drain. Div., A.S.C.E., 105(4), 439-441.
47. Rallison, R.E. and N. Miller (1982),'Past, present, and future,' Proc., Int. Symp. Rainfall-Runoff Relationship, (ed. V.P. Singh), Water Resources Pub., P.O. Box 2841, Littleton, Colorado 80161.
48. Ramasastri, K.S. and S.M. Seth (1985),'Rainfall-runoff relationships,' Rep. RN-20, National Institute of Hydrology, Roorkee-247 667, Uttar Pradesh, India.
49. Rawls, W.J. (1983),'Estimating soil bulk density from particle size analysis and organic matter content,' Soil Sci., Vol. 135, No. 2, pp. 123-125.
50. Rawls, W.J., L.R. Ahuja, D.L. Brakensiek, and A.D. Shirmohammadi (1992), 'Infiltration and soil water movement,' Chapter 5 in Handbook of Hydrology, D.R. Maidment (Ed. in Chief), Mc-Graw Hill, Inc., New York.
51. SCD (1972),'Handbook of Hydrology,' Soil Conservation Department, Ministry of Agriculture, New Delhi.
52. Schueler, T.R. (1987),'Controlling urban runoff: a practical manual for planning and designing urban BMPs,' Department of Environmental Programs, Metropolitan Washington Council of Governments, Washington, D.C., Publ. No. 87703.
53. SCS (1986),'Urban hydrology for small watersheds,' Tech. Release No. 55, Soil Conservation Service, U.S.D.A., Washington D.C.

54. Sherman, L.K. (1949),'The unit hydrograph method,' In: O.E. Meinzer (ed.) Physics of the Earth, Dover Publications, Inc., New York, N.Y., pp. 514-525.
55. Simanton, J.R., K.G. Renard, and N.G. Sutter (1973),'Procedures for identifying parameters affecting storm runoff volumes in a semiarid environment,' Rep. W-1, U.S.D.A., Agriculture Research Service, Washington D.C.
56. Singh, V.P. (1992),'Elementary Hydrology,' Prentice Hall, Englewood Cliffs, New Jersey 07632.
57. Singh, V.P. and F.X. Yu (1990),'Derivation of infiltration equation using systems approach,' J. Irrig. and Drain. Engrg., A.S.C.E., Vol. 116, No. 6, pp. 837-857.
58. Soil Conservation Service (SCS) (1956, 1964, 1971, 1972, 1985),'Hydrology, National Engineering Handbook, Supplement A, Section 4, Chapter 10, Soil Conservation Service, U.S.D.A., Washington, D.C.
59. Soni, B. and G.C. Mishra (1985),'Soil water accounting using SCS hydrologic soil classification,' Case Study, National Institute of Hydrology, Roorkee (India).
60. Springer, E.P., B.J. McGurk, R.H Hawkins, and G.B. Goltharp (1980),'Curve numbers from watershed data,' Proc., A.S.C.E. Irrigation and Drainage Symp. on Watershed Management, A.S.C.E., New York, N.Y., Vol. II, 938-950.
61. Stankowski, S.J. (1974),'Magnitude and frequency of floods in New Jersey with effects of urbanization,' Spec. Rep. 38, U.S. Geological Survey, Water Resources Division, Trenton, N.J.
62. Tarboton, D.G., R.L. Bras, and I. Rodriguez-Iturbe (1991),'On the extraction of channel networks from digital elevation data,' J. Irrig. and Drainage Engrg., A.S.C.E., Vol. 121, No. 3, pp. 81-100.
63. U.S. Environmental Protection Agency (EPA) (1983),'Results of the nationwide urban runoff program, Volume 1- final report,' NTIS No. PB84-185552, Washington, D.C.
64. USFS (1959),'Forest and Range Hydrology Handbook,' U.S. Forest Service Publication, Washington, D.C.
65. USGS (2000),'U.S. Geological Survey, EROS Data Center, http://edcwww.cr.usgs.gov/doc/edchome/ndcdb/ndcdb.html.
66. Van Mullem, J.A. (1989),'Runoff and peak discharges using Green-Ampt infiltration model,' J. Hydr. Engrg., A.S.C.E., 117(3), 354-370.
67. Williams, J.R and V. LaSeur (1976),'Water yield model using SCS curve numbers,' J. Hydraul. Engg., 102(HY9), pp. 1241-1253.
68. Zoch, R.T. (1934),'On the relation between rainfall and streamflow,' Monthly Weather Review, Vol. 62, pp. 315-322.
69. Zoch, R.T. (1936),'On the relation between rainfall and streamflow-II,' Monthly Weather Review, Vol. 64, pp. 105-121.

Chapter 3

1. Andrews, R.G. (1954),'The use of relative infiltration indices in computing runoff,' Paper presented at the meeting of the American Society of Agricultural Engineers (Unpublished), Soil Conservation Service, Fort Worth, Texas, p. 6.
2. Aron, G., A.C. Miller Jr., and D.F. Lakatos (1977),'Infiltration formula based on SCS curve number,' J. Irrig. and Draing. Div., A.S.C.E., Vol. 103, No. IR4, pp. 419-427.
3. Chen Cheng-lung (1982),'An evaluation of the mathematics and physical significance of the soil conservation service curve number procedure for estimating runoff volume,' in Rainfall-Runoff Relationship, edited by V.P. Singh, Water Resour. Pub., Littleton, Colo. 80161.
4. Cook, H.L. (1946),'The infiltration approach to the calculation of surface runoff,' EOS Trans. Am. Geophysical Union, 27, pp. 726-747.
5. Flanagan, D.C., G.R. Foster, and W.C. Moldenhauer (1988),'Storm pattern effect on infiltration, runoff, and erosion,' Trans. A.S.A.E., 31, pp. 414-420.
6. Fogel, M.M. and L. Duckstein (1970),'Prediction of convective storm runoff in semi-arid regions,' Proc., IAHS-UNESCO Symp. on the Results of Research on Representative Experimental Basins, Wellington, New Zealand.
7. Fok, Y. S. and S.H. Chiang (1984),'2-D infiltration equations for furrow irrigation,' J. Irrig. and Drainage Engrg., A.S.C.E., 110(2), 208-217.
8. Gray, D.D., P.G. Katz, S.M. deMonsabert, and N.P. Cogo (1982),'Antecedent moisture condition probabilities,' J. Irrig. and Draing. Div., A.S.C.E., 108(2), pp. 107-114.

9. Hawkins, R.H. (1982),'Interpretations of source area variability in rainfall-runoff relations,' Proc., Int. Symp. on Rainfall Runoff Modeling, Water Resources Pub., Littleton, Colo., 303-324.
10. Hawkins, R.H. A.T. Hjelmfelt, Jr., and A.W. Zevenbergen (1985), 'Runoff probability, storm depth, and curve numbers,' J. Irrig. and Drainage Engrg., A.S.C.E., 111(4), pp. 330-339.
11. Hjelmfelt, A.T. Jr. (1982), closure to 'Empirical investigation of the curve number technique,' J. Hydraulics Div., A.S.C.E., Vol. 108, No. HY4.
12. Hjelmfelt, A.T. Jr., K.A. Kramer, R.E. Burwell (1982),'Curve numbers as random variables,' Proc. Int. Symp. on Rainfall-Runoff Modeling, (Ed. V.P. Singh), Water Resources Publication, Littleton, Colo., 365-373.
13. Horton, R.I. (1933),'The role of infiltration in the hydrologic cycle,' Trans., Am. Geophysical Union, 14, 446-460.
14. Huggins L.F. and E.J. Monke (1966),'The mathematical simulation of the hydrology of small watersheds,' Tech. Rep. No. 1, Purdue Water Resources Research Centre, Lafayette.
15. Istanbulluoglu, E. (2000), Discussion on 'Theoretical justification of SCS method for runoff estimation', J. Irrig. And Drainage Engrg., A.S.C.E., Vol. 126, No. 1, p. 74.
16. McCuen, R.H. (1982),'A Guide to Hydrologic Analysis Using SCS Methods,' Prentice-Hall Inc., Englewood Cliffs, New Jersey.
17. Mishra, S.K. (1998),'Operation of a multipurpose reservoir,' Unpub. Ph.D. Thesis, University of Roorkee, Roorkee-247 667, Uttaranchal State, India.
18. Mishra, S.K. and V.P. Singh (1999a),'Behaviour of SCS-CN method in C-I_a^*-λ spectrum,' Proc., Int. Conf. on Water, Environment, Ecology, Socio-economics, and Health Engineering, Seol Nat. Uni., Korea, Oct. 18-21.
19. Mishra, S.K. and V.P. Singh (1999b), Another look at the SCS-CN method, J. Hydrologic Engineering, A.S.C.E., Vol. 4, No. 3, pp. 257-264.
20. Mishra, S.K. and V.P. Singh (2002), Chapter 13, 'SCS-CN-based hydrologic simulation package,' In: Mathematical Models in Small Watershed Hydrology, (eds.) V.P. Singh and D.K. Frevert, Water Resources Publications, P.O. Box 2841, Littleton, Colorado 80161.
21. Mockus, V. (1949),'Estimation of total (peak rates of) surface runoff for individual storms,' Exhibit A of Appendix B, Interim Survey Report Grand (Neosho) River Watershed, U.S.D.A., Dec. 1.
22. Moldenhauer, W.C., W.C. Burrows, and D. Swartzendruber (1960),'Influence of rainfall storm characteristics on infiltration measurements,' 7th Int. Congr. Soil Sci., Int. Soc. of Soil Sci., 1, pp. 426-432.
23. Moore, R.J. (1985),'The probability-distributed principle and runoff production at point and basin scales,' Hydrological Sciences J., 30, pp. 273-297.
24. Murai, H. and Y. Iwasaki (1975),'Studies on function of water and soil conservation on forest land (I)-Influence of difference in forest condition upon water run-off, infiltration and erosion,' Bull. No. 274, Gov. Forest Exp. Sta., Tokyo, Japan, pp. 23-24.
25. Ponce, V.M. (1989),'Engineering Hydrology: Principles and Practice,' Prentice Hall, Englewood Cliffs, New Jersey 07632.
26. Ponce, V.M., and R.H. Hawkins (1996), Runoff curve number: Has it reached maturity?, J. Hydrolog. Engrg., A.S.C.E., Vol. 1, No. 1, pp. 11-19.
27. Rallison, R.E. and N. Miller (1982),'Past, present, and future SCS runoff procedure,' In: Rainfall-Runoff Relationship (ed. V.P. Singh), Water Resources Publications, PO Box 2841, Littleton, Colorado 80161.
28. SCD (1972),'Handbook of Hydrology,' Soil Conservation Department, Ministry of Agriculture, New Delhi, India.
29. Schaake, J.C, V.I. Koren, Qing-Yun Duan, K. Mitchell, and F. Chen (1996),'Simple water balance model for estimating runoff at different spatial and temporal scales,' J. Geophysical Research, Vol. 101, No. D3, pp. 7461-7475.
30. SCS (1986),'Urban hydrology for small watersheds,' Tech. Release No. 55, Soil Conservation Service, U.S.D.A., Washington D.C.
31. Sherman, L.K. (1949),'The unit hydrograph method,' Physics of the Earth, IX, Hydrology, O.E., Meinzer, ed., McGraw-Hill, New York, N.Y.
32. Sobhani, G. (1975),'A review of selected small watershed design methods for possible adoption to Iranian conditions,' M.S. thesis, Utah State Univ., Logan, Utah.

33. Soil Conservation Service (SCS) (1956, 1964, 1971, 1985),'Hydrology, National Engineering Handbook,' Supplement A, Section 4, Chapter 10, Soil Conservation Service, U.S.D.A., Washington, D.C.
34. Steenhuis, T., S. M. Winchell, J. Rossing, J.A Zollweg, and M.F. Walter (1995),' SCS runoff equation revisited for variable-source runoff areas,' J. Irrig. & Drainage Engrg., A.S.C.E., 121(3), pp. 234-238.
35. Williams, J.R and V. LaSeur(1976),'Water yield model using SCS curve numbers,' J. Hydraul. Engg., 102(HY9), pp. 1241-1253.
36. Yu, B. (1998),'Theoretical Justification of SCS method for runoff estimation,' J. Irrigation and Drainage Engineering, A.S.C.E., Vol. 124, No. 6, pp. 306-310.
37. Yu, B. (2000), Closing discussion on 'Theoretical justification of SCS method for runoff estimation', J. Irrig. and Drainage Engrg., A.S.C.E., Vol. 126, No. 1, p. 76.
38. Yu, B., C.W. Rose, K.J. Coughlan, and B. Fentie (1997),'Plot-scale rainfall-runoff characteristics and modeling at six sites in Australia and South-East Asia.' Trans. A.S.A.E., 40, pp. 1295-1303.
39. Zoch, R.T. (1934),'On the relation between rainfall and streamflow,' Monthly Weather Review, Vol. 62, pp. 315-322.
40. Zoch, R.T. (1936),'On the relation between rainfall and streamflow-II,' Monthly Weather Review, Vol. 64, pp. 105-121.

Chapter 4

1. Bonta, J.V. (1997),"Determination of watershed curve number using derived distributions", J. Irrig. and Draing. Engrg., A.S.C.E., 123(1), pp. 28-36.
2. Black, T.A., Gardner, W.R., and Thurtell, G.W. (1969),'The prediction of evaporation, drainage, and soil water storage for a bare soil,' Soil Sci. Soc. Am. Proc. 33, pp. 655-660.
3. Elrick, D.E. and Bowman, D.H. (1964),'Note on an improved apparatus for soil moisture flow measurements,' Soil Sci. Am. Proc. 28, pp. 450-453.
4. Green, R.E.(1962),'Infiltration of water into soils as influenced by antecedent moisture,' Unpub. Ph.D. Thesis, Iowa State Univ., Iowa.
5. Haverkamp, R. and M. Vauclin (1979),'A note on estimating finite difference interblock hydraulic conductivity values for transient unsaturated flow problems,' Water Resources Research, Vol. 15, No. 1, pp. 181-187.
6. Hawkins, R.H. (1993),'Asymptotic determination of runoff curve numbers from data, J. Irrig. and Drain. Engrg., A.S.C.E., Vol. 119, No. 2, pp. 334-345.
7. Hawkins, R.H. (1996), Discussion on 'SCS-runoff equation revisited for variable-source runoff areas,' J. Irrig. and Drainage, Sept.-Oct., p. 319.
8. Hjelmfelt, A.T. Jr. (1980),'Empirical investigation of curve number technique,' J. Hydr. Div., A.S.C.E., 106, 9, 1471-1476.
9. Laliberte, G.E., A.T. Corey, and R.H. Brooks (1966),'Properties of unsaturated porous media,' Hydrol. Paper 17, Colorado State Uni., Fort Collins, Colorado.
10. McCuen, R.H. (1982),'A Guide to Hydrologic Analysis using SCS Methods,' Prentice Hall, Englewood Cliffs, New Jersey 07632.
11. McCuen, R.H. (2002),'Approach to confidence interval estimation for curve numbers,' J. Hydrologic Engrg., Vol. 7, No. 1, 43-48.
12. Mein, R.G. and C.L. Larson (1971),'Modeling the infiltration component of the rainfall-runoff process,' Water Resources Research Centre, University of Minnesota, Graduate School, Minneapolis, Minnesota.
13. Mishra, S.K., and V.P. Singh (1999),'Another look at the SCS-CN method,' J. Hydrologic. Engrg., A.S.C.E., Vol. 4, No. 3, pp. 257-264.
14. Mockus, V. (1964), Letter to Orrin Ferris, March 5, 6p, In: Rallison, R.E, Origin and evolution of the SCS runoff equation, Proc., A.S.C.E. Symp. Watershed Management, Boise, Idaho, July, 1980.
15. Moore, R.E. (1939),'Water conduction from shallow water tables,' Hilgardia 12, (6), pp. 383-426.
16. SCD (1972),'Handbook of Hydrology,' Soil Conservation Department, Ministry of Agriculture, New Delhi, India.
17. Schaake, J.C, V.I. Koren, Qing-Yun Duan, K. Mitchell, and F. Chen (1996),'Simple water balance model for estimating runoff at different spatial and temporal scales,' J. Geophysical Research, Vol. 101, No. D3, pp. 7461-7475.

18. Soil Conservation Service (SCS) (1956, 1964, 1971, 1985),'Hydrology, National Engineering Handbook,' Supplement A, Section 4, Chapter 10, Soil Conservation Service, U.S.D.A., Washington, D.C.
19. Shuttleworth, W.J. (1992),'Evaporation,' Chapter 4 in Handbook of Hydrology, D.R. Maidment (ed. in chief), Mc-Graw Hill, Inc., New York.
20. Young, E.G. (1964),'An infiltration method measuring the hydraulic conductivity of unsaturated porous materials,' Soil Sci., Vol. 97, pp. 307-311.
21. Yu, B. (1998),'Theoretical justification of SCS-CN method for runoff estimation,' J. Irrig. & Drain. Engrg., Vol. 124, No. 6, pp. 306-310.
22. Yu, B. (2000), Closure by author on 'Theoretical justification of SCS method for runoff estimation,' J. Irrig. and Draing. Engrg., A.S.C.E., January/February, p. 76.

Chapter 5

1. Campbell, G.S. (1974),'A simple method for determining unsaturated conductivity from moisture retention data,' Soil Science, Vol. 117, No. 6, pp. 311-314.
2. Chang Kun Park and Jung Ho Sonu (1993),'Capillary hysteresis model in unsaturated flow: State of the art,' Korean J. of Hydrosciences, Vol. 4, June, pp. 33-49.
3. Chen Cheng-lung (1982),'An evaluation of the mathematics and physical significance of the soil conservation service curve number procedure for estimating runoff volume,' In: V.P. Singh (ed.) Rainfall-Runoff Relationship, Water Resources Publications, Littleton, CO 80161.
4. Collis-George, N. (1977),'Infiltration equations for simple soil systems,' J. Water Resour. Res., Vol. 13, No. 2, pp. 395-403.
5. Crawford, F.S. (1965),'Waves: Berkeley Physics Course,' Vol. 3, McGraw-Hill, New York.
6. De Souza, P.A. and R.K. Price (1980),'The Saint Venant equations: A linearised analysis,' Report No. IT 203, Institute of Hydrology, Wallingford, Sept.
7. Feddes, R.A., P. Kabat, P.J.T. Van Bakel, J.J.B. Bronswijk, and J. Halbertsma (1988),'Modelling soil water dynamics in the unsaturated zone-State of the art,' J. Hydrology, 100, 69-111.
8. Gardener, W.R., D. Hillel, and Y. Benzamin (1970),'Post irrigation movement of soil water: I. Redistribution,' Water Resour. Res., 6(3), 851-61; 'II. Simultaneous redistribution and evaporation,' Water Resour. Res., 6(4), 1148-53.
9. Hillel, D. (1976),'Soil and Water: Physical Properties and Processes,' Academic Press, New York.
10. Jarvis, N.J., P.E. Jansson, P.E. Dik, and I. Messing (1991),'Modeling water and solute transport in macroporous soil, I. Model description and sensitivity analysis,'J. Soil Sci., 42, 59-70.
11. Knisel, W.G. (1980),'CREAMS: a field scale model for chemicals, runoff, and erosion from agricultural management systems,' Conser. Res. Rep., USDA, Vol. 26, 643 pp.
12. Kirkham, D. and W. L. Powers (1972),'Advanced Soil Physics,' Interscience, New York.
13. Kundzewicz, Z.W. and J.C.I. Dooge (1989),'Attenuation and phase shift in linear flood routing,"J. Hydrological Sciences, Vol. 34, No. 1, pp. 21-40.
14. Lee, H.W. (1983),'Determination of infiltration characteristics of a frozen palouse silt loam soil under simulated rainfall,' Ph.D. Dissertation, University of Idaho Graduate School.
15. Mallants, D., Peng-Hsiang Tseng, M. Vanclooster, J. Feyen (1998),'Predicted drainage for a sandy loam soil: sensitivity to hydraulic property description,' J. Hydrology, 206, 136-148.
16. Marshall, T.J., J.W. Holmes, and C.W. Rose (1996),'Soil Physics,' Cambridge University Press, New York.
17. McCuen, R.H. (1982),'A Guide to Hydrologic Analysis using SCS Methods,' Prentice Hall, Englewood Cliffs, New Jersey 07632.
18. Mein, R.G. and C.L. Larson (1971),'Modelling the infiltration component of the rainfall-runoff process,' Bull. 43, Water Resources Res. Centre, Uni. of Minnesota, Minneapolis, Minnesota.
19. Mishra, S.K. and S.M. Seth (1996),'Use of hysteresis for defining the nature of flood wave propagation in natural channels,' Hydrological Sciences Journal,41(2), April.
20. Mishra, S.K., and V. P. Singh (1999),'Hysteresis-based flood wave analysis,' J. Hydrologic Engineering, A.S.C.E., Vol. 4, No. 4, pp. 358-365.
21. Mishra, S.K., and V. P. Singh (2001a),'Hysteresis-based flood wave analysis using the concept of strain,' J. Hydrological Processes, Vol. 15, 1635-1651.

22. Mishra, S.K. and Vijay P. Singh (2001b),'On the Seddon speed formula,' J. Hydrological Sciences, IAHS, Vol. 46, No. 3, pp. 333-347.
23. Mishra, S.K. and Vijay P. Singh (In Press),'Role of dimensionless numbers in wave analysis,' J. Hydrological Processes.
24. Mishra, S.K., S.R. Kumar, and V. P. Singh (1999),'Calibration of a general infiltration model,' J. Hydrological Processes, Vol. 13, pp. 1691-1718.
25. Mishra, S.K. and M.K. Jain and S.M. Seth (1997),'Characterization of Flood Waves by Rating Curves,'J. Nordic Hydrology, Vol. 28(1).
26. Mockus, V. (1964), Letter to Orrin Ferris, March 5, 6p, In: Rallison, R.E, Origin and evolution of the SCS runoff equation, Proc., A.S.C.E. Symp. Watershed Management, Boise, Idaho, July, 1980.
27. Moore, R.E. (1939),'Water conduction from shallow water tables,' Hilgardia 12, 383-426.
28. Novotny, V. and H. Olem (1994),'Water Quality: Prevention, Identification, and Management of Diffuse Pollution,' John Wiley & Sons, New York, N.Y.
29. Pedlosky, J. (1979),'Geophysical Fluid Dynamics,' Springer Verlag, Berlin.
30. Philip, J.R. (1957),'Theory of Infiltration,' Chapters 1 and 4, Soil Sci., 83(5), pp. 345-357.
31. Philip, J.R. (1969),'Theory of Infiltration,' Advances in Hydrosciences (ed. V.T. Chow), Academic Press, New York.
32. Philip, J.R. (1974),'Recent progress in the solution of nonlinear diffusion equations,' Soil Sci., 117, 257-264.
33. Ponce, V.M. and D.B. Simons (1977),'Shallow wave propagation in open channel flow,' J. Hydraul. Div., A.S.C.E., 103 (HY120, pp. 1461-1476.
34. Ponce, V.M., and R.H. Hawkins (1996),'Runoff curve number: Has it reached maturity?,' J. Hydrolog. Engrg., ASCE, Vol. 1, No. 1, pp. 11-19.
35. Ponce, V.M, R.M. Li, and D.B. Simons (1978),'Applicability of kinematic and diffusion models,' J. Hydraul. Div., A.S.C.E., 104(HY3), pp. 353-360.
36. Rankine, M. (1870),'On the thermodynamic theory of waves of finite longitudinal disturbance,' Philosophical Trans. of the Royal Society of London, 160, pp. 277-288.
37. Singh, V.P. and F.X. Yu (1990),'Derivation of infiltration equation using systems approach,' J. Irrig. and Drainage Engrg., A.S.C.E., Vol. 116, No. 6, pp. 837-857.
38. SCS (1956, 1964, 1971, 1985),'Hydrology,' National Engineering Handbook, Supplement A, Section 4, Chapter 10, Soil Conservation Service, U.S.D.A., Washington, D.C.
39. Todd, D.K. (1980),'Groundwater Hydrology,' John Wiley & Sons, New York.
40. Vanclooster, M., P. Viane, J. Diels, and K. Chrismans (1994),'Water and agrochemicals in the soil and vadose environment,'Reference and user's manual, Release 2.0, Institute for Land and Water Management, KULeuven, Belgium.
41. Visser, W.C. (1969),'An empirical expression for the desorption curve,' In: Water in the Unsaturated Zone, eds., P.E. Rijtema and H. Wassnik, UNESCO, Paris/IAHS, Gentbrugge, pp. 329-35.
42. Vreugdenhill, C.B. (1972),'Mathematical Methods for Flood Waves,' Delft Hydraulics Laboratory, Research Report S89-IV, August.
43. Wagenet, R.J. and J.L. Hutson (1987),'LEACHM: leaching estimation and chemistry model,' Continuum Ser. Pub., Vol. 2, Water Resource Inst., Cornell Univ., Ithaca, N.Y., pp. 1-80.
44. Young, E.G. (1957),'Moisture profiles during vertical infiltration,' Soil Sc. 84, 283-290.
45. Young, E.G. (1964),'An infiltration method measuring the hydraulic conductivity of unsaturated porous materials,' Soil Sci., Vol. 97, pp. 307-311.

Chapter 6

1. Aron, G., A.C. Miller Jr., and D.F. Lakatos (1977),'Infiltration formula based on SCS curve number,' J. Irrig. and Draing. Div., A.S.C.E., Vol. 103, No. IR4, pp. 419-427.
2. Chen Cheng-lung (1982), An evaluation of the mathematics and physical significance of the soil conservation service curve number procedure for estimating runoff volume, in Rainfall-Runoff Relationship, edited by V.P. Singh, Water Resour. Pub., Littleton, Colo. 80161.
3. Collis-George, N. (1977),'Infiltration equations for simple soil systems,' J. Water Resour. Res., Vol. 13, No. 2, pp. 395-403.
4. Fok, Y. S. and S.H. Chiang (1984),'2-D infiltration equations for furrow irrigation,' J. Irrig. and Drainage Engrg., A.S.C.E., 110(2), 208-217.

5. Gray, D.D., P.G. Katz, S.M. deMonsabert, and N.P. Cogo (1982),'Antecedent moisture condition probabilities,' J. Irrig. and Draing. Div., A.S.C.E., 108(2), pp. 107-114.
6. Hawkins, R.H. (1982),'Interpretations of variable source area variability in rainfall-runoff relationships,' Int. Symp., Rainfall-Runoff Modelling, Mississippi State Uni., Water Resources Publications, Littleton, Colo.
7. Hawkins, R.H. (1984),'A comparison of predicted and observed runoff curve numbers,' Proc., Spec. Conf., Irrig. and Drainage Div., A.S.C.E., New York, N.Y., 702-709.
8. Hawkins, R.H. (1993),'Asymptotic determination of runoff curve numbers from data, J. Irrig. and Drain. Engrg., A.S.C.E., Vol. 119, No. 2, pp. 334-345.
9. HEC (1981, 1990),'HEC-1 Flood Hydrograph Package,' Users Manual, U.S. Army Corps of Engineers, Hydrologic Engineering Center, Davis, California.
10. Hewlett, J.D. and A.R. Hibbert (1967),'Factors affecting the response of small watersheds to precipitation in humid area,' Proc., Int. Symp. on Forest Hydrology, edited by W.E. Sopper and H.W. Lull, Pergamon Press, Oxford, England, 275-290.
11. McCuen, R.H. (1982),'A Guide to Hydrologic Analysis using SCS Methods,' Prentice Hall, Englewood Cliffs, New Jersey 07632.
12. McCuen, R.H. (2002),'Approach to confidence interval estimation for curve numbers,' J. Hydrologic Engrg., Vol. 7, No. 1, 43-48.
13. Mein, R.G. and C.L. Larson (1971),'Modelling the infiltration component of the rainfall-runoff process,' Water Resources Research Centre, University of Minnesota, Graduate School, Minneapolis, Minnesota.
14. Mishra, S.K. (1998),'Operation of a multipurpose reservoir,' Unpub. Ph.D. Thesis, University of Roorkee, Roorkee-247 667, Uttaranchal, India.
15. Mishra, S.K. and V. Garg (2000),'An SCS-CN-based time distributed model,' J. Hydrology, Indian Association of Hydrologists, Roorkee, Uttaranchal, India.
16. Mockus, V. (1964), Letter to Orrin Ferris, March 5, 6p, In: Rallison, R.E, Origin and evolution of the SCS runoff equation, Proc., A.S.C.E. Symp. Watershed Management, Boise, Idaho, July, 1980.
17. Ponce, V.M. (1989),'Engineering Hydrology: Principles and Practices,' Prentice Hall, Englewood Cliffs, New Jersey.
18. Ponce, V.M., and R.H. Hawkins (1996), Runoff curve number: Has it reached maturity?, J. Hydrolog. Engrg., A.S.C.E., Vol. 1, No. 1, pp. 11-19.
19. Rallison, R.E and N. Miller (1982), Past, present, and future SCS runoff procedure, in Rainfall-Runoff Relationship, (ed.) V.P. Singh, Water Resour. Pub., Littleton, Colorado 80161.
20. Sansalone, John J. and S.G. Buchberger (1997),'Partitioning and first flush of metals in urban roadway storm water,' J. Environmental Engineering, A.S.C.E., Vol. 123, No. 2, pp. 134-143.
21. Simanton, J.R., K.G. Renard, and N.G. Sutter (1973),'Procedures for identifying parameters affecting storm runoff volumes in a semiarid environment,' Rep. W-1, U.S.D.A. Agricultural Research Service, Washington, D.C.
22. Singh, V.P. (1988a),'Hydrologic Systems: Vol. I. Rainfall-Runoff Modelling,' Prentice Hall, Englewood Cliffs, New Jersey.
23. Singh, V.P. (1988b),'Hydrologic Systems: Vol. II. Watershed Modelling,' Prentice Hall, Englewood Cliffs, New Jersey.
24. Singh, V.P. (1992),'Elementary Hydrology,' Prentice Hall, Englewood Cliffs, New Jersey, 973 pp.
25. Smith, R.E. (1972),'The infiltration envelope: Results from a theoretical infiltrometer,' J. Hydrology, 17, pp. 1-21.
26. Soil Conservation Service (SCS) (1956, 1964, 1965, 1971, 1972, 1985, 1993), 'Hydrology, National Engineering Handbook, Supplement A, Section 4, Chapter 10, Soil Conservation Service, U.S.D.A., Washington, D.C.
27. SCS (1975),'Urban hydrology for small watersheds,' Tech. Release No. 55, Conservation Service, U.S.D.A., Washington, D.C.

Chapter 7

1. Burton, J. S. Compiler (1993), Proceedings of the Federal Interagency Workshop on Hydrologic Modeling Demands for the 90's, U. S. Geological Survey Water Resources Investigations Report 93-4018, Federal Center, Denver, Colorado.

2. Haan, C.T. (1975),'Evaluation of a model for simulating monthly water yields from small watersheds,' Southern Cooperative Series Bulletin 201, Agricultural Experiment Stations of Kentucky, Mississippi, North Carolina, South Carolina, Tennessee, and Virginia, July.
3. Hawkins, R.H. (1978),'Runoff curve numbers with varying site moisture,' J. Irrig. and Draing. Div., A.S.C.E., Vol. 104 (IR4), pp.389-398.
4. HEC (2000),'Hydrologic modeling system HEC-HMS,' User's manual, version 2, Hydrologic Engineering Center, U. S. Army Corps of Engineers, Davis, California.
5. Holtan, H.N. and N.C. Lopez (1971),'USDHAL-70 model of watershed hydrology,' U.S.D.A., Tech. Bull. 1435.
6. Huber, W.C., J.P. Heaney, B.P. Bedient, and J.P. Bender (1976),'Environmental resources management studies in the Kissimmee river basin,' Rep. No. ENV-05-76-3, Dept. of Envir. Engrg., Univ. of Florida, Gainesville, Fla, May.
7. Knisel, W.G. (1980),'CREAMS, A field-scale model for chemicals, runoff, and erosion from agricultural management systems,' Conservation Rep. No. 26, U.S.D.A. Agriculture Research Service, Washington D.C.
8. Mishra, S.K., N.K. Goel, S.M. Seth, and D.K. Srivastava (1998),'An SCS-CN-based long-term daily flow simulation model for a hilly catchment,' Int. Symp. 'Hydrology of Ungauged Streams in Hilly Regions for Small Hydro Power Development,' Mar. 9-10, New Delhi, India.
9. Nash, J.E. and J.V. Sutcliffe (1970),'River flow forecasting through conceptual models, Part I-A discussion of principles,' J. Hydrology, Vol. 10, pp. 282-290.
10. Pandit, A. and G. Gopalakrishnan (1996),'Estimation of annual storm runoff coefficients by continuous simulation,' J. Irrigation and Drainage Engineering, Vol. 122, No. 4, pp. 211-220.
11. Ponce, V.M. (1989),'Engineering Hydrology: Principles and Practices,' Prentice Hall, Englewood Cliffs, New Jersey.
12. Ponce, V.M. and R.H. Hawkins (1996),'Runoff curve number: Has it reached maturity?,' J. Hydrologic Engrg., A.S.C.E., Vol. 1, No. 1, pp. 11-19.
13. Schaake, J.C., V.I. Koren, Qing-Yun Duan, K.Mitchell, and F. Chen (1996),'Simple water balance model for estimating runoff at different spatial and temporal scales,' J. Geophysical Research, Vol. 101, No. D3, pp. 7461-7475.
14. Singh, V. P. (editor) (1995),'Computer models of watershed hydrology,' Water Resources Publications, Littleton, Colorado.
15. Singh, P. and V.P. Singh (2001),'Snow and Glacier Hydrology,' Kluwer Academic Publishers, P.O. Box 322, Dordrecht, The Netherlands.
16. Singh, R.D., S.K. Mishra, and H. Chowdhary (2001),'Regional flow duration models for 1200 ungauged Himalayan watersheds for planning micro-hydro projects,' J. Hydrologic Engineering, Technical Note, A.S.C.E., Vol. 6, No. 4, July/August.
17. Singh, V.P. (1992),'Elementary Hydrology,' Prentice Hall, Englewood Cliffs, New Jersey.
18. Singh, V.P. and F.X. Yu (1990),'Derivation of infiltration equation using systems approach,' J. Irrigation and Drainage Engineering, A.S.C.E., Vol. 116, No. 6, pp. 837-857.
19. Soni, B. and G.C. Mishra (1985),'Soil water accounting using SCS hydrologic soil classification,' Case Study, National Institute of Hydrology, Roorkee (India).
20. Subcommittee on Hydrology (1998), Proceedings of the first federal interagency hydrologic modeling conference, Interagency Advisory Committee on Water, U. S. Geological Survey Reston, Virginia.
21. Williams, J.R and V. LaSeur (1976),'Water yield model using SCS curve numbers,' J. Hydraulic Engineering, 102(HY9), pp. 1241-1253.
22. Woodward, D.E. and W.J. Gburek (1992),'Progress report ARS/SCS curve number work group,' Proc., A.S.C.E. Water Forum '92, ASCE, New York, N.Y., 378-382.
23. Wurbs, R. A. (1998), Dissemination of 'Generalized water resources models in the United States,' Water International, 23, pp. 190-198.
24. Yu, B. (1998),'Theoretical justification of SCS-CN method for runoff estimation,' J. Irrig. and Drain. Engrg., Vol. 124, No. 6, pp. 306-310.

Chapter 8

1. Armstrong, L.J. (1994),'Contribution of heavy metals to storm water from automotive disc brake pad wear,' Report for Santa Clara Valley Nonpoint Source Pollution Control Program, Woodward-Clyde Consultants, pp. 1-37.
2. Ball, D., Hamilton, R., and R. Harrison (1991),'The influence of highway-related pollutants on environmental quality,' In: R. Hamilton and R. Harrison, eds., Highway Pollution, Elsevier Science Publishing Company, Inc., New York, NY., pp. 1-47.
3. Chen Cheng-lung (1982),'An evaluation of the mathematics and physical significance of the soil conservation service curve number procedure for estimating runoff volume,' In: V.P. Singh (ed.) Rainfall-Runoff Relationship, Water Resources Publications, Littleton, CO 80161,USA
4. Cunge, J.A., F.M. Holly Jr., and A. Verwey (1980),'Practical Aspects of Computational River Hydraulics,' Pitman Advanced Publishing Program, Boston.
5. Driscoll, E., P. Shelley, and E. Strecker (1990),'Pollutant loadings and impacts from highway storm water runoff, volume IV: research data appendix,' FHWA-RD-88-009, Federal Highway Administration, U.S. Dept. of Transport, Washington, D.C., 1-139.
6. Ferrick, M.G. (1985),'Analysis for river wave types,' Water Resources Res., Vol. 21, No. 2, pp. 209-220.
7. Fread, D.L. (1984,85),'Dam break flood forecasting model (DAMBRK),' User Manual, National Weather Service, Maryland, U.S.A.
8. French, R.H. (1985),'Open Channel Hydraulics,' McGraw-Hill Book Co., Inc., New York, NY, 551.
9. Foster, I.D.L. and S.M. Charlesworth (1996),'Heavy metals in the hydrological cycle: Trends and explanation,' Hydrological Processes, Vol. 10, pp. 227-261.
10. Glenn, D.W., D. Liu, and J.J. Sansalone (2001),'Influence of highway runoff chemistry, hydrology, and residence time on non-equilibrium partitioning of heavy metals- Implications for treatment at the highway shoulder,' J. of Transportation Research Record, 1775, 129-140, November.
11. Hamilton, R.S. and R.M. Harrison (1991),'Highway Pollution,' Elsevier Science Publishing Co. Inc., New York.
12. Hayden, H.W., W.G. Moffatt, and J. Wulff (1986),'The Structure and Properties of Materials: Mechanical Behaviour,' Vol. III, Wiley Eastern Limited, New Delhi, India.
13. Henderson, F.M. (1966),'Open Channel Flow,' Macmillan, New York, U.S.A.
14. Jain, M.K., S.K. Mishra and S.M. Seth (1996),'Hysteresis criteria for the applicability of CPC and CPMC methods of flood routing,' Proc. Int. Conf. on Disasters and Mitigation, Anna University, Madras (India), Jan. 19-22.
15. Klein, L.A., M. Lang, N. Nash, and S.L. Kirschner (1974),'Sources of metals in New York city wastewater,' J. Water Pollution Control Federation, Vol. 46, No. 12, pp. 2653-2662.
16. Kobriger, N.P. and A. Geinopolos (1984),'Sources and migration of highway runoff pollutants,' Research Report, Vol. III, Report FHWA/RD-84/059 (PB86-227915) FHWA, U.S. Department of Transportation.
17. Kundzewicz, Z.W. and Dooge, J.C.I. (1989),'Attenuation and phase shift in linear flood routing,' Hydrological Sciences Journal, Vol. 34, No. 1, pp. 21-40.
18. Lapin, L.L. (1983),'Probability and statistics for modern engineering,' Brooks/Cole Engineering Division, Montrey, California.
19. Li, Y., S.G. Buchberger, and J.J. Sansalone (1999),'Variably saturated flow in storm-water partial exfiltration trench,' J. Environmental Engineering, A.S.C.E., Vol. 125, No. 6, pp. 556-565.
20. Lygren, E., Gjessing, E., and L. Berglind (1984),'Pollution Transport from a Highway,' The Science of the Total Environment, Vol. 33, pp. 147-159.
21. Martin, J.L. and S.C. McCutcheon (1999),'Hydrodynamics and transport for water quality modeling,' Lewis Publishers, New York, N.Y.
22. McBride, M. (1994),'Environmental Chemistry of Soils,' Oxford University Press, New York, pp. 406.
23. Menendez, A.N. and R. Norscini (1982),'Spectrum of shallow water waves: An analysis,' J. Hydr. Div., A.S.C.E., 108(HY1), 75-94.
24. Menendez, A.N. and R. Norscini (1983), discussion on 'Nature of flood wave attenuation in open channel flow,' J. Hydr. Div., A.S.C.E., 109(5), pp.786-789.
25. Mishra, S.K., S.M. Seth (1994),'Effect of downstream boundary conditions on dam break flood wave propagation characteristics,' Tech. Rep., National Institute of Hydrology, Roorkee (India).

26. Mishra, S.K. and S.M. Seth (1996),'Use of hysteresis for defining the nature of flood wave propagation in natural channels,' Hydrological Sciences-Journal-des Sciences Hydrologiques, 41(2), pp. 153-170.
27. Mishra, S.K., M.K. Jain and S.M. Seth (1997),'Characterization of flood waves by rating curves,' J. Nordic Hydrology, Vol. 28(1), pp. 51-64.
28. Mishra, S.K. and V.P. Singh (1999),'Hysteresis-based flood wave analysis,' J. Hydrologic Engineering, A.S.C.E., Vol. 4, No. 4, pp. 358-365.
29. Mishra, S.K. and V.P. Singh (2001),'On Seddon speed formula,' J. Hydrological Sciences, IAHS, Vol. 46, No. 2.
30. Mockus, V. (1949),'Estimation of total (peak rates of) surface runoff for individual storms,' Exhibit A of Appendix B, Interim Survey Report Grand (Neosho) River Watershed, U.S. Dept. of Agri., Dec. 1.
31. Muschack, W. (1990),'Pollution of street runoff by traffic and local conditions,' The Science of the Total Environment, Vol. 93, pp. 419-431.
32. Novotny, V. and H. Olem (1994),'Water Quality Prevention, Identification, Management of Diffuse Pollution,' Van Nostrand Reinhold, NewYork, NY.
33. O'Connor, D.J. (1985),'Models of sorptive toxic substances in freshwater systems, I: Basic equations,' J. Env. Engrg., A.S.C.E., Vol. 114, No. 3, pp. 507-532.
34. Pandit, A. and G. Gopalakrishnan (1996),'Estimation of annual storm runoff coefficients by continuous simulation,' J. Irrigation and Drainage Engineering, A.S.C.E., Vol. 122, No. 4, pp. 211-220.
35. Pandit, A. and G. Gopalakrishnan (1997),'Estimation of annual pollutant loads under wet-weather conditions,' J. Hydrologic Engineering, A.S.C.E., Vol. 2, No. 4, pp. 211-218.
36. Perumal, M. (1994),'Hydrodynamic derivation of a variable parameter Muskingum method: 2. Verification,' Hydrol. Sci. J., 39(5), 443-458.
37. Ponce, V.M. (1989),'Engineering Hydrology: Principles and Practices,' Prentice Hall, Englewood Cliffs, New Jersey.
38. Ponce V.M. and D.B. Simons (1977),'Shallow wave propagation in open channel flow,' J. Hydr. Div., A.S.C.E., 103 (HY12), pp.1461-1476.
39. Ponce, V.M., A.K. Lohani, and C. Scheyhing (1996),'Analytical verification of Muskingum-Cunge routing,' J. Hydrology, Vol. 174, 235-241.
40. Ponce, V.M. and A. Lugo (2001),'Modeling looped ratings in Muskingum-Cunge routing,' J. Hydrologic Engineering, A.S.C.E., Vol. 6, No. 2, pp. 119-124.
41. Price, R.K. (1973),'Flood routing methods in British rivers,' Proc. Instn. Civil Engrs. (London), Part 2, Vol. 57, 913-930, paper 7674.
42. Price, R.K. (1985),'Flood routing,' Ch. 4 of Developments in Hydraulic Engineering, E.J.P. Novak, Elsevier Applied Science.
43. Rutschmann, P. and W.H. Hager (1996),'Diffusion of flood waves,' J. Hydrology, Elsevier, Vol. 178 (1996), pp. 19-32.
44. Sansalone, J.J. and D.W. Glenn III (2000),'Temporal variations in heavy metal partitioning and loading in urban pavement sheet flow: Implications for in-situ treatment design,' Transportation Research Board, 1720, Paper No. OO-0354.
45. Sansalone, J.J and S.G. Buchberger (1997),'Partitioning and first flush of metals in urban roadway storm water', J. Environmental Engineering, A.S.C.E., Vol.123, No. 2, pp. 134-143.
46. Sansalone, J.J., J.M. Koran, J.A. Smithson, and S.G. Buchberger (1998),'Physical characteristics of urban roadway solids transported during rain events,' J. Environmental Engineering, A.S.C.E., Vol. 125, No. 6, pp. 556-565.
47. Sansalone, J.J. and T. Tribouillard (1999),'Variation in characteristics of abraded roadway particles as a function of particle size – Implications for water quality and drainage,' In: Transportation Research Record 1690, TRB, National Research Council, Washington, D.C., pp. 153-163.
48. SCS (1956, 1971),'Hydrology,' National Engineering Handbook, Supplement A, Section 4, Chapter 10, Soil Conservation Service, U.S.D.A., Washington, D.C.
49. SCS (1986),'Urban hydrology for small watersheds,' Tech. Release No. 55, Soil Conservation Service, U.S.D.A., Washington D.C.
50. Singh, V.P (1996),'Kinematic Wave Modelling in Water Resources: Surface Water Hydrology,' John Wiley & Sons, New York, p. 1399.
51. Smoley, C.K. (1993),'NPDES storm water sampling guidance manual,' U.S. Environmental Protection Agency, Office of Water, publishers CRC Press, Boca Raton, Florida.
52. "Soil survey of Hamilton County" (1982), U.S. Department of Agriculture, Washington D.C., pp. 1-220.

53. 'Standard methods for the examination of water and wastewater,' 18th Ed. (1992),' A. Greenberg, L. Clesceri, and A. Eaton, eds., Am. Public Health Association, Washington, D.C.
54. Stumm, W. and J.J. Morgan (1981),'Aquatic Chemistry', 2nd edition, Wiley, New York.
55. U.S. Environmental Protection Agency (EPA) (1990),'National water quality inventory-1988 report to congress,' EPA 440-4-90-003, U.S. EPA, Office of Water, Wahington, D.C.
56. Warren, L.A. and A.P. Zimmerman (1993),'Trace metal/SPM associations in a fluvial system: physical and chemical influences. In: Particulate Matter and Aquatic Contaminants (ed. S. Rao), CRC Press, Boca Raton, Fla, pp. 127-155.
57. Warren, L.A. and A.P. Zimmerman (1994),'The influence of temperature and NaCl on cadmium, copper, and zinc partitioning among suspended particulate and dissolved phases in an urban river,' Water Research, Vol. 28, No. 9, pp. 1921-31.

Chapter 9

1. Beasley, D.B., L.F. Huggins, and E.J. Monk (1980),'ANSWERS: A model for watershed planning,' Trans., Am. Soc. Agric. Engrs., 23, 938-944.
2. Chen Cheng-lung (1982), An evaluation of the mathematics and physical significance of the soil conservation service curve number procedure for estimating runoff volume, in Rainfall-Runoff Relationship, edited by V.P. Singh, Water Resour. Pub., Littleton, Colo. 80161.
3. Clark, E.H., J.A. Haverkamp, and W. Chapman (1985),'Eroding soils- The Off-Farm Impacts,' The Conservation Foundation, Washington D.C.
4. Cook, H.L. (1936),'The nature and controlling variables of the water erosion process,' Soil Science Society of America Proceedings, 1: 60-64.
5. Dabney, S.M., C.E. Murphee, L.D. Meyer, E.H. Grissinger, W.C. Harmon, and G.B. Triplett (1997),'Runoff and sediment from conventional and conservation cropping systems,' In: S.S.Y. Wang, E.J. Langendoen, and F.D. Jr. Shields (eds.), Management of Landscapes Disturbed by Channel Incision (MLDCI), Conf. Proc., May 20-22, 1997, Oxford, MS, pp. 983-989.
6. Ellison, W.D. (1944),'Studies of raindrop erosion. Agricultural Engineering,' 25: 131-136, 181-182.
7. Ellison, W.D. (1947),'Soil erosion studies. Agricultural Engineering,' 28: 145-146, 197-201, 245-248, 297-300, 349-351, 402-405, 442-444.
8. Foster, I.D.L. and S.M. Charlesworth (1996),'Heavy metals in the hydrological cycle: Trends and explanation,' Hydrological Processes, Vol. 10, pp. 227-261.
9. Gottschalk, L.C. (1964),'Sedimentation, Part I: Reservoir sedimentation,' In: Handbook of Applied Hydrology, V.T. Cow (ed), McGraw Hill, New York.
10. Huntington, T.G., J.W. Hardon, S.M. Dabney, D.A. Marion, C. Alonso, J.M. Sharpe, and T.L. Fries (1998),'Soil, Environment, and Watershed Measurements in Support of Carbon Cycling Studies in Northwestern Mississippi,' U.S. Geological Survey, Open File Report 98-501, Version 1, Washington, D.C.
11. Jain, M.K. and U.C. Kothyari (2000),'Estimation of soil erosion and sediment yield using GIS,' J. Hydrol. Sci., 45(5), 771-786.
12. Knisel, W.G. (1980),'CREAMS: A field scale model for chemicals, runoff, and erosion from agricultural management systems,' Cons. Res. Report No. 26, USDA-SEA, Washington, D.C., 643p
13. Kothyari, U.C., A.K. Tiwari, and R. Singh (1997),'Estimation of temporal variation of sediment yield from small catchments through the kinematic method,' J. Hydrol., 203, 39-57.
14. Langbein, W.B. and S.A. Schumm (1958),'Yield of sediment in relation to mean annual precipitation,' Trans. Am. Geophys. Union 39, pp. 1076-84.
15. Leonard, R.A., W.G. Knisel, and D.A. Still (1987),'GLEAMS: Groundwater loading effects on agricultural management systems,' Trans. Am. Soc. Agri. Engrs, 30(5), pp. 1403-1428.
16. Leopold, L.B., M.G. Wolman, and J.P. Miller (1964),'Fluvial Processes in Geomorphology,' W.H. Freeman, San Francisco.
17. Meyer, L.D. and W.H. Wischmeier (1969),'Mathematical simulation of the processes of soil erosion by water,' Trans, Am. Soc. Agri. Engrs, 12: 754-759.
18. Musgrave, G.W. (1947),'The quantitative evaluation of factors in water erosion, a first approximation,'

J. Soil Wat. Conserv., 2(3), 133-138.

19. Nash J.E. and J.V. Sutcliffe (1970),'River flow forecasting through conceptual models, Part I-A discussion of principles,' J. Hydrology, Vol. 10, pp. 282-290.

20. Nearing, M.A., G.R. Foster, L.J. Lane, and S.C. Finkner (1989),'A process based soil erosion model for USDA water erosion prediction project technology,' Trans., Am. Soc. Agric. Engrs., 32 (5), 1587-1593.

21. Negev, M. (1967),'A sediment model on a digital computer,' Technical Report 76, Department of Civil Engineering, Stanford University, Stanford, CA, 109pp.

22. Novotny, V. (1980),'Delivery of suspended sediment and pollutatnts from non-point sources during overland flow,' Water Resources Bull., 16 (6), pp. 1057-65.

23. Novotny, V. and H. Olem (1994),'Water Quality: Prevention, Identification, and Management of Diffuse Pollution,' John Wiley & Sons, New York, N.Y.

24. Novotny, V. et al. (1979),'Simulation of Pollutant Loadings and Runoff Quality,' EPA 905/4-79-029, U.S. Environmental Protection Agency, Chicago, IL.

25. Novotny, V. and G. Chesters (1989),'Handbook of Nonpoint Pollution: Sources and Management,' Van Nostrand Reinhold, New York.

26. Novotny, V., G.V. Simsima, and G. Chesters (1986),'Delivery of pollutants from nonpoint sources,' In: Proc., Symp. on Drainage Basin Sediment Delivery, Int. Assoc. Hydrol. Sci., Publ. No. 159, Wallingford, England.

27. Ponce, V.M (1989),'Engineering Hydrology: Principles and Practices' Prentice Hall, Englewood Cliffs, New Jersey.

28. Renard, K. G., G. R. Foster, G. A. Weesies, and J. P. Porter (1991),'RUSLE, Revised universal soil loss equation,' J. Soil Wat. Conserv., 46(1), 30–33.

29. Rode, M. and H.G. Frede (1997),'Modification of AGNPS for agricultural land and climate conditions central Germany,' J. Environ. Qual., Vol. 26, No. 1, pp. 165-172.

30. Roehl, J.W. (1962),'Sediment Source Areas, Delivery Ratios, and Influencing Morphological Factors,' Publ. No. 59, Int. Assoc. Hydrol. Sci., pp. 202-213.

31. SCD (1972),'Handbook of Hydrology,' Soil Conservation Department, Ministry of Agriculture, New Delhi.

32. Sansalone, J.J and S.G. Buchberger (1997),'Partitioning and first flush of metals in urban roadway storm water', J. Env. Engrg., ASCE, Vol. 123, No. 2, pp. 134-143.

33. Schultz, G.A. (1994),'Meso-scale modelling of runoff and water balance using remote sensing and other GIS data,' J. Hydrol. Sci., 39(2), 121-142.

34. Shen, H.W. and P.Y. Julien (1992),'Erosion and Sediment Transport,' Chap. 12 In: Maidment, D.R. (ed.) Handbook of Hydrology, McGraw Hill, New York, 12.1-12.61.

35. Singh, R.D., S.K. Mishra, and H. Chowdhary (2001),'Regional flow duration models for 1200 ungauged Himalayan watersheds for planning micro-hydro projects,' J. Hydrologic Engineering, Technical Note, ASCE, Vol. 6, No. 4, July/August.

36. Singh, V.P. (1985),'A mathematical study of erosion from upland areas,' Water resources Report, Dept. of Civil and Env. Engrg., Louisiana State University, Baton Rouge, LA 70803-6405.

37. Singh, V.P. (1992),'Elementary Hydrology,' Prentice Hall, Englewood Cliffs, New Jersey.

38. Singh, V.P. and V.J. Chen (1983),'The relationship between storm runoff and sediment yield,' Tech. Rep. 3, Water Resources Program, Dept. of Civil and Environmental Engineering, Louisiana State University, Baton Rouge, LA 70803.

39. Smith, D.D. (1941),'Interpretation of soil conservation data for field use,' Agricultural Engineering, 22: 173-175.

40. Smith, R.E. (1977),'Field test of a distributed watershed erosion/sedimentation model,' Proc. Nat. Conf., Soil Erosion: Prediction and Control, May 24-26, 1976, Special Publication No. 21, Soil Conservation Society of America, 7515 Northeast Ankeny Road, Ankeny, IOWA 50021, pp. 201-209.

41. Smith, R. E., Goodrich, D. C., Woolhiser, D. A. and Unkrich, C. L. (1995), 'KINEROS-A kinematic runoff and erosion model,' Chapter 20 in Computer Models of Watershed Hydrology, edited by V. P. Singh, Water resources Publications, Littleton, Colorado.

42. Tideman, E.M. (1996),'Watershed Management: Guidelines for Indian Catchments,' Omega Scientific Publishers, New Delhi, India.

43. Tripathi, M.P. (1999),'Hydrological modeling for effective management of a small watershed,' Unpub. Ph.D. thesis, Dept. of Agri. and Food Engineering, Indian Institute of Technology, Kharagpur 721302, India.

44. Walling, D.E. and P. Kane (1983),'Temporal variation of suspended sediment properties,' Proc. of Exter Symp. on Recent Developments in the Explanation and Prediction of Erosion and Sediment Yield,' IAHS Pub. No. 137, Washington D.C., pp. 409-419.

45. Walling, D.E. and B.W. Webb (1983),'Patterns of sediment yield,' In: Background to Paleohydrology, K.J. Gregory (ed.), Wiley, New York.

46. Wicks, J.M. and J.C. Bathurst (1996),'SHESED: A physically based, distributed erosion and sediment yield component for the SHE hydrological modelling system,' J Hydrol., 175, 213-238.

47. Williams, J.R. (1978),'A sediment graph model based on instantaneous unit sediment graph,' Water Resources Research, 14(4): 659-664.

48. Wischmeier, W.H. and D.D. Smith (1978),'Predicting rainfall erosion losses,' Agriculture Handbook No. 537, USDA, Science and Education Administration, 58 pp.

49. Wolman, M.G. (1977),'Changing needs and opportunities in the sediment field,' Water Resources Res., 13, pp. 50-54.

50. Woolhiser, D. A., Smith, R. E. and Goodrich, D. C. (1990), 'KINEROS-A kinematic runoff and eroion model, ' Documentation and User Manual. U. S. Department of Agriculture, Agriculktural Research Service, ARS-77, 130 pp.

51. Wu, T.H., J.A. Hall, and J.V. Bonta (1993),'Evaluation of runoff and erosion models,' J. Irrig. Drain. Div., 119(4), 364-382.

52. Young, R.A., C.A. Onstad, D.D. Bosch, W.P. Anderson (1987),'AGNPS:An agricultural non-point source pollution model: A large water analysis model,' U.S. Dept. of Agri., Cons. Res. Report No. 35, p. 77.

53. Zingg, A.W. (1940),'Degree and length of land slope as it affects soil loss in runoff,' Agricultural Engineering, 21: 59-64.

Appendix A

1. Aron, G., A.C. Miller Jr., and D.F. Lakatos (1977),'Infiltration formula based on SCS curve number,' J. Irrig. and Draing. Div., A.S.C.E., Vol. 103, No. IR4, pp. 419-427.
2. Hawkins, R.H. (1976),'Runoff curve numbers for Northern Arizona watersheds,' Project Report Contract No. 171-475, Utah State University, Logan, Utah.
3. Hawkins, R.H. (1978a),'Effects of rainfall intensity on runoff curve numbers: Hydrology and water resources in Arizona and the Southwest,' Vol. 8, pp. 53-64.
4. Hawkins, R.H. (1978b),'Runoff curve numbers with varying site moisture,' J. Irrig. and Draing. Div., Proc., A.S.C.E., Vol. 104, No. IR4, pp. 389-398.
5. Hawkins, R.H. (1979),'Runoff curve numbers from partial area watersheds, J. Irrig. and Draing. Div., Proc., A.S.C.E., Vol. 105, No. IR4, pp. 375-389.
6. Soil Conservation Service (SCS) (1971),'Hydrology, National Engineering Handbook,' Supplement A, Section 4, Chapter 10, Soil Conservation Service, U.S.D.A., Washington, D.C.

Appendix B

1. Levenberg, K. (1944),'A method for the solution of certain non-linear problems in least squares,' Quart. Appl. Math., 2, pp. 164-168.
2. Marquardt, D.W. (1963),'An algorithm for least-squares estimation of nonlinear parameters,' J. Soc. Indust. Appl. Math., 11(2), 431-441.
3. Nash, J.E. and J.V. Sutcliffe (1970),'River flow forecasting through conceptual models, Part I-A discussion of principles,' J. Hydrology, Vol. 10, pp. 282-290.

4. Press, W.H., S.A. Teukolsky, W.T. Vellerling, and B.P. Flannery (1998),'Numerical Recipes in Fortran: The Art of Scientific Computing,' Second Edition, Cambridge University Press, New York.

Appendix C

1. Mahmood, K. and V. Yevjevich (1975),'Unsteady Flow in Open Channels,' Vol. I, Water Resources Publications, PO Box 303, Fort Collins, Colorado 80522.
2. Menendez, A.N. and R. Norscini (1982),'Spectrum of shallow water waves: An analysis,' J. Hydr. Div., A.S.C.E., 108(HY1), 75-94.
3. Menendez, A.N. and R. Norscini (1983), discussion on 'Nature of flood wave attenuation in open channel flow,' J. Hydr. Div., A.S.C.E., 109(5), pp.786-789
4. Mishra, S.K. and V. P. Singh (1999),'Hysteresis-based flood wave analysis,' J. Hydrologic Engineering, A.S.C.E., Vol. 4, No. 4, pp. 358-365.
5. Mishra, S.K. and Vijay P. Singh (2001),'Hysteresis-based flood wave analysis using the concept of strain,' J. Hydrological Processes, Vol. 15, pp. 1635-1651.
6. Murphy, G. (1957),'Properties of Engineering Materials,' International textbook Company, Scranton, Pa.
7. Ponce V.M. and D.B. Simons (1977),'Shallow wave propagation in open channel flow,' J. Hydr. Div., A.S.C.E., 103(HY12), 1461-1476.
8. Price, R.K. (1973),'Flood routing method in British rivers,' Proc., Institution of Engineers (London), Part 2, Vol. 57, pp. 913-930, paper 7674.
9. Price, R.K. (1985),'Flood Routing,' Ch. 4: Development in hydraulic engineering, E.J.P. Novak, Elsevier Applied Science.
10. Rankine, M. (1870),'On the thermodynamic theory of waves of finite longitudinal disturbance,' Philosophical Trans. of the Royal Society of London, 160, pp. 277-288.
11. Seddon, J.A. (1900),'River hydraulics,' Trans., A.S.C.E., 43, 179-229.
12. Singh, V.P (1996),'Kinematic Wave Modelling in Water Resources: Surface Water Hydrology,' John Wiley & Sons, New York.

Appendix D

1. Foster, G.R., L.D. Meyer, and C.A. Onstad (1977),'An erosion equation devised from basic erosion principles,' Trans. A.S.A.E., 20, pp. 678-682.
2. Haan, C.T., B.J. Barfield, and J.C. Heyes (1994),'Design Hydrology and Sedimentology for Small Catchments,' Academic Press, A Division of Harcourt Brace & Company, New York.
3. Musgrave, G.W. (1947),'The quantitative evaluation of factors in water erosion: a first approximation,' J. Soil and Water Conservation, Vol. 2, pp. 133-138.
4. Novotny, V. and H. Olem (1994),'Water Quality: Prevention, Identification, and Management of Diffuse Pollution,' John Wiley & Sons, New York, N.Y.
5. Ponce, V.M (1989),'Engineering Hydrology: Principles and Practices' Prentice Hall, Englewood Cliffs, New Jersey.
6. Singh, P. and V.P. Singh (2001),'Snow and Glacier Hydrology,' Kluwer Academic Publishers, P.O. Box 322, Dordrecht, The Netherlands.
7. Singh, V.P. (1992),'Elementary Hydrology,' Prentice Hall, Englewood Cliffs, New Jersey.
8. Williams, J.R. and H.D. Berndt (1977),'Sediment yield prediction based on watershed hydrology,' Trans. A.S.A.E. 20, pp. 1100-04.
9. Wischmeier, W.H. and D.D. Smith (1965),' Predicting Rainfall-Erosion Losses from Cropland East of Rocky Mountains,' U.S.D.A. Agricultural Handbook No. 282, Washington D.C.

AUTHORS INDEX

SUBJECT INDEX

D

Water Science and Technology Library

16/3: Water-quality hydrology ISBN 0-7923-3652-6
16/4: Water resources planning and management ISBN 0-7923-3653-4
Set 16/1–16/4 ISBN 0-7923-3654-2

17. V.P. Singh: *Dam Breach Modeling Technology.* 1996 ISBN 0-7923-3925-8
18. Z. Kaczmarek, K.M. Strzepek, L. Somlyódy and V. Priazhinskaya (eds.): *Water Resources Management in the Face of Climatic/Hydrologic Uncertainties.* 1996 ISBN 0-7923-3927-4
19. V.P. Singh and W.H. Hager (eds.): *Environmental Hydraulics.* 1996 ISBN 0-7923-3983-5
20. G.B. Engelen and F.H. Kloosterman: *Hydrological Systems Analysis.* Methods and Applications. 1996 ISBN 0-7923-3986-X
21. A.S. Issar and S.D. Resnick (eds.): *Runoff, Infiltration and Subsurface Flow of Water in Arid and Semi-Arid Regions.* 1996 ISBN 0-7923-4034-5
22. M.B. Abbott and J.C. Refsgaard (eds.): *Distributed Hydrological Modelling.* 1996 ISBN 0-7923-4042-6
23. J. Gottlieb and P. DuChateau (eds.): *Parameter Identification and Inverse Problems in Hydrology, Geology and Ecology.* 1996 ISBN 0-7923-4089-2
24. V.P. Singh (ed.): *Hydrology of Disasters.* 1996 ISBN 0-7923-4092-2
25. A. Gianguzza, E. Pelizzetti and S. Sammartano (eds.): *Marine Chemistry.* An Environmental Analytical Chemistry Approach. 1997 ISBN 0-7923-4622-X
26. V.P. Singh and M. Fiorentino (eds.): *Geographical Information Systems in Hydrology.* 1996 ISBN 0-7923-4226-7
27. N.B. Harmancioglu, V.P. Singh and M.N. Alpaslan (eds.): *Environmental Data Management.* 1998 ISBN 0-7923-4857-5
28. G. Gambolati (ed.): *CENAS. Coastline Evolution of the Upper Adriatic Sea Due to Sea Level Rise and Natural and Anthropogenic Land Subsidence.* 1998 ISBN 0-7923-5119-3
29. D. Stephenson: *Water Supply Management.* 1998 ISBN 0-7923-5136-3
30. V.P. Singh: *Entropy-Based Parameter Estimation in Hydrology.* 1998 ISBN 0-7923-5224-6
31. A.S. Issar and N. Brown (eds.): *Water, Environment and Society in Times of Climatic Change.* 1998 ISBN 0-7923-5282-3
32. E. Cabrera and J. García-Serra (eds.): *Drought Management Planning in Water Supply Systems.* 1999 ISBN 0-7923-5294-7
33. N.B. Harmancioglu, O. Fistikoglu, S.D. Ozkul, V.P. Singh and M.N. Alpaslan: *Water Quality Monitoring Network Design.* 1999 ISBN 0-7923-5506-7
34. I. Stober and K. Bucher (eds): *Hydrogeology of Crystalline Rocks.* 2000 ISBN 0-7923-6082-6
35. J.S. Whitmore: *Drought Management on Farmland.* 2000 ISBN 0-7923-5998-4
36. R.S. Govindaraju and A. Ramachandra Rao (eds.): *Artificial Neural Networks in Hydrology.* 2000 ISBN 0-7923-6226-8
37. P. Singh and V.P. Singh: *Snow and Glacier Hydrology.* 2001 ISBN 0-7923-6767-7
38. B.E. Vieux: *Distributed Hydrologic Modeling Using GIS.* 2001 ISBN 0-7923-7002-3

Water Science and Technology Library

39. I.V. Nagy, K. Asante-Duah and I. Zsuffa: *Hydrological Dimensioning and Operation of Reservoirs*. Practical Design Concepts and Principles. 2002 ISBN 1-4020-0438-9
40. I. Stober and K. Bucher (eds.): *Water-Rock Interaction*. 2002 ISBN 1-4020-0497-4
41. M. Shahin: *Hydrology and Water Resources of Africa*. 2002 ISBN 1-4020-0866-X
42. S.K. Mishra and V.P. Singh: *Soil Conservation Service Curve Number (SCS-CN) Methodology*. 2003 ISBN 1-4020-1132-6
43. C. Ray, G. Melin and R.B. Linsky (eds.): *Riverbank Filtration*. Improving Source-Water Quality. 2003 ISBN 1-4020-1133-4
44. G. Rossi, A. Cancelliere, L.S. Pereira, T. Oweis, M. Shatanawi and A. Zairi (eds.): *Tools for Drought Mitigation in Mediterranean Regions*. 2003 ISBN 1-4020-1140-7

Kluwer Academic Publishers – Dordrecht / Boston / London

CPSIA information can be obtained at www.ICGtesting.com
Printed in the USA
LVOW10*0827240214

374926LV00012B/128/P